NATO ASI Series

Advanced Science Institutes Series

A series presenting the results of activities sponsored by the NATO Science Committee, which aims at the dissemination of advanced scientific and technological knowledge, with a view to strengthening links between scientific communities.

The Series is published by an international board of publishers in conjunction with the NATO Scientific Affairs Division

A Life Sciences B Physics	Plenum Publishing Corporation London and New York
C Mathematical and Physical Sciences D Behavioural and Social Sciences E Applied Sciences	Kluwer Academic Publishers Dordrecht, Boston and London
F Computer and Systems Sciences G Ecological Sciences H Cell Biology I Global Environmental Change	Springer-Verlag Berlin Heidelberg New York London Paris Tokyo Hong Kong Barcelona Budapest

NATO-PCO DATABASE

The electronic index to the NATO ASI Series provides full bibliographical references (with keywords and/or abstracts) to more than 30000 contributions from international scientists published in all sections of the NATO ASI Series. Access to the NATO-PCO DATABASE is possible in two ways:

- via online FILE 128 (NATO-PCO DATABASE) hosted by ESRIN, Via Galileo Galilei, I-00044 Frascati, Italy.
- via CD-ROM "NATO-PCO DATABASE" with user-friendly retrieval software in English, French and German (© WTV GmbH and DATAWARE Technologies Inc. 1989).

The CD-ROM can be ordered through any member of the Board of Publishers or through NATO-PCO, Overijse, Belgium.

Series H: Cell Biology Vol. 51

The ASI Series Books Published as a Result of Activities of the Special Programme on CELL TO CELL SIGNALS IN PLANTS AND ANIMALS

The books published as a result of the activities of the Special Programme are:

Vol. 1: **Biology and Molecular Biology of Plant-Pathogen Interactions.** Edited by J.A. Bailey. 1986.

Vol. 2: **Glial-Neuronal Communication in Development and Regeneration.** Edited by H.H. Althaus and W. Seifert. 1987.

Vol. 3: **Nicotinic Acetylcholine Receptor: Structure and Function.** Edited by A. Maelicke. 1986.

Vol. 4: **Recognition in Microbe-Plant Symbiotic and Pathogenic Interactions.** Edited by B. Lugtenberg. 1986.

Vol. 5: **Mesenchymal-Epithelial Interactions in Neural Development.** Edited by J.R. Wolff, J. Sievers, and M. Berry. 1987.

Vol. 6: **Molecular Mechanisms of Desensitization to Signal Molecules.** Edited by T.M. Konjin, P.J.M. Van Haastert, H. Van der Starre, H. Van der Wel, and M.D. Houslay. 1987.

Vol. 7: **Gangliosides and Modulation of Neuronal Functions.** Edited by H. Rahmann. 1987.

Vol. 9: **Modification of Cell to Cell Signals During Normal and Pathological Aging.** Edited By S. Govoni and F. Battaini. 1987.

Vol. 10: **Plant Hormone Receptors.** Edited by D. Klämbt. 1987.

Vol. 11: **Host-Parasite Cellular and Molecular Interactions in Protozoal Infections.** Edited by K.-P. Chang and D. Snary. 1987.

Vol. 12: **The Cell Surface in Signal Transduction.** Edited by E. Wagner, H. Greppin, and B. Millet. 1987.

Vol. 19: **Modulation of Synaptic Transmission and Plasticity in Nervous Systems.** Edited by G. Hertting and H.-C. Spatz. 1988.

Vol. 20: **Amino Acid Availability and Brain Function in Health and Disease.** Edited by G. Huether. 1988.

Vol. 21: **Cellular and Molecular Basis of Synaptic Transmission.** Edited by H. Zimmermann. 1988.

Vol. 23: **The Semiotics of Cellular Communication in the Immune System.** Edited by E.E. Sercarz, F. Celada, N.A. Mitchison, and T. Tada. 1988.

Vol. 24: **Bacteria, Complement and the Phagocytic Cell.** Edited by F.C. Cabello and C. Pruzzo. 1988.

Vol. 25: **Nicotinic Acetylcholine Receptors in the Nervous System.** Edited by F. Celementi, C. Gotti, and E. Sher. 1988.

Vol. 26: **Cell to Cell Signals in Mammalian Development.** Edited by S.W. de Laat, J.G. Bluemink, and C.L. Mummery. 1989.

Vol. 27: **Phytotoxins and Plant Pathogenesis.** Edited by A. Graniti, R.D. Durbin, and A. Ballio. 1989.

Vol. 31: **Neurobiology of the Inner Retina.** Edited by R. Weiler and N.N. Osborne. 1989.

Vol. 32: **Molecular Biology of Neuroreceptors and Ion Channels.** Edited by A. Maelicke. 1989.

Vol. 33: **Regulatory Mechanisms of Neuron to Vessel Communication in the Brain.** Edited by F. Battaini, S. Govoni, M.S. Magnoni, and M. Trabucchi. 1989.

Vol. 35: **Cell Separation in Plants: Physiology, Biochemistry and Molecular Biology.** Edited by D.J. Osborne and M.B. Jackson. 1989.

Vol. 36: **Signal Molecules in Plants and Plant-Microbe Interactions.** Edited by B.J.J. Lugtenberg. 1989.

Vol. 39: **Chemosensory Information Processing.** Edited by D. Schild. 1990.

Vol. 41: **Recognition and Response in Plant-Virus Interactions.** Edited by R.S.S. Fraser. 1990.

Vol. 43: **Cellular and Molecular Biology of Myelination.** Edited by G. Jeserich, H.H. Althaus, and T.V. Waehneldt. 1990.

Vol. 44: **Activation and Desensitization of Transducing Pathways.** Edited by T.M. Konijn, M.D. Houslay, and P.J.M. Van Haastert. 1990.

Vol. 45: **Mechanism of Fertilization: Plants to Humans.** Edited by B. Dale. 1990.

Vol. 46: **Parallels in Cell to Cell Junctions in Plants and Animals.** Edited by A.W. Robards, W.J. Lucas, J.D. Pitts, H.J. Jongsma, and D.C. Spray, 1990.

Vol. 50: **Phytochrome Properties and Biological Action.** Edited by B. Thomas and C.B. Johnson, 1991.

Vol. 51: **Cell to Cell Signals in Plants and Animals.** Edited by V. Neuhoff and J. Friend, 1991.

Cell to Cell Signals in Plants and Animals

Progress Report

Edited by

Volker Neuhoff

MPI für Experimentelle Medizin
Hermann Rein Str. 3
3400 Göttingen

and

John Friend

University of Hull
School of Life Sciences
Dept. of Biology
Hull, HU6 7RX
England

Springer-Verlag
Berlin Heidelberg New York London Paris Tokyo
Hong Kong Barcelona Budapest
Published in cooperation with NATO Scientific Affairs Division

The final report of the NATO Special Programme on Cell to Cell Signals in Plants and Animals, running from 1984 to 1990 under the auspices of the NATO Science Committee.

ISBN 3-540-53739-2 Springer-Verlag Berlin Heidelberg New York
ISBN 0-387-53739-2 Springer-Verlag New York Berlin Heidelberg

Printed in Germany

31/3140-543210 – Printed on acid-free-paper

PREFACE

Cell to cell signals are the most sophisticated and, at the same time, most elementary signs of life in the whole biological kingdom. A detailed analysis of these signals is one of the most challenging research topics in the broad subject of biology, including theoretical and applied biology.

It was therefore with great foresight that the NATO Science Committee established in March 1984 a Special Programme Panel on Cell to Cell Signals in Plants and Animals.

At its first meeting the Panel decided that its activities should be focused on the three main areas described below:

Recognition in Parasitic and Symbiotic Systems

This topic covers the molecular biology and genetics of susceptibility and resistance of plants and animals to pathogens, parasites and symbionts. Recognition indicates interaction between definable constituents of the different cell surfaces and/or molecular signals and the ability of an organism to detect and translate signals into response. Emphasis will be placed on the early events in these interactions and on the nature of the ensuing signals and particularly on the structure and mechanism of transmission of the signalling molecules involved in compatibility or incompatibility which lead to the establishment of either parasitism or symbiosis or the ultimate rejections of both these possibilities. Symbiotic systems are understood to be a combination of two or more organisms co-existing in close relationship involving cell or tissue penetration, without immediate damage to the host cells, and generally leading to overall advantages for both organisms involved.

Cell to Cell Recognition and Differentiation

These activities are focused on one of the most challenging problems in developmental biology of plants and animals, i.e. how systems of positional information are generated in pattern information. Of special interest are short-range cell to cell communications at a molecular level and their consequences in the formation of specialised morphological structures, as well as signals traceable to direct contacts between cell surface constituents. Characterisation at a molecular level of mechanisms which govern or inhibit cell to cell interactions are crucial for the understanding of morphogenetic events, e.g. cell proliferation, migration and aggregation, fertilization, graft compatibility, formation and elimination of synapses and synaptic contacts, cell differentiation and death.

Plasticity in Cell to Cell Communication

Plasticity in cell to cell communications is thought to play the major role in cell differentiation and function. The underlying mechanisms need elucidation at a molecular level. Although many signals are already identified their mode of action is still poorly understood, e.g. modulation of receptor concentration, their transition from low to high affinity forms, regulation of state of activity, inhibition or termination of transduction in adaptation processes.

Among the goals of this Special Programme was the encouragement of interaction and cooperation between scientists working in different fields, e.g. plant and animal biology as well as microbiology, and thereby to foster the transfer of concepts and techniques, especially those of molecular biology between the different disciplines.

This goal was achieved in the 37 Advanced Research Workshops sponsored by the Special Programme Panel. The selected topics for the workshops sponsored by the Panel covered investigations on a diverse array of cells and organisms, including bacteria, fungi, higher plants and signalling systems are of crucial importance to growth, differentiation, adaptation and survival.

Molecular biology has made a major impact in this area over the past few years. It has demonstrated the high degree of conservation of the signalling systems; and also the importance of modification for particular organisms and situations.

It has been known for some time that a diverse array of proteins is involved in signalling processes. Molecular biology has indicated, however, that the true diversity of such systems is far greater than has ever been envisaged. It is to be expected that mathematics, along with computer sciences, will assist molecular biologists to achieve a better understanding of the dynamics of complex signalling systems. The functional significance of these complexities, and the mechanisms whereby signalling systems interact to form networks provide a considerable challenge for the future.

Differential gene expression is often a major step in signal transducing pathways. Progress has been made in understanding which proteins are involved, how they interact with genes and how they are modulated as a result of signalling processes. The combination of molecular genetics including recombinant DNA techniques and physico-chemical techniques, will allow us to analyse the structure and function of the individual proteins that are involved. It is to be expected that this area will be also expand rapidly in the near future.

Many of the workshops were devoted to neurobiology and neurochemistry which is an indication of the continuing advances made in the past few years. Large numbers of novel neuropeptide transmitters have been isolated, identified, sequenced and in many cases the genes isolated and cloned. Genes for receptor proteins have also been cloned which has enabled a detailed structural analysis of the interaction between the signalling molecules and the cell surface receptors.

In plants, as well as in animals, several types of receptor proteins such as auxin receptors, and different types of phytochrome, have been isolated and purified and in many cases the genes for these receptor proteins have been cloned and tested in functional systems.

One of the major advances in the area of plant-microbe communications has been the chemical identification of low molecular weight phenolic compounds produced by the host plants which act as signalling molecules for gene expression by either *Rhizobium* or *Agrobacterium*. Also receptor molecules in the bacteria have been isolated as products of the virulence genes. Viruses have been shown to induce the synthesis of new proteins which alter plasmodesmatal transport.

Similar approaches have also been valuable in other studies of short range communication in both plant and animal cells and of cell abscission in higher plants.

The early stages of these advances were documented in the original report of each of the Workshops. However because there have been such considerable advances in many of these areas since the Workshops were held, we have taken the opportunity to ask Workshop directors to provide an update, where appropriate, of their own Workshops and the twenty four chapters in the book are the result. They also serve as a memento of the work of the Special Programme Panel.

On the Special Programme Panel the following NATO countries were represented by the following scientists:

Professor M.A.B. De Sousa	Portugal	1987—1989
Professor J.A. Fortin	Canada	1984—1986
Professor J. Friend (Chairman 1987—89)	U.K.	1984—1989
Professor K.R. Libbenga	Netherlands	1984—1989
Professor C. Martinez Alonso	Spain	1987—1989
Professor J.P. Merlie	U.S.A.	1987—1989
Professor M. van Montagu	Belgium	1986—1987
Professor V. Neuhoff (Chairman 1984—86)	Germany	1984—1989
Professor E. Seeberg	Norway	1988—1989
Doctor J. Tempé	France	1988—1989
Professor G.H.N. Towers	Canada	1988—1989
Professor M. Trabucchi	Italy	1984—1986
Doctor S. Tzartos	Greece	1984—1985

We very much hope that this book will on the one hand make a contribution to a better understanding of cell to cell signals and on the other hand be a document which demonstrates the beneficial activities of the NATO Science Committee in supporting both basic and applied science.

Göttingen/Leeds im December 1990

Volker Neuhoff John Friend

TABLE OF CONTENTS

RECOGNITION IN PARASITIC AND SYMBIONTIC SYSTEMS

PLASTICITY IN CELL TO CELL COMMUNICATION

PLASTICITY, RECOGNITION AND DIFFERENTIATION IN CELLULAR SYSTEMS

RECOGNITION
IN PARASITIC AND SYMBIONTIC SYSTEMS

PHYTOTOXINS AND PLANT PATHOGENESIS

A. Graniti[1], R. D. Durbin[2] and A. Ballio[3]

[1]Dipartimento di Patologia vegetale,
Universita di Bari, Bari, Italy

[2]ARS/USDA and Department of Plant Pathology,
University of Wisconsin, Madison, Wisconsin, USA, and

[3]Departimento di Scienze biochimiche,
Universita 'La Sapienza', Rome, Italy

INTRODUCTION

In a broad sense, "phytotoxins" can be considered as microbial metabolites, other than enzymes, that damage or are harmful to plants at very low concentrations (1—3, 23, 25). Many plant pathogenic bacteria and fungi produce phytotoxins both in culture and in their hosts during the infection process. In several cases — especially if they are produced during the early stages of plant disease development — these compounds have a function in pathogenesis and cause part or even all of the symptoms of the disease (4, 5, 20). For microbial products which are not phytotoxic *per se* but have a role in pathogenesis, the term aggressions has been proposed (1).

A limited number of phytotoxins, the so-called host-selective (or host-specific) toxins, are toxic to plant species or cultivars that are susceptible to the pathogens producing these toxins; resistant plants are not damaged, at least at physiological concentrations of the toxin. Sometimes, virulence of the pathogen is correlated with its ability to be toxigenic, *e. g.*, to produce one or more toxins and to use them as a main chemical weapon (6—10, 18, 22).

The larger class of non-selective phytotoxins includes compounds which do not reproduce the patterns of resistance and susceptibility of the hosts to the pathogen. Their total impact on agriculture also may be greater because they are broad-spectrum toxins which can affect a variety of important crops (13, 21, 23).

The ability of pathogenic fungi and bacteria to excrete substances which reproduce symptoms on test plants in the absence of the pathogen was recognized as far back as the second half of the 19th century. However, studies on toxins produced by plant pathogens and their effects on plants first really began to accelerate around the middle of the present century and were especially developed after World War II (11, 12).

For long time the role of toxins in plant pathogenesis remained controversial. Although the culture filtrates of many pathogenic bacteria and fungi were shown to contain metabolites able to mimic plant disease symptoms, the inability to purify and chemically characterize the active substances has limited the interpretation of biological experiments designed to elucidate their role in disease production.

NATO ASI Series, Vol. H 51
Cell to Cell Signals in Plants and Animals
Edited by V. Neuhoff and J. Friend

However, much of the scientific skepticism about the role of toxins in plant disease, chiefly based on the data on fusarial toxins in tomato wilt caused by *Fusarium oxysporum* f. sp. *lycopersici*, were overcome when the production of toxins (victorin) by *Cochliobolus victoriae*, cause of the disastrous "Victoria blight" of oats, was demonstrated in the mid 1940's. The toxin affected only oat varieties susceptible to the pathogen and at low concentrations reproduced all the symptoms of the natural infection. Since this demonstration, which confirmed earlier work by Japanese plant pathologists on the black spot of Japanese pear caused by a pathotype of *Alternaria alternata*, about 20 such host-selective toxins have been recognized for diseases of considerable world-wide importance (10, 22, 24).

The role in plant diseases of non-selective phytotoxins was more difficult to establish, but in the last several decades progress on their knowledge has been impressive, mainly because of the development of highly sensitive analytical techniques, which has led to greater acceptance for a role of these compounds in disease production.

The first comprehensive attempt to bring together our knowledge on phytotoxins was made in 1970, when a "NATO Advanced Study Institute" entitled "Phytotoxins in Plant Diseases" was held in Italy (4). This, in turn, greatly stimulated research in many countries on chemical factors in plant diseases and was followed by several reviews and books on the subject, *e.g.*, 3—5, 8, 9, 14—19, 21—25, 29.

Recent years have seen substantial progress in our understanding of several aspects of phytotoxins. More scientists have been attracted to the area, especially as part of cooperative efforts among different laboratories; there have been significant advancements in analytical instrumentation and in new means for structural and biological determinations; and the chemical nature and structure of many bacterial and fungal phytotoxins have been elucidated. For example, the first structure of a selective toxin was reported in 1975, although this kind of toxins had been known 40 years previously. Some insights on how toxins act, *i.e.*, targets and mechanisms of action, made a substantial advancement of our knowledge about these biologically active substances. Finally, after having used traditional approaches to the genetic analysis of toxin production, the powerful and precise technologies of recombinant DNA have been applied to toxin systems.

Up-to-date information on current progress of the research on phytotoxins has been provided at the NATO Advanced Research Workshop on "Phytotoxins and Plant Pathogenesis" held in Italy in 1988 (26). The purpose of this workshop was to present and critically analyze the most important problems that remain and how to approach them experimentally.

The present discussion, although mainly based on the information given at the Workshop, also contains references to some more recent work on phytotoxins and their role in plant disease.

GENETICS

The genetical approach for defining the role of toxins in plant diseases have proven to be very fruitful. All the more so since the advent of molecular genetics with its powerful and concise techniques. We are currently on the threshold of an exciting era whose promise is to define the relevant genes and gene products. By so doing we should be able to finally understand toxin biosynthesis and its regulation, the role of toxins in disease production and the economy of the pathogens, and their evolutionary development.

Analysis of toxin genetics in both fungal and bacterial pathogens is proceeding apace; transformation and complementation systems have been developed, synthetic genes have been cloned and characterized, and a detoxification gene has been transferred into tobacco (27).

Particular emphasis is being placed on the molecular genetics of *Cochliobolus heterostrophus*, *C. carbonum* and *C. victoriae* by Yoder's group. T-toxin production by the first pathogen is controlled by a single gene (Tox1) which conveys virulence on T-cytoplasm corn. A second gene (Tox2) is responsible for HC-toxin production in *C. carbonum*, and yet a third gene locus (Tox3) controls HV-toxin production in *C. victoriae*. Various strategies for cloning these Tox genes are being pursued.

Similar approaches are being taken with tabtoxin and phaseolotoxin which are produced by *Pseudomonas syringae* pathovars. Here, however, regions of DNA have been cloned and now site-directed mutagenesis is being used to define the primary genes for toxin production. Willis' group has found that resistance to tabtoxin in *P. syringae* pv. *tabaci* as well as other tabtoxin-producing pathovars appears to reside in a common DNA region genetically linked to toxin production. It is usually lost, in a very precise matter, when toxin production is lost. Toxin resistance in *P. syringae* pv. *phaseolicola* appears to be conditioned in part by a second form of the toxin's target, ornithine carbamoyltransferase (OCTase) which is resistant rather than sensitive. Interestingly, this second OCTase gene is located in the genome where the toxin production genes reside.

Toxins in relation to plant resistance; evolution of pathogens

Some toxins are known to be able to activate metabolic pathways leading to the formation and accumulation of substances involved in plant defense reactions such as phytoalexins; in this respect they can act as "elicitors", causing localized death of plant cells and the hypersensitive response of the host. Cell death may restrict growth of biotrophic pathogens, but for non-obligate pathogens, plant disease resistance and suspectibility are apparently more determined by the timing and extent of plant cell death (28, 29).

In specific recognition phenomena, certain host-selective toxins may cause an alteration of cell membranes which makes the host tissues "accessible" to invasion by the pathogen. The toxins may act as signal factors which bind to primary receptor sites in the host cells; these in turn may transduct the signal, thus suppressing the host's defense mechanisms (10, 30).

Resistance of plants to disease is often expressed as tolerance or insensitivity to the pathogen's metabolites. In certain diseases, as shown by early work with the host-selective *Alternaria* toxins (31) and by the following literature, toxins act as "suppressors" of induced resistance. In such systems, disease development and resistance may be explained on a molecular level (32, 33).

The role of toxins in the evolution and ecology of plant pathogenic microorganisms has been discussed with particular reference to fungal pathogens which seem to have evolved in response to changes in agriculture (31). A number of genera, *e.g. Cochliobolus*, *Pyrenophora*, *Alternaria*, include highly pathogenic species together with weakly pathogenic or even saprophytic species or populations. Many of the avirulent populations possess mycelial structures for colonizing and invading living plants as parasites, but lack the ability to become pathogenic. However, they may evolve to virulent or specialized forms by acquiring the ability to synthesize toxins by mutation, gene transfer or genetic recombination. Under the selection pressure of new genotypes of agricultural crops, these fungi may increase in number and cause epidemics. When susceptibility of the host, virulence, or specificity of the pathogen, and its ability to form a toxin is controlled by single genes, as in the *Cochliobolus victoriae* — Victoria oats system, a single gene change may result in a devastating epidemic.

Structure and mode of action

N-(N'-sulpho-diaminophosphinyl)-L-ornithine (PSorn), a potent inactivator of ornithine transcarbomoylase, plays a role in the mode of action of phaseolotoxin, the chlorosis-inducing toxin produced by *Pseudomonas syringae* pv. *phaseolicola*. PSorn is known to be formed by the action of peptidases in leaves on the native toxin, phaseolotoxin. Progress towards synthesis of PSorn has been reported (34); its successful completion should make accessible amino acid analogues of the native toxin which may be useful by virtue of altered biochemical activity, or which may be used in studies of structure-activity relationships.

The elucidation of toxin biosynthetic pathways and the study of their regulation can yield new knowledge useful for understanding various aspects of plant disease. The results of a preliminary investigation of rhizobitoxine biosynthesis, carried out by feeding potential, ^{14}C-labelled precursors to *Pseudomonas andropogonis*, suggests that L-aspartic acid is a precursor and that hydroxythreonine is an intermediate in the pathway (35, 36).

The low molecular weight pseudomonad toxic metabolites continue to attract the attention of scientists interested in the study of bacterial plant pathogens. Syringomycin (SR) was reported for the first time in 1971 as a toxic metabolite of many strains of *P. syringae* pv. *syringae*, but amazingly its structure remained unassigned until very recently. First, it was shown that the SR produced in diverse media (37) is a mixture of a structurally related compounds and not a single substance (38, 39), and then, through the joint efforts of two groups who had separately reported a partial structure in preliminary communications (40, 41), the toxin was found to be a new lipodepsipeptide (42). SR appears *in vitro* to be complexed with a high molecular weight, proteinaceous component (43); the commonly used isolation procedure of Gross and DeVay releases SR at the first step in purification: acidified acetone. What, if any, significance this phenomenon has *in planta* remains to be seen.

Studies on the mode of action of SR, all carried out with impure preparations, have suggested that the primary effects of the toxin in plant and yeast cells is at the plasma membrane where ion transport, ATPase and polypeptide phosphorylation are affected. The latter process might involve calcium ions whose influx across the plasma membrane is stimulated by SR (44). More recently it has been demonstrated that SR activates the cellular efflux of orthophosphate and brings about cytoplasmic acidification (45). By perturbing the membrane transport process SR also causes guard cells to bring about stomatal closure (46).

The number of newly discovered fungal phytotoxins has increased remarkably during recent years and several novel structures have been identified with the help of several new, powerful instrumentation techniques. In the course of studies directed towards the search for natural products toxic to weeds, the group of Strobel (47) has reported the structures and biological activities of toxins isolated from weed pathogens such as several species of *Drechslera* (terpenes such as eremophilanes (47) and ophiobolins (48, 49); acetogenins such as curvulins (47), de-O-methyldiaporthin (50), resorcylides (47) and triticones (51, 52); amino acid-derived metabolites such as tryptophol (47)) and *Alternaria alternate* ((maculosins (53) and tetramic acid and perylenquinones (54)). These studies have made large use of X-ray crystallography for structural determinations.

Another group, also working on the identification of the toxins produced by weed pathogens, has isolated prehelminthosporol and dihydroprehelminthosporol from a species of *Bipolaris* pathogenic to Johnson grass, one of the worst weeds in tropical and subtropical areas of the world; these two substances were toxic towards Johnson grass in leaf-spot assays, while four other structurally related compounds, isolated from the same fungus, were inactive (55).

The chemistry of host-selective fungal phytotoxins produced by *Cochliobolus heterostrophus* (*Drechslera maydis*) and *Mycosphaerella zeae-maydis* (*Phyllosticta maydis*), both responsible of corn blight epidemics in USA, and by strains of *Alternaria citri* casual agents of brown spot of rough lemon and Dancy tangerine and other mandarin cultivars, has been reviewed by Kono (56). The studies of *M. maydis* have recently included an analysis of conformation-toxicity relationships (57), which has prompted the synthesis of several analogues of the natural bioactive products (58). The above investigations on host-selective toxins represent positive examples of successful collaborations between USA and Japan laboratories.

Extensive structure-activity studies on the victorins, the host-selective toxins of *Cochliobolus victoriae* (*Bipolaris victoriae*) (the causal agent of Victoria blight of oats) has become feasible when the fractionation of the culture filtrate afforded several related metabolites, all cyclic peptides of a novel type (59, 60), which in turn could be derivatized to yield modified victorins. Among the derivatives, some were capable of preventing or reducing the effects of victorin C, the predominant toxin in the culture filtrate, and another one (the Bulton-Hunter derivative) provided a substance which after labelling with ^{125}I was used in the search for host receptors. Indeed, this radioactive probe bound covalently *in vivo* and in genotype-specific, ligand-specific manner to a 100 kDa protein which may represent the victorin receptor (61). A previously unreported minor component of the toxin mixture produced by *C. victoriae* was shown by tandem FAB-mass spectrometry to differ from victorin C by having a glycine residue in place of the glytoxylic acid moiety (62).

Two new host-selective toxins have been isolated from culture filtrates of *Cochliobolus carbonum*, the causal agent of a leaf spot disease of maize (63). These toxins are analogues of the previously known HC-toxin and, together with other structurally similar metabolites, have permitted an examination of structure-activity relationships.

A novel type of fungal phytotoxin, drechserol B, has been obtained from the culture filtrate of a *C. maydis* strain pathogenic to *Costus speciosus*; its phytotoxicity towards twelve plants (hosts and non-hosts of the fungus) correlates with the susceptibility of the plants to the pathogen (64).

Among nonhost-selective fungal toxins, fusicoccin (FC) is probably the one most extensively studied. Its mode of action has been recently investigated by several groups. The hypothesis that its direct and primary action is at the plasma membrane, where it activates the H^+-ATPase responsible for electrogenic extrusion, which consequently influences the activity of a number of metabolic and physiological processes, has received strong support by recent work (65). Evidence was brought forth that this action is triggered by a signal originating from the interaction of FC with specific, high affinity receptors detected in plasma membrane-enriched fractions of a large number of higher plants and located on the apoplastic side of the plasma membrane (66, 67). These receptors have been characterized in tissues of several mono and dicots (66—70), and partially purified, either as such (71) or after photoaffinity labelling (72). Recently, the existence at the plasma membrane level of a functional relationship between FC receptors and the H^+-ATPase was supported by observing a specific stimulation of the ATPase activity and H^+-pumping in isolated membrane vesibles from radish (73) and broad bean (74). This functional relationship was confirmed by showing that FC can stimulate H^+-pumping in a system obtained by incorporating into liposomes a partially purified plasma membrane ATPase from maize roots or shoots together with a crude preparation of FC receptors from maize shoots (66, 75). The dual, reconstituted system provides a useful tool for the study of the signal transduction following FC binding to its receptor. The search for a physiological significance for FS receptors, which are ubiquitous in plant tissues, has led to the detection of endogenous ligands in several tissues: promising results for their isolation have been obtained by immunoaffinity chromatography (76, 77).

Interesting information on the mode of action of zinniol, a well known phytotoxin produced by several pathogenic *Alternaria* species, has been recently reported. It appears that this substance influences calcium fluxes in plant cells, an effect which might perturb a number of physiological processes in the host. In fact, zinniol binds specifically to protoplasts and membranes from zinniol-sensitive carrot cell lines (the binding is reversible and saturable) and, on receptor occupancy, the entry of calcium ions into protoplasts is stimulated with partial reversion of the inhibitory effect of calcium channel blockers (78).

The electrophysiological effects of toxins on plant cells was summarized by Novacky (79), whose group has done much in this area. They and others have found that, excepting HC-toxin, all the host selective toxins depolarize membranes. This is a quite interesting finding, considering the wide diversity in the structures of these toxins. He discussed the results on HT- and HV-toxins in detail.

Preliminary work on the toxins of three interesting fungi has shown that: (a) culture filtrates of *Ascochyta rabiei*, the causal agent of chickpea blight, contain two toxins which have been purified and partially characterized (80); (b) culture filtrates of *Septoria nodorum* have toxic activity in a wheat leaf necrosis assay which is not due to the mellein content but to some other toxins (81); and (c) culture filtrates of *Hypoxylon mammatum*, the organism which causes a stem canker on aspen, contains several diterpene sulfates, called hymatoxins, and trihydroxytetralones; various clones of popular have been tested for sensitivity towards the purified compounds and culture filtrates (82).

Applications

Usually research on phytotoxins is primarily directed towards determining their chemical nature, occurrence in diseased plants, role in pathogenesis, and selectivity for plant species or genotypes. With this information — sometimes before having a complete picture of what the toxin is and how acts — some practical applications become possible (83, 84).

Perhaps the oldest application is to screen large populations of host material, *e.g.*, seedlings, plant parts, protoplasts or tissue cultures, for resistant individuals using toxin preparations in place of the pathogen. The practice has extended to the selection of novel, useful plant genotypes or variants as a step *in vitro* breeding programs before the regeneration of plants. First applied with host-selective toxins, the practice has spread to non host-selective toxins where it has proved useful for selecting species, cultivars or clones showing lower sensitivity to the toxic effects of the pathogen in question.

Toxin-degrading microorganisms have been successfully used to control certain fungal plant diseases, even in the case of Fusarium wilt of tomato, where the role of toxins in pathogenesis has never been clearly demonstrated (85). On the basis of several findings, it seems possible that toxin detoxifying genes from microorganisms can be expressed in cultivars susceptible to species of *Alternaria*, *Cochliobolus* and other fungi that produce host-selective toxins.

The possibility that phytotoxins produced by plant pathogens may be used to control noxious weeds has attracted more and more attention in recent years. This is mainly due to the great economic importance of losses caused by weeds to agricultural crops, to the undesirable effects of weed-killers on the environment, and to the uncertain results of using biological herbicides. Thus, the search for selective, non-persistent herbicides to be used as agrochemicals has turned to phytotoxins, a class of natural compounds which could provide several substances having a tremendous potential as candidate herbicides. Another advantage would be that many genotypes of cultivated crops are not affected by selective phytotoxins which are active on weeds; if not, incorporation of genes for insensitivity to those toxins in their genomes might be possible.

Being biologically active substances, phytotoxins may have utility in areas other than plant pathology. For example, the fusicoccins and other related fungal metabolites increase the rate of plant transpiration by opening stomata; this favors the uptake of nutrients (and other agrochemicals including pesticides) from the soil or the absorption of CO_2 by crops, especially under glass. The increased water loss induced by these toxins can be utilized for speeding plant drying, *e.g.*, in making hay (86). Germination of dormant seeds, *e.g.*, lettuce, is also promoted by fusicoccin (84).

Phytotoxins may have applications in many other areas, such as in the taxonomy of pathogenic microorganisms, physiological studies of plant hormones and membrane receptors. The high degree of biochemical specificity exhibited by some toxins make them useful metabolic probes, not restricted to the plant kingdom. New approaches to use them in cancer therapy are now also being explored (84).

CONCLUSIONS

Taken together, there has been a great deal of exciting advances made in understanding phytotoxins during the 18 years between the two NATO Advanced Research Workshops. Currently, more people are being attracted to their study for a variety of reasons. We anticipate that results from the near future will not only yield information of relevance to plant diseases, but also provide fundamental knowledge and applications to the life sciences in general.

REFERENCES

1) GRANITI, A. (1972). The evolution of the toxin concepts in plant pathology. In: Ref. 4: 1—15.

2) SCHEFFER, R.P., BRIGGS, S.P. (1981). Introduction: a perspective of toxin studies in plant pathology. In. Ref. 5: 1—20.

3) TURNER, J.G., (1984). Role of toxins in plant disease. In: Plant diseases: infection. (Wood, R.K.S., Jellis, G.J., eds), Blackwell, Oxford, pp. 3—12.

4) WOOD R.K.S, BALLIO, A., GRANITI, A. (ed.) (1972). Phytotoxins in plant diseases. Academic Press, London, 530 p.

5) DURBIN, R.D. (ed.) (1981). Toxins in plant diseases. Academic Press, New York, 515 p.

6) SCHEFFER, R.P., YODER, O.C. (1972). Host specific toxins and selective toxicity. In: Ref. 4: 251—269.

7) KONO, Y., KNOCHE, H.W., DALY, J.M. (1981). Structure: fungal host-specific. In: Ref. 5: 221—257.

8) WHEELER, H. (1976). The role of phytotoxins in specificity. In: Specificity in plant diseases. (Wood, R.K.S., Graniti, A., eds.), Plenum Press, New York, pp. 217—235.

9) DALY, J.M., KONO, Y., SUZUKI, Y., KNOCHE, H.W. (1983). Biological activities and structures of host-selective pathotoxins. In: IUPAC Pesticide Chemistry. (Miyamoto, J., ed.), Pergamon Press, New York, p. 11.

10) KOHMOTO, K., DURBIN, R.D. (ed.) (1989). Host-specific toxins: recognition and specificity factors in plant disease. Tattori University, Tattori, Japan, 230 p.

11) GÄUMANN, E. (1950). Principles of plant infection. Crosby Lockwood, London, 543 p.

12) GÄUMANN, E. (1954). Toxins and plant diseases. Endeavor 13: 198—204.

13) RUDOLPH, K. (1976). Non-specific toxins. In: Physiological plant pathology. (Heitefuss, R., Williams, P.H., eds), Encyclopedia of Plant Physiology, Vol. 4, Springer-Verlag, Berlin, pp. 270—315.

14) STROBEL, G.A. (1977). Bacterial phytotoxins. Annu. Rev. Microbiol. 31: 205—224.

15) FRIEND, J. (1977). Biochemistry of plant pathogens. In: Plant Biochemistry II. International Review of Biochemistry, Vol. 13. (Northcote, D.H., ed.), University Park Press, Baltimore, p. 151—155.

16) YODER, O.C. (1980). Toxins in pathogenesis. Annu. Rev. Phytopathol. 18: 103—129.

17) STAPLES, R.C., TOENNIESSEN, G.H. (ed.) (1981). Plant disease control. Resistance and susceptibility. J. Wiley, New York, 339 p.

18) DALY, J.M., KNOCHE, H.W. (1982). The chemistry and biology of pathotoxins exhibiting host-selectivity. Adv. Plant Pathology 1: 84—138.

19) DALY, J.M., DEVERALL, B.J. (ed.) (1983). Toxins and plant pathogenesis. Academic Press, New York, 181 p.

20) DURBIN, R.D. (1983). The biochemistry of fungal and bacterial toxins and their modes of action. In: Biochemical Plant Pathology. (Callow, J.A., ed.), Wiley, Chichester, p. 137—162.

21) MICHELL, R.E. (1984). The relevance of non-host-specific toxins in the expression of virulence by pathogens. Annu. Rev. Phytopathol. 22: 215—245.

22) SCHEFFER, R.P., LIVINGSTON, R.S. (1984). Host-selective toxins and their role in plant diseases. Science 223: 17—21.

23) GOODMAN, R.N. KIRALY, Z., WOOD, K.R. (1986). The biochemistry and physiology of plant disease. Univ. Missouri PREss, Columbia, 433 p.

24) NISHIMURA, S., VANCE, C.P., DOKE, N. (ed.) (1987). Molecular determinants of plant diseases. Japan Sci. Societies Press, Tokyo — Springer-Verlag, Berlin, 293 p.

25) MISAGHI, I.J. (1982). Physiology and biochemistry of plant-pathogen interactions. Plenum Press, New York, 387 p.

26) GRANITI, A., DURBIN, R.D., BALLIO, A. (ed.) (1989). Phytotoxins and plant pathogenesis. Springer-Verlag, Berlin, 508 p.

27) ANZAI, H., YONEYAMA, K., YAMAQUCHI, I. (1990). Transgenic tobacco resistant to a bacterial disease by the detoxification of a pathogenic toxin. Mol. Gen. Genet (in press).

28) WOOD, R.K.S. (1989). Toxins and diseases resistance. In: Ref. 26, p. 267—273.

29) BAILEY, J.A., O'CONNELL, R.J. (1989). Plant cell death: a determinant of disease resistance and susceptibility. In: Ref. 26, p. 275—283.

30) KOHMOTO, K., OTANI, H., KODAMA, M., NISHIMURA, S. (1989). Host recognition: can accessibility to fungal invasion be induced by host-specific toxins without necessitating necrotic cell death? In: Ref. 26, p. 249—265.

31) HAYAMI, C., OTANI, H., NISHIMURA, S., KOHMOTO, K. (1982). Induced resistance in pear leaves by spore germination fluids of non pathogens to *Alternaria alternata,* Japanese pear pathotype, and suppression of the induction by AK toxin. J. Fac. Agric, Tottori Univ. 17: 9—18.

32) SCHEFFER, R.P. (1989). Host-specific toxins in phytopathology: origin and evolution of the concept. In: Ref. 10, p. 1—17.

33) NISHIMURA, S., NAKATSUKA, S. (1989). Trends in host-selective toxin research in Japan. In: Ref. 10, p. 19—31.

34) MITCHELL, R.E. (1989). Current research into chemical synthesis of phaseolotoxin. In: Ref. 26, p. 1—6.

35) MITCHELL, R.E. (1989). Biosynthesis and regulation of toxins produced by *Pseudomonas syringae* pv. *glycineas* (coronatine) and *Pseudomonas andropogonis* (rhizobitoxine). In: Ref. 26, p. 23—29.

36) MITCHELL, R.E. (1989). Biosynthesis of rhizobitoxine from L-aspartic acid and L-*threo*-hydroxythreonine by *Pseudomonas andropogonis*. Phytochemistry 28: 1617—1620.

37) SURICO, G., LAVERMICOCCA, P., IACOBELLIS, N.S. (1989). Syringomycin and syringotoxin production in cultures of *Pseudomonas syringae* pv. *syringae*. In: Ref. 26, p. 469—471.

38) BALLIO, A., BARRA, D., BOSSA, R., DEVAY, J.E., GRGURINA, I. IACOBELLIS, N.S., MARINO, G., PUCCI, P., SIMMACO, M., SURICO, G. (1988). Occurrence of multiple forms of syringomycin. In: Ref. 26, p. 363—366.

39) BALLIO, A., BARRA, D., BOSSA, F., DEVAY, J.E., GRGURINA, I. IACOBELLIS, N.S., MARINO, G., PUCCI, P., SIMMACO, M., SURICO, G. (1988). Multiple forms of syringomycin. Physio. Mol. Plant Pathol. 33: 493—496.

40) BACHMANN, R.C., TAKEMOTO, J.Y. (1989). Structure of syringomycin. A progress report. In: Ref. 26, p. 259—361.

41) SEGRE, A., BALLIO, A., BARRA, D., BOSSA, F., GRGURINA, I., IACOBELLIS, N.S., MARINO, G., PUCCI, P., SIMMACO, M., SURICO, G. (1989). Structural studies on syringomycin. In: Ref. 26, p. 367—369.

42) SEGRE, A., BACHMANN, R.C., BALLIO, A., BOSSA, F., GRGURINA, I., IACOBELLIS, N.S., MARINO, G., PUCCI, P., SIMMACO, M., TAKEMOTO, J.Y. (1989). The structure of syringomycins A, F. and G. FEBS Lett 255: 27—31.

43) IACOBELLIS, N.S., LAVERMICOCCA, P., SURICO, G., DURBIN, R.D. (1989). Occurrence of a syringomycin-high molecular weight complex in *Pseudomonas syringae* pv. *syringae*. In: Ref. 26, p. 429—431.

44) TAKEMOTO, J.Y., GIANNINI, J.L., VASSEY, T., BRISKIN, D.P. (1989). Syringomycin effects on plasma membrane Ca^{+2} transport. In: Ref. 26, p. 167—175.

45) REIDL, H.H., GROVER, T.A., TAKEMOTO, J.Y. (1989). ^{31}P-NMR evidence for cytoplasmic acidification and phosphate extrusion in syringomycin-treated cells of *Rhodotorula pilimanae*. Biochim Biophys. Acta 1010: 325—329.

46) MOTT, K.A., TAKEMOTO, J.Y. (1989). Syringomycin, a bacterial phytotoxin, closes stomata. Plant. Physiol. 90: 1435—1439.

47) KENFIELD, D., BUNKERS, G. STROBEL, G., SUGAWARA, F. (1989). Fungal phytotoxins — potential new herbicides. In: Ref. 26, p. 319—335.

48) YUN, C.-H., SUGAWARA, F., STROBEL, G.A. (1988). The phytotoxic ophiobolins produced by *Drechslera oryzae*, their structures and biological activity on rice. Plant Sci. 54: 237—243.

49) SUGAWARA, F., TAKAHASHI, N., STROBEL, G., YUN, C.-H., GRAY, G., FU, Y., CLARDY, J. (1988). Some new phytotoxic ophiobolins produced by *Drechslera oryzae*. J. Org. Chem. 53: 2170—2172.

50) HALLOCK, Y.F., CLARDY, J., KENFIELD, D.S., STROBEL, G. (1988). De-O-methyl diaporthin, a phytotoxin from *Drechslera siccans*. Phytochemistry 27: 3123—3125.

51) SUGAWARA, F., TAKAHASHI, N., STROBEL, G.A., STROBEL, S.A., LU, H.S.M., CLARDY, J. (1988). Tritocones A and B, novel phytotoxins from the plant pathogenic fungus *Drechslera tritici-repentis*. J. Am. Chem. Soc. 110: 4086.

52) KENFIELD, D., STROBEL, S., SUGAWARA, F., BERGLUND, D. STROBEL, G. (1988). Triticone A: a novel bioactive lactam with potential as a molecular probe. Biochem. Biophys. Comm. 157: 174—182.

53) STIERLE, A.C., CARDELLINA II, J.H., STROBEL, G.A. (1989). Maculosin, a host-specific phytotoxin for spotted knapweed from *Alternaria alternata*. Proc. Natl. Acad. Sci. 85: 8008—8011.

54) STIERLE, A.C., CARDELLINA II, J.H. STROBEL, G.A. (1989). Phytotoxins from *Alternaria alternata*, a pathogen of spotted knapweed. J. Nat. Prod. 52: 42—47.

55) PENA-RODRIGUEZ, L.M., ARMINGEON, N.A., CHILTON, W.S. (1988). Toxins from weed pathogens, I. Phytotoxins from a *Bipolaris* pathogen of Johnson grass. J. Nat. Prod. 51: 821—828.

56) KONO, Y. (1989). Structural studies on host-specific phytotoxins in corn blight and citrus brown spot disease. In: Ref. 26, p. 7—21.

57) SUZUKI, Y., DANKO, S.J., KONO, Y., DALY, J.M., KNOCHE, H.W., TAKEUCHI, S. (1988). Studies on the conformations of PM-toxin, the host-specific corn pathotoxin produced by *Phyllosticta maydis*. Agric. Biol. Chem. 52: 15—24.

58) SUZUKI, Y., DANKO, S.J., KONO, Y., DALY, J.M., KNOCHE, H.W., TAKEUCHI, S. (1989). Synthesis and biological activity of a new type of PM-toxin analogue containing oxygen atoms in the carbon-chain skeleton. Agric Biol. Chem. 53: 453—459.

59) MACKO, V., WOLPERT, T.J., ACKLIN, W., ARIGONI, D. (1989). Biological activities of structural variants of host-selective toxins from *Cochliobolus victoriae*. In: Ref. 26, p. 31—41.

60) WOLPERT, T.J., MACKO, V., ACKLIN, W., ARIGONI, D. (1988). Molecular features affecting the biological activity of the host-selective toxins from *Cochliobolus victoriae*. Plant Physiol. 88: 37—41.

61) WOLPERT, T.J., MACKO, V. (1989). Specific binding of victorin to a 100-kDa protein from oats. Proc. Nat. Acad. Sci. 86: 4092—4096.

62) KINOSHITA, T., KONO, Y, TAKEUCHI, S., DALY, J.M. (1989). Structure of HV-toxin M, a host-specific toxin-related compound produced by *Helminthosporium victoriae*. Agric Biol. Chem. 53: 1283—1290.

63) RASMUSSEN, J.B., SCHEFFER, R.P. (1988). Isolation and biological activities of four selective toxins from *Helminthosporium carbonum*. Plant Physiol. 86: 187—191.

64) SHUKLA, R.S., AGRAWAL, P.K., THAKUR, R.S., HUSAIN, A. (1989). Drechserol-B, a host-selective phytotoxin produced by *Drechslera maydis*. Phytochemistry 28: 2089—2091.

65) MARRE, E., MARRE, MT, ROMANI, G. (1989). Action of fusicoccin *in vivo*: physiological and biochemical consequences. In: Ref. 26, p. 131—141.

66) ADUCCI, P., BALLIO, A. (1989). Mode of action of fusicoccin: the role of specific receptors. In: Ref. 26, p. 143—150.

67) FEYERABEND, M., WEILER, E.W. (1988). Characterization and localization of fusicoccin-binding sites in leaf tissues of *Vicia faba* L probed with a novel radioligand. Planta 174: 115—122.

68) STOUT, R.G. (1988). Fusicoccin activity and binding in *Arabidopsis thaliana*. Plant Physiol. 88: 999—1001.

69) MEYER, C., FEYERABEND, M., WEILER, E.W. (1989). Fusicoccin-binding proteins in *Arabidopsis thaliana* (L) Heynh. Characterization, solubilization and photoaffinity labeling. Plant Physiol. 89: 692—699.

70) DE MICHELIS, M.I., PUGLIARELLO, M.C., RASI-CALDOGNO, F. (1989). Fusicoccin binding to its plasma membrane receptors and the activation of the plasma membrane H^+-ATPase. I. Characteristics and intracellular localization of the fusicoccin receptor in microsomes from radish seedlings. Plant Physiol. 90: 133—139.

71) DE BOER, A.H., WATSON, B.A., CLELAND, R.E. (1989). Purifications and identification of the fusicoccin binding protein from oat root plasma membrane. Plant Physiol. 89: 250—259.

72) FEYERABEND, M., WEILER E.W. (1989). Photoaffinity labeling and partial purification of the putative plant receptor for the fungal wilt-inducing toxin, fusicoccin. Planta 178: 282—290.

73) RASI-CALDOGNO, F., DE MICHELIS, M.I., PUGLIARELLO, M.C. (1989). Fusicoccin-receptor. Interaction and activation of plasma-membrane H^+-ATPase in native membrane vesicles. In: Ref. 26, p. 123—129.

74) BLUM, W., KEY, G., WEILER, E.W. (1988). ATPase activity in plasmalemma-rich vesicles isolated by aqueous two-phase partitioning from *Vicia faba,* mesophyll and epidermis: Characterization and influence of abscisic acid and fusicoccin. Physiol. Plant 72: 279—287.

75) ADUCCI, P., BALLIO, A., BLEIN, J.-P., FULLONE, M.R., ROSSIGNOL, M., SCALLA, R. (1989). Functional reconstitution of a proton-translocating system responsive to fusicoccin. Proc. Natl. Acad. sci. 85: 7849—7851.

76) MARRA, M., ADUCCI, P., BALLIO, A. (1989). Immunoaffinity chromatography of endogenous ligands for fusicoccin binding sites. In: Ref. 26, p. 357—358.

77) MARRA, M., BALLIO, A., ADUCCI, P. (1988). Immunoaffinity chromatograph of fusicoccin. J. Chromatogr. 440: 47—51.

78) THULEAU, P., GRAZIANA, A., ROSSIGNOL, M., KAUSS, H., AURIOL, P., RANJEVA, R. (1988). Binding of the phytotoxin zinniol stimulates the entry of calcium into plant protoplasts. Proc. Natl. Acad. Sci. 85: 5932—5935.

79) NOVACKY, A., ULLRICH-EBERIUS, C.I., BALL, E. (1989). Interactions of phytotoxins with plant cell membranes: electrophysiology and ion flux-induced pH changes. In: Ref. 26, p. 151—166.

80) ALAM, S.S., SIRANGE, R.N. (1989). Isolation and properties of two toxins from culture filtrates *Ascochyta rabiei*. In: Ref. 26, p. 385—386.

81) KURZYCA, H.P., FRIEND, J. (1989). Leaf spotting toxins from *Septoria nodorum.* In: Ref. 26, p. 387—388.

82) GENETET, I., PINON, J., BODO, B., REBUFFAT, S. (1989). *Hypoxylon mammatum* toxins. Their nature and their role in host parasite relationships. In: Ref. 26, p. 389—391.

83) DURBIN, R.D. (1981). Applications. In: Ref. 5, p. 495—505.

84) DURBIN, R.D., GRANITI, A. (1989). Possible applications of phytotoxins. In: Ref. 26, p. 337—355.

85) VAN ALFEN, N.K. (1989). Reassessment of plant wilt toxins. Annu. Rev. Phytopathol. 27: 533—550.

86) GRANITI, A. (1989). Fusicoccin and stomatal transpiration. In: Ref. 10, p. 143—152.

RECENT ADVANCES IN PLANT-MICROBE INTERACTIONS

Ben J.J. Lugtenberg and Ruud A. de Maagd

Leiden University
Department of Plant Molecular Biology
Nonnensteeg 3, 2311 VJ Leiden, The Netherlands

INTRODUCTION

Living plants interact extensively with microorganisms in their environment. Such interactions can be benificial, neutral or harmful to the plant, depending on the microbe and on the exact conditions.

Pathogenic microbes, virusses, bacteria and fungi are stil resposible for huge losses of crop. Scientists and farmers have since long taken measures like rotating crops, treatment with pesticides and breeding of resistant crops. Understanding of plant-microbe interactions, which may lead to development of new protection strategies, has gained momentum in the last decade. The strategy of the NATO Science Committee to develop a special program "Cell to Cell Signals in Plants and Animals" was very timely. By sponsoring two major workshops (1,2) in the field of plant-microbe interactions it has significantly stimulated the interactions and even collaborations between scientists involved in this field.

Using molecular genetic techniques exquisite tools have been developed to study plant-microbe interactions at the molecular level (e.g. ref. 1). Using these tools, eg appropriate promoters linked to reporter genes, it became possible to detect and purify signal molecules involved in plant microbe interaction (e.g. refs. 1 and 2). Subsequent studies are clearly necessary for understanding the mechanism of action of these signal molecules at the molecular level.

The extent of progress obtained in the understanding of different plant-microbe interaction depends on the microbe involved. Microbes that long ago became genetically accessible and that because of their importance have received the interest of many scientists, appeared to provide the best tools for the elucidation of plant-microbe interactions. Currently the interaction between Rhizobium bacteria and their leguminous host plants is best understood. This interaction will receive most attention in this review because it illustrates best the scientific strategies that are currently available to study plant-microbe interactions. Much less knowledge is available about interactions of plants with pathogenic fungi that are not genetically accessible. The ultimate scientific goal of these studies is to understand interactions between plants and microbes at the molecular level. This understanding will lead to applications that improve crop quality and yield. In that respect molecules and elements involved in regulation and in signalling pathways are of prime interest.

NATO ASI Series, Vol. H 51
Cell to Cell Signals in Plants and Animals
Edited by V. Neuhoff and J. Friend

INTERACTIONS BETWEEN PLANTS AND MICROBES: BACKGROUND INFORMATION

Plants have physical barriers, e.g. cutin, suberin, lignin, cellulose, hemicellulose, callose and certain cell wall proteins, that prevent entry of most microbes. As a response to the recognition of a pathogen plants can synthesize (increased amounts of) molecules that actively contribute to the elimination of microbes, of which phytoalexins (substances that are toxic to microbes) and pathogenesis related (PR)proteins (some of which have later been identified, e.g. as chitinases or glucanases) are best known (3). This response of the plant is triggered by so-called elicitors, molecules which can either be of microbial or of plant origin.

Some microbes enter the plant but are eliminated by the so-called hypersensitive reaction (HR) of the plant. This is the rapid and localized death of challenged plant cells associated with the restriction of microbial colonization.

Bacteria of the genus *Rhizobium* can enter via the root hairs and establish themselves in leguminous plants, apparently without eliciting HR or another form of plant defence. In the last stage of the symbiotic process the bacteria localized within plant cells fix atmospheric nitrogen and provide the plant with NH_{4+}. Since lack of an appropriate nitrogen-source is after lack of available water the second limiting factor for plant growth, this process is of prime economic importance.

A group of microbes that have gained considerable interest in the last few years are organisms that can be used for biological control of harmful insects, fungi and bacteria. *Trichoderma* and certain bacteria can produce antifungal toxins (4). A particular group of toxins are siderophores, Fe^{3+}-binding peptides which control pathogens by limiting the amount of Fe^{3+} available for their growth (5). The applied aims of studying these interactions are (i) to developed novel ways of control, and (ii) to partly replace the often harmful chemical pesticides.

An interesting group of microbes are those which stimulate plant growth by increased development of the root system or by promoting early emergence of the root system. These properties can be decisive for growth in arid regions or in regions with a brief growth season.

SIGNALS IN THE INTERACTION OF RHIZOBIUM AND LEGUMINOUS PLANTS

Rhizobium bacteria induce nodules on leguminous plants. In these nodules a differentiated form of the bacterium, the bacteroid, converts atmospheric nitrogen into ammonia ions, which are used as a nitrogen source by the plant. In return the plant cell provides the bacteroid with a carbon source, most likely a dicarboxylic acid. Nodulation is a host-specific process in that a certain *Rhizobium* species can only nodulate a restricted number of leguminous host plants, e.g. *R. leguminosarum* bv. *viciae* nodulates vetch and pea, *R. leguminosarum* bv, *trifolii* nodulates clover, *R. meliloti* nodulates alfalfa and *Bradyrhizobium japonicum* nodulates soybean.

The processes of nodule formation (nodulation) and nitrogen fixation are very complex. The early stages are the best known ones. Several stages will be discussed here in the context of molecules involved in plant-microbe communication.

1. Chemotaxis

It is likely that the ability of Rhizobium to swim and carry out chemotaxis to root exudate, plays a role in nodulation, at least under some circumstances. Indeed, El-Haloui et al. (6) found that flagella-containing rhizobia have a competitive advantage under field conditions. Similarly it has been found that chemotaxis is conditionally required for tumor formation by the plant pathogenic bacterium *Agrobacterium tumefaciens* (7).

2. Attachment

Attachment to the plant cell is assumed to be an essential step in nodulation by *Rhizobium* (8) and tumor formation by *A.tumefaciens* (9,10). Presently the attachment process can be divided in the following molecular interactions (11).

a) A bacterial cell surface protein, termed rhicadhesin, interacts with a common plant root hair tip component (12). Rhicadhesin is a Ca^{2+}-binding protein which shares properties with calmodulin. Purified rhizobial rhicadhesin is able to restore attachment of *chvB* mutant cells of *A. tumefaciens* to pea root hair tips. Even more spectacular, it restores virulence of these cells towards plants (13). Since *Rhizobium* and *Agrobacterium* share many properties, it is surprizing that the molecules assumed to be responsible for attachment to plant cells are so different, namely a protein in *Rhizobium* and β1,2-glucan in *Agrobacterium* (14). In fact the real situation may be a combination of both viewpoints, namely that both the protein and the glucan are required for receptor activity. Such is the case for many phage receptors in *Escherichia coli* (15).

b) Cellulose fibrils of *Rhizobium* are partly or completely responsible for the accumulation of *Rhizobium* bacteria on the tips of pea root hairs (cap formation). The fibrils seem to promote interactions between bacteria. Their receptor is unknown but may well be the same fibril (16).

c) The bacterial lectin receptor, which has not been identified so far, can also be involved in cap formation. When this is the case, attachment is followed by infection thread formation, in contrast to the situation when only cellulose fibrils are responsible for cap formation. Involvements of these two bacterial components can be influenced by growth conditions. The present state of the art indicates that recognition by the bacterial lectin receptor of plant lectin at the tip of the root hair is an important step in the initiation of infection thread formation (17). It should be noted that lectin is probably also involved in a later stage of the modulation process (17).

3. Activation of nodulation genes

Fusion of *nod* gene promoters to reporter genes provided a tool for the elucidation of the regulation of *nod* genes. Of the Sym plasmid localized *nod* genes only *nodD*, the positive regulatory protein, is transcribed constitutively, whereas the other *nod* genes are induced by plant substances (18,19,20). The identification of the latter components as flavonoids by four different research groups (from Canberra, Leiden, Norwich and Stanford) was first reported at a NATO Advanced Research Workshop in Biddinghuizen in 1986 under the auspices of the Scientific Affairs Division of NATO as part of the programme "Cell to cell signals in Plants and Animals". The fact that many scientists working on the biosynthesis of flavonoids had been invited to the same workshop brought this observation immediately in a broader perspective and induced collaboration and exchange of materials.

Later evidence was provided for a direct interaction between flavonoid inducers and NodD protein (21), which takes place in the cytoplasmic membrane of the *Rhizobium* cell (22,23). A certain class of hybrid NodD proteins does not require flavonoids for activation. Such cells have a largely extended host range, an increased nitrogen fixing ability and are resistant against the action of inhibitors of *nod* gene activation (24,25).

Bacterial *nod* genes may not be necessary during the whole period of symbiosis. The levels of *nod*-specific mRNA's and Nod proteins in bacteroids is extremely low compared to levels in induced bacteria (Schlaman et al., unpublished). Under field conditions the activity of *nod* gene inducer in exudate can be limiting for nodulation (26).

4. Inhibition of *nod* gene activation

Plants not only secrete activators of nodulation but also substances that inhibit activation, the most powerful ones being umbelliferone (27,28).

5. *nod* gene products

Upon activation of the *nod* genes the bacterium causes a number of changes in the plant like hair curling, hair deformation, thick and short roots (20,29), cortical cell division (30,31) and the induction of increased production and/or secretion of inducers (32). Several of these phenomena can be brought about by soluble secreted factors in the absence of bacteria (29,31,32). Several of these factors are being purified and their structures will be known soon (32,33,34). Since mutants in the production of extra cellular polysaccharide (EPS) are also defective in hair curling one or more of the soluble factors may be related to (precursors of) EPS (35).

6. Infection thread formation

Proteins involved in the initiation of infection thread formation or in their stabilization are NodE protein (36), NodO protein (37) and plant lectin (17,38). Bacterial NodE protein and the plant lectin are major determinants of host specificity (17,36).

7. Release of bacteria from the infection thread

Mutants lacking the O-antigen part of lipopolysaccharide (LPS) (39,40) form small ineffective nodules, apparently due to a block in the release of bacteria from the infection thread into the plant cell. Several possible explanations for this defect have been proposed (41) including a role of the O-antigen in surface hydrophobicity and/or surface charge (40).

8. Bacteroid development

Compared with free-living bacteria, bacteroids show a large numver of changes at the molecular level, the most important one being the production of nitrogenase components, required for nitrogen fixation (see below). The (plant) signal that is responsible for these developmental changes are largely unknown. Promoters are currently being sought which respond to this signal. Future research is required to identify the signald and to establish whether these changes are essential for effective nitrogen fixation.

9. Effective nitrogen fixation

During the development of bacteroids the components of the nitrogenase enzym complex are induced. A central role, in the induction of the structural genes of this complex is played by the regulatory *nifA* gene. Oxygen limitation has been identified as the signal required for the induction by the *nifA* gene (42).

10. Induction of nodulins

Nodulins are plant proteins specifically present in nodules. More than 25 of those proteins have been found. Based on the time of expression after infection of the plant with Rhizobium bacteria, they are divided into early (ENOD) and late nodulins (43). The late nodulin genes are first expressed shortly before or concomitantly with the onset of nitogen fixation.

Rhizobium nod mutants have been used to see which bacterial genes are required for the induction of a specific nodulin, e.g. the common *nod* genes *ABC* as well as the host specific gene *nodE* are required for the expression of ENOD12 (44). A soluble factor produced by the bacterium is presumably responsible for this effect.

Using *in situ* hybridization it could be shown that different ENOD transcripts are present in successive zones of the central nodule tissue. ENOD12 transcripts are found in cells that become prepared for infection thread growth and in cells containing the infection thread tip. Therefore ENOD12 may have a role in the preparation of cells for infection thread growth and perhaps also in the information of the infection thread (43). Several early nodulin genes have been sequenced. The prediction is that several nodulins are rich in (hydroxy-)proline. Therefore these may be cell wall proteins (44).

PLANT PATHOGENS

From an economic point of view, fungi are the most important group of pathogens. Basically, there are two ways in which organisms can be pathogenic for plants. Class I pathogens damage the host through secretion of toxins and/or enzymes. Such pathogens often elicit host defence mechanisms non-specifically (e.g. induction of phytoalexins), but are not always inhibited because they are equippod with enzymes or other devices that inactivate and/or avoid the defence. Class II pathogens do not seriously damage the host in the early stages and avoid non-specific initation of host defence mechanisms. If elicited, defense reactions are effective (45).

Enzymes that can be produced by class I pathogens include pectic enzymes, celluloses, phosphatidases, proteases, galactosidases, xylanases, arabanases, cutinases and nucleases.

Cutinase-dificient mutants of *Fusarium Solani* f.sp. *pisi* are non-pathogenic. Pathogenicity is restored by the addition of purified cutinase to the inoculated mutant spores (46). The role of cutinase became known in more detail be recent experiments which showed that the introduction of the cutinase gene into *Mycosphaerella,* a fungus that normally requires wounds in order to cause damage, gives the fungus the ability to infect intact papaya fruits (47).

Several enzymes of class I pathogens can degrade host phytoalexins, such as kievitone (48) and pisavin (49). Class I pathogens frequently produce toxins that damage plant cells, e.g. victorin, syringomycin, coronatine, phaseolotoxin and tabtoxin.

PLANT DEFENCE MECHANISMS

The major physical barrier to potential pathogens of plant tissues is the polysaccharide-rich cell wall. In response to invading microorganisms, plant cell walls undergo structural changes, e.g. lignification, accumulation of hydroxyproline-rich glycoproteins, deposition of callose and addition of cutin or suberin. For further details the reader is referred to ref. 50.

Among the preformed chemical plant factors which may prevent infection by a potentially pathogenic fungus, are several compounds with antibiotic activity. Tomatine is toxic to *Fusarium solami*, whereas tomatine-restistant mutants of *Fusarium* are pathogenic (51). High levels of an antifungal diene in the peel of avocado fruits is correlated with resistance of unripe fruits to colonization by the fungus *Colletotrichum gloesperioides*.

Systemic induced disease resistance in whole plant has been observed with a number of plant-pathogen systems (52).

The best resistance effects against viral, bacterial and fungal diseases have been achieved by pre-infection with the same or different microorganisms. Systemic signals transported over long distances in the plant have to be proposed to explain these effects (53).

HYPERSENSITIVE REACTION (HR)

The hypersensitive response reaction is the defence reaction of the plant after recognition of an avirulent micro-organism. HR restricts the further development of the micro-organism in a process that is correlated with the induction of 30 or more plant genes.

Biochemical events associated with the expression of HR include the production of phytoalexins, chitinases, antiviral proteins, β-1,3-glucanases, lignin and hydroxyproline-rich glycoproteins (54,55,56).

ELICITORS OF PLANT RESPONSES

Research on avirulence genes is aiding our understanding of plant-pathogen specificity. Genetic evidence now links an *avirulence* gene, *avrD*, with the initiation of the plant's hypersensitive response by causing the bacterium to produce a specific elicitor, even when *Escherichia coli* is the bacterial host of *avrD* (55).

Various substances have been proposed as elicitors of the hypersensitive response. These includes cell wall preparations and lipids of fungal origin as well as a small peptide and a glycan from fungus-infected plants (57,58). Cell wall fragments or either pathogen or plant appear to be able to induce the same defense reaction as the attack by the pathogen itself.

The best characterized elicitor isolated from fungal walls is a branched hepta-β-glucoside. It is capable of eliciting phytoalexin accumulation in soy bean cotylodons at concentrations as low as 10^{-8} to 10^{-9}M (50). Chitosan, a fungal β(1->4)-linked polymer of glucosamine, is also an elicitor of phytoalexin accumulation. Chitosan oligosaccharides also elicited the accumulation of protease inhibitors and induced callose synthesis.

Several pathogens, e.g. *Erwinia chrysanthemi* and *Fusarium solani* f.sp.*pisi* secrete pectin-degrading enzymes. Plant cell wall pectic fragments can elicit plant defense responses. Linear oligosaccharides of α(1->4)-linked galacturonic acid residues, especially those between 10 and 13 residues in length, elicit phytoalexin accumulation. This activity is destroyed by treatment of the oligosaccharides with purified endopolygalacturonase (50). Interestingly, plant cell walls contain proteins that specifically inhibit fungal polygalacturonases. When the hepta-β-glucoside elicitor of fungal origin and the plant-derived oligogalacturonide elicitor are simultaneously applied on plant tissues, they act synergistically (59). This discovery will be relevant for plant defense.

Killing fragments are plant cell wall fragments that kill plant cells. They are distinct from the oligogalacturonide elicitors of phytoalexins in that they are not elicitors of phytoalexin accumulation, although they eventually cause this effect through reactions triggered by cell death. They also can be generated by enzymes secreted by pathogens. Hahn et al. (50) hypothesize that the killing fragments are the triggers of the hypersensitive resistance response. They speculated that the key to the understanding of race-cultivar-specific olant-pathogen interactions may be that race-cultivar-specific pathogens may have the rare combination of abilities that enables them to invade their host while avoiding the accumulation of an effective concentration of host-derived killing fragments.

ELICITOR SIGNAL PERCEPTION AND TRANSDUCTION TO PLANT GENE RESPONSE

A wide variety of defense responses has been characterized, and several genes and gene products of the plant, involved in these responses have been identified (60). Our knowledge of the transduction events from pathogen recognition or elicitor to the activations of defense gene is poor. It has been suggested that several cascades are induced by elicitors, and that elicitor recognition induces membrane depolarization, which in turn affects several membrane-related events, especially ethylene biosynthesis. Ethylene and membrane depolarization might play the role of rapidly transduced cell-to-cell signals. Intracellular signals might involve phospholipid metabolites, e.g. derived from lipoxygenase action or from inositol phosphates (58).

Remarkable progress in research on the structure and function of receptors has been made in animal systems. Signal receptors involved in the transduction pathways of plants are presently forcefully being pursued. For a critical review see (61). Similarly, research on second messengers in plants (e.g. cyclic nucleotides, products of phospholipid metabolism and Ca^{2+}-ions) has become intensive (62). Phosphorylation of plant plasma membrane proteins in response to oligosaccharides that regulate proteinase inhibitor gene expression has been reported (63).

Ca^{2+} has been implicated in the regulation of a number of basic processes associated with plant growth and development, including cell elongation, cell division, hormone action and turgor regulation (63). Substantial evidence has been provided for Ca^{2+} involvement in the activation of components that activate plant defence during a plant-pathogen interaction (64,65).

CONCLUDING REMARKS

The last decade has provided the beginning of the understanding of genes and signal molecules involved in plant—microbe interactions. This has provided a firm basis for further unravelling these interactions at the molecular level, with a spin off to practical applications.

REFERENCES

1) Lugtenberg, B.J.J. 1986. Recognition in Microbe-Plant Symbiotic and Pathogenic Interactions. NATO ASO Series H. p. 449. Springer Verlag, Heidelberg.

2) Lugtenberg, B.J.J. 1989. Signal molecules in Plant and Plant-Microbe Interactions. NATO ASI Series H. p. 425. Springer Verlag, Heidelberg.

3) Hahlbrock, K., B. Cuypers, C. Douglas, K.H. Fritzemeier, H. Hoffman, F. Rohwer, D. Scheel, and W. Schulz. 1986. Biochemical interactions of plants with potentially pathogenic fungi. pp. 311—-323. In: ref. 1.

4) Chet, I. 1987. Innovative approaches to plant disease control. p. 372. Wiley Interscience, New York.

5) De Weger, L.A., B. Schippers, and B.J.J. Lugtenberg. 1987. Plant growth stimulation by biological interference in iron metabolism in the rhizosphere. pp. 387—400. In: G. Winkelman, D. Van der Helm, and J.B. Neilands (eds.), Iron transport in microbes, plants and animals. VCH Verlaagsgesellschaft, Weinheim.

6) El-Haloui, N.E., D. Ochin, and R. Tailliez. 1986. Competitivité pour l'infection entre souches de *Rhizobium meliloti:* role de la mobilité. Plant Soil 95: 337—344.

7) Hawes, M.C. 1989. Conditional requirement for chemotaxis in induction of rooty tumors by *Agrobacterium tumefaciens* on roots of *Pisum sativum*. pp. 123—128. In: ref. 2.

8) Kijne, J.W., G. Smit, C.L. Díaz, and B.J.J. Lugtenberg. 1986. Attachment of *Rhizobium leguminosarum* to pea root hair tips. pp. 101—111. In: ref. 1.

9) Douglas, C.J., W. Halperin, and E.W. Nester. 1982. *Agrobacterium tumefaciens* mutants affected in attachment to plant cells. J. Bacteriol. 152: 1265—1275.

10) Matthysse, A.G. 1986. Attachment of *Agrobacterium tumefaciens* to plant host cells. pp. 219—227. In: ref. 1.

11) Lugtenberg, B.J.J., G. Smit, C. Díaz, and J.W. Kijne. 1989. role of attachment of *Rhizobium leguminosarum* cells to pea root hair tips in targeting signals for early symbiotic steps. pp. 129—-136. In: ref. 2.

12) Smit, G., T.J.J. Logman, M.E.T.I. Boerrigter, J.W. Kijne, and B.J.J. Lugtenberg. 1989. Purification and partial characterization of the *Rhizobium leguminosarum* biovar *viciae* Ca^{2+}-dependent adhesin, which mediates the first step in attachment of cells of the family Rhizobiaceae to plant root hair tips. J. Bacteriol. 171: 4054—4062.

13) Smit, G. 1988. Adhesins from *Rhizobiaceae* and their role in plant-bacterium interactions. Ph.D. thesis, Leiden University, The Netherlands.

14) Puvanesarajah, V., F.M. Schell, G. Stacey, C.J. Douglas, and E.W. Nester. 1985. Role for 2-linked-beta-D-glucan in the virulence of *Agrobacterium tumefaciens*. J. Bacteriol. 164: 102—106.

15) Lugtenberg, B., and L. Van Alphen. 1983. Molecular architecture and functioning of the outer membrane of *Escherichia coli* and other Gram-negative bacteria. Biochim. Biophys. Acta 737: 51—115.

16) Kijne, J.W., C.L.Díaz, and B.J.J. Lugtenberg. 1989. Role of lectin in the pea-*Rhizobium* symbiosis. pp. 351—358. In: ref. 2.

17) Díaz, C.L., L.S. Melchers, P.J.J. Hooykaas, B.J.J. Lugtenberg, and J.W. Kijne. 1989. Root lectin as a determinant of host-plant specificity in the *Rhizobium* legume symbiosis. Nature (London) 338: 579—581.

18) Innes, R.W., P. L. Kuempel, J. Plazinski, H.C.J. Canter Cremers, B.G. Rolfe, and M.A. Djordjevic. 1985. Plant factors induce expression of nodulation and host-range genes in *R. trifolii*. Mol. Gen. Genet. 201: 426—432.

19) Mulligan, J.T., and S.R. Long. 1985. Induction of *Rhizobium meliloti nodC* expression by plant root exudate requires *nodD*. Proc. Natl. Acad. Sci. USA 82: 6609—6613.

20) Van Brussel, A.A.N., S.A.J. Zaat, H.C.J. Canter Cremers, C.A. Wijffelman, E. Pees, T. Tak, and B.J.J. Lugtenberg. 1986. Role of plant root exudate and sym plasmid-localized nodulation genes in the synthesis by *Rhizobium leguminosarum* of Tsr factor which causes thick and short roots on common vetch. J. Bacteriol. 165: 517—522.

21) Spaink, H.P., C.A. Wijffelman, E. Pees, R.J.H. Okker, and B.J.J. Lugtenberg. 1987. *Rhizobium* nodulation gene *nodD* as a determinant of host specificity. Nature (London) 328: 337—340.

22) Recourt, K., A.A.N. Van Brussel, A.J.M. Driessen, and B.J.J. Lugtenberg. 1989. Accumulation of a *nod* gene inducer, the flavonoid naringenin, in the cytoplasmic membrane of *Rhizobium leguminosarum* biovar *viciae* is caused by the pH-dependent hydrophobicity of naringenin. J. Bacteriol. 171: 4370—4377.

23) Schlaman, H.R., H.P. Spaink, R.J.H. Okker, and B.J.J. Lugtenberg. 1989. Subcellular localization of the *nodD* gene product in *Rhizobium leguminosarum*. J. Bacteriol. 171: 4686—4693.

24) Spaink, H.P., R.J.H. Okker, C.A. Wijffelman, T. Tak, L.Goosen-de Roo, E. Pees, A.A.N. Van Brussel, and B.J.J. Lugtenberg. 1989. Symbiotic properties of Rhizobia containing a flavanoid-independent hybrid *nodD* product. J. Bacteriol. 171: 4045—4053.

25) Spaink, H.P., C.A. Wijffelman, R.J.H. Okker, and B.J.J. Lugtenberg. 1989. Localization of functional regions of the *Rhizobium nodD* product using hybrid *nodD* genes. Plant Molec. Biol. 12: 59—73.

26) Kapulnik, Y., C.M. Joseph, and D.A. Phillips. 1987. Flavone limitations to root nodulation and symbiotic nitrogen fixation in Alfalfa. Plant Physiol. 84: 1193—1196.

27) Firmin, J.L., K.E. Wilson, L. Rossen, and A.W.B. Johnston. 1986. Induction of *Rhizobium leguminosarum nod* genes by flavonoids is antagonized by other plant-specified aromatic compounds. pp. 137—148. In: ref. 1.

28) Djordjevic, M.A., J.W. Redmond, M. Batley, and B.G. Rolfe. 1987. Clovers secrete specific phenolic compounds which either stimulate or repress *nod* gene expression in *Rhizobium trifolii.* EMBO J. *6*: 1173—1179.

29) Van Brussel, A.A.N., T. Tak, A. Wetselaar, E. Pees, and C.A. Wijffelman. 1982. Small leguminosae as test plants for nodulation of *Rhizobium leguminosarum* and other *Rhizobia* and *Agrobacteria* harbouring a leguminosarum plasmid. Plant Sci.Lett. 27: 317—325.

30) Long, S.R., N.K. Peters, J.T. Mulligan, M.E. Dudley, and R.F. Fisher. 1986. Genetic analysis of *Rhizobium*-plant interactions. pp. 1—16. *In:* ref. 1.

31) Schmidt, J., R. Wingender, M. John, U. Wieneke, and J. Schell. 1988. *Rhizobium meliloti nodA* and *nodB* genes are involved in generating compounds that stimulate mitosis of plant cells. Proc. Natl. Acad. Sci. USA 85: 8578—8582.

32) Van Brussel, A.A.N. 1990. Symbiotic signals in early stages of the morphogenesis of *Rhizobium* induced root nodules. Symbiosis: in press.

33) Faucher, C., P. Lerouge, P. Roche, C. Rosenberg, F. Debellé, J. Vasse, E. Cervantès, S.B. Sharma, G. Truchet, J.-C. Promé, and J. Dénarié. 1989. The common *nodABC* genes and the *nodH* and *nodQ* host-range genes of *Rhizobium meliloit* determine the production of low molecular weight extracellular signals. pp. 379—386. In: ref. 2.

34) Hollingsworth, R., A. Squartini, S. Philip-Hollingsworth, and F. Dazzo. 1989. Root hair deforming and nodule initiating factors from *Rhizobium trifolii.* pp. 387—394. In: ref. 2.

35) Leigh, J.A., E.R. Signer, and G.C. Walker. 1985. Exopoly-saccharide deficient mutants of *Rhizobium meliloti* that form ineffective nodules. Proc. Natl. Acad. Sci. USA 82: 6231—6235.

36) Spaink, H.P., J. Weinman, M.A. Djordjevic, C.A. Wijffelman, R.J.H. Okker, and B.J.J. Lugtenberg. 1989. Genetic analysis and cellular localization of the *Rhizobium* host-specificity-determining nodE protein. EMBO J. 8: 2811—2818.

37) De Maagd, R.A., H.P. Spaink, E. Pees, I.H.M. Mulders, A. Wijfjes, C.A. Wijffelman, R.J.H. Okker, and B.J.J. Lugtenberg. 1989. Localization and symbiotic function of a region on the *Rhizobium leguminosarum* Sym plasmid pRL1JI responsible for a secreted flavonoid-inducible 50 kD protein. J. Bacteriol. 171: 1151—1157.

38) Kijne, J.W., C.L. Díaz, and B.J.J. Lugtenberg. 1989. Role of lectin in the pea-*Rhizobium* symbiosis. pp. 351—358. In: ref. 2.

39) De Maagd, R.A., A.S. Rao, I.H.M. Mulders, L. Goosen-de Roo, M. van Loosdrecht, C.A. Wijffelman, and B.J.J. Lugtenberg. 1989. Isolation and characterization of mutants of *Rhizobium leguminosarum* biovar *viciae* strain 248 with altered lipopolysaccharides: role of surface charge or hydrophobicity in bacterial release from the infection thread?. J. Bacteriol. 171: 1143—1150.

40) Priefer, U.B. 1989. Genes involved in lipopolysaccharide production and symbiosis are clustered on the chromosome of *Rhizobium leguminosarum* biovar *viciae* VF39. J. Bacteriol. 171: 6161—6168.

41) De Maagd, R.A., and B.J.J. Lugtenberg. 1989. Lipopolysaccharide: a signal in the extablishment of the Rhizobium/legume symbiosis?. pp. 337—344. In: ref. 2.

42) Fischer, H.M., and H. Hennecke. 1987. Direct response of *Bradyrhizobium japonicum nifA*-mediated *nif* gene regulation to cellular oxygen status. Mol. Gen. Genet. 207: 621—626.

43) Nap, J.P., and T. Bisseling. 1990. Nodulin function and nodulin gene regulation in root nodule development. In press. In: P. M. Gresshoff (ed.), The Molecular Biology of Nitrogen Fixation. CRC Press, Florida.

44) Scheres, B., C. Van de Wiel, A. Zalensky, A. Hirsch, A. Van Kammen, and T. Bisseling. 1989. Identification of *Rhizobium leguminosarum* genes and signal compounds involved in the induction of early nodulin gene expression. pp. 367—377. In: ref. 2.

45) Keen, N. 1986. Pathogenic strategies of fungi. pp. 171—188. In: ref. 1.

46) Kolattukudy, P. E. 1985. Enzymatic ponotration of the plant cuticle by fungal pathogens. Ann. Rev. Phytopathol. 23: 223—250.

47) Dickman, M.B., G.K. Podila, and P.E. Kolattukudy. 1989. Insertation of cutinase gene into a wound pathogen enables it to infect intact host. Nature (London) 342: 446—448.

48) Smith, D.A., H.E. Wheeler, S.W. Banks, and T.E. Cleveland. 1984. Association between lowered kievitone hydratase activity and reduced virulence to bean in variants of *Fusarium solani* f.sp. *phaseoli*. Physiol. Plant Pathol. 25: 135—147.

49) Van Etten, H., D. Matthews, P. Matthews, V. Miao, A. Maloney, and D. Straney. 1989. A family of genes for phytoalexin detoxification in the plant pathogen *Nectria haematococca*. pp. 219—228. In: ref. 2.

50) Hahn, M.G., P. Bucheli, F. Cervone, S.H. Doares, R.A. O'Neill, A. Darvill, and P. Albersheim. 1989. Roles of cell wall constituents in plant-pathogen interactions. pp. 131—181. In: T. Kosuge, and E.W. Nester (eds.), Plant-Microbe Interactions, Molecular and Genetic perspectives. McGraw-Hill, New York.

51) Defago, G., H. Kern, and L. Sedlar. 1983. Genetic analysis of tomatine sensitivity, sterol content and pathogenicity for green tomato fruits in mutants of *Fusarium solanii*. Physiol. Plant Pathol. 22: 39—43.

52) Kuc, J. 1987. Plant immunization and its applicability for disease control. pp. 225—274. In: ref. 4.

53) Binder, A., S. Baer, C. Hofmann, and K. Kovats. 1989. Mechanisms in systemic induced disease resistance. pp. 267—272. In: ref. 2.

54) Mansfield, J.W. 1986. Recognition, eleicitors and the hypersensitive reaction. pp. 433—437. In: ref. 1.

55) Keen, N.T., S. Tamaki, D. Kobayashi, M. Stayton, D. Gerhold, H. Shen, S. Gold, J. Lorang, and H. Thordal-Christensen. 1989. Characterization and function of avirulence genes from *Pseudomonas syringae* pv. *tomato*. pp. 184—188. In: ref. 2.

56) Hahlbrock, K., N. Arabatzis, M. Becker-Andre, H.J. Joos, E. Kombrink, M. Schroder, G. Strittmatter, and J. Taylor. 1989. Local and systemic gene activation in fungus infected patato leaves. pp. 241—249. In: ref. 2.

57) Doares, S.H., P. Bucheli, P. Albersheim, and A.G. Darvill. 1989. Fungal enzymes, plant cell wall fragments, and plant cell death. A possible mechanism for hypersensitivity in the rice blast system. pp. 229—233. In: ref. 2.

58) Esquerre-Tugaye, M.T., J. Fournier, M. Rickauer, and M.L. Pouenat. 1989. Signals and defense responses associated with the race-cultivar specific interaction between tobacco and *Phytophthora parasitica nicotianae*. pp. 409—414. In: ref. 2.

59) Davis, K.R., A.G. Darvill, and P. Albersheim. 1986. Host-pathogen interactions XXXI. Several biotic and abiotic elicitors act synergistically in the induction of phytoalexin accumulation in soybean. Plant Molec. Biol. 6: 23—32.

60) Collinge, D.B., and A.J. Slusarenko. 1987. Plant gene expression in response to pathogens. Plant Molec. Biol. 9: 389—410.

61) Hahn, M.G. 1989. Animal receptors — Examples of cellular signal perception molecules. pp. 1—26. In: ref. 2.

62) West, C.A., R. Bruce, and Y.Y. Ren. 1989. Second messengers in animals and their possible relevance for plants. pp. 27—40. In: ref. 2.

63) Farmer, E.A., and C.A. Ryan. 1989. Phosphorylation of plant plasma membrane proteins in response to oligosaccharides that regulate proteinase inhibtior gene expression. pp. 49—56. In: ref. 2.

64) Kauss, H. 1987. Some aspects of calcium-dependent regulation in plant metabolism. Ann. Rev. Plant Physiol. 38: 47—72.

65) Kauss, H., T .Waldmann, W. Jeblick, G. Euler, R. Ranjeva, and A. Domard. 1989. Ca^{2+} is an important but not the only signal in callose synthesis induced by chitosan, saponins and polyene antibiotics. pp. 107—116. In: ref. 2.

SIGNALLING BETWEEN PLANTS AND VIRUSES

R. S. S. Fraser

AFRC Institute of Horticultural Research
Worthing Road
Littlehampton, West Sussex BN17 6BD, U.K.

INTRODUCTION

The interactions between plants and their viruses form a special model system within the Nato Programme on Cell-to-cell Signalling in Plants and Animals. Clearly, the virus side of the pairing is not cellular. But plants and viruses recognize each other by the interactions between virus- and plant-coded molecules. These can directly affect pathogenesis, or can send signals which condition the subsequent development of the infection and the host response. The subject was discussed at a Nato Advanced Research Workshop on 'Recognition and Response in Plant-Virus Interactions' in April 1989 (1).

The particular attraction of plant viruses for research on signalling is that they can now be completely defined in molecular terms. For several viruses, the complete sequences of the nucleic acid and protein components are known, and much is known about the control and pathway of expression of the genetic information. Physically, the particle structure is well understood as a result of x-ray diffraction and other methods. In contrast, the immensely more complex host is less well understood in molecular terms, and the processes of signal reception and transduction in higher plants are just beginning to be explored. Plant viruses, by their apparent simplicity, and yet by the diversity of their interactions with the host, offer excellent opportunities for studying the phenomena of recognition, signalling and response (2).

The virus replicative cycle, considered in the widest sense, offers a number of stages at which signalling phenomena may influence the development of the interaction between plant and virus. Some of these stages are shown in Figure 1. Recognition and signalling events can be considered in two main areas: those occurring outside the plant, affecting the establishment of infection, and those occurring within the plant after infection. The latter include effects on virus multiplication and spread throughout the plant, symptom formation by the host, and complex interactions resulting from the operation of host resistance mechanisms. Some of these signalling processes may be interfered with, to allow development of methods of crop protection.

In this Chapter I will try to review the major stages where recognition and signalling processes occur between plants and viruses. Literature citations will be to key original papers or recent reviews. Further details can of course be found in the relevant chapters by individual contributors in the workshop proceedings (1); for reasons of space these are not cited individually in the present paper, except where they constitute the only comprehensive recent review of the particular subject area.

NATO ASI Series, Vol. H 51
Cell to Cell Signals in Plants and Animals
Edited by V. Neuhoff and J. Friend

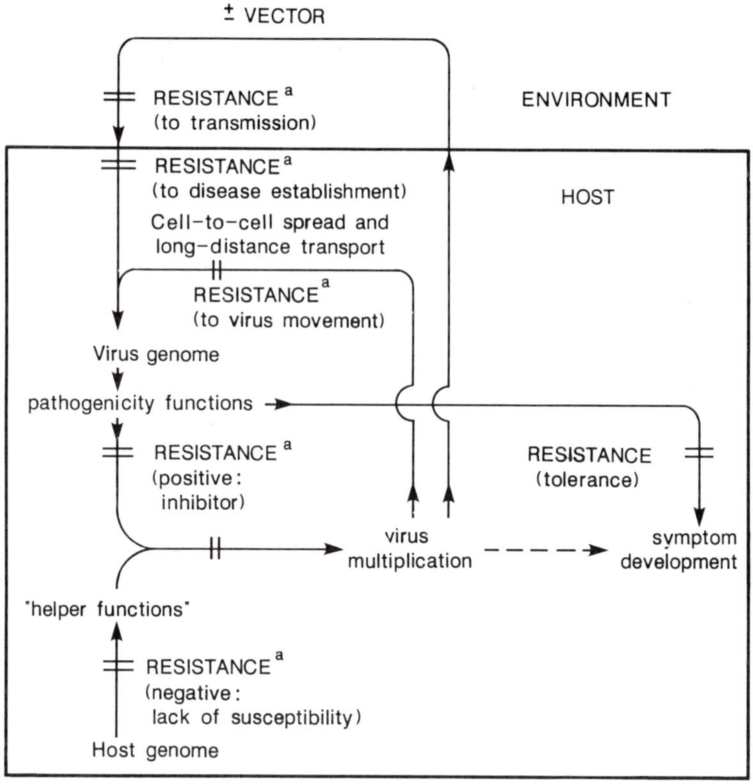

FIGURE 1
The replicative cycle of a plant virus, showing various possible targets of cultivar resistance mechanisms. [a]Indicates resistance mechanisms which might also be involved in the determination of host range. Reproduced from (1).

VIRUS HOST RANGE AND TRANSMISSION

Host range

Most plant species are not hosts for most plant viruses — or are apparently not so. The non-host species are defined as those in which no detectable virus multiplication occurs, and in which no symptoms form. What mechanisms determine whether a species is a host or non-host for a particular virus?

In a classical series of papers, Bald and Tinsley (3—5) introduced the idea that host range was controlled by pathogenicity factors in the virus and susceptibility factors in the host. When the two sets of factors 'matched', successful pathogenesis was established; when they did not, the species was a non-host for that particular virus. Clearly, this model involves recognition events between plant- and virus-specified molecules, and it is likely that a 'successful' recognition, from the virus standpoint, would signal changes facilitating virus multiplication and pathogenesis. The Bald and Tinsley model is attractive in that it offers explanations of many features of host range control, such as how two viruses may have overlapping host ranges, or how the host range of one virus may be entirely contained within that of another. However, there is as yet very little direct experimental evidence for the model. This reflects the paucity of knowledge of which host- and virus-coded molecules interact during disease establishment. One possible example is the interaction of host- and virus-coded subunits to form a replicase responsible for multiplication of the viral genome. But with one partial exception, that of turnip yellow mosaic virus (TYMV) (6), the nature of the true replicase and the importance of host and virus-coded subunits are not well understood.

Experiments with protoplasts have shown, in the small number of cases examined, that they may support multiplication of ostensibly non-hosted viruses (7). Furthermore, when some apparent non-host plants are mechanically-inoculated by rubbing with the non-hosted virus, multiplication occurs in the few directly-inoculated cells (8). It is possible, therefore, that the barrier to successful establishment of infection in at least some cases of host-range control occurs at the stage of cell-to-cell transport of the virus, and does not represent an inhibition of virus multiplication as such. The involvement of recognition and signalling in control of cell-to-cell movement of viruses is important in normal pathogenesis, and will be considered later.

Transmission

Many plant viruses require a plant-feeding vector such as an insect, nematode or motile fungal zoospore to transfer them between individuals of the host plant species; insect vectors are commonest and economically the most important. The relationship between a virus and its vector is highly specific, in terms both of vector species and mechanism of transmission (9). Non-circulative transmission occurs when the virus is associated with the external surfaces (including those of the mouthparts), while circulative transmission occurs when the virus passes through the gut wall into the haemocoel of the insect, then reaches the salivary glands and is secreted. There are clear recognition events between viral proteins and sites on some types of vector, which determine whether transmission will occur. Most involve the virus coat protein, which determines whether a virus particle is adsorbed to a retention site. However, there are more complex examples in which one virus produces a non-structural protein ('helper component') which can assist transmission of other viruses as well as itself. Little is known about the biochemistry of the receptor surface in the vector, and of the mechanism by which circulative viruses are transferred through the gut wall to the haemocoel.

INFECTION, REPLICATION AND SPREAD

Initiation of infection

Once a virus has reached a host plant — by vector or mechanical transmission — the processes of establishment of infection can begin. Infection occurs in the cytoplasm, and it is necessary for the virus to gain access to this, by a wounding process. This occurs during feeding of the vector, or mechanically for viruses which do not rely on a vector. The latter process appears to be non-specific.

The next stage is for the information contained in the virus genome to be expressed, so that viral direction of pathogenic processes can commence. This generally involves removal of the virus capsid protein, although some of the rare, more complex plant viruses can express their genetic information without completely exposing it. The paradox is that many plant virus have particles which are extremely stable *in vitro,* requiring extreme pH or highly anionic detergents for disassembly, yet can obviously be disassembled under the mild physiological conditions present *in vivo.*

The processes of disassembly *in vivo* have been studied most for tobacco mosaic virus (TMV), although similar principles appear to operate for at least some other viruses. The TMV genome consists of a single RNA molecule, which is messenger-sense and contains the sequences specifying four polypeptides. The first two of these, of 126 and 183 kDa, are 5' co-terminal: their AUG start codon is at positions nt 69—71. Evidence from Wilson and co-workers (10) indicates that the first 68-residue 'leader sequence' of TMV RNA is loosely or not encapsidated by viral coat protein under physiological conditions. This appears to be because the coat protein-RNA binding in this region is particularly weak, because of the absence of G-containing triplets. The leader sequence functions as an efficient recruiter of host ribosomes, which disassemble the virus particle in a 5'-3' direction as they translate the 126 kDa polypeptide, thought to be part of the viral replicase. This 'co-translational disassembly' is highly efficient for virus replication for two reasons: the high translational efficiency conferred by the leader sequence, and because the viral RNA is protected by coat protein from nuclease degradation until the moment of translation. Furthermore, co-translational disassembly is clearly able to expose the viral genome under mild physiological conditions.

The recognition event between the TMV RNA 5' leader sequence and host ribosomes is therefore important in the establishment of pathogenesis: the response in the form of synthesis of a replicase or replicase component opens up the way to the next important stage. The means by which the remainder of the viral RNA is exposed and expressed are not yet clear, and much remains to be discovered about how other types of virus expose and express their genes. Finally, it is important to stress that co-translational disassembly of TMV RNA does not appear to involve any determination of host specificity, as the virus particles are efficiently translated in highly heterologous systems such as rabbit reticulocyte and *Escherichia coli* cell-free lysates, and *Xenopus* oocytes. Whether exposure of the genomes of other types of virus occurs by mechanisms expressing host specificity or not remains to be discovered.

Replication of the viral genome

The vast majority of plant viruses characterized so far have single-stranded RNA genomes of plus polarity. The greatest research effort on genomic multiplication has been devoted to this type, although considerable progress has also been made for viroids and for plant DNA viruses of the caulimovirus group; the last resemble animal retroviruses in that DNA replication involves an RNA intermediate.

Replication of (+) strand RNA viruses requires synthesis of a complementary minus strand, from which the progeny plus strands are copied. At some stage, sub-genomic messenger RNAs for such products as the virus coat protein may be produced. These processes require one or more viral replicases, which are RNA-dependent RNA polymerases but which appear to be different from the RNA-dependent RNA polymerase activity present in healthy plants. The significance of this last enzyme in plant metabolism is unknown.

The viral replicase(s) must be dependent on the presence of a viral RNA template, must be able to initiate transcription, to elongate and to terminate chains; they must be able to copy plus to minus, and minus to plus, to produce progeny viral genomes. They may also be required to initiate internally to produce subgenomic mRNAs, although the mechanism of this is still in doubt. So far, no viral replicase has been purified and shown to possess all of these properties *in vitro,* although some will perform some of the required tasks (11).

As far as recognition events are concerned, it is argued that the functional replicase may contain host- and virus-coded subunits; partly by analogy with RNA bacteriophages. The evidence for the virus-coded components is stronger than for the host-coded components, and is based on two main types of approach.

Firstly, although the RNA viruses of eukaryotes show a wide range of types of genome structure, some close relationships have been revealed by nucleotide sequence analysis. Thus many viruses can be classified into two 'supergroups' based on similarities to the animal poliovirus and sindbisviruses (12). Coding regions which have been clearly identified as specifying replicase components in well understood viruses show sequence homology to regions in less-well understood viruses — such as the majority of plant viruses. The involvement of these plant virus genes in replication is inferred. Secondly, partially purified replicase preparations from infected plants can be shown to contain virus-specified polypeptides (11).

The purified replicase from TYMV-infected Chinese cabbage leaves is the best characterized to date (6, 11). It contains a virus-coded polypeptide of 115 kDa and a host-coded subunit of 45 kDa. The latter is of unknown function in the host cell; in addition, the 'true' replicase *in planta* may contain additional virus- and host-coded subunits.

Many viruses cause formation of characteristic ultrastructural modifications in cells of infected plants; they can be shown to be associated with viral RNA synthesis by the presence of double-stranded RNAs and replicase complex (11). These ultrastructural modifications most probably involve further host- and perhaps virus-encoded molecules, but the relationship to viral RNA replication has still to be established.

Viroids are the smallest known pathogenic agents of plants, consisting of a single-stranded circular RNA molecules of 246—375 nucleotides. They do not code for any proteins, and are completely dependent on the host plant for RNA replication. Current evidence indicates that the host DNA-dependent RNA polymerase II is responsible for all steps of viroid RNA replication (13, 14). Excision of monomeric viroid RNA from the oligomeric linear form produced by the 'rolling circle' type replication mechanism, and joining of the ends of the linear monomer to form the circular form, appears also to involve host proteins, perhaps only one ribonuclease performing both cleavage and reverse-cleavage. Thus viroid RNA molecules are in themselves particularly interesting recognition entities, in that they appear to associate with host catalytic agents which then perform activities quite different from their role in the healthy plant.

Virus movement

Although replication of a single virus particle can produce millions of progeny in the initially-infected cell, the full pathogenic potential can only be realized if the infection spreads from cell to cell, and to other parts of the plants (15). The two types of movement probably involve different types of mechanism.

Cell-to-cell spread is interesting in that the intercellular connections — plasmodesmata — appear to be too small to allow transit of virus particles, whether judged by electron microscopy or by the passage of solutes of known molecular mass. It appears that infection by some viruses may cause modification of plasmodesmata, making them suitable for movement of an infectious entity. For a small number of viruses, of which TMV is the best characterized example, a virus coded-protein has been shown to be involved in cell-to-cell transport (16). The implication is that the viral movement protein is involved in modification of the plasmodesmata; this is supported by the immunocytochemical localization of the movement protein in the cell wall fraction (17). The nature of the interaction between the virus-coded movement protein and the host components involved — perhaps some membrane fractions — are not yet clear. For some viruses, the interaction and transport mechanism probably involve components other than the plasmodesmata. However, it seems likely that the virus/host membrane interaction is a critical one in the development of pathogenesis; speculatively, it may also be a major stage in signalling which heralds the onset of particular host reactions.

The nature of the infectious entity which moves from cell to cell remains to be fully established. Some viruses, such as cowpea mosaic virus, require their coat proteins for movement, whereas other such as TMV can move as naked RNAs. It has been suggested that there may be two distinct types of transport mechanism for different types of viruses; one involving viral particles, and the other viral nucleic acid (18, 19). These two mechanisms are likely to involve quite different types of recognition and signalling events between host- and virus-coded molecules.

The mechanisms by which viruses may enter the vascular system for systemic spread remain to be established. They could share features with those involved in cell-to-cell spread, but there appear to be some differences. Some viruses, such as the luteoviruses — transmitted by phloem-feeding insects — are normally restricted to the phloem tissue, although they can move from cell to cell. There is also evidence that certain types of host resistance mechanism may operate by preventing movement of virus into the vascular system, while permitting cell-to-cell spread.

PATHOGENESIS AND SYMPTOM FORMATION

The previous sections have dealt with recognition between plant- and virus-coded molecules which have been predominantly related to the immediate requirements of virus multiplication, and which have had little wider implication for signalling and host response. In this and the following section, the interest shifts to interactions which are primarily of interest because of the host response, and which may display quite complex signal transduction pathways.

We may distinguish between quantitative aspects — inhibition of host growth or loss of yield — and qualititative aspects such as the mechanisms involved in formation of mosaic symptoms, epinasty, leaf distortion and necrosis. Generally, infection by a plant virus causes both inhibition of growth and formation of visible symptoms. However, the cryptic viruses cause no visible symptoms, although they can cause severe reductions in yield (20). It is not known whether the failure of these viruses to cause symptoms is related to their unusual pathway of transmission — entirely through pollen and seed, without any vector or mechanical system — or whether it stems from the fact that they may be mycoviruses which have become adapted to multiply in plants, but have not developed symptom-causing capability.

Control of plant growth

Virus inhibition of plant growth could involve competitive or non-competitive mechanisms. In the former, the amount of virus accumulated would have to be large enough for virus protein and nucleic acid synthesis to be sufficiently active sinks for amino acids and nucleoside triphosphates, that host protein and nucleic acid synthesis would be inhibited. Evidence from quantitative experiments indicates that such competitive inhibition of host growth only occurs with those viruses, such as TMV, which accumulate to very high concentrations (21).

It is in non-competitive mechanisms of growth inhibition that opportunities for signalling are found. For TMV-infected tobacco plants, virus-induced increase in the growth inhibiting plant hormone abscisic acid (ABA) appears to be important in inhibition of growth. In healthy, unstressed leaves, ABA is normally sequestered in the chloroplasts. TMV infection causes an increased rate of synthesis of ABA, and increased rate of metabolism to conjugated forms such as the glucosyl ester and to oxidised forms such as phaseic acid, and a change in compartmentation resulting in large increases in the concentration of ABA outside the chloroplasts. The receptor for ABA-mediated growth inhibition is thought to be extrachloroplastic. The means by which TMV infection signals increased ABA synthesis are unclear — as indeed are several aspects of ABA synthesis in the healthy plant — but appear to be correlated with symptom severity and possible damage to the chloroplasts (22). There is also evidence that transport of virus-induced ABA to other parts of the plant is involved in the control of development of infected plants at the whole-plant level.

In other systems, virus-induced changes in ABA metabolism are probably not central to control of growth inhibition (23). Other plant hormones show various changes after infection, which could be involved in growth inhibition (24), but few studies have been as detailed as those with ABA. Stunting may be associated with alterations in gibberellic acid metabolism, and leaf epinasty and inhibition of growth seem clearly linked with ethylene synthesis. This is a feature of virus infections causing a necrotic response in the host, but also of some viroid infections without necrosis (25).

Formation of symptoms

The visible symptoms formed by virus infections are very diverse, and depend on the particular host and virus involved. Different mechanisms of symptom formation may operate in parallel in a single combination of host and virus. Processes at the cell, tissue, organ and whole plant level have to be considered to form the complete picture.

Virus-induced changes in the chloroplasts are likely to be involved in formation of mosaic symptom, in yellowing reactions, and probably in mechanisms involving changes in plant growth regulators. By what mechanisms do viruses affect chloroplasts so as to bring about symptom formation? The TMV coat protein accumulates in the chloroplasts and specifically inhibits electron transport in Photosystem II (26). Genomic RNA of TMV, but not the monocistronic coat protein messenger, has been detected within the chloroplast (27). It remains to be discovered whether the TMV RNA is translated into coat protein by the 70S chloroplast ribosomes, or whether coat protein synthesized in the cytoplasm is transported across the chloroplast membrane. Furthermore, it is not known how the coat protein in the chloroplast may cause symptoms, but different strains of TMV, with coat proteins having different properties, do cause different types of symptom.

Another ultrastructural alteration which appears to be associated with symptom formation concerns the so-called X-bodies in TMV-infected tobacco. These contain the 126 kDa protein which is thought to be a replicase component, and may be virus replication complexes. TMV strains which cause severe symptoms cause formation of large X-bodies which associate with the nucleus, whereas strains causing mild symptoms cause no or only small X-bodies, which do not associate with the nucleus. The implication is that the X-bodies of severe symptom strains may in some way influence gene activity leading to symptom formation, but the means by which this might happen remain to be demonstrated (28).

Signalling in symptom formation at a molecular level is perhaps best understood in the special case of plant virus satellites. These small RNA molecules are dependent on the helper virus for their multiplication and transmission, but are not required by the helper virus, and do not apparently derive from it. They can ameliorate the visible symptoms caused by the helper virus and may reduce its multiplication, or can intensify symptom severity.

The best studied example is the satellite of cucumber mosaic virus, although several others are known. RNA sequencing studies on variants of satellites which have different effects on virus symptoms, and mutational analysis, have suggested that there are separate domains for different effects: one is involved in induction of yellow mosaic symptoms, and the other in lethal necrosis (29). The means by which these effects are signalled are not yet clear; it is possible that the satellite RNA can interact both with the mechanism of symptom production by the virus, and with the mechanism of response in the plant. A further domain on the satellite RNA may affect satellite multiplication or transport, or virus multiplication.

RESISTANCE AND VIRULENCE

In plant species which are normally susceptible to a particular virus, some individuals or cultivars may contain a gene or genes conferring resistance to that virus. In many cases, isolates or mutants of the virus have the ability to overcome a specific host resistance gene. Like plants and their microbial pathogens, plants and viruses have co-evolved gene-for-gene interactions between resistance genes in the host, and matching virulence genes in the virus. These genes, and their allelic forms for susceptibility and avirulence, appear to involve a number of recognition and signalling mechanisms, from which resistance or susceptibility may result. Analysis of the genetics gives some clues to the mechanisms involved (30).

Resistance controlled by dominant genes appears to involve a recognition event between a molecule coded by the resistance gene, and one coded by a viral avirluence gene. This recognition results in the induction of the resistance mechanism which eventually prevents development of pathogenesis. Commonly, the phenotypic expression of

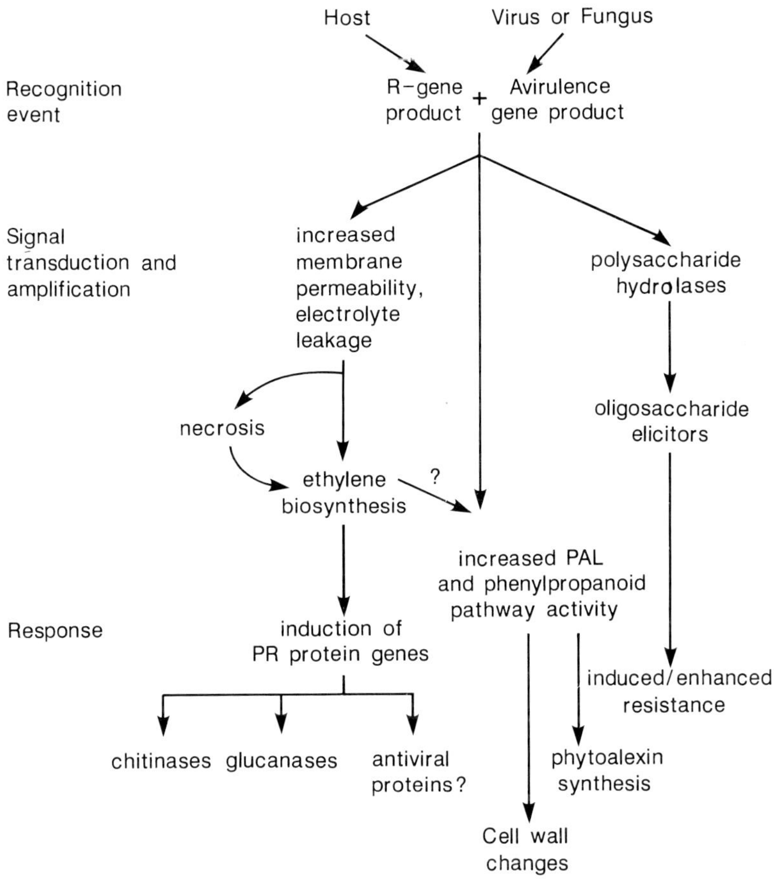

FIGURE 2
Cascade control metabolic changes after an initial recognition event between a resistant plant host and an avirulent strain of virus or fungus. Reproduced from (1).

this type of resistance is the localization of the virus to a small cluster of infected cells, normally followed by the formation of a necrotic local lesion (2).

The induction of resistance may involve a cascade control, with several signal transduction pathways, and several induced responses, as is shown in Figure 2. Enhanced ethylene synthesis appears to be an important component in necrotic responses, although signal transduction pathways which do not involve ethylene also operate. The pathogenesis-related (PR) proteins are amongst the best understood examples of final induction of the host response to viruses (31), but paradoxically, their significance appears to be as potential agents of resistance to fungal and bacterial pathogens rather than to viruses (32). The actual mechanism which localizes the virus infection in hypersensitively-responding plants remains to be discovered. A protein which inhibits virus replication has been isolated and may be involved in the resistance mechanism (33). Alternatively, resistance might operate against the viral function controlling cell-to-cell movement.

Resistance controlled by gene-dosage dependent alleles may be constitutive in at least some cases, and may inhibit virus multiplication throughout the plant. This could involve the direct interaction between an inhibitor of some stage of virus multiplication, and the relevant viral molecule such as a replicase.

Resistance involving recessive genes appears to be comparatively rare, and may involve a negative type of mechanism in that the resistant plant lacks some susceptibility function required by the virus for pathogenesis. Resistance here is the failure of a required recognition event, and any dependent signalling.

While the host components of the recognition reactions leading to resistance are unknown, molecular analysis of virus strains and artificial mutants which overcome particular resistance genes has now allowed mapping of the determinants of virulence/avirulence to single bases or defined regions of the viral genome. Virulence against different resistance genes maps to genes for different viral functions. Thus for TMV, virulence against the tomato gene *Tm-1,* which inhibits multiplication, maps to the putative TMV replicase (34); virulence against *Tm-2,* which prevents spread from the initially-infected cell, maps to the viral protein involved in cell-to-cell transport (35), and virulence against the *N'* gene in tobacco, which causes necrotic local lesions, maps to the viral coat protein (36).

EXPLOITING MIS-SIGNALLING IN CROP PROTECTION

If the normal recognition, signalling and response processes involved in virus multiplication and pathogenesis can be interfered with, there may be opportunities for development of methods of practical crop protection. One of these is already widely used in the field; others are at the development stage.

In 'cross protection', a plant is deliberately inoculated with a mild strain of a virus, and may become partly resistant to subsequent infection by a severe strain (37). This has been widely used in citrus against tristeza virus, and in tomato against TMV. The mechanism appears to involve the protecting strain coat protein, which may interfere with the uncoating of the invading severe strain.

Plants have also been transformed with DNA copies of the coat protein gene, and this has been shown to confer a useful degree of resistance for a number of virus groups (38). DNA copies of benign viral satellite RNAs have also been used to transform plants, and shown to confer a useful degree of resistance to the helper virus (39). These mechanisms of mis-signalling by genetic engineering have great potential for future cuse in crop protection, but environmental safety aspects will have to be carefully assessed before the virus-derived genes are released into commercial useage (40).

REFERENCES

1) Fraser, R.S.S. (ed) (1990) Recognition and response in plant virus interactions. Springer-Verlag, Heidelberg, pp. 467.

2) Fraser, R.S.S. (1987) The biochemistry of virus-infected plants. Research Studies Press/John Wiley & Sons, Chichester. pp. 259.

3) Bald, J.G., Tinsley, T.W. (1967) A quasi-genetic model for plant virus host ranges. I. Group reactions within taxonomic boundaries. Virology 31: 616—624.

4) Bald, J.G., Tinsley, T.W. (1967) A quasi-genetic model for plant virus host ranges. II. Differentiation between host ranges. Virology 32: 321—327.

5) Bald, J.G., Tinsley, T.W. 1967. A quasi-genetic model for plant virus host ranges. III. Congruence and relatedness. Virology 32: 328—336.

6) Mouches, C., Candresse, T., Bove, J.M. (1984) Turnip yellow mosaic virus RNA-replicase contains host and virus-encoded subunits. Virology 134: 78—90.

7) Huber, R., Hontelez, J., Van Kammen, A. Infection of cowpea protoplasts with both the common strain and the cowpea strain of TMV. J Gen Virol 55: 241—245.

8) Sulzinski, M.A., Zaitlin, M. (1982) Tobacco mosaic virus replication in resistant and susceptible plants: in some resistant species virus is confined to a small number of initially infected cells. Virology 121: 12—19.

9) Harrison, B.D., Murant, A.F. (1984) Involvement of virus-coded proteins in transmission of plant viruses by vectors. In: Mayo MA, Harrap KA (eds) Vectors in virus biology. Academic Press, London, p. 1.

10) Wilson, T.M.A., Shaw, J.G. (1987) Cotranslational disassembly of filamentous plant virus nucleocapsids *in vitro* and *in vivo.* In: Brinton MA, Rueckert RR (eds) Positive strand RNA viruses. UCLA Symp Molec Cell Biol Vol 54. Alan Liss, New York, p. 159.

11) Gargouri, R., Haenni, A.-L. (1990) Replication of plant RNA viruses *in vivo* in relation to virus-host interactions. In: Fraser RSS (ed) Recognition and response in plant virus interactions. Springer-Verlag, Heidelberg, p. 163.

12) Goldbach, R.W., Wellink, J. (1988) Evolution of plus-strand RNA virus. Intervirology 29: 260—267.

13) Sänger, H.L. (1987) Viroid replication. In: Diener TO (ed) The viroids. Plenum Press, New York, p. 117.

14) Tabler, M., Tsagris, M. 1990. Viroid replication mechanisms. In: Fraser RSS (ed) Recognition and response in plant virus interactions. Springer-Verlag, Heidelberg, p. 185.

15) Hull, R. (1989) The movement of viruses in plants. Annu Rev Phytopathol, 27: 213—240.

16) Meshi, T., Watanabe, Y., Saito, T., Sugimoto, A., Maeda, T., Okada, Y. (1987) Function of the 30K protein of tobacco mosaic virus: involvement in cell-to-cell movement and dispensability for replication. EMBO J 6: 2557—2563.

17) Stussi-Garaud, C., Garaud, J.C., Berna, A., Godefroy-Colburn, T. (1987) In situ location of an alfalfa mosaic virus non structural protein in plant cell walls: correlation with virus transport. J Gen Virol 68: 1779—1784.

18) Goldbach, R., Eggen, R., de Jager, C., van Kammen, A., van Lent, J., Rezelman, G., Wellink, J. (1990) Genetic organization, evolution and expression of plant viral RNA genomes. In: Fraser RSS (ed) Recognition and response in plant virus interactions. Springer-Verlag, Heidelberg, p. 147.

19) Harrison, B.D., Barker, H., Derrick, P.M. (1990) Intercellular spread of potato leafroll luteovirus: effects of co-infection and plant resistance. In: Fraser RSS (ed) Recognition and response in plant virus interactions. Springer-Verlag, Heidelberg, p. 405.

20) Boccardo, G., Lisa, V., Luisoni, E., Milne, R.G. (1987) Cryptic plant viruses. Adv Virus Res 32: 171—214.

21) Fraser, R.S.S., Gerwitz, A., Morris, G.E.L. (1986) Multiple regression analysis of the relationships between tobacco mosaic virus multiplication, the severity of mosaic symptoms, and the growth of tobacco and tomato. Physiol Mol Plant Pathol 29: 239—249.

22) Whenham, R.J., Fraser, R.S.S., Snow, A. (1985) Tobacco mosaic virus-induced increase in abscisic acid concentration in tobacco leaves: intracellular location and relationship to symptom severity and to extent of virus multiplication. Physiol Plant Pathol, 26: 379—387.

23) Fraser, R.S.S., Whenham, R.J. (1989) Abscisic acid metabolism in tomato plants infected with tobacco mosaic virus: relationships with growth, symptoms and the *Tm-1* gene for TMV resistance. Physiol Mol Plant Pathol 34: 215—226.

24) Fraser, R.S.S., Whenham, R.J. (1982) Plant growth regulators and virus infection: a critical review. Plant Growth Regulation 1: 37—59.

25) Bellés, J.M., Granell, A., Durán-Vila, N., Conejero, V. (1989) ACC synthesis as the activated step responsible for the rise of ethylene production accompanying citrus exocortis viroid infection in tomato plants. J Phytopathol 125: 198—208.

26) Hodgson, R.A.J., Beachy, R.N., Pakrasi, H.B. (1989) Selective inhibition of photosystem II in spinach by tobacco mosaic virus: an effect of the viral coat protein. FEBS Lett 245: 267—270.

27) Schoelz, J.E., Zaitlin, M. (1989) Tobacco mosaic virus RNA enters chloroplasts *in vivo.* Proc Natl Acad Sci USA 86: 4496—4500.

28) Van Loon, L.C., van Kooten, O., Linders, E.G.A., Meurs, C., Wijdeveld, M.M.G. (1990) Recognition and disease development in the tobacco-TMV system. In: Fraser RSS (ed) Recognition and response in plant virus interactions. Springer-Verlag, Heidelberg, p. 311.

29) Devic, M., Jaegle, M., Baulcombe, D.C. (1989) Symptom production on tobacco and tomato is determined by two distinct domains of the satellite RNA of cucumber mosaic virus (strain Y). J Gen Virol, in press.

30) Fraser, R.S.S. (1986) Genes for resistance to plant viruses. CRC Crit Rev Plant Sci, 3: 257—294.

31) Bol, J.F., Linthorst, H.J.M., Meuwissen, R.L.J. (1989) Molecular responses of tobacco to virus infection. In: Smith CJ (ed) Biochemistry and molecular biology of plant-pathogen interactions. Oxford University Press, Oxford, in press.

32) Fritig, B., Kauffmann, S., Rouster, J., Dumas, B., Geoffroy, P., Knopp, M., Legrand, M. (1990) Defence proteins, glycanhydrolases and oligosaccharide signals in plant-virus interactions. In: Fraser RSS (ed) Recognition and response in plant virus interactions. Springer-Verlag, Heidelberg, p. 375.

33) Gera, A., Loebenstein, G., Salomon, R., Frank, A. (1990) An inhibitor of virus replication from tobacco mosaic virus infected protoplasts of a hypersensitive tobacco cultivar is associated with a 23K protein species. Phytopathology, in press.

34) Meshi, T., Motoyoshi, F., Adachi, A., Watanabe, Y., Takamatsu, N., Okada, Y. (1988) Two concomitant base substitutions in the putative replicase genes of tobacco mosaic virus confer the ability to overcome the effects of a tomato resistance gene, *Tm-1,* EMBO J 7: 1575—1580.

35) Meshi, T., Motoyoshi, F., Maeda, T., Yoshiwoka, S., Watanabe, H., Okada, Y. (1989) Mutations in the tobacco mosaic virus 30-kD protein gene overcome *Tm-2* resistance in tomato. Plant Cell 1: 515—522.

36) Culver, J.N., Dawson, W.O. (1989) Point mutations in the coat protein gene of tobacco mosaic virus responsible for the induction of hypersensitivity in Nicotiana sylvestris. Mol Plant Microbe Interact 2: 209—213.

37) Sherwood, J.L. (1987) Mechanisms of cross-protection between plant virus strains. In: Evered D, Harnett S (eds) Plant resistance to virues. John Wiley & Sons, Chichester, New York, p. 136.

38) Nelson, R.S., McCormick, S.M. Delannay, X., Dubé, P., Layton, J., Anderson, E.J., Kaniewska, M., Proksch, R.K., Horsch, R.B., Rogers, S.G., Fraley, R.T., Beachy, R.N. (1988) Virus tolerance, plant growth, and field performance of transgenic tomato plants expressing coat protein from tobacco mosaic virus. Bio/Technology 6: 403—409.

39) Harrison, B.D., Mayo, M.A., Baulcombe, D.C. (1987) Virus resistance in transgenic plants that express cucumber mosaic virus satellite RNA. Nature 328: 802—805.

40) Hull, R. (1989) Non-conventional resistance to viruses in plants — concepts and risks. Proc 19th Stadler Conference, in press.

RECOGNITION AND CELL SIGNALS IN MUTUALISTIC SYMBIOSES

Silvano Scannerini

Dipartimento di Biologia Vegetale dell'Universita di Torino
V. le P. A. Mattioli, 25. I-10125 Torino. Italia

MUTUALISTIC SYMBIOSIS: BASIC BIOLOGY AND CONCEPTUAL FRAMEWORK

The definition of mutualistic symbioses is still a matter of dispute, as are their landmarks and cataloguing. The many semantic, topological and functional problems involved have been reviewed on several occasions (1, 2, 3) and will not discussed in this paper. There is none the less a general consensus of opinion that a mutualistic symbiont system, usually referred to as a symbiont system *tout court*, can be regarded as a combination of two or more organisms of different species that share a very close common existence for long periods and gain a reciprocal advantage therefrom that is reflected in their biological cycle, availability of energy and metabolites, and growth. The overall result is an increase in fitness that is open to assessment through clearly defined cost-benefit algorithms (4).

Symbiosis of this kind is a very widespread phenomenon. It can be found in all five kingdoms: Monera, Protoctista, Fungi, Plants and Animals according to Margulis (5). In Plants and Fungi, indeed, it would seem to be the rule — see e.g. (6).

Symbiont systems display some common features:

1) at the population level, they constitute a very efficient way of colonising extreme environments — e.g. symbiosis between worms and sulfurating bacteria (7), or those in which nutrient substances are scarce — e.g. lichens (8), mycorrhizae (9), reef corals (10).

2) at the organism level, they control both the reproduction of the partners — e.g. bacteria-coleoptera (11), lichens (12) — as well as their growth — e.g. mycorrhizae (9) — and morphogenesis — e.g. luminescent bacteria and the Leiognathid fishes (13), *Rhizobium*-Leguminosae (14), bacteria-insects (15), mycorrhizae (9, 16), lichens (12).

3) at the cell level, interaction between the partners involves penetration of the symbiont into the tissues and or cells of the host, and hence the long-term establishment of compatibility (17, 3). The partners may regulate both the number of intracellular symbionts — e.g. *Paramecium-* (18) and *Hydra-* (19) — *Chlorella-,* the rate of the cell division — e.g. vesicular-arbuscular mycorrhizae (16) — and cell polarity — e.g. bacteria-insects (15).

Yet the mechanisms of integration between the partners and its degree may vary considerably from one symbiosis to another. In some cases, the host has to be reinfected at each new cycle — e.g. mycorrhizae (9) — in others the symbiont is inheritable — e.g. bacteria-coleoptera (11), *Hydra-Chlorella* (20) — and in some cases it is not only inheritable, but also essential for the survival of the host — bacteria-*Amoeba* (21) some bacteria-leafhoppers (22).

NATO ASI Series, Vol. H 51
Cell to Cell Signals in Plants and Animals
Edited by V. Neuhoff and J. Friend

Mutualistic symbioses are thus a complex of highly differentiated systems whose common denominator is a cost-benefit equilibrium that benefits both partners and is reached by putting their properties to common use. Such systems are a source of biological innovation influencing both ecological equilibria (4) and phylogenetic mechanisms (23, 24). Innovation dependent on symbiosis is particularly significant at the cell level: intracellular symbioses (endocytobioses) sustained by prokaryotes are now regarded as responsible for the origin of the eukaryote (5, 25, 26). Endocytobiosis provides a pathway for the acquisition of new exogenous replicons or activation of specific genic segments of the partners — e.g. bacteria-amoeba (27) bacteria-insects (28), *Rhizobium*-Leguminosae (29), algae-algae (30) resulting in the expression of new metabolic capabilities — e.g. luminescent bacteria and nematodes (31). Symbiont systems can thus be regarded as responsible for horizontal gene transfers other than through sexual reproduction and hence as a source of totally new information for the receiving organism (32).

Biological systems which characteristics of this kind are obviously of interest in cytology and molecular biology for two reasons:

1) In terms of the biology of development, they enable an analysis to be made of the effect of heterologous signals on the morphogenesis of the partners. Examples include both short range signals — e.g. lichens (12), VAM Fungi (33, 34) — and long-range signals — e.g. root production in VAM (16).

2) In terms of cell pathology, they allow a study to be made of the signals responsible for compatibility between different organisms and for maintenance and blockade of the two-way relationships typical of mutualistic symbiosis — *Paramecium-Chlorella* (35), *Tetraselmis-Convoluta* (36), VAM (37). This is of particular importance, since mutualistic and pathogenetic symbioses can be seen as a variously modulated continuum in topological and functional terms (1) at the cell level (3, 37, 38, 39).

Mutualistic symbioses thus provide a large number of experimental systems through which to gain a better understanding of the signals regulating cell morphogenesis, differentiation and death, including those that govern cell-recognition, and hence of the mechanisms controlling susceptibility and resistance to pathogens and to mutualistic symbionts in animals and plants.

Mention must, however, be made of certain drawbacks associated with mutualistic symbioses:

1) They have long been studied in isolation. There has been a lack of communication between specialists in different branches and hence a dearth of comparisons in different systems. The state of the art is thus very uneven and the best known systems are not necessarily those with the greatest prospective interest.

2) Mutualistic symbiont systems are complex entities. Integration of the partners is so intimate that there is little scope for a reductionistic approach. Except for *Rhizobium,* e.g. (29), and *Amoeba* of Jeon (40) their molecular biology and molecular genetics are virtually unknown. Moreover, often the biochemical techniques usually employed to investigate molecular signals are unsuitable and in most cases it has been necessary to rely on the indirect evidence offered by ultrastructural, cytochemical, cytoimmunological and genetical analyses.

3) Some systems are so closely integrated that their partners cannot be cultured separately in vitro, nor can their dissociation and synthesis be undertaken under controlled conditions, e.g. : bacteria-cyanobacteria-*Azolla* (41), amoeba-bacteria (40), bacteria-like organisms (BLOs)-VAM fungi (42).

Under these conditions, therefore, examination of what is currently known with regard to cell signalling in mutualistic symbioses must necessarily be preceded by comparison of the cell structures involved in starting and maintaining different symbioses to see what processes are involved in the host-symbiont cell contact. The following discussion is based on the work of the NATO ARW "Cell-to-Cell Signals in Plant. Animal and Microbial Symbiosis" held at the Villa Gualino, Turin, from 19 to 22 May 1987. Account is also taken of more recent data observed in various systems and published or presented at congresses during the last two years.

FUNCTIONAL MORPHOLOGY OF CELL-CELL INTERACTIONS IN MUTUALISTIC SYMBIOSIS

It would seem that the best approach to an understanding of the compatibility and specificity mechanisms involved in a mutualistic symbiosis is to compare three cytologies, namely that of the partners during their separate growth, that of the natural symbiont system, and that of a similar system obtained by resynthesis under controlled conditions. This procedure provides a precise catalogue of the cell changes induced by symbiosis in each partner, For the reasons mentioned early, however two major groups cannot be studied in this way:

1) Natural symbioses not yet reproduced in the laboratory because of the impossibility of obtaining host aposymbiotic lines as *Amoeba*-bacteria (40).

2) Symbioses whose partners are not subculturable, see e.g. Glomaceae mycelia (37, 42) or are not able to infect their host in vitro, see e.g. bacteria partners of *Anabaena-Azolla* symbiosis (41).

3) Inherited symbioses in which symbiont is transmitted by cell reproduction and hence infection cannot be obtained in vitro, e.g. BLO-Glomaceae (42), *Amoeba*-bacteria (40), some bacteria-leafhoppers (22).

It is clear from what has so far been learned about the functional morphology of mutualistic symbioses, however, that they have three common cellular features (3)

1) absence of cytopathological symptoms in the active phase of mutualism;

2) presence of complex interfaces implying a great variety in their contact structures;

3) presence of several types of structures reminiscent of the stages of phagocytic interactions.

1) Absence of cytopathological symptoms

Both animal and plant cells acting as hosts to parasites display symptoms typical of a defence reaction, followed by signs of "cell damage". All these symptoms are familiar feature of both animal cell pathology (43) and plant cell pathology (44) and have been discussed in other specific NATO ARW and ASI programmes. In the case of the other kingdoms the data are fewer. None the less, it is said that the cell pathology of the Protozoa is comparable to that of animals and those of Algae and Fungi to that of Plants. A reference can thus be made to common models. Yet the initial defence reactions of animal cells are usually "mediated by antibodies", whereas this form of defence is lacking in plant cells. Resistance induced as the result of preimmunization (45) on the part of non virulent strains of fungal pathogen, too, appears to be due to the production of non specific elicitors (46). More important, a plant's defence barriers are linked to cell changes in its walls: formation of papillae, layers of callose, cell wall lignification, together with secondary modification induced by metabolism of phenylpropanoid compounds (46, 47). Such are reactions very rarely immediate however. Even though some parasites may kill an invaded cell straightaway (48), most are biotrophic. A defence reaction is thus preceded by a period of compatibility, during which infection is established in a living host both in plants (44) and in animals (43). Once this period is over, symptoms characteristic of disease appear in the event of pathogenic interaction: membrane damage, changes in semiautonomous organelles, lysosome systems, nuclei, and cell surfaces. none of these are observed in mutualistic interactions. In other words, a cell hosting a mutualistic symbiont often displays changes, especially on its surface (see later): None the less, during the active phases of the interaction it behaves as though the compatibility period was of indefinite duration. This does not rule out membrane flow and organelle modification: nuclear hypertrophy and chromatin despiralization, with possible activation of transcription in VAM host cells (49); production of specialized vesicules in Paramecium due to differentiation of the perisymbiotic vacuole around *Chlorella* (18); production of specialized membranes around Rhizobium in dinitrogen-fixing root nodules (50); changes in the lysosome cycle of Hydra hosting *Chlorella* (20) or in *Convoluta rescoffensis (Turbellarium acoelum)* as the host of the Alga *Tetraselmis* (36); absence of host membranes around the BLOs of Glomaceae (42). As far as the symbiont is concerned, however, the situation is different, since the compatibility stage may be followed by stages during which the single-cell symbiont population is controlled by the host cell in a variety of ways: extrusion of the excess of symbionts (36) or more commonly their harvesting (i. e. digestion) as in *Hydra-Chlorella* (20, 51) and *Paramecium-Chlorella* (18) or lysis through mechanisms not yet clearly demonstrated as e.g. in the collapsed arbuscule phases in VAMs (34) and in the final phases of leguminosae-*Rhizobium* symbioses (14). Gradual alteration of the hosted prokaryotes and eukaryotes obviously appears during these stages. Nevertheless, signs of pathological degeneration are not found in the symbiont during the compatibility period, which can be regarded as the active phase of mutualism. On the other hand, there are progressive indications of its simplification, especially on the cell surface: Examples include loss of its theca by *Tetraselmis* following its internalization by *Convoluta* (36), and gradual thinning and simplification of the wall structures of VAM Fungi on transition from the extraradical to the arbuscule stage (33, 34, 37). These alterations imply reduction of intersymbiont barriers during two-way exchanges. They may, however, have a precise morphogenetic meaning, particularly in Plants and Fungi, where morphogenesis is closely associated with wall changes (see later).

2) Presence of complex interfaces

It has long been known that cell surfaces undergo extensive alteration when cells of different organisms come into contact and establish a stable relationship. This is true of both pathogenic and mutualistic eukaryotes (52, 37, 38) and prokayotes (53, 14). Not surprisingly, therefore, the interface between organisms in mutualistic and pathogentic symbiosis has long been regarded as the focal point for understanding the relations between the cells of both plant pathogens (54) and mutualists (55). More recently this interpretation have been extended to the relations between the cells of animal pathogens and

TABLE 1

This table compare the types of interface established between symbionts and the steps in phagocytosis resulting in the creation of similar interfaces in conventional phagocytic interactions. All symbioses can be regarded as the outcome of a "choice" by the symbiont of one or more of the four stages of a delayed phagocytosis as its ecological niche. Step 5, of course, is confined to organisms that actually cross the perisymbiotic membrane barrier and become cytoplasmic endocytobionts (11) or are inheritable not dissociable endocytobionts (21, 42).

CELL TO CELL INTERACTIONS BETWEEN MUTUALISTIC SYMBIONTS			PHAGOCYTOSIS
INTERFACES	ASSOCIATIONS	STEPS OF INTERACTION	STEPS OF PHAGOCYTOSIS
1 WALL-WALL (and/or Interface Material)	Lichens, Plant-Cyanobacteria,Plant--Dinitrogen fixing bacteria, Ectomy-corrhizae, Endomycorrhizae, Nematode--luminescent bacteria	Recognition and entrapment of symbiont on host surface	Recognition and entrapment on the cell surface
2 PERYSYMBIOTIC MEMBRANE - INTERFACE MATERIAL - WALL	Lichens, Plant-Dinitrogen fixing bacteria, Endomycorrhizae	First steps of internalization	Invagination of the plasmalemma and formation of the open phagocytic vacuole
3 PERISYMBIOTIC MEMBRANE - WALL	Plant-Dinitrogen fixing bacteroids, Endomycorrhizae, animal-algae, protozoa--algae, protozoa-bacteria, insect-bacteria, insect-yeasts	Complete internalization, Symbiosome differentiation	Closure of the vacuole
PERISYMBIOTIC MEMBRANE - PLASMAMEMBRANE	Animal-Algae		
4 PERISYMBIOTIC MEMBRANE - SYMBIONT REMNANTS	"Harvesting phase" of intracellular associations	"Harvesting phase" lysis of the symbiont *	differentiation into a heterophagocytic vacuole, activation of the lysosomal system through membranes and lytic enzymes, lysis of the segregated material by enzymes
5 CYTOPLASM - WALL (and/or PLASMAMEMBRANE)	Ameba-Bacteria, Insect-bacteroids VA Fungi - bacteria-like organisms (BLOs)	Escape from symbiosome into cytoplasm	

* may be substituted by control of symbiont population by extrusion.

mutualists (3). The term "interface" has been applied to the "complex of contact structures between the partners", including both membrane and wall ultrastructures, and any other materials (matrix) that may deposited. Table 1 sets out the types of interface between mutualistic symbionts as defined in terms of their ultrastructural morphology. This classification, however, is by no means exhaustive. Three points have still to be established: 1) the architecture and chemical composition of the extra-membrane materials (glycocalix, walls, matrix materials) and 2) their origin (host and/or symbiont), and hence the domains proper to each partner, and 3) the architecture, chemical composition, and biochemical and chemical-physical properties of the membrane units involved in the interface, as well as the membrane flow that has constructed them. Apart from the morphological and topological ultrastructural references, little information is available.

As far as the "non membrane" material is concerned, the bulk of information is related to mycorrhizae and dinitrogen fixing root-nodules, It is only very recently that some experimental systems have been examined cytochemically using affinity techniques, such as monoclonal antibodies, see for technical discussion (56) and or Au conjugated lectins (57). Very precise data on the supramolecular organization of the host walls, the matrix material, and changes in the fungus wall have been obtained for the VAM (37, 38) and Ericales (57) mycorrhizae composition and topology of VAM interfaces. The model of the VAM interface structures displays progressive simplification of the supramolecular organization and the polysaccharide and aminopolysaccharide composition of the fungus wall, and a parallel simplification of the polysaccharide architecture in the host cell domain (37, 58, 59, 60) such as to minimise the apoplastic barrier and to differentiate a matrix material, which is produced by the host and contains pectins and non-cristalline cellulose (58). A number of problems remain, however, particularly that of the origin of the "simplification" of the matrix material (partial enzymatic lysis of an already differentiated wall, or blockade of cellulose during deposition of new wall material?). Similar results, supplied by immunoblotting and by electron microscopy using Au-conjugated monoclonal antibodies, have also be obtained for the infection thread of Rhizobia in the leguminosae (56): this matrix material however is built up by the host and the symbiont together (56).

Even less precise are the findings concerning the membranes that bound the symbiont when an endocytobiosis (intracellular relation)is established. This uncertainty of definition is reflected in the proposed consideration of the space containing the intracellular symbionts as a specialized symbiosome (61), leaving the simple denomination "perisymbiotic membrane" for their surrounding membrane (3, 61). Another point to borne in mind is that the data for the different symbiont systems are in partial disagreement. In VAM, for example, it is agreed that it is an extension of the host plasmalemma on the strength of the presence of vanadate repressed ATPase (34). None the less the cell potentials measured in maycorrhizae are significantly greater than those of axenic root cells (59). In *Hydra-Chlorella* (20), *Tetraselmis-Convoluta* (36), *Paramecium-Chlorella* (35) and perhaps marine invertebrate-algae (62) it is regarded as an endoplasmic reticulum membrane bounding a space (the symbiotic vacuole) in which hydrolase transport is inhibited. It can also been seen as a conventional eterophagosome membrane in *Hydra-Chlorella,* however (51). Precise data on the *Rhizobium*-Leguminosae membrane provided by genetic, cytochemistry and electron microscopy point to an origin from differentiated membrane flows (50, 63). Namely a membrane flow involving endoplasmic reticulum, plasmalemma and Golgi variously modulated at different times during internalization of *Rhizobium* into the host cell.

In general terms, therefore, the interface data demonstrate the presence of structures which, in all the symbiont systems investigated so far, are referable to cell-cell adhesion mechanisms with various degrees of "introduction of the symbiont microorganism into the host cell". It is on these albeit extremely incomplete findings that one must rely in attempting to create an entirely hypothetical model aimed at identifying a single phenomenomenon for the host and symbiont cell-cell interactions of any and every taxonomic category and, at allowing a comparative approach to be made for symbiotic interactions.

Comparison of the interface data in the table 1 clearly shows that there are three types of cell-cell topological relations:

1) extracellular, where the two partners are in very close contact via surfaces protected by various materials aimed (inter alia) at ensuring adhesion;

2) intracellular with a differentiation of a symbiosome, within which, after an initial extracellular stage, the symbiont is introduced into the cell, maintained throughout the active phase of interaction, and then eliminated by farming or extrusion at the end of the mutualism; 3) intracellular, without interposition of the perisymbiotic membrane.

3) Structures that can be correlated with phagocytosis or related phenomena.

These interfaces are the result of interactions that involve the activities of the cells of the partners *in toto.* If one wishes to determine the existence of a common model in the different symbiotic interactions, it is essential to undertake a comparative analysis of the cell biology of the partners in the various interaction stages. For the reason stated earlier, most of the information available on the cell biology of symbionts is merely ultrastructural and ultracytochemical: it is thus possible to make a comparative ultrastructural analysis of all the symbioses mentioned.

This has led to establishment of a common denominator for all animal or protoctistic cells hosting intracellular symbionts closed in a symbiosome: their behaviour is that of "blocked" professional phagocytes. The data collected for *Paramecium-Chlorella* (18), *Hydra-Chlorella* (19, 20, 51), *Tetraselmis-Convoluta* (36), dinoflagellates-corals (62. 64), and some bacteria-insect interactions (22, 65) are in line with classic phagocytosis models — see for a general review (66). The series of phenomena described in the part 2) of this chapter, in fact, corresponds to the stages of a delayed heterophagocytic phenomenon. This result is obtained through mechanisms documented in professional phagocytes in which phagocytosis is inhibited experimentally or blocked by pathogenic organisms — for a complete review see (66), namely: 1) inhibition of hydrolase transport in the perisymbiotic space as in *Paramecium-Chlorella* (18), *Tetraselmis-Convoluta* (36), dinoflagellates-marine invertebrates (62, 64), a part of bacteria-insects (22, 65), Hydra-*Chlorella* according to McAuley (20); 2) selection of Chlorella strains resistant to the phagolysosomes environment, *Hydra-Chlorella* according to Rahat (51). The cell of plants hosting the symbiont closed in a symbiosome also display structures reminiscent of phagocytic interactions both in VAMs (67) and dinitrogen-fixing root-nodules (50, 63), though something more must be said about this phenomenon. It was long asserted, in fact, that the plant cells were incapable of phagocytosis and endocytosis in general. They can however, engage in endocytosis provided their wall is removed. Protoplasts can take in both lectin gold conjugates (68) and cationized ferritin (69). Moreover, a lysosome system fully comparable with that of of the animal cell is present in plant cells (70) and it has recently been shown that receptor mediated endocytosis is also possible in plant cells in energy term (71). A frustrated phagocytosis model for type 2) symbiotic interaction in plants, too, has already been proposed (67) on the strength of ultrastructural evidence of symbiont adhesion-introduction-segregation sequences investigated by perisymbiotic membrane in VAM (34, 37) and in *Rhizobium*-Leguminosae (14), and the demonstration of lysis inside the symbiosome of both *Rhizobium* and VA mycelia. The recent data provide further support for this view. In VAM, highly specific cytochemical markers have shown that the matrix material consists of the wall (reduced to the minimum) of the host cell, which becomes functionally similar to a protoplast (58); in the same way, control of peroxidase activity during the development of infection displays an initial stimulation when the relations are primarily extracellular (72), as in the adhesion stage of professional phagocytosis (66, 73). In Rhizobia-Leguminosae systems, the immunocytochemical data recently reported (74) point to a reduced wall (apoplastic barrier) in the host cell as in VAMs, whereas those for the perisymbiotic membrane are indicative of endocytosis during the *Rhizobium* internalization (infection thread) stage and an origin from different integrated endomembrane systems of the symbiosome (63). Type 2) symbioses with plant cells can thus be referred to the previously stated phagocytic interaction model. The fact that in VAMs the symbiosome does not completely surround the hyphae can be attributed to the well-known phenomenon of frustration of phagocytosis on account of geometrical encumbrance of the ingested microorganism and possible emission of the lytic enzymes (66, 73, 75). The final stages — lysis of the microorganism at the end of the interaction — can only tentatively assigned to this phenomenon, because there is no incontrovertible evidence of the release of the host's lytic enzymes into the symbiosome to obtain the harvesting of the symbiont and autolysis cannot be ruled out.

Yet this is not all: both type 1) (extracellular) symbioses and those of type 3) (symbiont in the host cytoplasm without interposition of a perisymbiotic membrane) can be tentatively correlated with type 2), and hence regarded as a continuum including phagocytic stages. A type 3) symbiosis is often inherited as in bacteria-insects (65), BLOs-Glomaceae (42) and bacteria-Amoeba (40). It is formed of symbionts that are so closely integrated that they can be regarded as cell organelles, and has been widely discussed in terms of the endosymbiotic origin of the eukaryotic cells (5, 24 pro parte). It can be seen as the result of the "flight" from the symbiosome of the symbiont, which gradually changes into an organule according to the serial enosymbiotic theory (SET) (5, 76). Indirect confirmation of this possibility is provided by the presence of residual organelles (chloroplasts) of algae phagocytosed and digested in the cytoplasm of marine invertebrate cells and their maintenance after artificial introduction (77) and, at the opposite extreme, the fact that some intracellular symbionts can cross the symbiosome barrier and survive to host degenerating cell at the end of mutualistic interaction, or pass into non specific hosts where they act as weak pathogens (78).

Type 1) symbioses (extracellular symbioses) in which the symbiont is accomodated closely adhering to the cell surfaces or in introflections of its wall — ectomycorrhizae (79), lichens (12), *Anabaena*-cyanobacteria-Bacteria (41) involve the production of specific metabolites perhaps on the part of the host, as neoproteins in ectomycorrhizae — ectomycorrhizins — (79, 80) and in lichens (12), polysaccharide glues in *Anabaena*-Cyanobacteria-Bacteria (41, 81). The same is true in the extracellular phases of intracellular symbionts, they involve the production of specific structure and/or metabolites as lectins on the surface of dinitrogen-fixing nodule (82). Some of these products could be involved in adhesion. Extracellular interactions could be seen as an initial "entrapment" step, whereas the subsequent "engulfment" step is blocked by the presence of wall material and the inhability of the symbiont to reach the membrane receptors. Indirect corroboration of this hypothesis is the fact that there are several types of interface in lichen systems, ranging from simple wall-to-wall juxtaposition to so-called "intraparietal" and "intracellular" haustoria (12), or transfer cells in relation with the fungus in some ectomycorrhizae (83).

Functional morphology tests can thus be employed to build a unifying model for cell-cell interactions in symbiont systems based on the steps of a delayed "conventional" or "frustrated" phagocytosis ensuring the maintainance of long periods of compatibility. A scheme of this type (Tab. 1) brings into correspondence the stages of heterophagocytosis according to classic patterns (66, 73) and the different mutualistic symbiotic systems during their cell-cell interactions by identifying the location of the partners and specific compatibility stages (3). From a cytological standpoint, this model is compatible with the behaviour of both biotrophic parasites and the mutualistic symbionts studied so far.

The model is admittedly hypothetical and requires further in-depth studies to confirm its correctness, particularly with regard to type 1) and type 3) symbioses. Even so, it enables a single approach to be made to the question of cell signals required for the establishment and maintenance of both mutualistic and biotrophic pathogenic symbioses, The signals in question are those for: adhesion and introduction of the symbiont; control of the membrane flow and functioning of symbiosome as in equilibrium between lysosomes and symbiosomes; blockade of compatibility and commencements of the stages of control through lysis or extrusion of the intracellular symbiont; alternatively,

TABLE 2
Types of stimulus occurring in the better known stages of a host-symbiont interaction. Some interactions have been determined in outline only or are subject of conflicting interpretations (after SMITH, 1988, slightly modified).

Type of Interaction	Type of Symbiosis	Nature of Stimulus	Response
CONTACT OF EARLY PHASES OF ESTABLISHMENT			
Specific Signals	legume/rhizobium	lectins	attachment
		flavonoids	regulator gene activated
	V.A. Mychorrizae	?host chitinase	?fungal "infectivity"
Recognition cues	legume/rhizobium	root surface	adhesion of symbiont
	V.A. Mychorrizae	root exudate	hyphal branching
	" "	root surface	appressoria formed
	" "	pecto-cellulosic walls	intercellular penetration
	lichen	algal cell surface	hyphal extension
	Convoluta roscoffensis	algal cell wall	entry to host
Non-specific interactions	gut symbionts	anaerobic atmosphere	anaerobic metabolism
	green hydra	contact with digestive cell surface	phagocytosis
LATER PHASES OF ESTABLISHMENT			
Specific Signals	legume/rhizobium	?host + symbiont gene products	?specific cell-to-cell interactions
	mychorrizae	?wall proteins	binding of symbiont to host wall
	V.A. Mychorrizae	?wall saccharides	intracellular penetration
	lichens	?symbiont product	synthesis of secondary metabolic products to host
	green hydra	vacuolar pH	maltose release by symbiont
	" "	maltose release	avoidance of lysosome fusion
	" "	?'division factor'	symbiont cell division
	anemones, corals	?'host factor (s)'	photosynthate release from symbionts
	N_2-fixing symbioses	?	Inhibition of symbiont glutamine synthetase
Recognition cues	Ruminants	release of VFA by symbionts	Input of saliva to mantain pH
	V.A. Mychorrizae	no encrousted cell wall	intracellular penetration

crossing of the perisymbiotic membrane and "evasion" of the symbiont resulting in its integration in the cytoplasm or its behaviour as a pathogen.

This is equivalent to considering as a pattern for the search for cell signals in mutualistic symbioses, the mechanisms responsible for: 1) cell adhesion and 2) modulation of the behaviour of the lysosome system.

Both phenomena are poorly understood and in present state of the art not open to correlation with a particular specialized mechanism in a given organism. In the case of adhesion mechanisms, the determinants appear to be primarily linked to proteins and glycides (84). Yet the determinant recognized by an anti-*Candida* monoclonal antibody and involved in the adhesion of Candida to epithelial animal cells (85), for example, is also found on *Hymenoscyphus ericae,* where it appears to be involved in adhesion to Ericaceae roots for mycorrhization (38). Modulation: of the lysosomes appears to be the task of a "system" involving Ca++ and other ions, matabolites of arachidonic acid, cytoskeletal structures, soluble cytoplasmic proteins, cyclic nucleotides, etc. (66). Some of this factor (e.g. soluble proteins, modified ion flows, variation of membrane component metabolites) have also been demonstrated in the interactions between mutualistic symbionts as shown in Tab. 2 and in (20, 59) and between biotrophic pathogens (86).

However, in parallel with the several phases of symbiotic interaction, beginning with recognition, mechanisms that greatly alter the morphogenesis of the partners are set up in almost all systems: dinitrogen-fixing nodules (14), mycorrhizae (9, 16, 17, 34), lichens (12) bacteria-insects (15, 65), animal-algae (20, 36, 62), protozoa-algae (18). Moreover a complex two-way exchange of nutrients take place between the symbiotic partners requiring modifications of their membrane activity and ion fluxes. Under these conditions, there can be no certainty that recognition, "the set of phenomena resulting in the expression of specificity or selectivity in associations between hosts and symbionts" in the words of D. C. Smith (87), and the subsequent course of the interaction, are solely dependent on molecular signals.

STIMULI, CUES AND SIGNALS IN SYMBIOTIC CELL TO CELL INTERACTIONS

Cell to cell interactions between mutualistic symbionts pass through three, theoretical, stages: recognition and adhesion, engulfment and establishment (maintenance) and control. In nature however, many inherited interactions as weevils-bacteria (11) *Hydra-Chlorella* (19, 20), *Paramecium-Chlorella* (18) do not go through the recognition stage, and this is solely reconstructed by laboratory experiments. By contrast in interactions with obligate reinfection as mycorrhizae (9) and dinitrogen-fixing symbioses (14), repeat all three stages at each cycle. In both cases the interfaces must be continuously kept under control the interaction via stimuli that regulate its compatibility and nutritional exchanges.

The stimuli so far documented vary considerably both in different systems and in different moments of the interaction (see Tab. 2 for a general survey): nutrients — e.g. maltose released by Chlorella for Hydra (19, 20); glycerol and aminoacids by *Symbiodinium* for marine invertebrate (62); saccharose by roots for VA fungi and phosphate by VA Fungi for roots (34) —; molecules constituting the interface and contained in it — e.g. fungus wall proteins in the Ericales mycorrhizae (37), 1, 4 glucanes and glucuronyls of the host (58) and N-acetylglucosamine oligomers of the fungal wall (38) in VAMs, glucuronyl and glucose in the developing infection thread of dinitrogen-fixing nodules (74, 88); translocating molecules: chitinase secreted by the host at the beginning of the interaction in VAMs (89) and Leguminosae flavonoids that activate the nod genes of *Rhizobium* at the beginning of the plant-*Rhizobium* interaction (29, 90) miscellaneous variations in physico-chemical cell parameters: in the electric charges of the cell surfaces as in *Hydra-Chlorella* (20), in cell potential (or membrane potential) as in VAMs (59), in the pH of the symbiosome as in *Tetraselmis-Convoluta* (36); nonspecific factors or factors whose chemical composition has not yet been determined — factors stimulating or inhibiting the nutrient exchange in Zooxanthellae (64), anaerobic envinroment in nematodes-*Xenorhabdus* symbioses (31) and in dinitrogen-fixing nodules (14), altered or simplified wall materials as in lichens (12, 91), *Paramecium-Chlorella* (18), *Tetraselmis-Convoluta* (36), *Hydra-Chlorella* (20). The significance of these stimuli and their correlation with intersymbiont signalling mechanisms are poorly understood for the most part, one reason being the lack of data on molecular biology of nearly all symbiotic systems, except *Rhizobium*-Leguminosae and mutualistic amoeba-bacteria of Jeon.

Moreover, cell signalling mechanisms in vertebrates are quite well known. They could perhaps be extrapolated to invertebrates, but in plants they are subject to debate.

In animals, cell signalling depends on molecular signals and is well established, particularly for systems activated by water-soluble peptides, namely hormones and neurotransmitters, which are signals received by specific proteins or glycoprotein membrane receptors that act via second messengers (cyclic AMP, Ca++, inositol phosphatides) responsible for activating proteins that produce the response to the signal. More recent work has shown that the second messengers include diacylglycerol, leukotrienes, lipoxins, prostanoids, superoxide anions, and oligoadenylates. Receptors, signals and second messengers belonging to these categories may be involved in the regulation of endocytosis and phagocytosis (66, 84).

In plants, while it may be inferred that the cell-cell signalling mechanism moves along the same general lines as that of animal cells, the experimental evidence concerning molecular signals is far from concordant. The most likely candidates for the function of molecular signals are oligosaccharins, for a complete discussion (92). These originate from the demolition of wall material and are involved in both the morphogenesis of plants and their resistance to pathogens (86). The question of the receptors and second messengers (if any), however is far from being solved. There is substantial evidence that the Ca++ cation plays an important part in plant cell regulation (93), whereas nothing is known for certain with regard to receptor molecules, nor the function of cyclic AMP as a second messenger, though the importance of the lipoperoxidation reactions that govern the balance of the lysosome system in professional phagocytosis (66) is now suspected to be a predominant factor in cell control of membrane function (94). In a recent pattern for cell signalling proposed for the host-pathogen relationships in plants, the primary signal is converted in a messenger outside the cell and then internalized to act as an elicitor of the resistance response (86). This arrangement is in line with the commonly accepted fact that the cell language of plants is essentially composed of glycide molecules (by contrast with the protein language in animals) and with gene by gene mechanism for the control of resistance to pathogens (95). However, it lacks sufficient proofs and is still for the most part the result of working hypotheses or even speculations. This state of frank backwardness in our knowledge of cell-signalling in plants is also determined as Smith pointed out in our NATO ARW by the technical complication of the existence of cell walls that render it difficult to locate the receptors and separate them, and complicate the work of the experimenter, forcing him to resort to indirect techniques, such as affinity reactions with enzymes, lectins or monoclonal antibodies conjugated with gold or fluorochromes — see for technical discussions and examples in different symbiotic systems (56, 57, 89) — whose quantification is subjected to intriguing approach by means of static and flow cytometry and image analysis.

In view of the molecular and structural complexity of the interfaces, the difficulties involved in the demonstration of signals, receptors and messengers already mentioned when dealing with plants recur and are compounded by the fact that here two different cells come into contact in mutualistic symbiont systems.

Furthermore, not all stimuli with a significant function in the establishment of a symbiosis can be regarded as signals in the strict sense. Examination of the stimuli so far documented in the systems set out in Table 2 — from Smith (87) modified — shows that they can be divided into three categories:

1) stimuli whose sole function is to induce a specific response from the partner resulting from activation of a gene or group of genes (specific stimuli determining a specific response) signals in the strict sense;

2) stimuli that are the normal consequence of one of the partner's ordinary characteristics, but to which the other organism responds in a specific manner, namely by activating gene, structures or metabolic pathways permitting establishment of the symbiosis: recognition cues;

3) Nonspecific stimuli that do not determine specific responses: non specific interactions.

By "signals" according to D. C. Smith (87), we mean stimuli formed of specific molecules whose only function is that of inducing a specific cell response. In our opinion, what is more, a supramolecular structure modified during the interaction whose only function is to induce a specific response could be a *bona fide* signal and hence regarded as a "new" molecule.

Reconstruction of cell-cell signalling in a symbiotic system thus means that an answer must be found to the following serie of questions:

1) What is the molecular nature of the signals and receptors responsible for adhesion between the partners ?

2) What is the molecular nature of the signals and receptor responsible for the specificity of recognition and the establishment of compatibility?

3) What is the molecular nature of the signals and receptors ensuring the exchange of nutrients and the maintenance of compatibility?

4) Do signals, once internalized by the partner cell, regulate the gene(s) responsible for the response directly, or do they give rise to secondary signals that act directly or through second messengers?

Finally the following goals must be reached:

1) The genes responsible for production of the signal and those that control the response must be determined and possibly cloned.

2) In the case of the recognition cues, an answer must be found to the question of their variability among different individuals, clones, or strains of the same species, since this is closely linked to the problem of the "preadaptation" as an addition or supplement to conventional "recognition", and to the problem of the specificity of the recognition in symbiotic partners.

3) The correlations between the control of metabolism and the non specific interactions must be established. No symbiotic system has as yet provided the information needed to answer the questions and to reach the goals necessary for a full understanding of the mechanism(s) by which information is exchanged between cells in symbiosis. None the less, data have been collected, especially with regard to the *Rhizobium*-Leguminosae, since their agrotechnological significance has made them the subject of considerable interest (88), some work has been performed on the location of signals and cues in mycorrhizae and with regard to the biochemistry of some animal-algae, protoctista-algae, bacteria-insects, protoctista-bacteria, animal-bacteria systems. Two fundamental facts emerge from the work carried out so far (87).

1) initial recognition, which corresponds to the adhesion stage, appears to be substantially nonspecific, and this promotes the infectability of the host, with obvious advantages with respect to its fitness,

2) the following stages ensure compatibility, and exchanges are regulated by more specific stimuli.

The first recognition stages are primarily linked to common recognition cues: algal wall (36), secreted sugars from the roots in the mycorrhizae (34, 96), surface polymers in *Paramecium-Chlorella* (18), pecto-cellulosic not encrusted host wall components in

VAMs (97), or to nonspecific interactions anaerobic environment for *Xenorhabdus-nematodes* (31). Recognition signals are by no means lacking however: flavonoid produced by the host root that activate the nod genes of Rhizobia (29), specific metabolites and proteins produced by lichenized fungi (12, 91). D. C. Smith (87) views the non-specificity or low specificity of recognition as a selective advantage that allows the host to be ready for reinfection whenever the symbiont happens to be in its surroundings. This correlation would appear to be indirectly confirmed by the great diffusion of molecules serving as adhesion signals in totally distant systems: the monoclonal antibody that recognises a determinant for *Candida* pathogenic for human epithelia can also recognise infectious strains of the fungus symbiont of ericoid mycorrhizae (47). The lectin involved in the adhesion of *Rhizobium* to some leguminosae is probably a host cues (87) and the flavonoids themselves (29) are not strictly specific in their selection of the symbiont species, since they can bind various *Rhizobium* strains. Furthermore, there is nothing to show that the initial stages of recognition are definitely dependent on signals or recognition cues. Experiments with laboratory maintained aposymbiotic strains of *Hydra fusca* (51) have shown that the uptake of *Chlorella* of various species or strains is not selective, and that it is only after introduction into the cell that a distinction is made between what is and is not digested. Rahat (51) therefore suggests that there are preadaptation mechanisms on which selection acts at random.

The initial, a specific stage is followed by a selective stage to ensure the compatibility of the system. The progress of *Rhizobium* infection is accompanied by specific modifications of the interface architecture (61, 74) and by the activation of genes other than nod (90). Transformation of VA fungus into an intracellular arbuscule involves a change in the cristallization of the fungus wall amino saccharides components (37) and in the supramolecular organization of the interface material, even if the single monosaccharide and disaccharide component correspond to those of the host wall (57, 60). Moreover the walls of the cells hosting arbuscules are modified in their texture (57, 60). In the same way there is a marked despiralization of the chromatin of the cells hosting the arbuscule. This can be demonstrated with static and flow cytometry and points to enhanced transcriptional activity (49). Specific compounds produced during mycorrhizic interactions — ectomycorrhizin (79, 80), perhaps endomycorrhizin (96) — suggest that it results in new proteins induced by the symbiosis or at least in new products. The same is true of the interaction between *Xenorhabdus* and Nematodes — production of antibiotics by the symbiont — (31), and between lichenized fungi and algae or cyanobacteria (12). Likewise, in the well-known *Chlorella-Hydra* system the vacuolar pH governs the release of maltose of the alga, and this in turn inhibits the fusion of the symbiosome with the lysosomes (19). A system of the same kind appears to operate in *Paramecium-Chlorella* (18) and *Tetraselmis-Convoluta* (36), The position with regard to the anemone or coral Zooxanthellae symbioses seems to be more complicated than had been supposed by previous authors (62). Experiments with cell extracts have revealed the existence of both a host releasing factor that stimulates the two-ways transport, and substances with an inhibiting effect (64).

As the interaction proceeds, there is in all the studied systems an increase in the number of interactive processes. Recognition mechanisms are joined by morphogenetic mechanisms and two-way nutrient transfers that require their own control systems. The task of interpretation is thus rendered more complicated. . As a result, patterns illustrating the nutritional interactions and their relation to control of the symbiont population have been worked out for no more than a handful of symbiotic systems as in *Hydra-Chlorella*

(19, 20), *Tetraselmis*-Convoluta (36) whereas for others there is indirect evidence that allows changes in the cytochemical, ultrastructural and bioelectric characteristic of the interface to be correlated with the exchange of nutrients (49, 59) but cannot be used to build reliable models of the signals controlling host and symbiont morphogenesis or of the blockade of the interaction.

CONCLUSIONS

That the history of science has had its full share of fallacious predictions and conclusions which failed to stand the test of time is a pronouncement bordering on the pontifical. The field of symbiosis itself provides a fine example of an assessment of poor success in the comment whereby E. B. Smith, some 60 years ago, elegantly disposed of the basic ideas of endocytobiology with the remark that the origin of Eukaryotes by means of symbiosis was "too fantastic for present mention in polite biological society" — quoted in (5) preface —. With such a shining dictum before me, I hope that — "si parva licet componere magnis" — I may be allowed to steer clear of over-compromising conclusions, and that such perspectives and conclusions as may be put forward will be none the less borne by the reader in the event of their being belied by future research and theoretical schemes.

Symbioses have now been admitted to the polite society of biological research. There is evidence, too, of their being a source of specific "capacities" exploitable in terms of application. This matter, however, lies outside the scope of this review and will not be discussed. Suffice it to recall that mutualistic symbioses are already being put to work in agriculture and afforestation, and there are good prospects for the application of both symbiont organisms and their genes as innovative "information" in various fields of biotechnology through genetic engineering (88).

This review of what is currently known of the cell and molecular biology of mutualistic symbiotic systems has shown that none of them has so far been provided with a satisfactory model of the mechanisms governing its recognition mechanism and blockade stages. Even so, some experimental models have advanced a long way towards the clarification of certain particular points. This is the case with regard to *Rhizobium*-Leguminosae among the non inherited systems (29, 88, 98), perhaps *Paramecium*-*Chlorella* (18, 99) and *Sytophylus* — bacteria gram — (11) among those inherited and experimentally reproducible by reinfection, and bacteria-amoeba (27) among those that are inherited, but not dissociable and not reinfectable. Moreover different systems are approached with different techniques ranging from electron microscopy to biochemistry from molecular biology to classic genetic. In this way a lack of comparable results in different systems is always possible.

The threads of Ariadne on which we must rely to negotiate the labyrinth of results obtained is composed of three strands:

1) Organization of our knowledge of the cell biology of symbioses must be preceded by a thorough investigation of the structure and dynamics of their interfaces, so that correct assessment can be made of their molecular and supramolecular organisation throughout the interaction. A unifying reference to the mutualistic symbiosis analogy known as delayed or "frustrated phagocytosis" may be tentatively proposed for this purpose.

2) Understanding of the signalling between the partners demands recognition of the existence of stimuli that are significant for the establishment of reciprocal "comprehen-

sion", but cannot be classed as signals in the true sense of specific molecules inducing a specific response. Examples of these "non-signals" are recognition cues, i. e. normal host structures or molecules that induce a specific response on the part of the symbiont, and nonspecific interactions, i. e. the sum of factors that are nonspecific in terms of both their origin and their effect on the partner.

3) In view of the complexity of their integration, less ambiguous information concerning cell signalling in symbiotic systems can only be obtained by combining conventional biochemical and cytochemical techniques with the employment of affinity reactions and monoclonal antibodies for specific signals, receptors and messengers, so as to arrive at their identification *in situ,* and genetic engineering techniques with which to clone the genes responsible for signals and cues, and build up a picture of their expression.

Two major questions fundamental to the understanding of cell signalling in symbioses still remain to be solved:

1) Is symbiotic recognition reducible to simply the sum of different cell to cell signalling, or are other mechanisms involved?

It is true that cues and nonspecific interactions are also involved, it is equally true that our knowledge of the subject is still primitive, and there is evidence to show that some symbiotic stimuli may perhaps be referable to the sum of as yet unknown cell-to-cell signals. It has, in fact, been claimed that *Rhizobium*-Leguminosae system, whose molecular biology is the best known, makes use of gene-to-gene host-pathogen like recognition mechanism, and that *Rhizobium* is able to bypass the host's defence mechanism by active suppression or avoidance, thus acting as a refined parasite (98). Indirect support for this contention is provided by the complex and variable surface antigen systems of Paramecium (99) in the *Paramecium*-Chlorella symbiosis, and in VAM, where ultrastructural, biochemical and cytometric studies have revealed an initial interaction stage marked by enhanced peroxidase and chitinase activity in the host reminiscent of incompatibility reaction (72, 89) as well as a progressive chromatin despiralization in host nuclei, with possible activation of transcription (49).

2) Are the recognition signals responsible for symbiont compatibility and the restricted periods of biotrophic parasite compatible interactions completely lacking in correlation, or do they share a common base?

It is indeed true that control of pathogens leads to rigidly specific recognition, whereas the recognition of symbionts is non-specific in the initial stages, the observation, however, of a different degrees of efficiency and infectiveness of different symbiont strains in both plants (ectomycorrhizae, endomycorrhizae of Ericales) and animals (symbiotic and aposymbiotic strains of *Hydra* cultured in the laboratory and, though controversial, in symbioses between marine invertebrates and algae, Zooxanthellae) implies discrimination phenomena that can be explained either as specific signals, or as the preadaptation of strains resulting in differentiated cues. This conclusion appears justifiable as a working hypothesis even though in the present state of the art it can plausibly be supposed that in nature symbiotic recognition is more complicated than the simple sum of cell-cell recognitions, and that even though from the standpoint of cell biology the compatibility of biotrophic pathogens and that of mutualistic symbionts are similar, the stimuli that regulate them are different. Even so, the opposite hypothesis cannot be completely ruled out. Signalling for compatibility through an initial incompatibility response that is depressed during the later stages of the interaction, is by no means an unreasonable mechanism at least for plant symbioses. Phytoalexin detoxication by the symbiont is well known as the method adopted by some fungal pathogens to bypass the host's defences (100). VAM Fungi for example in the initial infection stage elicit a nonspecific incompatibility response from the host (72, 89). It has not yet been shown that the subsequent depression of these activities — (89), see also (97) and quoted references — is attributable to a fall in production or to degradation. If a mechanism of this kind does exist, the ecological evidence clearly points to nonspecific sorting and recognition of mutualistic symbionts. Some of the laboratory findings, on the other hand, suggest that the effective establishment of symbiosis may be controlled by specific symbiont genes (29, 88).

Investigation of the phenomena of symbiont recognition and of the establishment and maintenance of a symbiosis thus relates to a particularly complex experimental system that raises considerable complications in the laboratory, and imposes, if the reductionistic working hypothesis previously stated cannot be substantiated, a holistic approach towards understanding the system (recognition, control of nutrient transfer and morphogenesis are only different aspects superimposed on a single interactive phenomenon) and a reductionistic approach to the experimental reconstruction of its mechanisms at the molecular level.

The decision to address such an unusual level of complexity and a state of the art that is still so fragmentary and chaotic is justified by the fact that symbiotic systems are the result of an equilibrium that simultaneously controls nutrition, morphogenesis and resistance to a non-self. It is thus not unlikely that all the cell regulation phenomena on which the biology of development and the pathology of both animal and plant depend, can be better understood through a satisfactory knowledge of the cell signalling mechanisms in mutualistic symbioses.

REFERENCES

1) LEWIS, D.H., 1985. Symbiosis and Mutualism. Crisp Concepts and Soggy Semantics. In "The Biology of Mutualism. Ecology and Evolution". Edited by Boucher, D.H. (Crown Helm. London & Sidney). pp. 29—39.

2) SMITH, D.C., DOUGLAS, A.E., 1987. The biology of mutualism. (Edward Arnold. London).

3) SCANNERINI, S., 1988. The Cell structures in Plant, Animal and Microbial Symbionts. Their differences and similarities. In "Cell to Cell Signals in Plant, Animal and Microbial Symbiosis". Edited by SCANNERINI, S., SMITH, D.C., BONFANTE-FASOLO, P., GIANINAZZI-PEARSON, V., (Springer-Verlag, Berlin, Germany). pp. 143—157.

4) BOUCHER, D.H. (ed.) 1985. The Biology of Mutualism. Ecology and Evolution. (Crown Helm. London-Sidney).

5) MARGULIS, L., 1981, Symbiosis in Cell Evolution. (Freeman & Co. San Francisco).

6) LEWIS, D.H., in press. Fungi and the Flora which inherited the Earth. In "Evolution and Speciation: Symbiosis as a Source of Evolutionary Innovation". Edited by MARGULIS, L. & FESTER, R., (MIT Press. Boston Ma.) in Press.

7) VETTER, R. in press. Symbioses and novel energy sources: chemoautotrophic organism at idrotermal vent. In "Evolution and speciation: Symbiosis as a source of Evolutionary Innovation". Edited by MARGULIS, L. and FESTER, R. (MIT PRESS BOSTON Ma.).

8) AHMADIJAN, V. and HALE, M.E. (eds). 1974. The lichens (Academic Press, London-New York).

9) HARLEY, J.L. SMITH, S.E., 1983. Mycorrhizal symbiosis. (Academic Press. London).

10) MUSCATINE, L., 1973. Nutrtion of corals. In “Biology and Geology of Coral Reefs”. Edited by JONES, O.A. and ENDEAN, R. (Academic Press N. York and London). 2 (1): 77—155.

11) NARDON, P., GRENIER, A.M., 1988. Genetical and biochemical interactions between the host and its endocytobiotes in the weevils. Sytophilus (Coleoptere, Curculionidae) and other related species. In “Cell to cell signals in Plant, Animal and Microbial Symbiosis”. Edited by SCANNERINI, S., SMITH, D.C., BONFANTE-FASOLO, P., GIANINAZZI-PEARSON, V. (Springer Verlag, Berlin, Germany). pp. 255—270.

12) HONNEGER, R., 1988. The functional morphology of cell to cell interactions in lichens. In “Cell to Cell Signals in Plant, Animal and Microbial Symbiosis”. Edited by SCANNERINI, S, SMITH, D.C., BONFANTE-FASOLO, P., GIANINAZZI-PERSON, V. (Springer-Verlag. Berlin, Germany) pp. 36—53.

13) MC FALL-NGAY, M.J., in press. Luminous bacterial symbiosis in fish evolution: adaptative radiation in Leiognathid Fishes. In “Evolution and speciation: Symbiosis as a source of evolutionary innovation”. Edited by MARGULIS, L. and FESTER, R., (MIT Press. Boston Ma.).

14) TORREY, J.G., 1988. Cellular interactions between host and endosymbiont in Dinitrogen-fixing root-nodules of woody plants. In “Cell to Cell Signals in Plant, Animal and Microbial Symbiosis”. Edited by SCANNERINI, S., SMITH, D.C., BONFANTE-FASOLO, P., GIANINAZZI-PEARSON, V. (Springer-Verlag, Berlin. Germany) pp. 1—25.

15) SCWEMMLER, W. in press. Analysis of the gas metabolism of fertilized and unfertilized leafhopper eggs with endocytobionts. In “Endocytobiology IV”. Edited by NARDON, P., GRENIER, A.M., (INRA Presse, Paris, France).

16) BERTA, G., FUSCONI, A., TROTTA, A. and SCANNERINI, S., in press. Morphogenetic modifications induced by the mycorrhizal fungus Glomus strain E3 on the root system of *Allium porrum* L. New Phytol.

17) SMITH, D.C., 1979. From extracellular to intracellular: the establishment of a symbiosis. Proc. R. Soc. London, B 204: 115—130.

18) REISSER, W. 1988. Signals in Paramecium bursaria-Chlorella sp. Association. In “Cell to Cell Signals in Plant, Animal and Microbial Symbiosis”. Edited by SCANNERINI, S., SMITH, D.C., BONFANTE-FASOLO, P., GIANINAZZI-PEARSON, V. (Springer-Verlag, Berlin) pp. 271—282.

19) DOUGLAS, A. 1988. Nutrition interactions as signals in the Green Hydra symbiosis. In “Cell to Cell Signals in Plant, Animal and Microbial Symbiosis”. Edited by SCANNERINI, S., SMITH, D.C., BONFANTE-FASOLO, P., GIANINAZZI-PEARSON, V., (Springer-Verlag, Berlin, Germany), pp. 283—310.

20) McAULEY, P, J., 1988. Cell to cell interactions during the establishment of the *Hydra-Chlorella* symbiosis. In “Cell to Cell Signals in Plant, Animal and Microbial Symbiosis”. Edited by SCANNERINI, S., SMITH, D.C., BONFANTE-FASOLO, P., GIANINAZZI-PEARSON, V., (Springer Verlag, Berlin) pp. 115—130 ATTENZIONE.

21) JEON, K.W., Integration of bacterial endosymbionts in *Amoeba*. In International Review of Cytology. Suppl. 14. Intracellular Symbiosis. Edited by JEON, K. W. pp. 29—47.

22) TIIVEL, T. 1987. Leafhopper endocytobiosis. Endocyt. C. Res. 4: 25—39.

23) MARGULIS, L., BERMUDES, 1988. Symbiosis and Evolution: a brief guide to recent literature. In “Cell to Cell Signals in Plant, Animal and Microbial Symbiosis”. Edited by SCANNERINI, S., SMITH, D.C., BONFANTE-FASOLO, P, GIANINAZZI-PEARSON, V., (Springer-Verlag, Berlin, Germany) pp. 159—165.

24) MARGULIS, L. and FESTER, R., (Eds.). in press. Evolution and Speciation: Symbiosis as a source of Evolutionary Innovation. (MIT Press. Boston Ma.).

25) TAYLOR, F.J.R., 1983. Some eco-evolutionary aspects of intracellular symbiosis. In International Review of Cytology. Suppl. 14: Intracellular Symbiosis. Edited by JEON, K.W., pp. 1—28.

26) SCHWEMMLER, W., 1989. Symbiogenesis a Macro-Mechanism of Evolution. (W. de Gruyter, Berlin, Germany).

27) PARK, M.S., JEON, K.W., in press. A symbiont gene coding for a protein required for the host Amoeba: Cloning and expression in phage-transformed *E. coli.* In "Endocytobiology IV". Edited by NARDON, P., GRENIER, A.M. (INRA Presse, Paris, France).

28) HARA, E., ISHIKAWA, E., in press. Characterization of symbionin only one protein produced by an aphid endosymbiont in vivo. In "Endocytobiology IV". Edited by NARDON, P., GRENIER, A.M. (INRA Presse, Paris, France).

29) ZAAT, S.A.J., SPAINK, H.P. WIJFFELMAN, C.A., Van BRUSSEL, A.A.N., OKKER, R.J.H. and LUGTENBERG, B.J.I., 1988. Flavonoid compounds as molecular signals in *Rhizobium*-Legume Symbiosis. In "Cell to Cell Signals in Plant, Animal and Microbial Symbiosis". Edited by SCANNERINI, S., SMITH, D.C., BONFANTE-FASOLO, P., GIANINAZZI-PEARSON, V., (Springer-Verlag, Berlin, Germany) pp. 189—205.

30) GOFF, L. in press. Symbiosis, interspecific gene transfer and the evolution of new species: A case study in the parasitic Red Algae. In "Evolution and Speciation: Symbiosis as a source of Evolutionary innovation". Edited by MARGULIS, L., FESTER, R. (MIT Press, Boston Ma).

31) NEALSON, K., SCHMIDT, T.M., and BLEAKLEY, B., 1988. Luminescent bacteria symbionts of Nematodes and Pathogen of Insects. In "Cell to Cell Signals in Plant, Animal and Microbial Symbiosis". Edited by SCANNERINI, S., SMITH, D.C., BONFANTE-FASOLO, P., GIANINAZZI-PEARSON, V. (Springer-Verlag, Berlin, Germany). pp. 101—113.

32) KORDYUM, V.A., 1988. Information channels of pro- and eukaryotes. Comparative analysis. In "Lectures in Theoretical Biology". Edited by KULL, K., TIIVEL, T., (Valgus, Tallinn, Estonia) pp. 84—101.

33) SCANNERINI, S., BONFANTE-FASOLO, P., 1983. Comparative Ultrastructural Analysis of mycorrhizal associations. Can. J. Bot. 61: 917—943.

34) GIANINAZZI-PEARSON, V., GIANINAZZI, S., 1988. Morphological Integration and functional compatibility between symbionts in vesicular-arbuscular endomycorrhizal associations. In "Cell to Cell Signals in Plant, Animal and Microbial Symbiosis". Edited by SCANNERINI, S., SMITH, D.C., BONFANTE-FASOLO, P., GIANINAZZI-PEARSON, V., (Springer-Verlag, Berlin, Germany). pp. 73—84.

35) REISSER, W., in press. Endocytobiotic ciliate-algae associations: Model system for the study of basic principles ofsymbiosis formation. In "Endocytobiology IV": Edited by NARDON, P., GRENIER, A.M., (INRA, Presse, Paris, France).

36) DOUGLAS, A., 1988. Specificity in the Convoluta roscoffensis-Tetraselmis Symbiosis. In "Cell to Cell Signals in Plant, Animal and Microbial Symbiosis". Edited by SCANNERINI, S., SMITH, D.C., BONFANTE-FASOLO, P., GIANINAZZI-PEARSON, V., (Springer-Verlag, Berlin, Germany). pp. 219—233.

37) BONFANTE-FASOLO, P., 1988. The role of cell wall as a signal in mycorrhizal associations. In "Cell to Cell Signals in Plant, Animal and Microbial Symbiosis". Edited by SCANNERINI, S., SMITH, D.C., BONFANTE-FASOLO, P, GIANINAZZI-PEARSON, V., (Springer-Verlag, Berlin, Germany). pp. 219—233.

38) BONFANTE-FASOLO, P. and PEROTTO, S., in press. Mycorrhizal and Pathogenic Fungi. Do they share any common features? In "Electron Microscopy and Plant Pathology". Edited by MENDGEN, K., (Springer-Verlag, Berlin, Germany).

39) HALVESTON, G. and STACEY, S., 1986. Signal exchange in plant-microbe interactions. Microbiol. Rev. 50: 193—225.

40) JEON, K.W., 1986. Change of cellular pathogen into required Cell component. In "Endocytobioogy III". Edited by LEE, J.I. and FREDERICK, J.F., Ann. N.Y. Acad. Sci. 503: 359—371.

41) GRILLI-CAIOLA, M., ALBERTANO, P., 1988. Recognition Mechanisms in the *Azolla-Anabaena* Symbiosis. In "Cell to Cell Signals in Plant, Animal and Microbial Symbiosis". Edited by SCANNERINI, S., SMITH, D.C., BONFANTE-FASOLO, P., GIANINAZZI-PEARSON, V. (Springer-Verlag, Berlin). pp. 27—38.

42) SCANNERINI, S., BONFANTE-FASOLO, P., in press. Glomaceae, Endocytobiosis and Fungal Evolution: A lost opportunity? In "Evolution and Speciation: Symbiosis as a Source of Evolutionary Innovation". Edited by MARGULIS, L., FESTER, R., (MIT Press, Boston Ma).

43) PALADE, G.E., FARQUAR, M.G., 1981. Cell Biology and Pathophysiology. In "The Biological Principles of disease". Edited by SMITH, L.H., THIER, S.O., (Sanders, Phyladelphia, USA). pp. 1—56.

44) OUCHI, S., 1983. Induction of resistance or susceptibility. Ann. Rev. Phytopathol. 21: 289—315.

45) MATTA, A., 1971. Microbial penetration and immunization of uncongenial host plants. Ann. Rev. Phytopathol. 9: 387—410.

46) MATTA, A., 1989. Induced Resistance to Fusarium wilt Disease. In "Vascular wilt Diseases of Plants". Edited by TJAMOS, J., BECKMANN, C., (Springer-Verlag, Berlin, Germany) pp. 175—196.

47) HAHLBROCK, K., CUYPERS, B., DOUGLAS, C., FRITZMEIER, K.H., HOFFMANN, H., ROHVER, F., SCHEEL, D. and SCHULTZ, W., 1986. Biochemical interactions of plants with potentially pathogen Fungi. In "Recognition in Microbe-Plant symbiotic and pathogenic interactions". Edited by LUGTENBERG, B., (Springer-Verlag, Berlin, Germany). pp. 311—324.

48) KEEN, N.T., 1986. Pathogenic strategies in Fungi. In "Recogniton in Microbe-Plant symbiotic and pathogenic interactions". Edited by LUGTENBERG, B., (Springer-Verlag, Berlin, Germany). pp. 171—188.

49) BERTA, G., SGORBATI, S., SOLERI, V., FUSCONI, A., TROTTA, A., CITTERIO, M.G., SPARVOLI, E. and SCANNERINI, S., in press. Chromatin structure variations in host nuclei of a VA mycorrhiza. New Phytol.

50) ROTH, L.E., STACEY, G., 1989. Bacterium release into host cells of nitrogen-fixing soybean nodules: the symbiosome membrane come from three sources. Europ. J. Cell Biol. 49: 13—23.

51) RAHAT, M., REICH, V. 1988. The establishment of *Algal/Hydra* symbioses — A case of recognition or preadaptation? In "Cell to Cell Signals in Plant, Animal and Microbial Symbiosis". Edited by SCANNERINI, S., SMITH, D.C., BONFANTE-FASOLO, P., GIANINAZZI-PEARSON, V. (Springer-Verlag, Berlin, Germany). pp. 297—310.

52) ABU-EL TEEN, K., GHANNUM., M., and STRETTON, R.J., 1989, Effects of sub-inhibitory Agents on the aderence of *Candida* spp. to buccal epithelial cells in vitro, Mykosen. 32: 552—561.

53) SHAPER, W.M., REST, R.E., 1989. Interaction of Gonococci with phagocytic cells. Ann. Rev. Microbiol. 43: 121—145.

54) BRACKER, C.E., LITTLEFIELD, L.J., 1973. Structural concepts of Host-Pathogen interfaces. In "Fungal Pathogenecity and the Plant's Response". Edited by BYDE, R.J.W. and CUTTING, C.V., (Academic Press, London UK) pp. 159—318.

55) SCANNERINI S., BONFANTE-FASOLO, P., FONTANA, A., 1975, An ultrastructural model for the host-symbiont interactions in the endotrophic mycorrhiza of Ornithogalum umbellatum L. In "Endomycorrhizae". Edited by SANDERS, F.E., MOSSE, B. and TINKER, P.B., (Academic Press, London, UK) pp. 313—324.

56) GREEN, J.R., JONES, J.L., and CALLOW, J.A. 1988. The application of Monoclonal Antibody Technology to the study of Cell-Cell interactions. In "Cell to Cell Signals in Plant, Animal and Microbial Symbiosis". Edited by SCANNERINI, S., SMITH, D.C., BONFANTE-FASOLO, P., GIANINAZZI-PEARSON, V., (Springer Verlag, Berlin, Germany) pp. 362—371.

57) BONANTE-FASOLO, P., PEROTTO, S., TESTA, B., FACCIO, A., 1987. Ultrastructural localization of Cell surface sugar residues in ericoid mycorrhizal fungi by gold labelled lectins. Protoplasma. 137: 25—35.

58) BONFANTE-FASOLO, P., VIAN, A., PEROTTO, S., FACCIO, A. and KNOX, J.P., in press. Cellulose and pectin localization in roots of mycorrhizal Allium porrum: labelling continuity between host cell wall and interfacial material. Planta.

59) BONFANTE-FASOLO, P. and SCANNERINI, S., in press. The cellular basis of Plant-Fungus interchanges in mycorrhizal associations. In "Mycorrhizal Functioning". Edited by ALLEN, M.F. (Chapmann and Hall, London UK).

60) BONFANTE-FASOLO, P., VIAN, B., 1989. Cell wall architecture in mycorrhizal roots of *Allium porrum* L, Ann. Sci. Nat. Bot. 10: 93—109.

61) ROTH, L.E., JEON, K.W., STACEY, G., 1988. Homology in endosymbiontic systems. The term symbiosome. In "Molecular Genetics of Plant-Microbe interactions". Edited by PALACIOS, R., VERNA, D.P.S. (APS Press, Saint Paul, Minnesota) pp. 220—225.

62) TRENCH, R.K., 1988 Specificty in dinomastigote-marine invertebrate symbioses: An evaluation of hypotheses of mechanisms involved in producing specificity. In "Cell to Cell Signals in Plant, Animal and Microbial Symbiosis". Edited by SCANNERINI, S., SMITH, D.C., BONFANTE-FASOLO, P., GIANINAZZI-PEARSON, V., (Springer-Verlag. Berlin Germany) pp. 325—346.

63) ROTH, L.E., STACEY, G., 1989. Cytoplasmic membrane systems involved in bacterium release into soybean nodule cells as studied in two *Bradyrhizobium japonicum* mutant strains. Eur. J. Cell Biol. 49: 24—32.

64) HINDE, R., 1988 Factors produced by symbiotic marine invertebrate which affect traslocation between the symbiont. In "Cell to Cell Signals in Plant, Animal and Microbial Symbiosis". Edited by SCANNERINI, S., SMITH, D.C., BONFANTE-FASOLO, P., GIANINAZZI-PEARSON, V., (Springer-Verlag, Berlin, Germany) pp. 311—324.

65) NARDON, P., 1988. Cell-to-Cell interactions in insect endocytobiosis. In "Cell to Cell Signals in Plant, Animal and Microbial Symbiosis". Edited by SCANNERINI, S., SMITH, D.C., BONFANTE-FASOLO, P., GIANINAZZI-PEARSON, V. (Springer-Verlag, Berlin, Germany) pp. 85—100.

66) HOLTZMAN, E., 1989. Lysosomes. (Plenum Press) New York. USA.

67) SCANNERINI, S., 1985. Mycorrhizal Symbiosis. 2: The Process. Riv. Biol. 78: 546—553.

68) HILMER, S., DEPTA, H., ROBINSON, D.G., 1986. Confirmation of endocytosis in higher plant protoplast using lectin-gold conjugates. Eur. J. Cell Biol. 41: 142—149.

69) TANCHAK, W.A., GRIFFING, L.R., MERSEY, B.G. and FOWKES, C.C. 1984. Endocytosis of cationized ferritin by coated vesicles of soybean protoplasts. Planta, 162: 481—486.

70) BOLLER, T., WIEMKEN, A., 1987. Dynamics of lysosomal functions in plant vacuoles. In "Plant Vacuoles". Edited by MARIN, B. (Plenum Press, New York USA) pp. 361—368.

71) SAXTON, M.J., BREIDENBACH, R, W., 1988. Receptor mediated endocytosis in plants is energetically possible. Plant Physiol. 86: 993—995.

72) SPANU, P., BONFANTE-FASOLO, P., 1988. Cell wall bound peroxidase activity in roots of mycorrhizal Allium porrum. New Phytol. 109: 119—124.

73) HOFFSTEIN, S.T., 1980. Intra- and extracellular secretion from polymorphonuclear leukocytes. In "The Cell Biology of inflammation" (Elsevier-North Holland, Amsterdam The Nederland) pp. 387—480.

74) VANDENBOSCH, K.A., BRADLEY, D.J., PEROTTO, S., KNOX, J.P., BUTCHER, G.W., BREWIN, N.J., 1989. Common components of the infection thread matrix and the intercellular space identified by immunicytochemical analysis of pea nodules and uninfected roots. EMBO J. 8: 335—342.

75) VAN DEURS, B., PETERSEN, O.W., OLSENS, S. and SANDVIG, K., 1989. The ways of Endocytosis. International Review of Cytology; A survey of Cell Biology. 117: 131—1t5.

76) CAVALIER-SMITH, T., 1988. Eukaryote Cell Evolution. In "Proceedings of XIV International Botanical Congress", Edited by RAYNER, A.D.M., BRASIER, C.M. and ZIMMER, B., (Koeltz Scientific Publications, Koeningstein, Germany). pp. 203—233.

77) TRENCH, K.R., 1980. Uptake, retention and function of chloroplasts in animal cells. In "Endocytobiology I". Edited by SCHWEMMLER, W. and SCHENK, H.A.E., (W. De Gruyter, Berlin, Germany) pp. 703—727.

78) BONFANTE-FASOLO, P., GIANINAZZI-PEARSON, V., MARTINENGO, L., 1984. Ultrastructural aspects of endomycorrhizae in the Ericaceae. IV: Comparison of infection by *Pezizella ericae* in host and non host plants. New Phytol. 98: 329—333.

79) PICHE, Y., PETERSON, R.L. and MASSICOTTE, H.B., 1988. Host-Fungus interactions in ectomycorrhzae. In "Cell to Cell Signals in Plants, Animal and Microbial Symbiosis". Edited by SCANNERINI, S., SMITH, D.C., BONFANTE-FASOLO, P., GIANINAZZI-PEARSON, V., (Springer-Verlag, Berlin, Germany) pp. 55—71.

80) HILBERT, J.L., MARTIN, F., 1988. Regulation of gene expression in ectomycorrhizae, 1. Protein changes and the presence of ectomycorhiza specific polypeptides in the Pisolithus-Eucalyptus symbiosis. New Phytol. 110: 339—346.

81) PETERS, G.A., MEECKS, J.C. 1989. The *Azolla-Anabaena* Symbiosis. Basic Biology. Ann. Rev. Plant Physiol. Mol. Biol. 40: 193—210.

82) DAZZO, F.R., HOLLINGSWORTH, R.I., PHILIP-HOLLINGSWORTH, S., SMITH, K.B., WELSCH, M.A., DJORDJEVIC, M., and B.G. ROLFE, 1988. Early recognition signals in *Rhizobium*-legume Symbiosis. In "Cell to Cell Signals in Plant, Animal and Microbial Symbiosis". Edited by SCANNERINI, S., SMITH, D.C., BONFANTE-FASOLO, P., GIANINAZZI-PEARSON, V, (Springer-Verlag, Berlin, Germany) pp. 189—200.

83) MASSICOTTE, H.B., ACKERLEY, C.A. and PETERSON, R.L., 1989. Ontogeny of *Alnus rubra-Alpova* diplophaeus ectomycorrhizae. II. Transmission Electron Microscopy. Can. J. Bot. 65: 1940—1947.

84) MONSIGNY, M., ROCHE, A.C., KIEDA, C., MAYER, R. and MIDOUX, P. 1988. Peptide and carbohydrate moieties as molecular signals in animal cell recognition. In "Cell to Cell Signals in Plant, Animal and Microbial Symbiosis". Edited by SCANNERINI, S., SMITH, D.C., BONFANTE-FASOLO, P., GIANINAZZI-PEARSON, V., (Springer-Verlag, Berlin, Germany) pp. 237—254.

85) CASSONE, R., TORASANTUCCI, A., BOCCANERA, M., PELLEGRINI, G., PALMA, C., MALAVASI, F., 1988. Production and characterization of a monoclonal antibody to a cell surfacee, glucomannoprotein constituent of *Candida albicans* and other pathogenic *Candida* species. J. Med. Microbiol. 27: 233—238.

86) BOLLER, T., 1989. Primary Signals and second messengers in the reactions of Plants to Pathogen. In "Second Messengers in Plant growth and Development" pp. 227—255.

87) SMITH, D.C. 1988. Concepts leading to an understanding of recognition and signalling between hosts and symbionts. In "Cell to Cell Signals in Plant, Animal and Microbial Symbiosis". Edited by SCANNERINI, S., SMITH, D.C., BONFANTE-FASOLO, P., GIANINAZZI-PEARSON, V. (Springer-Verlag, Berlin, Germany) pp. 347—359.

88) NUTI, M.P., PASTI, M.B. and SQUARTINI, A., 1988. Application of genetic engineering to symbiontology in Agriculture. In "Cell to Cell Signals in Plant, Animal and Microbial Symbiosis". Edited by SCANNERINI, S., SMITH, D.C., BONFANTE-FASOLO, P., GIANINAZZI-PEARSON, V, (Springer-Verlag, Berlin, Germany) pp. 347—359.

89) SPANU, P., BOLLER, T., LUDWIG, A., VIEMKEN, A., FACCIO, A., BONFANTE-FASOLO, P., 1989. Chitinase in roots of mycorrhizal *Allium porrum* regulation and localization. Planta 177: 447 456.

90) GALUN, M. and GERTY, J. 1988 Soredia formation of compatible and incompatible lichen symbiosis. In "Cell to Cell Signals in Plant, Animal and Microbial, Symbiosis". Edited by SCANNERINI, S., SMITH, D.C., BONFANTE-FASOLO, P., GIANINAZZI-PEARSON, V., (Springer-Verlag, Berlin, Germany). pp. 208—217.

91) BREWIN, N.J., BRADLEY, D.J., WOOD, E.A., KANNENBERG, E.L., VANDENBOSCH, K.A., BUTCHER, G.W. 1988. The use of monoclonal antibodies to investigate Plant-Microbe interactions in pea root nodules containing *Rhizobium leguminosarum.* In "Cell to Cell Signals in Plant, Animal and Microbial Symbiosis". Edited by SCANNERINI, S., SMITH, D.C., BONFANTE-FASOLO, P., GIANINAZZI-PEARSON, V., (Springer-Verlag, Berlin) pp. 374—383.

92) CALLOW, J.A., RAY, T, ESTRADA-GARCIA, T.M. and GREEN, J.R., 1988. Molecular Signals in Plant Cell Recognition. In "Cell to Cell Signals in Plant, Animal and Microbial Symbiosis". Edited by SCANNERINI, S., SMITH, D.C., BONFANTE-FASOLO, P., GIANINAZZI-PEARSON, V., (Springer-Verlag, Berlin, Germany), pp. 167—182.

93) TREWAVAS, J, A., (Ed.) 1986. Molecular and cellular aspects of Calcium in Plants develpoment. (Plenum Press. New York, USA)

94) CASTORIA, R., FABBRI, A.A. and FANELLI, C., 1989 Hypersensitive response in potato tubers induced by arachidonic and eicosapentaenoic acid. Giorn. Bot. Ital. 123 (suppl. 2): 47.

95) LUGTENBERG, B., (ed.) 1986. Recognition in microbe-plant symbiotic and pathogenic interactions. (Springer-Verlag, Berlin, Germany).

96) SMITH, S.E, and GIANINAZZI-PEARSON, V., 1988. Physiological interactions between symbionts in vesicular arbuscular mycorrhizal plants. Ann. Rev. Plant Physiol. Mol. Biol. 39: 221—244.

97) CODIGNOLA, A., VEROTTA, L., SPANU, P., MAFFEI, M., SCANNERINI, S., and BONFANTE-FASOLO, P., 1989. Cell wall bound phenols in roots of vesicular-arbuscular mycorrhizal plants. New Phytol. 112: 221—228.

98) DJORDJEVIC, M.A GABRIEL, D.W., ROLFE, B.G., 1987. Rhizobium the refined parasite of legumes. Ann. Rev. Phytopathol. 25: 145—168.

99) CARON, F., MAYER, E., 1989. Molecular basis of surface antigen variation in Paramecia. Ann. Rev. Microbiol. 43: 33—42.

100) VAN ETTEN, H.D., MATTHEWS, D.E. and MATTHEWS, P.S. 1989. Phytoalexins detoxication. Importance for pathogenicity and practical implications. Ann. Rev. Phytopathol. 27: 143—164.

CELL TO CELL RECOGNITION AND DIFFERENTIATION

PARALLELS IN CELL TO CELL COMMUNICATION IN PLANTS AND ANIMALS

A.W. Robards[1] and J.D. Pitts[2]

[1]Institute for Applied Biology
University of York
York YO1 5DD, England

[2]Beatson Institute for Cancer Research
Bearsden, Glasgow G61 1BD, Scotland

INTRODUCTION

The concept of junctional communication first arose more than a hundred years ago (1) and the theoretical consequences of this form of cell-cell interaction were widely discussed by plant physiologists in the 1930s (e.g. 2). Intercellular bridges in plants, it was suggested, produced a continuous cytoplasmic network (symplast) separated by a continuous plasma membrane from the extracellular space. The cellular state within the symplast was seen as intermediate between the independence of complete isolation and the suppression of individuality seen in a true syncytium. This partial syncytial state was thought to provide a mechanism for the synchronization of cellular activities at the tissue level.

More recent structural and functional studies have confirmed the presence of permeable intercellular junctions in both plants and animals. Plasmodesmata are found in plants and gap junctions in animals. They are structurally unrelated but functionally very similar. They both provide direct cytoplasmic continuity *via* aqueous channels between adjacent cells. The channels are size selective and the junctions act like intercellular sieves allowing the free intercellular movement of small ions and molecules but not macromolecules. The permeability limit of both is based primarily (or solely) on molecular size and the cut-off, which is normally expressed for convenience in terms of molecular weight rather than molecular dimensions, is equivalent to about Mr 900. There is no active transport and molecules passing through the junctional channels move down gradients of chemical and/or electrical potential.

Botanists may have foreseen in general terms the importance of junctional communication but in recent years major advances have been made in the detailed understanding of this phenomenon through studies on animal tissues. This has mainly been due to the development of procedures for the isolation of gap junctions which has allowed the identification and characterization of component proteins and led to the introduction of the powerful techniques of molecular genetics.

Plasmodesmata and gap junctions provide a striking example of convergent evolution. Their functional similarity belies their structural differences. Junctional communica-

NATO ASI Series, Vol. H 51
Cell to Cell Signals in Plants and Animals
Edited by V. Neuhoff and J. Friend

tion appears to have been a pre-requisite for the evolution of complex multicellular organisms in both the plant and animal kingdoms but different structural solutions were required in the two systems. The presence of structurally important materials in the extra-cellular spaces in plants results in a relatively large separation of neighbouring cell membranes which must be bridged by a long tube-like structure. To maintain the integrity of the cells this is coated with an extension of the plasma membrane. In animals, on the other hand, the membranes of adjacent cells can be brought close together so that the small residual space can be bridged and sealed by much simpler protein structures.

It seems likely that junctional communication in plants and animals will share common features and provide similar functions despite the differences in the mediating structures. It is appropriate therefore to try and draw the two fields together, as was done at a NATO Advanced Research Workshop, held in York during July 1989, which discussed "Parallels in Cell to Cell Communication in Plants and Animals" (3). Much of the work reported here was discussed at that workshop and we are grateful to the contributors for allowing us access to their work, some of which is unpublished.

CELL INTERACTIONS IN PLANT TISSUES

As reviewed by Robards and Lucas (4), there is now a substantial body of information on plasmodesmatal formation, structure, and tissue-specific variation in substructure. However, there remains a strong need to obtain further information on dynamic aspects of plasmodesmata. When this has been done, together with an improvement in our understanding of the physiological roles played by plasmodesmata, it will be possible, among other things, to establish the necessary framework for future studies on the role of cell-to-cell communication in the control of plant development.

Structure of plasmodesmata

A simple plasmodesma is illustrated in Fig. 1. The points of general agreement are: that the pore through the wall is lined by the plasmalemma; that an axial component passes through the plasmalemma-lined tube and that this axial strand often connects in some way with the endoplasmic reticulum (ER) in the neighbouring cells; and that there is a space, the "cytoplasmic sleeve", between the axial strand and the plasmalemma. This cytoplasmic sleeve may be more or less occluded in the neck region of the plasmodesmata where the axial strand and the plasmalemma are in close contact with each other. The basic structure of a plasmodesma is, therefore, rather simple although the dimensions and relative dispositions at the molecular level — so important to understanding how plasmodesmata actually work — remain inadequately resolved (5).

It is now generally thought that intercellular transport must occur via the cytoplasmic sleeve. More specifically, the channel(s) for communication would be through the spaces between a ring of 5 nm subunits that partly occlude this sleeve. The cross-sectional area of each of these spaces has been calculated to be 2.6 — 3.6 nm^2 for plasmodesmata in *Azolla* roots. This is remarkably similar to the 3.1 nm^2 cross-sectional area of the pores in the macromolecular components of animal gap junctions, the connexons.

There appears to be a growing consensus that plasmodesmatal regulation may take place at the level of the neck region where particulate subunits are closely packed

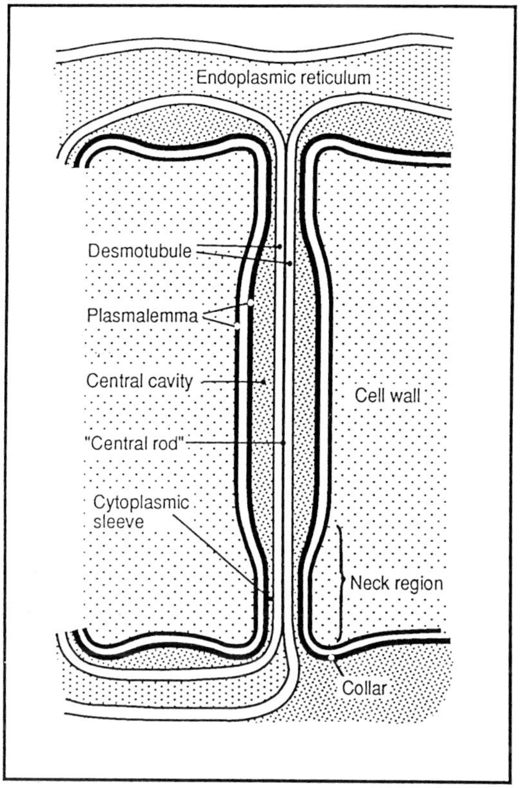

FIGURE 1
Diagram illustrating the component parts of a simple plasmodesma.
[Taken from Olesen and Robards (5)]

between the plasmalemma and the desmotubule. It is relatively simple to envisage a mechanism for using such a structure to control transport, even if the precise signalling and transduction components remain to be elucidated. One model for plasmodesmatal regulation at the sphincter has been forwarded by Olesen and Robards (5).

Evidence has been presented to suggest that some polypeptides in plant cells are immunologically related to rat liver-type connexin — the animal gap junctional protein. Immunolocalisation studies suggest that the polypeptide is both widely distributed in plants and their tissues and is localised at the cell periphery. It is too early to comment on the generality or significance of this result but it is clearly an area open for further study. Conkling et al. (6) have reported that *Arabidopsis thaliana* and *Nicotiana tabacum* each contain a gene coding for a 26 kDa polypeptide that is approximately 30—40% related by amino acid sequence to membrane intrinsic protein (MIP) 26 isolated from animal eye lens. MIP 26 has been suggested to be a gap junctional protein while the plant 26 kDa protein appears to be confined exclusively to roots.

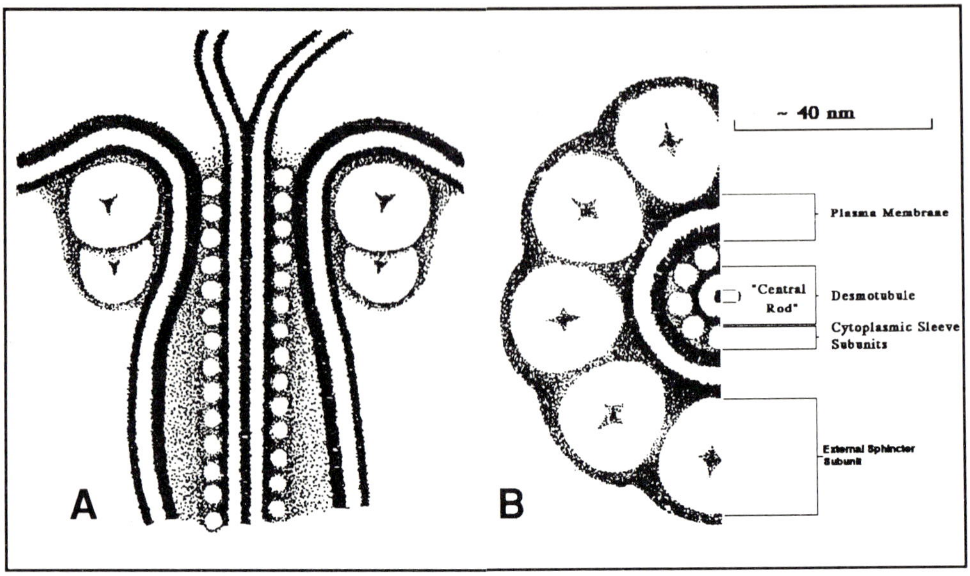

FIGURE 2

Diagrammatic representation of a plasmodesma drawn approximately to scale, based on electron micrographs of plasmodesmata from the mesophyll-bundle sheath interface in *Salsola kali* leaves.

Components are shown in longitudinal (A) and transverse sections (B). Taken from (5).

TABLE 1
Dimensions (all values in nm)
of the various components of plasmodesmatal neck regions [Taken from (5)]

	Range
Central rod:	2—4
Desmotubule:	10—13
Ring of subunits in cytoplasmic sleeve:	17—22
Average subunit size:	4—5
Space between subunits:	1—2
Plasma membrane, inner diameter:	21—44
Plasma membrane, outer diameter:	35—60
External sphincter, outer diameter:	97—117[a]
Particulate subunits of external sphincter:	27[b]

[a] Variation in *Salsola kali* only
[b] *Salsola kali*

Future work on plasmodesmatal structure will need to resolve better the nature of the axial component (desmotubule). This component clearly forms a membranous connection between the ER of neighbouring cells but whether there is continuity with the lumen of the ER remains conjectural. Modulation of ER luminal contact may be an essential attribute for physiological function, and this may be the basis for the different electron microscopic images obtained of the axial component (i.e. open tubule or tightly appressed lipid cylinder).

Although molecular size exclusion limit studies are consistent with the hypothesis that spatially arranged globular entities, located within the cytoplasmic sleeve of the sphincter, restrict molecular movements between cells, almost nothing is known about these putative structures. Studies on rapidly frozen material may provide valuable information, and here freeze-fracture of rapidly-frozen tissue culture cells may prove to be an experimental system worthy of further study. Furthermore, if these structures are proteinaceous in nature, it will be important to establish their synthetic pathway and method of targeting of the cytoplasmic sleeve. In this regard it may be fruitful to investigate likely parallels between the sub-surface cisternae of animal cells.

Isolation of plasmodesmata. Isolation of plasmodesmata from cell walls may provide a useful avenue for investigation, especially if such experiments are conducted as coordinated biochemical, immunocytochemical and ultrastructural studies. If successful, this approach could enable the isolation and purification of the major proteins involved in plasmodesmata function. The development of antibodies directed against these putative proteins would provide a major breakthrough in terms of studying the substructure and dynamics of plasmodesmata.

Membrane dynamics. It is now possible to study he dynamic nature of the plasma membrane and ER-axial component of plasmodesmata by conducting further photobleaching studies (7,8). Ideally, these experiments should utilize fluorescent molecules that can be inserted into the membranes from the cytosol. Obviously, if a fluorescent molecule could be targeted to the lumen of the ER, photobleaching experiments could provide valuable information on the dynamics, or otherwise, of lumen of the desmotubule.

Plasmodesmatal formation

It has long been known and accepted that most plasmodesmata are formed from the entrapment of strands of endoplasmic reticulum in the developing cell plate. This leaves open, however, questions such as how plasmodesmata come to be arranged in regular patterns (e.g. pit fields) within the cell wall and how specific cell types consistently insert more or less the same frequency of plasmodesmata across a specific interface, this frequency sometimes varying in relation to age/number of divisions of the parent cell.

While most plasmodesmata in mature cells have arisen from the process described above, it is now beyond argument that plants also have the capacity, under appropriate circumstances, to form plasmodesmata secondarily in "non-division" walls. A recent review of this topic concentrated on the secondary formation of plasmodesmata in cultured cells, leading to a model for secondary plasmodesmata formation (9). Tubular cisternae of ER are constricted to the size of a desmotubule. Golgi vesicles are involved in the enclosure, and exclusion, of protoplasmic strands containing the constricted ER cisternae. In general, secondary plasmodesmatal formation appears to arise in relatively undifferentiated cells: those forming connections, for example, between grafts and dedifferentiated cells. There is scope for considerably more work in this interesting area.

Plasmodesmatal distribution and frequency

Plasmodesmata are found between all young, living cells of higher plants (10,11). Their frequencies are broadly within the range 0.1 to 10.0 per μm^2 although exceptions can be found.

Improvements in the ability to measure plasmodesmatal frequencies with high accuracy, and then to relate these to observed rates of intercellular transport, has led to a number of publications concerned with the quantitative correlation of plasmodesmatal structure, frequency and transport capacity. The demands of this approach are that it is necessary to obtain high-grade ultrastructural data from a system that is also amenable to the experimental determination of rates of water and solute movement (e.g. 12; see [4] for review of this topic).

Plasmodesmata in development

While it now seems accepted that the secondary formation of plasmodesmata may have a role to play in the normal development of some relatively undifferentiated plant cells, the question of how many plasmodesmata are inserted into a given cell plate and what subsequent losses and/or dilution there may subsequently be during development are important aspects of the regulation of cell growth and development. Different numbers of plasmodesmata are inserted into the cell plates of different cell types. In the few detailed experiments carried out in this area, it appears that a good case may be made for the frequency of plasmodesmata between cells being a determinant of the ability to maintain cell division (13). The ultimate extent of plasmodesmatal dilution is, of course, total symplastic isolation. This occurs, for example, in stomatal guard cells.

Plasmodesmata are likely to play a central role in terms of establishing local tissue-specific symplasmic domains during embryogeny. It should now be possible to conduct biophysical and cell-to-cell dye coupling experiments on higher plant embryos to test this hypothesis. In addition, studies on the cellular mechanism(s) responsible for controlling plasmodesmatal frequencies, in various tissues, are also now possible so that experiments on the relationships between environmental and/or physical parameters, such as light and pressure gradients, and the establishment of symplasmic domains can be carried out.

Biophysical investigations on plasmodesmata

Electrophysiological experiments have been conducted on a range of plant cell types in order to determine whether plasmodesmata truly represent a low resistance pathway for the movement of ions and solutes. Spanswick and Costerton (14) were able to show that the specific resistance of the node was approximately 50 times lower than the specific resistance of the plasma membranes of the two internodal cells.

Ultrastructural studies conducted on the same experimental tissue revealed that numerous plasmodesmata were present within the cell walls that connected the internodal cells with the cells of the nodal complex . The desmotubule is absent from the plasmodesmata of characean cells and so Spanswick and Costerton were able to model this low electrical resistance pathway as being represented by simple pores filled with 100 mM KCl. From this analysis they deduced that the measured resistance should have been even lower than they actually measured. Thus, these electrophysiological and ultrastructural results were consistent with the hypothesis that plasmodesmata constitute a low resistance pathway. However, the discrepancy between experimental and theoretically determined plasmodesma resistance led Spanswick & Costerton (14) to conclude that there must be some restriction within the plasmodesma that limits the free diffusion of ions.

Tyree (15) used the principles of irreversible thermodynamics to evaluate various transport processes that may function at the level of the plasmodesma and concluded that plasmodesmata constitute the pathway of least resistance for the cell-to-cell movement of all small solutes. Furthermore, he concluded that whereas complex transport phenomena involving active mechanisms cannot be completely discounted, diffusion will be the predominant mechanism of transport through plasmodesmata.

Drake (16) used a biophysical approach to investigate possible energy requirements for plasmodesmatal function. In the presence of metabolic inhibitors like azide and cyanide, the electrical coupling between parenchyma cells in the oat coleoptile was almost completely eliminated. Although some part of this reduction could be attributed to the deposition of callose, this could not account for the loss of electrical coupling in the presence of azide. Whether these results can be explained on the basis of an inhibition of the histochemically identified ATPase activity within plasmodesmata remains to be established. However, Tucker (17,18) used high levels of cyanide and azide to inhibit cytoplasmic streaming in staminal hairs of *Setcreasea purpurea,* and under these conditions the plasmodesmata appeared normal, at least in terms of dye coupling.

The structural and biophysical studies performed on the plasmodesmata of *Azolla* roots by Overall and coworkers (19,20) is perhaps the best example available of an attempt to integrate ultrastructural details into a biophysical model to test the predicted characteristics of plasmodesmata against experimental data. The *Azolla* root provides an interesting experimental system for electrical coupling studies, since it is relatively simple in structure and, as it ages, the apical cell of this root lays down walls with progressively fewer plasmodesmata. An additional cytological feature of importance, in terms of analyzing the electrical coupling data, is that young cells near the apex have very small vacuoles, thus making it easy to insert microelectrodes into the cytoplasm.

From the combined electrical, anatomical and ultrastructural data obtained on the *Azolla* root system, Overall and Gunning (20) established that a quantitative relationship does exist between electrical coupling and the number of plasmodesmata in the cell wall across which the injected current flows. No correlation between cell age and single plasmodesma resistance (average 4.4×10^{11}) was found. These workers also used their elegant ultrastructural data to compute the resistance of a single plasmodesma which they determined to be approximately 2.5×10^{9} Ohms, which is some 170 times lower than the experimentally derived value! They attributed this discrepancy to an "impediment factor", which might be a tortuosity factor if the particles do not form files, but rather are helically arranged, or a callose-mediated constriction of the plasmodesma due to wounding the cell upon electrode insertion. Since the 3.0 nm^2 cross-sectional area of the inter-particle spaces is close to the known dimensions of animal gap junctions (3.1 nm^2), and as there is close agreement between the experimental and theoretical resistances of the gap junction system, Robards and Lucas (4) used these equations to investigate this discrepancy further. The relevant plasmodesmatal parameters used in the gap junction equations were: resistivity of pore fluid, 91 Ohm-cm; pore radius, 8×10^{-8} cm; pore length, 1.47×10^{-5} cm; geometric packing of 9 particles within the cytoplasmic sleeve. Based on these calculations the theoretical resistance of a single plasmodesma would be approximately 6.7×10^9 Ohms (or a conductance of 149 pS) [see (4)], which is very close to the value obtained by Overall and Gunning. In that the two theoretical values have approximately the same value and are much lower than the experimental resistance, it would seem that the most likely basis for the discrepancy resides with the spatial arrangement of the particles within the cytoplasmic sleeve. However, it is worth noting that changing the resistivity of the pore from 91 to 200 Ohm-cm raised the single plasmodesma resistance to 1.5×10^{10}; which reduced the discrepancy to a factor of less than 30.

More work needs to be done to establish the unitary conductance of plasmodesmata in simple cellular systems: cells in callus or suspension culture would seem ideal. If a two-celled system can be developed, then it would be possible to use the voltage clamp approach that has been utilized for the characterisation of gap junctions. However, great care will need to be taken to reduce complications associated with wound-induced callose formation around the plasmodesmata. In addition, since there is preliminary evidence that plasmodesmata may be pressure-sensitive, care must also be taken to ensure that osmotic perturbations are avoided. If unambiguous measurements of single plasmodesmal conductance can be made, then this would help to resolve further the roles played by the particles in the cytoplasmic sleeve, and the region of the neck constriction, in terms of establishing the biophysical properties of plasmodesmata.

Molecular size exclusion limits

Electrical coupling experiments have indicated that ions can pass through plasmodesmata, with the most likely route being the cytoplasmic sleeve. The molecular properties of this pathway have been further probed by the use of fluorescent dyes, a technique adapted from those used for gap junctions. Tyree and coworkers (21,22) were the first to conduct a semiquantitative analysis on the diffusion of fluorescein through staminal hairs of *Tradescantia virginia* and the trichomes of *Lycopersicon esculentum*. Goodwin (23) and Tucker (24) subsequently established that the molecular size exclusion limits of plasmodesmata were from 700 to 874 Da, which is very close to the limits observed for gap junctions. The relationship between the molecular size exclusion limit, expressed in terms of molecular weight, and the actual physical dimensions of the fluorescent dyes has been addressed by several groups. Goodwin used a simple molecular modelling approach to predict that the physical size limits of plasmodesmata in *E. canadensis* would be within the range 3.0—5.0 nm. Tucker, on the other hand, relied on the earlier gap junction studies of Loewenstein (25) to predict that in *S. purpurea* this physical limit would be somewhat greater than 1.4 nm. Microinjection studies in which fluorescent dye molecules were introduced into the cytoplasm, or the nucleoplasm of the staminal hair cells of *Setcreasea purpurea*, indicated that the plasmodesmata which interconnect these cells constitute a rather complex molecular diffusion barrier. Tucker and colleagues (26,27) were later able to develop a numerical analysis model that allowed them to compute the junctional (intercellular) diffusion coefficient for carboxyfluorescein. Although dye transfer to the vacuole had previously been considered as unimportant, Tucker et al had to incorporate this transfer coefficient in their numerical model in order to obtain reasonable agreement between simulated and experimental data. The computed plasmodesma-specific diffusion coefficient for carboxyfluorescein calculated by Tucker et al was 5.37×10^{-8} cm^2 s^{-1}, which is approximately 80 times lower than the value of 4.5×10^{-6} cm^2 s^{-1} obtained for diffusion of carboxyfluorescein in water. This result provides the first experimental support for the hypothesis that plasmodesmata represent a significant barrier to the diffusion of small molecules. Perhaps the most extensive and rigorous evaluation of the physical properties responsible for establishing the size limits of plasmodesmata has been carried out by Terry and Robards (28). An extensive array of FITC-peptide conjugates was evaluated using space-filling models to determine the most energetically favourable conformation, and, thus, the most probable molecular dimensions. These probes were then employed in dye-coupling experiments performed on the nectary trichome cells of *Abutilon striatum* to determine the molecular size exclusion limits for plasmodesmata in this system. Terry and Robards then adapted the permeability equations, developed for permeation through the pores of the nuclear envelope, to their plasmodesmatal system to explore the relationship between the radius of the permeant dye and the rate of diffusion. From these studies they determined that, at least for the *Abutilon* system, the mobility of a probe is determined solely by its effective Stokes radius. In addition, the rates of diffusion observed for the different probes utilized in their study can best be explained if the cytoplasmic sleeve is sub-divided into a number of discrete channels (from 9 to 20) each having a diameter of approximately 3 nm.

Regulation of plasmodesmatal permeability

An extension of the concept that plasmodesmata are involved in regulating intercellular transport of solutes, on the basis of size (or effective Stokes radius), is that their actual size exclusion limits are dynamic, being under some form of cellular control. Erwee and Goodwin (29) found that pretreatment of *Egeria densa* (synonym *Elodea densa*) cells with Ca^{2+} caused a dramatic reduction in the subsequent movement of fluorescent dye molecules; similar results were also obtained with Mg^{2+} and Sr^{2+}. However, if they waited for 30 minutes after Ca^{2+} had been iontophoresed into the cell, normal kinetics for cell-to-cell dye movement were obtained. These experiments, along with others (see 4) support the contention that plasmodesmata are regulated by Ca^{2+}-mediated processes other than callose formation; a conclusion that is further supported by recent fluorescein photobleaching studies performed on soybean suspension culture cells. Subsequent findings indicated that plasmolysis can disrupt normal regulatory mechanism(s), the function of which appears to be the control of plasmodesmatal size exclusion limits.

Tucker (30) has investigated the effects of various D-myo-inositol metabolites on the movement of carboxyfluorescein in the *Setcreasea* staminal hair system and suggested that inositol trisphosphate-diacylglycerol (IP3-DG) second-messenger system may form part of the regulatory pathway.

Plasmodesmatal-virus interactions

It has long been known that plant viruses interact, in some way, to facilitate systemic infection (4,31). It has now been unequivocally established that in transgenic tobacco plants, in which the tobacco mosaic virus 30 kD movement protein is expressed, the plasmodesmatal molecular size exclusion limit is raised more than 10-fold compared with control plants. Further genetic, molecular and cellular studies into the mechanism responsible for causing this change in these transgenic plants may well provide an alternative route to obtaining the genes encoding for the putative plasmodesmatal proteins. In addition, since modulation in callose deposits is often observed in plasmodesmata of virally-infected tissue, molecular studies of plant-viral interaction may also provide essential information on the regulation of the β-1,3 D-glucan synthase and glucanase moieties. Immunological studies using antibodies raised against these moieties will also provide information on the spatial distribution of these enzymes along the plasmodesmatal membranes.

CELL INTERACTIONS IN ANIMAL TISSUES

Gap junctions are sieve-like structures which form in regions of close contact between adjacent cells. The basic unit of a gap junction is an aqueous channel which crosses both membranes and provides direct continuity between the cytoplasms of adjacent cells. Junctions contain varying numbers of channels which are normally packed together in loose hexagonal arrays. The channels are permeable to small ions and molecules (M.Wt. < 900) but not to macromolecules (32). As with plasmodesmata (28), the permeability limit is most likely governed by the shape and dimensions of the hydrated molecules but is normally expressed for convenience in terms of molecular weight. In practice most cytoplasmic molecules are either small enough to pass freely through the junctional channels or clearly too large; few are of a size which makes their transfer uncertain.

Structure of gap junctions

Gap junctions were first isolated by Benedetti and Emmelot (33) more than 20 years ago. They started with rat liver plasma membranes and exploited the relative detergent resistance of the junctions to selectively isolate plaque-like, double membrane structures packed with hexagonal arrays of 7-8 nm diameter particles. Since then other isolation procedures have been developed but all depend on the unique stability of gap junctions in various chaotropic conditions (detergent solutions, 6M urea, high pH), on their resistance to proteolysis and on their characteristic bouyant density which results from a composition of approximately equal amounts of protein and residual phospholipid in the double bilayer.

Isolated liver gap junctions retain the characteristic features seen *in situ* and can be identified by electron microscopy in sectioned pellets or negatively stained suspensions. Centrifugation of even crude preparations tends to produce stratified pellets with a layer rich in gap junctions. These layers, because junctions are flat structures which pack face to face, have sufficient organization for X-ray diffraction analysis and individual junctions,

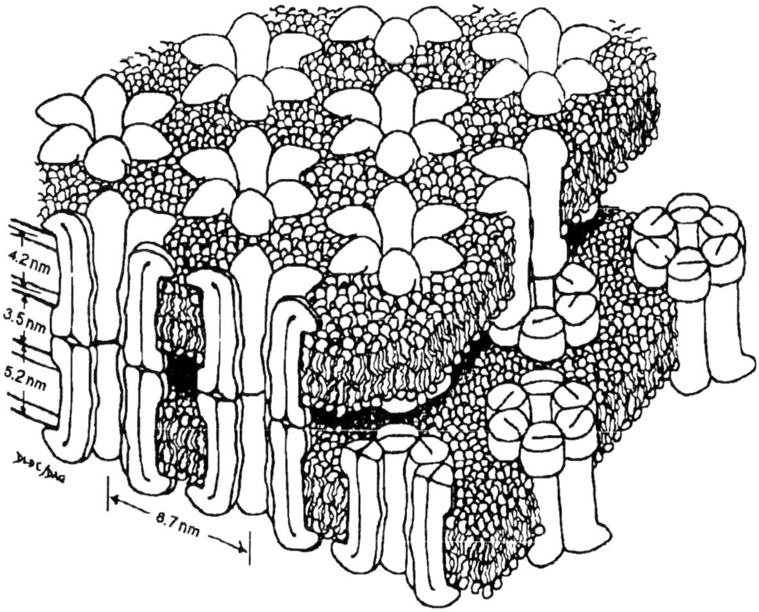

FIGURE 3
The Gap junction (from 36)

selected for their large size and uniform crystalline packing, can be examined by electron microscopy and electron diffraction (34,35). These imaging procedures all produce basically the same model (Fig. 3) and there is now general agreement that this is an acceptable representation of the main features of the isolated structures.

The cells of most tissues form gap junctions which are very similar to those found in liver but the junctions between lens fibres may be different (37). In thin sections they look similar but freeze fracture images show there are two forms of particle array in areas of close membrane apposition. In one the junctional particles are rather loosely packed and in the other they appear in close-packed orthogonal arrays. It is not clear how these structures are related to each other or to the model shown in Fig. 3 and there is some uncertainty about the role of these lens junctions in cell to cell communication.

Analysis of the proteins present in isolated junction preparations has provided additional information on the relationships between gap junctions in different tissues and it has also shown that these structures are more complex than was first thought. Purified liver junction fractions contain three proteins, all thought to be components of the junctional structure. Two show partial homology in their N-terminal regions and are now known to be members of the connexin family of proteins (38). These proteins vary in size from 26 kD to 46 kD and have a tissue specific distribution. The third protein is smaller, 16 kD, is not related to the connexins and is apparently expressed in all tissues where gap junctions are found (39). This protein has been highly conserved through evolution and is also found in arthropods which form gap junctions but do not express connexin-related proteins. Lens fibres express a 46 kD connexin (which shows N-terminal homology with a larger protein Mr 75,000) but the major component of isolated lens gap junctions in a 28 kD protein (MIP26 — mol. wt. by SDS-PAGE is 26,000) which is unrelated to either the connexins or the 16 kD protein (37).

Despite the application of sophisticated techniques, there are still divergent views on which of these proteins forms the subunits of the model shown in Fig. 3. There is better evidence (39,40) for the presence of the 16 kD protein in the isolated junctions (as used for imaging analyses) but better evidence (41) for the presence of the connexins in gap junctions *in situ*. Inhibition by intracellular injection of specific antibodies (42,43) shows that both types of protein are necessary for successful coupling and suggests that the active junction is a more complex structure than depicted by the model. Reconstitution of the purified connexins and/or 16 kD protein into double lipid bilayers to form morphologically identifiable junctions should help to resolve the issue of which proteins are required but this has not yet been achieved. MIP26, on the other hand, has been reconstituted into liposomes to form structures which closely resemble the lens fibre junctions seen *in situ* (37). Recent work has shown that the sequence of the junctional 16k protein is closely related to that of a 16 kD protein believed to be the proton channel component of the vacuolar ATPase (39). The 16k protein in gap junction preparations is insoluble in a variety of detergents but the protein in vacuolar ATPase preparations is readily solubilized in these same detergent solutions. In the gap junction, the N-and C-termini are cytoplasmic but in the vacuolar ATPase the protein is arranged with the opposite orientation. These observations suggest the protein can exist in different forms which provide different channel functions.

Molecular modelling has shown that the 16k protein can form a stable hexamer in which the subunits, each containing four extended trans-membrane alpha-helices, surround a central water-filled pore (39,44). The dimensions of this model are very similar to those of the gap junctional half-channel derived from the X-ray and diffraction studies. The connexins are reported to have four trans-membrane domains in the N-terminal region but it is not possible to accommodate the hydrophilic C-terminal tail, which varies widely in size, in the currently accepted junctional channel structure (35).

Control of junctional communication

The junctional pathway in animal tissues can be controlled at the level of channel formation and at the level of channel gating (interchange between open and closed states). There is some evidence for interchange between open and closed states of plasmodesmata but the two forms are less well understood and the response time appears to be longer than that for gap junctions (minutes or longer rather than nanoseconds or shorter).

Unlike plasmodesmata, which are formed (or left behind) during the process of cytokinesis, gap junctions form whenever the membranes of two mutually competent cells come close together. In culture, gap junctions form between cells from different species and phyla indicating a high degree of conservation.

Controls acting on gap junction formation affect the number of channels at the cellular interfaces. These controls may operate at the level of gene expression or at the level of assembly. As yet little is known about the control elements associated with genes for the structural (or control) elements of gap junctions and nothing is known about the mechanism of assembly. It is generally believed that half-channels exist in the cytoplasmic membrane in a precursor state and that these interact with counterparts on the approaching membrane when junctions are formed. Such hypothetical precursor half-channels have not yet been identified. Similarly, nothing is known about the pathway which leads from protein synthesis to the formation of the putative hexameric half-channels except that both the 16k protein (and the connexins) do not have cleavable signal sequences (Finbow and Pitts, unpublished;45). It is interesting to note however that the 16k protein must either be directed to the vacuolar membrane system in one orientation (N- and C-termini on the non-cytoplasmic face) or to the cytoplasmic membrane in the opposite orientation (the termini on the cytoplasmic face), a choice which may be governed by the presence of (a) chaperone protein(s).

Various diverse treatments and agents, such as low cytoplasmic pH, high Ca^{2+}, intermediate chain alkanols and retinoids are reported to reduce channel permeability (see 46). Electrical measurements and in particular single channel recordings, suggest that individual channels can adopt open or closed states and there is some evidence to suggest that intermediate conductance states may exist, at least transiently (see 47). Some concern must be expressed about data derived from single channel recordings however as gap junctions normally exist as aggregates of many channels and artifical conditions must be introduced to eliminate most of these channels. One approach has been to treat pairs of cells with uncoupling reagents and make the electrical measurements on residual channels just before coupling is finally lost (48). It is assumed that measurements made under these conditions are representative of the characteristics of channels in intact junctions in untreated cells.

Some junctional conductances also appear to be voltage dependent, varying with both trans-junctional and trans-membrane voltage changes (46). The physiological significance of these findings for junctional communication between non-excitable or even excitable cells is at the moment unclear.

The observation of two alternative conductance states is however tantalisingly equivalent to the open and closed conformations seen in isolated gap junctions by electron diffraction analysis (49). The switching of the channels from open to closed states in purified junctional preparations by increasing the calcium concentration suggests that the response is due to some direct effect of calcium on the channel protein. If this is correct and separate calcium binding proteins are not involved, it is important to identify the binding site which should presumably lie on the cytoplasmic surface.

Connexin expression has been shown to affect the level of junctional communication between cells. In principle, this could be due to the availability of extra channels, if connexins form such channels, or to increased channel formation from endogenous precursor half-channels if connexins play a role in the control of junction assembly. The effect of connexin expression on gap junctional conductance has been examined between pairs of *Xenopus* oocytes after microinjection of RNAs coding for different connexins (50). The picture which emerges is, however, complex and difficult to understand, partly due the presence of uncharacterized endogenous connexins. Expression of either connexin 32 or 43 in both cells in any combination results in large, voltage-insensitive increases in junctional conductance while injection of water into both cells produces small, symetrically voltage-dependent conductances. Expression of connexin 32 in only one cell has little effect on conductance but expression of connexin 45 in one cell results in a large, asymmetrically voltage-dependent increase. Clearly much will be learned from studies such as these but it will be important either to fully define the extent of endogenous expression of all relevant proteins or to find a cell system where gap junctions are not formed and none of the proteins potentially involved in junctional communication is endogenously expressed.

Gap junctions and cellular differentiation

Gap junctions are formed between all the cells at early stages of embryogenesis and communication persists in most tissues throughout development and adult life. They are lost in in a few cell types, as part of the programme of terminal differentiation, seemingly to avoid problems which would be caused by the presence of cell coupling. For example, in skeletal muscle gap junctions would result in short-circuiting between stimulated cells producing all or nothing responses (as in heart muscle) rather than movements of graded intensity. Similarly, in tissues subject to continuous renewal such as the epidermis, the fully differentiated dying cells become uncoupled from the underlying population to prevent the leakage of small ions and molecules from the live cells.

Gap junctions provide a general mechanism for intercellular homeostasis (51). Plasmodesmata probably provide the same homeostatic function but there is also a clear requirement in plants for bulk flow between cells, a phenomenon which is not generally considered in animal systems.

Cells joined by gap junctions share their pools of metabolites, second messengers, inorganic ions etc. which results in the coordination of cellular activities and a consequent suppression of cellular differences. This inhibition of difference presents a potential problem for the development of differentiated tissues within complex organisms and has led to the evolution of mechanisms of segregation. In animal tissues and in model systems in culture, cells are divided into communication compartments (51). Cells within a compartment are well coupled by gap junctions to each other but not to cells in adjacent compartments. Compartmental boundaries, which may or may not be associated with morphologically recognisable barriers (such as basement membranes), are marked by a reduced or undetectable rate of heterologous junction formation.

It is now becoming clear that intercellular homeostatic control operates within these compartments and compartmentation allows the expression of differences in different parts of an organism. During embryogenesis the compartments are first seen when the three germ layers become apparent and the compartmental boundaries appear to coincide with the morphological boundaries of the different tissues (52). If such compartmentation is necessary before groups of cells can embark on significantly different differentiation pathways the formation of the boundaries will be key events in development.

Junctional communication between cells within a compartment results in the coordination of cellular activities and a uniform tissue phenotype. Cells in different compartments can be subject to different homeostatic pressures so different activities are more likely to be associated with cells in different compartments. Furthermore, cells in different compartments can not only maintain different cytoplasmic concentrations of small ions and molecules but can also use the same small molecules for different purposes. One explanation for the evolution of compartmentation may therefore be related to the limited number of second messenger systems which are available and which mediate different activities in different cell types.

These general ideas on the roles of compartmentation have been supported and extended by studies on compartmentation and the patterns of junctional communication in skin (53). Proliferative control in the epidermis appears to operate within epidermal proliferative units, each composed of 5 or 6 basal cells and the overlying differentiating cells. Recent work indicates that these units are communication compartments and this observation has led to a model for epidermal proliferative control (54). Second messengers produced by growth factor stimulation of the basal cells will, it is suggested, dissipate into the unresponsive differentiating cells of the compartment. The concentration of second messengers in the basal cells will depend on the number (total cytoplasmic volume) of overlying cells. If a threshold value of second messengers is required to initiate cell division, junctional communication will result in a skin thickness dependent modulation of growth factor stimulated cell division. A decrease in compartment size through loss of terminally differentiated cells should stimulate division whereas an increase in size through cell division should be inhibitory.

There is an obvious defect in the model, however, and some mechanism is necessary to prevent uncontrolled division of additional basal cells during the period before the volume of the compartment increases as a consequence of the stochastic initiation and subsequent division of the first cell. Fortunately a solution to this problem is provided by the observation that the boundary between the epidermal compartments and the dermis breaks down soon after a hyperproliferative stimulus (55). If such a breakdown is a consequence of the initiation of the first cell, the resulting dissipation of second messengers from the epidermal compartment into what is in effect a dermal sink (the dermal compartment is apparently limitless) provides a simple mechanism for the inhibition of the unwanted divisions. Regeneration of the boundary then occurs after the first cell has completed its division and one of the daughter cells has moved up into the overlying layer to begin its programme of terminal differentation and the basal layer. This additional feature depends on the same basic principle of intercellular homeostasis but in this situation across a regulated boundary. Such regulation has now been observed in a variety of situations.

The repeated epilation (Er) mutant mouse has provided a convenient test for this growth control model (56). The observed hyperproliferation of the mutant epidermis is a predicted consequence of the observed tissue-specific loss of junctional communication in the dermis. In this case the ineffective dermal sink results in inadequate dissipation of second messengers and explains the observed excess epidermal proliferation.

The general idea that compartmentation allows the expression of difference and that homeostatic control operates within compartments have been further strengthened by recent work on the patterns of communication in hair follicles (Kam and Hodgins, unpublished) and molluscan embryos (57). Hair follicles are highly differentiated structures which are derived from invaginations of the epidermis into the dermis. Their development and growth is influenced by hormones and by interactions between the epithelial and dermal tissues. Analysis of the patterns of junctional communication in these structures has shown: (1) cells in the germinative region (near the base of the hair bulb) all communicate freely with each other; (2) cells immediately above this region sort-out intocommunication compartments according to their respective differentiation destiny; (3) the differentiated epithelial cells in the various layers of the inner und outer root sheath, derived from these compartments, remain either well coupled within their compartments or become poorly coupled. This loss of coupling presumably reflects the onset of terminal differentiation.

A similar picture has emerged from the more detailed study of early development in molluscan embryos (57). Again the same general principles emerge. Prior to commitment there is wide-spread coupling but when groups of cells differentiate along distinctive pathways they are either found to be in separate communication compartments or become uncoupled altogether. The loss of coupling is associated with embryonic cells which have become highly differentiated in the early larval stages. The cells of the anlagen however, remain coupled but form separate communication compartments. Interestingly, development of adult structures from the anlagen is more regulative and probably more dependent on cell-cell interactions.

Cellular proliferation and commitment to differentiation are normally controlled within coupled cell populations but as a prerequisite for differentiation selective uncoupling of groups of cells results in compartmentation. The controls leading to the creation of compartmental boundaries allow sub-populations of cells to establish separate identities and seem likely to be a critical processes in morphogenesis. Further advances in our understanding of the role of junctional communication in the control of proliferation and differentiation will be facilitated in future by an ability to genetically manipulate the boundary state and hence the levels of trans-boundary junctional communication.

Recent work (58) has revealed the presence of specific patterns of communication in plant tissues. Although it is premature to interpret these patterns in terms of defined, developmentally determined communication compartments, it is possible that the general principles of compartmentation, based on observation in animal systems, may also apply to plants.

CONCLUSIONS

Although plasmodesmata and animal gap junctions exhibit certain parallels in terms of their physiological size exclusion limits, they possess no structural similarities. However, it appears highly likely that these two divergent cell-to-cell communication systems have evolved, by completely separate pathways, in order to permit the molecular and biochemical coordination of cells within specialized tissues. It will be of great interest to determine the cellular mechanisms utilized by these two systems to control the size exclusion limits of their cell-to-cell junctions, and in particular to identify common attributes. Studies of this kind may well establish a new operational framework for comparative analysis of developmental systems in plants and animals.

ACKNOWLEDGEMENTS

We gratefully acknowledge the help and support of the other co-organisers of our NATO ASW: Dr. Habo Jongsma (Amsterdam), Prof. David Spray (New York) and, particularly Prof. Bill Lucas (Davis, California). Many of the ideas incorporated into this article arise from points made during the ARW and from subsequent discussions among the co-organisers. We are also grateful to NATO for its sponsorship of the Workshop; and to the Agricultural and Food Research Council and the Science and Engineering Research Council for grants to AWR for the study of plasmodesmata.

REFERENCES

1) Tangl, E., 1879. Ueber offene Communicationen zwischen den Zellen des Endosperms einiger Samen. *Jb. wiss Bot.* 12: 170—190.

2) Münch, E., 1930. *Die Stoffbewegung in der Pflanze.* Gustav Fischer, Jena.

3) Robards, A.W., Jongsma, H.,Lucas, W.J., Pitts, J., Spray, D. (eds.), 1990. *Parallels in Cell to Cell Junctions in Plants and Animals*, Springer-Verlag, Heidelberg, In Press.

4) Robards, A.W. and Lucas, W.J., 1990. Plasmodesmata. *Annu. Rev. Plant Physiol.* In Press.

5) Olesen, P., Robards, A.W., 1990. The neck region of plasmodesmata: general architecture and some functional aspects. In *Parallels in Cell to Cell Junctions in Plants and Animals,* eds, A.W. Robards, H. Jongsma, W.J. Lucas, J. Pitts, D. Spray, Springer-Verlag, Heidelberg, In Press.

6) Conkling, M.A. Yamamoto,Y.T., Acedo, G.N., 1990. A tobacco root-specific gene homologous to the mammalian lens major intrinsic in soybean nodulin 26 proteins. *Science* (submitted).

7) Baron-Epel, O., Hernandez, D., Jiang, L.-W., Meiners, S., Schindler, M., 1988. Dynamic continuity of cytoplasmic and membrane compartments between plant cells. *J. Cell Biol.* 106: 715—-721.

8) Wade, M.H., Trosko, J.E., Schindler, M., 1986. A fluorescence photobleaching assay of gap junction-mediated communication between human cells. *Science* 232: 525—528.

9) Monzer, J., 1990. Secondary formation of plasmodesmata in cultured cells. In *Parallels in Cell to Cell Junctions in Plants and Animals,* eds, A.W. Robards, H. Jongsma, W.J. Lucas, J. Pitts, D. Spray, Springer-Verlag, Heidelberg, In Press.

10) Robards, A.W., 1976. Plasmodesmata in higher plants. In *Intercellular Communications in Plants: Studies on Plasmodesmata*, pp. 15—57, Gunning B.E.S., Robards A.W. eds.Springer-Verlag, Heidelberg.

11) Fisher, D.G., 1990. Distribution of plasmodesmata in leaves. A comparison of *Cananga odorata* with other species using different measures of plasmodesmatal frequency. In *Parallels in Cell to Cell Junctions in Plants and Animals,* eds, A.W. Robards, H. Jongsma, W.J. Lucas, J. Pitts, D. Spray, Springer-Verlag, Heidelberg, In Press.

12) Gunning, B.E.S., Hughes, J.E., 1976. Quantitative assessment of symplastic transport of pre-nectar into trichomes of *Abutilon* nectaries. *Aust. J. Plant Physiol.* 3: 619—637.

13) Gunning, B.E.S., 1978. Age-related and origin-related control of the numbers of plasmodesmata in cell walls of developing *Azolla* roots. *Planta* 143: 181—190.

14) Spanswick, R.M., Costerton, J.W.F., 1967. Plasmodesmata in *Nitella translucens:* structure and electrical resistance. *J. Cell Sci.* 2: 451—464.

15) Tyree, M.T., 1970. The symplast concept. A general theory of symplastic transport according to the thermodynamics of irreversible processes. *J. Theor. Biol.* 26: 181—214.

16) Drake, G., 1979. Electrical coupling, potentials, and resistances in oat coleoptiles: effect of azide and cyanide. *J. Exp. Bot.* 30: 719—725.

17) Tucker, E.B., 1987. Cytoplasmic streaming does not drive intercellular passage in staminal hairs of *Setcreasea purpurea. Protoplasma* 137: 140—144.

18) Tucker, E.B. Analytical studies of dye-coupling between plant cells. In *Parallels in Cell to Cell Junctions in Plants and Animals,* eds, A.W. Robards, H. Jongsma, W.J. Lucas, J. Pitts, D. Spray, Springer-Verlag, Heidelberg, In Press.

19) Overall, R.L., Wolfe, J., Gunning, B.E.S., 1982. Intercellular communication in *Azolla* roots: I. Ultrastructure of plasmodesmata. *Protoplasma* 111: 134—150.

20) Overall, R.L., Gunning, B.E.S., 1982. Intercellular communication in *Azolla* roots: II. Electrical coupling. *Protoplasma* 111: 151—160.

21) Barclay, G.F., Peterson, C.A., 1982. Transport of fluorescein in trichomes of *Lycopersicon esculentum. Can. J. Bot.* 60: 397—402.

22) Tyree, M.T., Tammes, P.M.L., 1975. Translocation of uranin in the symplasm of staminal hairs of *Tradescantia. Can. J. Bot.* 53: 2038—2046.

23) Goodwin, P.B., 1983. Molecular size limit for movement in the symplast of the *Elodea* leaf. Planta 157: 124—130.

24) Tucker, E.B., 1982. Translocation in the staminal hairs of *Setcreasea purpurea*. I. Study of cell ultrastructure and cell-to-cell passage of molecular probes. *Protoplasma* 113: 193—201.

25) Loewenstein, W.R., 1979. Junctional intercellular communication and the control of growth. *Biochim. Biophys. Acta* 560: 1—65.

26) Tucker, E.B., Spanswick, R.M., 1985. Translocation in the staminal hairs of *Setcreasea purpurea*. II. Kinetics of intercellular transport. *Protoplasma* 128: 167—172.

27) Tucker, J.E., Mauzerall, D., Tucker, E.B., 1989. Symplastic transport of carboxyfluorescein in staminal hairs of *Setcreasea purpurea* is diffusive and includes loss to the vacuole. *Plant Physiol.* 90: 1143—1147.

28) Terry B.R., Robards A.W., 1987. Hydrodynamic radius alone governs the mobility of molecules through plasmodesmata. *Planta* 171: 145—157.

29) Erwee, M.G., Goodwin, P.B., 1983. Characterisation of the *Egeria densa* Planch. leaf symplast: inhibition of the intercellular movement of fluorescent probes by group II ions. *Planta* 158: 320—328.

30) Tucker, E.B., 1988. Inositol bisphosphate and inositol trisphosphate inhibit cell-to-cell passage of carboxyfluorescein in staminal hairs of *Setcreasea purpurea. Planta* 174: 358—363.

31) Lucas, W.J., 1990. Viruses and plasmodesmata. In *Parallels in Cell to Cell Junctions in Plants and Animals*, eds, A.W. Robards, H. Jongsma, W.J. Lucas, J. Pitts, D. Spray, Springer-Verlag, Heidelberg, In Press.

32) Pitts, J.D., Simms, L.W. Permeability of junctions between animal cells. *Exp. Cell Res., 1977*: 153—163.

33) Benedetti, E.L., Emmelot, P., 1968. Hexagonal array of subunits in tight junctions separated from rat liver plasma membranes. *J. Cell Biol* 38: 15—24.

34) Unwin, P.N.T., 1987. Gap junction structure and the control of cell to cell communication. *Ciba Foundn Symp.* 125: 78—91.

35) Makowski, L., 1988. X-ray diffraction studies of gap junction structure. *Adv. Cell Biol.* 2: 119—154.

36) Makowski, L., Casper, D.L.D., Phillips. W.C., Goodenough, D.A., 1977. Gap junction structures: II Analysis of X-ray diffraction data. *J. Cell Biol.* 74: 629—645.

37) Benedetti, E.L., Dunia, I., Manenti. S., Bloemendal, H., 1990. Biochemical and structural properties of the protein constituent of junctional domains in eye lens fiber plasma membranes. In *Parallels in Cell to Cell Junctions in Plants and Animals*, eds, A.W. Robards, H. Jongsma, W.J. Lucas, J. Pitts, D. Spray, Springer-Verlag, Heidelberg, In Press.

38) Willecke, K., Hennemann, H., Herbers, K., Heynkes, R., Kozjek, G., Look, J., Stutenkemper, R., Traub, O., Winterhager, E., Nicholson, B., 1990. Molecular heterogeneity of gap junctions in different mammalian cells. In *Parallels in Cell to Cell Junctions in Plants and Animals*, eds, A.W. Robards, H. Jongsma, W.J. Lucas, J. Pitts, D. Spray, Springer-Verlag, Heidelberg, In Press.

39) Finbow, M.E., Thompson. P., Keen. J., Jackson. P., Eliopoulos. E., Meaghe.R L., Findlay, J.B.C., 1990. A structural analysis of the molecules involved in gap junctional communication. In *Parallels in Cell to Cell Junctions in Plants and Animals*, eds, A.W. Robards, H. Jongsma, W.J. Lucas, J. Pitts, D. Spray, Springer-Verlag, Heidelberg, In Press.

40) Buultjens, T.E.J., Finbow, M.E., Lane, N.J., Pitts, J.D., 1988. Tissue and species conservation of the vertebrate and arthropod forms of the low molecular weight (16—18,000) proteins of gap junctions. *Cell Tissue Res* 251: 571—580.

41) Dermietzel, R., Hwang, T.K., 1990. Structural and molecular diversity of gap junctions. In *Parallels in Cell to Cell Junctions in Plants and Animals*, eds, A.W. Robards, H. Jongsma, W.J. Lucas, J. Pitts, D. Spray, Springer-Verlag, Heidelberg, In Press.

42) Hertzberg, E.L., Spray, D.C., Bennett, M.V.L., 1985. An antibody to gap junctions blocks gap junctional conductance. *Proc. Natn. Acad. Sci. U.S.* 82: 2412—2416.

43) Serras, F., Buultjens, T.E.J., Finbow, M.E., 1988. Inhibition of dye coupling in *Patella* (Mollusca) embryos by microinjection of antiserum against *Nephrops* (Arthropoda) gap junctions. *Exp. Cell. Res.* 179: 282—288.

44) Findlay, J.B.C., Thompson. P., Jackson, P.J., Keen, J.N., Eliopoulos, E., Attwood, T.K., Pitts, J.D., Finbow, M.E., 1990. Structural analysis of integral membrane proteins. In *Structure, Function and Biogenesis of Energy Transfer Systems* Elsevier, Amsterdam (in Press).

45) Paul, D., 1986. Molecular cloning of cDNA for rat liver gap junction protein. *J. Cell Biol.* 103: 123—134.

46) Spray, D.,1990. Electrophysiological properties of gap junction channels. In *Parallels in Cell to Cell Junctions in Plants and Animals*, eds, A.W. Robards, H. Jongsma, W.J. Lucas, J. Pitts, D. Spray, Springer-Verlag, Heidelberg, In Press.

47) Jongsma, H., Rook, M.B., 1990. Cardiac gap junctions: gating properties of single channels. In *Parallels in Cell to Cell Junctions in Plants and Animals*, eds, A.W. Robards, H. Jongsma, W.J. Lucas, J. Pitts, D. Spray, Springer-Verlag, Heidelberg, In Press.

48) Neyton, J., Trautmann, J., 1985. Single-channel current of an intercellular junction. *Nature* 317: 331—335.

49) Unwin, P.N.T., Ennis P.D., 1984. Two configurations of a channel forming membrane protein. *Nature* 307: 609—613.

50) Swenson, K.I., Jordon, J.R., Beyer, E.C., Paul, D.L., 1989. Formation of gap junctions by expression of connexins in *Xenopus* oocyte cell pairs. *Cell* 57: 145—155.

51) Pitts, J.D., 1990. Junctional communication: the role of communication compartments in complex multicellular organisms. In *Parallels in Cell to Cell Junctions in Plants and Animals*, eds, A.W. Robards, H. Jongsma, W.J. Lucas, J. Pitts, D. Spray, Springer-Verlag, Heidelberg, In Press.

53) Kam, E., Melville, L., Pitts, J.D., 1986. Patterns of junctional communication in skin. *J. Invest. Dermatol.* 87: 748—753.

53) Kam, E., Melville, L., Pitts, J. D., 1986. Patterns of junctional communication in skin. *J. Invest. Dermatol.* 87: 748—753.

54) Pitts, J.D., Finbow, M.E., Kam, E., 1988. Junctional communication and cellular differentiation. *Brit J. Cancer* 58: 52—-57.

55) Kam, E., Pitts, J.D., 1988. Affects of the tumour promotor TPA on junctional communication in intact mouse skin; persistance of homologous communication and increase in dermal-epidermal coupling. *Carcinogenesis* 9: 1389—1394.

56) Kam, E., Pitts, J.D., 1989. Tissue specific regulation of junctional communication in the skin of foetuses homozygous for the repeated epilation (Er) mutation. *Development* 107: 923—929.

57) Serras, F., van den Biggelaar, J.A.M., 1990. Progressive restrictions in gap junctional communication during development. In *Parallels in Cell to Cell Junctions in Plants and Animals*, eds, A.W. Robards, H. Jongsma, W.J. Lucas, J. Pitts, D. Spray, Springer-Verlag, Heidelberg, In Press.

58) Erwee, M.G., Goodwin, P.B., 1985. Symplastic domains in extrastelar tissues of *Egeria densa* Planch. *Planta* 163: 9—19.

MECHANISM OF FERTILIZATION: PLANTS TO HUMANS

Brian Dale

Stazione Zoologica
Villa Comunale
80121, Naples, Italy

INTRODUCTORY REMARKS

Most information on the process of fertilization has come from the study of a few selected invertebrates and lower vertebrates. Over the last decades, in vitro culture techniques have allowed comparable experimentation in mammals, including the human, while more recently access to plant material has permitted a start in this direction. In order to identify unifying concepts in the mechanism of fertilization in plants, invertebrates and vertebrates a Nato workshop was held in Sorrento, Italy in October, 1989 (see Dale, 1990). The traditional temporal limits of fertilization were extended in the meeting to cover late stages of gametogenesis and early stages of cytoplasmic segregation in the zygote. In this review, I will attempt to highlight some of the major advances made in this field and point out areas suggested for future study.

SURFACE RECEPTORS AND SPERM-EGG INTERACTION

Fertilization is a complex process of cell-cell interaction which starts with the specific recognition and binding of spermatozoa with eggs and ultimately leads to the fusion of the male and female pronuclei. Although the concept of receptor mediated gamete adhesion dates back more than 60 years to F. R. Lillie, the macromolecules involved in this process have been characterized only recently. The initial stages of fertilization depend on two structures, an extracellular glycoprotein sheet in the egg (called the vitelline coat in echinoderms, the zona pellucida in mammals and the chorion in other animals) and the acrosome in the sperm.

Plants

Eggs of the marine brown alga *Fucus* do not have an obvious extracellular coat, although their surface appears rough at the SEM (1). Sperm make contact with the egg membrane by probing the surface with the tips of their anterior flagella (2). There is no evidence of a process comparable to the acrosome reaction seen in animal sperm-egg interactions. The lectins Con A and RCA 120 bind to *Fucus* eggs, but not to sperm, and both inhibit

NATO ASI Series, Vol. H 51
Cell to Cell Signals in Plants and Animals
Edited by V. Neuhoff and J. Friend

fertilization. Fucose Binding Protein binds to both gametes and also inhibits fertilization (3). These results point to a role for mannose/glucose and galactose residues on *Fucus* eggs in the fertilization process. Inhibition type experiments with a variety of carbohydrases and polysaccharides confirm these findings and therefore it appears that mannose and fucose containing egg surface components interact with complementary carbohydrate binding structures on the sperm cell surface. Using monoclonal antibodies raised to cell surface antigens of *Fucus* sperm it has been shown that some antigens are distributed over the entire sperm surface, including a 200kDa glycoprotein. Others are restricted to the sperm body, to the mastigonemes or to the anterior flagellum plasma membrane. The 200kDa glycoprotein is an immunodominant antigen. It is possible that differences in the carbohydrate side chains of this molecule are responsible for the gamete, species and genus preferential binding of the antibodies raised (4).

Owing to new techniques, male gametes may now be collected from angiosperms in sufficient numbers to allow biochemical analysis and monoclonal antibody production. Female gamete isolation is still in its early days, however it is expected that *in vitro* fertilization techniques in plants will emerge shortly (5, 6).

Invertebrates

In the ascidian egg, receptors for the species specific recognition and binding of gametes are located in the chorion and are synthesized by the oocyte (7). Binding occurs between fibrils emerging from the chorion and the tip of the sperm plasma membrane. In *Ciona* fucosyl residues localized on these fibrils play a key role in sperm binding (8). In other species acetylglucosamine residues have the same function. A fucosyl protein of high molecular weight and a protein — carbohydrate ratio of 2:1, bearing the properties of a sperm receptor, has been partially characterized. The biological activity of this glycoprotein is similar to that of ZP3 found in the mouse. Hoshi (9) has suggested that sperm-egg binding in ascidians is in fact an enzyme substrate reaction, with the complimentary counterpart on the sperm head being the enzyme α-fucosidase.

In the freshwater bivalve *Unio* spermatozoa interact with and enter the egg at a specialized point of the surface, a crater, found at the vegetal pole. This area is in fact the only part of the egg surface that binds with lectins, suggesting that the sperm receptors in this case are also fucosyl-containing glycoproteins (10). As in other animals it appears that these sperm receptors are synthesised by the growing oocyte, not somatic cells (11).

The vitelline coat in echinoderms and frogs is in turn externally covered by a thick gelatinous layer. Lillie (12) was one of the first to observe that this coat had the capacity to alter sperm physiology and behaviour. Since then many investigators have isolated soluble factors from the jelly layer of the sea urchin that show activity. Suzuki and colleagues (13) purified a decapeptide from the jelly layer of the sea urchin that stimulates sperm respiration. So far over 40 different sperm activating peptides have been isolated from various sea urchins that may be classified into four groups according to their biological activity . These peptides either trigger sperm respiration, sperm motility, or effect sperm physiology in a more general fashion. They also act as specific co-factors with a second component of the jelly layer, a fucose rich glycoconjugate, in inducing the acrosome reaction (14). The receptor protein for jelly peptides on the spermatozoon has recently been isolated and it is expected that the DNA sequence will shortly be known (15).

Mammals

The fertilization process in mice may be sub-divided into several stages (16). First, loose attachment and then tight species specific binding of acrosome intact sperm to the zona pellucida (ZP). Second, completion of the acrosome reaction and penetration of the zona pellucida by the acrosome reacted sperm, and finally fusion of the plasma membrane of the sperm and egg. Binding occurs between sperm receptors in the ZP and egg binding proteins in the plasma membrane of the sperm overlying the anterior region of the head. The primary sperm receptor in the zona pellucida, called ZP3, is an 83, 000 M. Wt. glycoprotein (17). There are 402 amino acids in the polypeptide chain (18) and it probably consists of stretches of extended chain (5—29 residues) interrupted by short stretches of coils (4—10 residues). Purified ZP3 only binds to the head region of acrosome intact sperm and prevents fertilization. A galactose residue, located in the α-linkage at the non reducing terminus of ZP3 O-linked oligosaccharides serves as an essential determinant for binding of the glycoprotein to sperm (19). ZP3 also induces sperm to undergo the acrosome reaction and this effect depends on both the O-linked oligosaccharides and polypeptide chain. It has been suggested that induction of the acrosome reaction by ZP3 depends on multivalent interactions between the glycoprotien and a sperm plasma membrane component, which may induce capping of the sperm component (17, 20). ZP3 is synthesized by the growing oocyte and not the follicle cells and therefore is an example of sex-specific gene expression. It is probable that, gamete recognition and adhesion in many different mammals are carbohydrate mediated events and it will be interesting to see whether differences in sperm receptor oligosaccharide structure account for species specifity of gamete adhesion.

MECHANISM OF EGG ACTIVATION

Since the first stage of fertilization is in fact a receptor mediated interaction of two cells and there are many examples in cell biology of such signals being transduced across the cell membrane into an intracellular second messenger, it is not surprising that many investigators consider egg activation to be part and parcel of gamete binding. The contrasting idea is that while binding serves to ensure the adhesion and close apposition of the two gametes, egg activation is triggered at a subsequent stage of sperm-egg interaction.

Electrical Events

One of the first indications of egg activation is a change in the ionic permeability of the egg plasma membrane. This is followed by an increase in intracellular levels of Ca and a cascade of biochemical and morphological events (21). Over the past decade, owing partially to improved and new techniques, such as single channel recording, ion specific micro-electrodes and low light imaging, several breakthroughs into the mechanism of egg activation have been made.

The importance of ion fluxes through the plasma membrane was first noted in the 1950's by Monroy and colleagues (22). Since then electrical events at activation have been recorded in eggs from organisms as distant as algae to mammals (23). In 1978 it was shown that the fertilizing spermatozoon differed from supernumerary spermatozoa in its capacity to generate a discrete step like depolarization (24). This step event has since been observed in sea urchins, ascidians and amphibians. Much emphasis has been placed on elucidating the role of this event using the voltage clamp technique (25), however it is becoming increasingly apparent that the step reflects a stage in sperm-egg interaction rather than a trigger event per se.

At fertilization a new population of ion channels appear in the membrane and these generate the larger depolarization the fertilization potential. This may be considered a true activation event. Fertilization channels in the ascidian have a single channel conductance of 400pS, a reversal potential around zero and are most probably non specific for ions (25). In further studies we estimated that the fertilizing spermatozoon opens 200—1000 channels in an area localized to the site of gamete fusion. Fertilization channels are also aggregated around the point of sperm-egg fusion in the anuran *Discoglossus* (26). In the ascidian these channels are not gated by Ca, but are activated following the micro-injection of IP3 (27).

Soluble sperm factors vs Receptors and G-proteins

Two hypotheses prevail as to how a spermatozoon triggers an egg into metabolic activity (see 28 for references). The first points to a soluble trigger factor in the spermatozoon that is released into the egg cytoplasm following gamete fusion (29, 30). The contrasting idea favours an externally located receptor and G-protein transmembrane transduction mechanism (31, 32). Most of the evidence supporting these ideas is indirect and confirmation will require the identification and characterization of the trigger elements. As far as the former hypothesis is concerned an active soluble factor has been successfully isolated from sea urchin sperm (27, 29) and other indirect experiments have suggested the factor to be Ca (33), or cGMP (30). The second hypothesis is based on the following evidence. First, the injection of a non-hydrolyzable GTP analogue activates eggs, while GDP-β-S inhibits activation. To examine the possibilty that a receptor might be involved in the activation of a G-protein, exogenous G-protein related neurotransmitter receptors were introduced into the egg membrane. Adding the corresponding agonists triggered activation events (31, 32).

When does sperm-egg fusion occur?

Since the former hypothesis is based on a post-fusion event, information on the timing of sperm-egg plasma membrane fusion may help to further elucidate the mechanism of fertilization. Early experiments on oocytes suggested that the step event was in fact the result of sperm-egg fusion, the channels in the spermatozoon adding to the membrane conductance of the egg and causing a small depolarization (34). In a more direct attempt to determine when fusion actually occurs Longo and colleagues (35) fixed eggs for the EM following electrical recording and sectioned them. Their conclusion, and supported by others utilizing the concept of dye transfer between cells in continuity (36) was that fusion occured following the step event. Although preliminary, McCulloh and Chambers (37) have shown that the step occurs coincidently with an increase in membrane capacitance, supporting the idea that it is the result of sperm-egg fusion. If this turns out to be the case then the externally mediated receptor hypothesis loses weight.

In most animals there is a delay period from the moment the spermatozoon attaches to the egg surface until the appearence of activation events such as cortical granule exocytosis and the release of intracellular Ca. This period varies from species to species and is known as the latent period (38). There has been a recent surge of interest as to what is happening during the latent period. It seems probable that if the sperm activates a G-protein by interacting with a receptor on the outside of the plasma membrane, then the message would be transferred to the egg cytoplasm rather quickly, and therefore it is not easy to explain this delay period. Alternatively, the 10 second delay may simply represent the time required for a soluble factor to diffuse from the sperm cytoplasm into the egg. A microfilament dependent process occurs during the latent period and there appears to be little other change (23, 39).

Polyspermy prevention

A popular topic for students of fertilization is how eggs prevent polyspermy. Most eggs will not develop if more than one sperm nucleus enters the egg cytoplasm. Many texts talk of dynamic changes occurring at egg activation that "allow" the entry of one spermatozoon and exclude all others. A contrasting idea is that there is a limited number of pre-determined sperm entry sites on the egg surface and that the majority of attached spermatozoa are in fact incapable of penetration (see 28, for references). Proponents of the former idea have suggested that the electrical depolarization at activation serves as a fast block to polyspermy. Again precise timing of the moment of sperm-egg fusion will be determinative in resolving this problem. If the step depolarization, which in some cases leads to an regenerative overshooting depolarization, is the result of sperm-egg fusion then there must be a preceding event — that is not electrical — that discriminates the fertilizing capacity of all the attached spermatozoa. Conversely, if it turns out that fusion occurs after the electrical depolarization, as suggested by other groups (35, 36), then why does one and only one spermatozoon fuse to the egg in this depolarized state? New experimental approaches are required to further elucidate this rather interesting phenomenon.

CYTOPLASMIC SEGREGATION IN THE ZYGOTE

A fundamental process in the differentiation of an early embryo is the generation of asymmetry. Sperm entry in many animals, in addition to triggering the egg into metabolic activity, induces directional changes that lead to the formation of a polarized zygote or embryo.

The Role of Intracellular Ca

Shortly after the electrical changes at activation, intracellular levels of Ca have been shown to increase dramatically in a wide variety of eggs (40). Techniques of ion-selective electrodes, and low-light imaging have gone a long way to resolve these Ca changes both spatially and temporally. During the mid-eighties the inositol cascade was implicated in this Ca releasing mechanism (30), but as yet there are many unanswered questions.

Direct evidence for a calcium wave or indeed an elevation of cytoplasmic calcium in the activation of the marine alga *Fucus* is so far missing. *Fucus* eggs appear to contract at activation, and, in addition, secrete a cell wall, probably as the result of an exocytotic wave (41). Since Ca ionophores also stimulate cell wall secretion, this may well be a Ca dependent process. Calcium binding proteins such as calmodulin are also present.

The role of Ca in polarity formation is clearer. The polar axis is perpendicular to the plane of first cell division and can be determined by unilateral light. A gradient of free Ca in the growing rhizoid has been directly demonstrated using ion selective electrodes and fluorescent indicators (42, 43). In addition, polarization and future development of *Pelvetia* zygotes can be arrested by using various analogues of the calcium buffer BAPTA (44). Questions to be asked that will be of help in understanding the mechanisms of fertilization and development in plants generally are; What are the relative roles of G-proteins in the light transduction mechanism? Is there a Ca gradient in the polarizing egg and how is this polarising gradient setup?

A Ca wave following activation in the ascidian egg has been demonstrated in two independent studies using different techniques. This egg is a particularly useful model since within minutes of fertilization there is a massive re-organization of cytoplasmic components, called cytoplasmic segregation, that leads to the establishment of a highly polarized zygote. Both with Aequorin and Fura-2 it was calculated that the level of free Ca in the virgin egg is about 100nM (44, 45). About 10 sec after the initiation of the fertilization current a wave of high Ca rapidly traverses the egg and cytoplasmic levels reach 10uM. The Ca stays high for several minutes and then turns to resting levels at the time of ejection of the 1st polar body. Of importance is the fact that the direction of cortical contraction, cytoplasmic segregation and animal vegetal polarity follow the direction of this Ca wave. Increasing Ca levels uniformly in the egg by stimulating the opening of voltage-gated Ca channels did not induce a Ca wave or a contraction (45). Since ascidian eggs do not possess cortical granules it is probable that the Ca wave in this animal serves a different function to that in the sea urchin egg.

There are many unanswered questions as regards the Ca wave in eggs. For example, on the propagation mechanism itself; Is it a Ca induced Ca release mechanism or an IP3 induced propagative process? How does this signal trigger directional re-arrangement of the cytoskeleton, and finally, what is the location of the intracellular sources of Ca?

REFERENCES

1) Callow, J., Callow, M. and Evans L. 1985, Fertilization in *Fucus*. In "Biology of Fertilization". Vol. 2, pp. 389—407. Academic press.

2) Friedmann, I. 1961, Cinematography of spermatozoids and fertilization in Fucales. *Bull. Res. Counc. Isr.* 10D, 73—83.

3) Bolwell, G., Callow, J., Callow, M. and Evans, L. 1979, Fertilization in brown algae. 11 Evidence for lectin sensitive complementary receptors involved in gamete recognition in *Fucus serratus*. *J. Cell Sci.* 36: 19—30.

4) Green, J. Jones, J., Stafford, C. and Callow, J. 1990, Fertilization in *Fucus:* exploring the gamete cell surfaces with monoclonal antibodies. In "Mechanism of Fertilization: Plants to Humans". Ed. B. Dale, Nato ASI Cell Biology Series, Springer Verlag.

5) Russell, S. 1990, Isolation and characterization of the Angiosperm Gamete. In "Mechanism of Fertilization: Plants to Humans". Ed. B. Dale, Nato ASI Cell Biology Series, Springer Verlag.

6) Guilluy, C., Gaude, T., Digonnet-Kerhoas, C., Chaboud, A., Heizmann, P. and Dumas, C. 1990, New data and concepts in angiosperm fertilization. In "Mechanism of Fertilization: Plants to Humans". Ed. B. Dale, Nato ASI Cell Biology Series, Springer Verlag.

7) Rosati, F., Cotelli, F., De Santis, R., Monroy, A. and Pinto, M. 1982, Synthesis of fucosyl-containing glycoproteins of the vitelline coat in oocytes of *Ciona Intestinalis.* (Ascidia). *Proc. Nat. Acad. Sci.* USA 79: 1098—1911.

8) Pinto, M., DeSantis, R., D'Alessio, G., and Rosati, F. 1981, Studies on fertilization in ascidians. Fucosyl sites on the vitelline coat of *Ciona intestinalis. Exp. Cell Res.* 132: 289—295.

9) Hoshi, M. 1984, Roles of sperm glycosidases and proteases in the ascidian fertilization. In "Advances in Invertebrate reproduction", Engels et al. (eds.). Elsevier Science publishers., Vol 3, pp. 27—40.

10) Focarelli, R., Renieri, T., and Rosati, F. 1988, Polarized site of sperm entrance in the egg of a freshwater bivalve, *Unio elongatulus. Develop. Biol.* 127: 443—451.

11) Rosati, F. and Focarelli, R. 1990, Sperm egg interaction in bivalves. In "Mechanism of Fertilization: Plants to Humans". Ed. B. Dale, Nato ASI Cell Biology Series, Springer Verlag.

12) Lillie, F. 1913, Studies on fertilization. V. The behaviour of the spermatozoa of Nereis and Arbacia with special reference to egg extracts. *J. Exp. Zool.* 14: 515—574.

13) Suzuki, N., Nomura, K., Ohtake, H. and Isaka S. 1981, Purification and the primary structure of sperm activating peptides from the jelly coat of sea urchin eggs. *Biochem Biophys. Res. Commun.* 99: 1238—1244.

14) Yamaguchi, M. Niwa, T., Kurita, M., and Suzuki, M. 1988, The participation of speract in the acrosome reaction of *Hemicentrotus pulcherrimus. Develop. Growth Differ.* 30: 159—167.

15) Suzuki, N. 1990, Structure and Function of Egg Associated Peptides of Sea Urchins. In "Mechanism of Fertilization: Plants to Humans". Ed. B. Dale, Nato ASI Cell Biology Series, Springer Verlag.

16) Gwatkin, R. 1977, "Fertilization Mechanisms in Man and Mammals". Plenum Press, New York.

17) Wassarman, P. 1989, Profile of a mammalian sperm receptor, *Development* 106: (in press).

18) Kinloch, R., Roller, R., Fimiani, C. Wassarman, D., and Wassarman, P. 1988, Primary structure of the mouse sperm receptors polypeptide chain determined by genomic cloning. *Proc. Nat. Acad. Sci. USA.* 85: 6409—6413.

19) Bleil, J. and Wassarman, P. 1988, Galactose at the nonreducing terminus of O-linked oligosaccharides of mouse egg zona pellucida glycoprotein ZP3 is essential for the glycoproteins sperm receptor activity. *Proc. nat. Acad. Sci. USA* 85: 6778—6782.

20) Leyton, L. and Saling, P. 1989, Evidence that aggregation of mouse sperm receptors by ZP3 triggers the acrosome reaction. *J. Cell Biol.* 108: 2163—2168.

21) Dale, B. 1983, "Fertilization in animals". Edward Arnold, London.

22) Tyler, A., Monroy A., Kao, C. and Grundfest, H. 1956, Membrane potential and resistance of the starfich egg before and after fertilization *Biol. Bull.* 111: 153—177.

23) Dale, B. and DeFelice, L. 1990, Soluble sperm factors, electrical events and egg activation. In "Mechanism of Fertilization: Plants to Humans". Ed. B. Dale, Nato ASI Cell Biology Series, Springer Verlag.

24) Dale, B., DeFelice, L. and Taglietti, V. 1978, Membrane noise and conductance increase during single spermatozoon-egg interactions. *Nature Lond.* 275: 217—219.

25) Dale, B. and De Felice, L. 1984, Sperm activated channels in ascidian oocytes, *Develop. Biol.* 101: 235—239.

26) Talevi, R., Dale, B. and Campanella, C. 1985, Fertilization and activation potentials in *Discoglossus pictus* (Anura) eggs: A delayed response to activation by pricking. *Develop. Biol.* 111: 316—323.

27) Dale, B. 1988, Primary and secondary messengers in the activation of ascidian eggs, *Exp. Cell. Res.* 177: 205—211.

28) Dale, B. 1987, Mechanism of Fertilization. *Nature* Lond. 325: 762—763.

29) Dale, B., DeFelice, L. and Ehrenstein, G. 1985, Injection of a soluble sperm fraction into sea urchin eggs triggers the cortical reaction, *Experentia* 41: 1068—1070.

30) Whitaker, M., Swann, K., and Crossley, I. 1989, What happens during the latent period at fertilization. In "Mechanisms of Egg Activation". Eds. R. Nuccitelli, G. Cherr and W. Clark, Plenum Press, New York.

31) Kline, D., Simoncini, L., Mandel, G. Maue, R. Kado, R and Jaffe, L. 1988, Fertilization events induced by neurotransmitters after injection of mRNA in Xenopus Eggs, *Science* 241: 464—467.

32) Kline, D. Kado, R., Kopf, G., and Jaffe, . L. 1990, Receptors, G-proteins and activation of the amphibian egg. In "Mechanism of Fertilization: Plants to Humans". Ed. B. Dale, Nato ASI Cell Biology Series, Springer Verlag.

33) Jaffe, L. 1980, Calcium explosions as triggers of development. Ann. NY Acad. Sci. 339: 86—101.

34) Dale, B. and Santella, L. 1985, Sperm-oöcyte interaction in the sea-urchin, *J. Cell Sci.* 74: 153—167.

35) Longo, F., Lynn, J., McCulloh, D. and Chambers, E. 1986, Correlative ultrastructural and electrophysiological studies of sperm-egg interactions of the sea urchin Lytechinus variegatus, *Develop. Biol.* 118: 155—166.

36) Hinkley, R., Wright, B., and Lynn, J. 1986, Rapid visual detection of sperm-egg fusion using the DNA-specific fluorochrome Hoechst 33342. *Develop. Biol.* 118: 148—154.

37) McCulloh, D. and Chambers, E. 1986, When does the sperm fuse with the egg, *J. Gen. Physiol.* 88: 38a.

38) Ginsburg, A. 1988, Egg cortical reaction during fertilization and its role in block to polyspermy. *Sov. Sci. Rev. F. Physiol. Gen. Biol.* 1: 307—327.

39) Dale, B. and deSantis, A. 1981, The effect of cytochalasin B and D on the fertilization of sea urchins. *Develop. Biol.* 85: 474—484.

40) Jaffe, L. 1990, The roles of intermembrane Calcium in polarizing and activating eggs. In "Mechanism of Fertilization: Plants to Humans". Ed. B. Dale, Nato ASI Cell Biology Series, Springer Verlag.

41) Brawley, S., and Bell, E. 1987, Partial activation of *Fucus* eggs with calcium ionophore and low sodium sea water. *Develop. Biol.* 122: 217—226.

42) Brownlee, C. and Wood, J. 1986, A gradient of cytoplasmic free calcium in growing rhizoid cells of *Fucus serratus. Nature Lond.* 320: 624—626.

43) Brownlee, C. 1990, Intracellular signalling during fertilization and polarization in fucoid algae. In "Mechanism of Fertilization: Plants to Humans". Ed. B. Dale, Nato ASI Cell Biology Series, Springer Verlag.

44) Speksnijder, J., Corson, D., and Sardet, C., Jaffe, L. 1989, Free calcium pulses following fertilization in the ascidian egg. *Dev Biol.* 135: 182—190.

45) Brownlee, C. and Dale, B. 1990, Temporal and spatial correlation of fertilization current, calcium waves and cytoplasmic contraction in eggs of *Ciona intestinalis. Proc. R. Soc. B, Lond.* (in press).

SIGNALS FOR CELL SEPARATION IN PLANTS
PHYSIOLOGY, BIOCHEMISTRY AND MOLECULAR BIOLOGY

DAPHNE J. OSBORNE

Department of Plant Sciences
University of Oxford
Oxford, OX1 3RA, U.K.

The process of cell separation in plants is central to a wide range of differentiation and developmental events throughout the life-span of each individual organ and tissue. The NATO Advanced Research Workshop that met in Italy in September 1988 was the first international meeting to direct its activities to an overall assessment of the examples and range of physiological and biochemical interplay that could lead to the partial or complete separation of any particular cell from its neighbour (1).

This was not solely an academic exercise. Economic crop production relies heavily upon efficient harvesting, non-detrimental storage and the eventual marketing of a high quality product to the consumer. During each of these processes, cell separation events play a critical part. It was to an understanding of the underlying biochemical mechanisms that can determine and control cell separation, and to the practical means that can be used to implement such control that the 52 plant scientists from 13 different countries assembled for their deliberations in the Villa Gualino in the hills above Turin.

Plants, like animals, are comprised of many different cell types, each differentiating with a distinct anatomical or biochemical morphology with attendent specificities of gene expression and each exhibiting a precise positional differentiation in the body of the plant. The majority of organellar constituents and metabolic pathways are common to all cells and provide the basis for energy conversions and house-keeping biochemistry that are the essentials for life processes to continue. Superimposed upon these are highly coordinated and sequential chains of differential gene expression that impart the specificity to a cell and distinguish it as a distinct biological type.

It was immediately evident that an important part of cell specificity in relation to plant cell separation lay not only in the cytoplasmic components but to a very large extent in the character and composition of the enclosing cell walls. Once considered as essentially inert and functioning mainly as structural support to the plant body, cell walls are now assigned the role of the external organelle, sensing an outer ionic, enzymic and organic environment, filtering these signals through the electrostatically charged matrix of complex

NATO ASI Series, Vol. H 51
Cell to Cell Signals in Plants and Animals
Edited by V. Neuhoff and J. Friend

polysaccharide and proteins. In this way, the messages that pass then interface with the plasmamembrane surface of the cytosol. This system is seen concurrently functioning in reverse so that hormones and other effector molecules present within the cell transmit informational signals across the plasmamembrane to reach the wall.

Since many of the components of walls are in a constant state of turnover throughout the life of a cell, the dynamic potential for determining either cell separation or tenacious adhesion is high. Effective biochemical and molecular biological approaches to an understanding of the mechanisms involved have therefore centred upon those cells that can be obtained from tissues composed of a high proportion of a uniform cell type and which can also be obtained reproducibly and in quantity with relative ease. The flesh of ripening fruits provides such material and it is not surprising, therefore, that a molecular biological approach to cell separation has flourished in this area. Nor is it surprising that this Workshop deliberated long on the changes in polysaccharide composition associated with the changing texture of walls in the softening of fruit and on the enzymes that might control these processes.

Over recent years, cDNA clones for a number of mRNAs expressed in ripening tomato fruit have been characterized and one of them, pTom6, has been shown, by DNA and protein sequence analysis, to encode the enzyme polygalacturonase (PG). This enzyme is synthesised *de novo* during fruit ripening and is involved in the solubilization of the pectin fraction of the fruit cell walls. Production of polygalacturonase mRNA, as well as other less prominent mRNAs, is induced by ethylene during the normal ripening progression, but all are substantially reduced in tomato mutants that either fail to ripen their fruit or ripen very slowly.

The softening process of the fruit wall and the loosening of cell adhesion depends upon a number of factors, the first of which is the induction of new gene expression in the unripe fruit triggered by the considerable rise in ethylene synthesis that accompanies ripening.

Although all cells produce ethylene and all possess the PG gene, expression of the gene is specific to fruit cells and the protein coded for is synthesised in the cytosol and secreted into the fruit cell wall. Genomic cloning, sequencing and Southern hybridization indicate that in the tomato there may be only one PG gene of approximately 8000 base pairs interrupted by 8 introns. Although the gene is also present in mutant plants whose fruits are either non-ripening or slowly ripening, it fails to be expressed in the fruit even when additional ethylene is supplied. With reporter gene constructs, containing the PG structural gene and the bacterial chloramphenicol transferase (CAT), tomato plants have been transformed using the *Agrobacterium tumefaciens* T_1 plasmid technology. The fact that CAT activity was then found only in the ripening fruit of the transformed plants and not elsewhere showed that the PG gene-regulating sequences can control the bacterial CAT gene activity to the same extent, and in the same tissues, as the native PG gene to which it was attached. Inverting the PG coding sequence in the chimeric gene constructs can lead to a 90% reduction in PG enzyme activity in the fruit. Of special interest, therefore, is the finding that the lowering of the PG activity through the transcription of an antisense mRNA from the inverted gene did not, however, lead to a failure of fruit reddening or of the wall polyuronide degradation that has always been associated with normal fruit softening (2). Although the expression of the PG gene is so closely linked to the ripening of the tomato fruit, as so many past experiments confirm, this new research now casts doubts upon the absolute role of PG in the separation and softening events of the fruit cell wall (3). The unexpected result can, however, be put to what may or may not be considered as an economic spectacular in a few year's time complete suppression of the PG gene expression and the production on a commercial scale of the ripening, but non-squashy tomato.

The Workshop devoted wide ranging discussions to other wall-related enzymes; to cellulases, which degrade not cellulose but hemicelluloses, and to a variety of gene repressions and activations in senescing leaves and other ripening tissues. All are programmed by hormones (predominantly by ethylene) and by environmental stimuli that are transduced into changes in hormonal levels within the cells. The isolation of cDNA and genomic clones to the enzymes concerned in the regulation of hormonal levels (as well as the genes these levels control) will eventually permit the more precise experimentation required to elucidate hormonal function in the processes of cell separation and softening. A particularly important example will be referred to later with respect to ethylene.

Because the walls of a cell are always in a state of dynamic change with the synthesis and insertion of new polysaccharide and protein complexes, the enzymes achieving the loosening and softening also do not remain constant. The wall is a micro-composite of interpenetrating polymeric networks involving both covalent and non-covalent linkages. Although the mean saccharide and aminoacid compositions are known for many tissue types, the specific sequences and orientation of the crosslinks is known only for a limited number. A small family of basic hydroxyproline-rich glycoproteins, mostly O-glycosylated by short arabino-oligosaccharides and whose monomers exist as fluxuous (80 nm) rods, are seen to play an important role in maintaining the structure and rigidity of cell walls in the dicotyledon families that have been studied to date (4). These proteins — known as extensins — have now been detected in graminaceous monocotyledons (the grasses) as we heard at the Workshop. They are, however, present at a much attenuated level which is over an order of magnitude lower than in dicotyledons tested so far. Compared with dicotyledons, the monocotyledon wall differs also in being sparse in pectins. This raises the question, does the construction of the pectin-poor wall allow it to dispense with the extensin crosslinks or is there another structural substitute, such as phenolic cross-linking or even another specialised class of protein. So few walls have been as closely analysed as those of the sycamore or others raised in suspension culture that it is difficult to make firm conclusions. *In vivo,* the numbers of cell types in a plant are many and disperse and it is to the difficult task of establishing which enzymes are regulating which wall complexes, and how they interact, that future experimentation must be directed.

The quest is a very real one and particularly so for the cells of abscission zones. There, a plate of cells that positionally differentiates at the precise site of cell separation at the base of a leaf petiole or a fruit stalk, must have walls with a structure and composition that is specifically unlike those of their neighbours. If it were not so, the walls of all the surrounding cells would be caused to separate by the extracellular enzymes that are so specifically induced and secreted by the zone. Instead, cell separation is very rigidly restricted to those cells that constitute the differentiated abscission zone.

Immunological methods afford new tools for probing the specificity of wall complexes both in relation to cell growth and cell separation. Auxin-induced growth requires cell wall loosening and can, for example, be inhibited by anti-'-L-arabino-furanose in both oat coleoptiles and azuki bean segments. However, the antibodies raised against a-D-glucans inhibited auxin-induced growth in the monocotyledon (in this instance maize) but not in the dicotyledon azuki bean. All agreed that we do not yet understand the full nature of the cell wall loosening mechanism sufficiently well to account for these differences but much may depend upon the individual compositional specificities of the walls.

Other questions of the Workshop centred upon the sorting signals and trafficking of the enzymes concerned in loosening and separation processes. What mechanisms determine from where, and to where, each protein shall be transported? What determines the selective secretion to the cell wall? Glycosylation of polypeptides translating on the endoplasmic reticulum-bound polyribosomes, removal of the signal sequence followed by sequestration into the lumen of the reticulum and the further glycosylation and modification within the Golgi are seen as preludes to the delivery of the mature enzyme to the plasmamembrane and hence to the cell wall. Elucidation of these cellular postal codes is another area where research will be intensive in the years to come.

Whereas the enzyme-cell wall interactions endogenously orchestrated by the plant lead to predictable form and behaviour in the individual plant parts, an attack by fungal pathogens introduces a new array of enzymes under the genomic control of the invader. No longer are the walls partially or wholly dissociated in an organised and sequential fashion as in normal growth and differentiation, instead, destructive events of complete wall dissolution commonly occur.

Studies of the changes induced in the host by pathogenic enzyme secretions have provided a wide array of information on cell wall structure and composition and of the evocative and signal function of small oligosaccharide units cleaved from the cell wall. These matrix-derived 'elicitors' can activate host reponses to pathogens in relation to cell growth, cell separation or cell autolysis as well as to the induction of phytoalexins which form a major part of the plant's defense system against pathogenic attack. Of special significance is the role of these pectic oligomer fragments, particularly xyloglucans, in the enhancement of ethylene production. The specificity of certain hosts for their pathogens provides special opportunities for probing molecular interactions between the two organisms and we can expect much new information to derive from such studies which will be of value in re-evaluating methods for plant protection.

A search for cytoplasmic markers in order to identify specific cell types has been successful for cells of certain abscission zones. In the bean leaf for example, a 34kDa polypeptide has been found only in the zone cells and is undetectable elsewhere. This finding is important in relation to the target cell concept of specific induction and repression of genes. The presence of tissue-specific and developmentally regulated polypeptides, probably glycosylated, has important parallels with animal cell differentiation. In plant cells too, it seems apparent that each cell type can be seen as a target for particular hormonal or other signals which in turn evoke cell specific gene responses.

Wall loosening occuring during cell growth is primarily an auxin-induced response in immature cells, whereas the partial or complete separation of cells (as found most spectacularly in aerenchyma formation and abscission) is generally attributed to a perception and response to ethylene. The Workshop considered these and other hormonal signal mechanisms. For example, that for sepal separation occurring at low light when buds of the evening primrose (Oenothera) flowers open in the evening has been shown to be a light-regulated synthesis of wall separating enzymes determined by the low light-dependent synthesis of auxin. In pollinated Petunia flowers, corolla abscission, though speeded by pollination-induced ethylene, has a greatly enhanced speed of shedding in conjunction with certain C_8 to C_{10} fatty acids. The production of these fatty acids is now known to be the pollination-induced signal that increases the sensitivity of the corolla to ethylene. It is apparent, therefore, that the regulation of cell separation is not a single or simple inducible system in which only one signal operates. Repressors, inducers and activators/sensitizers all play their part and the precise nature of the functional control mechnism eventually lies with the dictates of the specific target nature of the cell. Much remains to be resolved in this area with many systems. The sloughing off of intact and viable root cap cells, for example (5), and the formation of free and separate sperm cells within the pollen germ-tube during sexual reproduction (6), still await the identification of the initiating signal molecule(s).

At present there seems no doubt that ethylene is the hormonal molecule most usually responsible for the initiation of cell separation in the many tissue types in which cell-cell dissociation may occur. It was appropriate, therefore, for the meeting to spend substantial time considering facets of ethylene biosynthesis and the limiting or regulatory steps in the pathway. In the conventional route from methionine to ethylene, 1-aminocyclopropane-1-carboxylic acid (ACC, the immediate ethylene precursor) and the enzyme ACC-synthase play a critical role. The regulation of the gene for the synthase enzyme is central to any growth or separation event that is modified by ethylene in any way. The Workshop heard of the newest attempts to isolate and clone the gene, and since the meeting, these goals have now been successfully achieved (7). The potential for the transformation of plants with respect to the ACC-synthase gene(s) now offers spectacular opportunities for plant manipulation, particularly in relation to cell separation and wall changes.

The exchange of ideas at meetings like the NATO Advanced Research Workshops and the basic science that leads to our improved understanding of how plants perform must be entirely praiseworthy. Whether this understanding be eventually implemented or used for good or ill must await for the future to unfold.

REFERENCES

1) Cell Separation in Plants. Physiology, Biochemistry and Molecular Biology. Eds. D.J. Osborne and M.B. Jackson. NATO ASI Series H. Cell Biology Vol. 35. Springer-Verlag Berlin, Heidelberg 1989.

2) Knapp, J., Moureau, P., Schuch, W., and Grierson, D. Organization and expression of polygalacturonase and other ripening related genes in Ailsa Craig 'Neverripe' and 'Ripening inhibitor' tomato mutants. Plant Molec Biol 12, 105—116, 1989.

3) Giovannoni, J.J., Della Penna, D., Bennett, A.B., and Fischer, R.L. Expression of a chimeric polygalacturonase gene in transgenic rin (Ripening inhibitor) tomato fruit results in polyuronide degradation but not fruit softening. Plant Cell 1, 53—63, 1989.

4) Heckman Jr., J.W., Terhune, B.T., and Lamport, D.T.A. Characterization of native and modified extensin monomers and oligomers by electron microscopy and gel filtration. Plant Physiol 86, 848—856, 1988.

5) Gochnauer, M.B., McCully, M.E., and Labba, H. Different populations of bacteria associated with sheathed and bare regions of roots of field-grown maize. Plant Soil 114, 107—120, 1989.

6) Wilms, H.J., and Keijzer, C.J., eds. Plant sperm cells as tools for biotechnology. Pudoc, Wageningen, The Netherlands, 1988.

7) Sato, T., and Theologis, A. Cloning the mRNA encoding 1-aminocyclopropane-1-carboxylate synthase, the key enzyme for ethylene biosynthesis in plants. Proc. Natl. Acad. Sci. 86, 6621—6625, 1989.

DETECTION AND CODING OF CHEMICAL SIGNALS: A COMPARISON BETWEEN ARTIFICIAL AND BIOLOGICAL SYSTEMS

D. Schild[1] and J.W. Gardner[2]

[1] Physiology Department,
University of Göttingen, FRG
[2] Department of Engineering,
University of Warwick,
Coventry CV4 7AL, U.K.

1. INTRODUCTION

Did you ever try to measure a smell? Can you tell whether one smell is just twice as strong as another? Can you measure the difference between one kind of smell and another? It is very obvious that we have very many different kinds of smells, all the way from the odour of violets and roses up to asafetida. But until you can measure their likeness and differences you can have no science of odour. If you are ambitious to found a new science, measure a smell. Alexander Graham Bell (1914)

Although there has been no lack of ambitious researchers in the past 75 years, the problems put forward by the inventor of the telephone still await an answer. Hundreds of thousands of stimuli can be detected and discriminated by the nose. However, the basic principles of odour detection and coding are not yet well understood, and the development of artificial chemosensitive devices is still in its "kindershoes". In this respect, the olfactory system is an exception among the sensory systems since there are excellent electrical analogues for all the other senses, above all the visual and auditory. Also, there are sophisticated methods of processing optically and acoustically recorded signals. On the other hand, there is an enormous need for intelligent chemosensitive devices: biochemical analyses in medicine and biology could be facilitated considerably. The control not only of food and beverage flavours, but also of poisonous and harmful substances in water and air would become feasible, if not easier. Although biological and artificial chemosensory systems show many obvious differences, there are also some principal and conspicuous similarities. E.g., some features of biological receptor cells such as a low stimulus selectivity resemble very much those of artificial sensors. Also, the second stage of both systems needs to process the stimuli in some ordered way which is simpler than the relatively complex stimulus response profiles of the primary receptor cells/ sensors. Research in both fields should profit from cross-fertilization. In this article, an attempt is made to highlight the common and different features in biological and artificial chemosensory systems.

NATO ASI Series, Vol. H 51
Cell to Cell Signals in Plants and Animals
Edited by V. Neuhoff and J. Friend

2. OLFACTORY RECEPTOR NEUROSES

In vertebrates, the olfactory receptor cells are situated in the epithelium in the nasal cavity, each receptor cell is tightly surrounded by sustentecular cells. In man, this epithelium is called the olfactory mucosa and covers an area of a few square centimetres. It is located in the upper part of the superior turbinate, the nasal septum, and the roof between. It is composed of three major cell types: (i) olfactory receptor neurones, (ii) basal cells, and (iii) sustentacular cells . Olfactory receptor cells have two unique properties: first, they are the only neurones which communicate directly with the environment. While other neurones receive neurotransmitters and modulators as input signals, olfactory neurones receive odours instead. Second, they regenerate throughout life (Farbman, 1986, Graziadei and Monti Graziadei, 1978, Moulton, 1974). Basal cells are progenitor cells of olfactory neurones; they divide, differentiate, and replace dying receptor neurones (Moulton et al., 197O, Moulton, 1974). The sustentacular cells separate olfactory receptor neurones, they have supporting, glia like, and secretory functions. The lamina propria underlying these cells contains the axons of the olfactory neurones and secretory glands. Secretion from these glands together with secretion from the sustentacular cells provide the mucus which forms a 20—30 μm thick layer above the olfactory tissue. The function of the mucus is not fully understood. It certainly acts as a barrier to air and water, and it also shows transport properties for lipophilic molecules (Vogt et al., 199O). Olfactory receptor cells are bipolar primary neurones with one long dendrite from the top of which, the so-called olfactory knob, several cilia issue (Fig. 1). In some

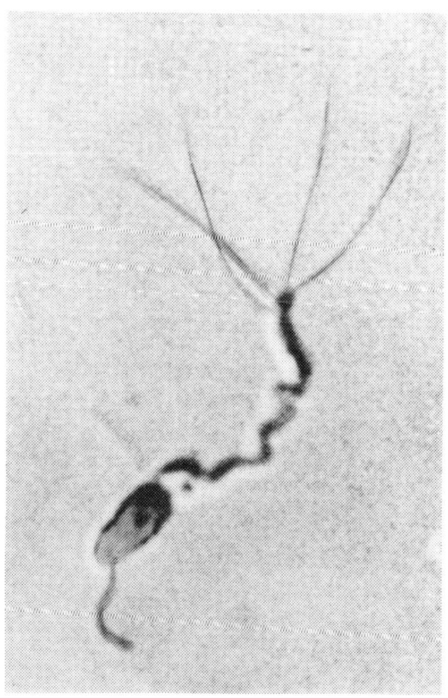

FIGURE 1
Olfactory receptor neurone. A piece of axon is attached to the cell's soma. At the top of the dendrite several cilia issue.

species, the cilia are relatively long as compared to the thickness of the mucus layer (up to 100 μm), and extend almost horizontally as a dense long-haired carpet in the mucus layer. The oval-shaped soma of an olfactory neurone is usually fairly small with diameters between 4 μm and 7 μm, even in the snail — a species which is well known for its large neurones. As primary neurones, the cells have an axon attached to the soma. The unmyelinated axons correspond to the olfactory nerve which passes through the cribriform plate into the olfactory bulb. Here the axon terminals make synaptic contact within the glomerula to dendrites of the mitral cells.

Electrophysiological studies on single olfactory receptor neurones have proved to be difficult because of the small size of these cells. Almost everything we know about vertebrate olfactory receptor cells is therefore based upon recordings from amphibian species which have relatively large olfactory receptor cells. However, no one really seems to know how amphibians use their sense of smell. Air-borne odorants enter the nose and reach the mucus of the olfactory mucosa. The odorous molecules enter the mucus according to a certain partition coefficient. Once in the mucus, there is a finite probability of the odorous molecule interacting with a cilium of an olfactory neurone. The overall responses to natural stimuli of olfactory neurones can be summarized as follows (for a review, see Getchell, 1986): (i) The neurones show a low spontaneous discharge rate which is well below one action potential (or spike) per second. This feature is advantageous because a relatively small change in firing rate indicates a significant response, i.e., a sufficiently high signal to noise ratio is easily achieved. (ii) When maximally stimulated by natural odorants, the neurones respond by producing 20 to 30 spikes per second. (iii) When the temporal intervals between spikes are analyzed, they show an interval distribution which suggests that the spikes are distributed randomly in time, i.e. only the mean firing rate of a response is characteristic. However, patch-clamp recordings from single cells suggest that firing might occur in perhaps irregular bursts (Frings and Lindemann, 1988). (iv) Responses to natural stimuli show a characteristic and well reproducible pattern: the spike rate increases within a few seconds to a maximum and then decreases to a fairly constant niveau (phasic-tonic response). (v) A typical vertebrate olfactory receptor neurone responds to a large number of stimuli, whereby every stimulus brings about a different response spike rate. Another olfactory receptor neurone would also respond to a large number of stimuli, maybe partly overlapping with those of the first mentioned cell, but one stimulus would usually cause different responses in two cells. From the experiments up to now, it is difficult to establish whether there are receptor neurones that respond identically to any sequence of stimuli. Such cells would form a receptor class, comparable to those responsive e.g. for blue light in vision, but unfortunately at present we don't know whether receptor classes exist. The response of only a few out of the several million olfactory neurones per mucosa has been recorded in animals to only a few of the thousands of possible stimuli. The events occurring between the binding of stimulus molecules to the cilia and the generation of action potentials, i.e. the details of stimulus transduction, are only partially known at present. One wide-spread hypothesis is that binding of a stimulus molecule to a ciliar receptor protein seems to trigger one or more intracellular enzyme cascades which eventually lead to a net inward current into the cilia, the so-called generator current. There is evidence that various G-proteins and cyclo-adenosinemonophosphate (cAMP), cycloguanosinemonophosphate (cGMP), as well as kinase C are involved in the transduction processes (for a review, see Vogt et al., 1990). However, the precise roles of these second messenger pathways remain to be established in detail. The generator current loads the cell's capacitance thereby depolarizing the membrane potential to the cell's firing threshold. It has been shown that olfactory receptor cells are extremely sensitive, responding with spikes to current injections of only a few (<1O pA) picoampères (Schild et al., 1990). The generator current is modulated and processed by voltage-gated currents (Schild, 1989). There seem to be two voltage-gated inward currents and three different outward currents in all studied species: (i) a fast inward current carried by Na+ ions and blocked by tetrodotoxin (TTX): this current is the classical Na^+-current responsible for the initiation of action potentials; (ii) a small Ca^{2+} current that partially inactivates; (iii) a Ca^{2+}-dependent outward current carried by K^+ ions; (iv) a fast K^+-current that inactivates within some hundreds of millisecond, and (v) a third K^+-current that inactivates slowly in the range of tens of seconds The physiological role of the outward currents are partially the repolarization and hyperpolarization of action potentials. The Ca^{2+} current and the Ca^{2+}-dependent K^+-current form a negative feedback loop, which is likely to be involved in the adaptive regulation of the cell's impedance and thereby its sensitivity as a function of its activity state and recent activity history.

3. CHEMOSENSITIVE SENSOR ELEMENTS

The detection and coding of gaseous chemical signals has been of great interest over the past thirty years since the discovery by Brattain and Bardeen (1953) that certain semiconductor materials were sensitive to the presence of gases. This important dis-

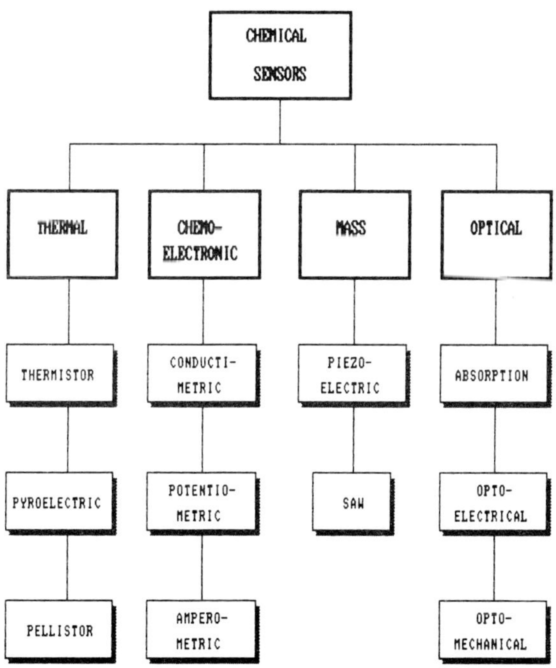

FIGURE 2
Classes and types of chemical sensors that have been applied to odour and gas detection

covery has led to the development of a wide range of chemical sensors that may be employed within an artificial or electronic nose (Gardner et al., 1990). Chemosensitive sensors may be conveniently classified into four groups that relate to their basic sensing principle, namely, thermal, chemoelectronic, mass and optical. Figure 2 shows the main types of sensors in each class that have been employed in gas or odour detection. They are the first stage in any artificial chemosensory system that transduces a chemical signal into, usually, an electrical DC signal. We shall now consider the relative performance of these chemosensitive sensors and compare their behaviour to that of an olfactory receptor cell.

3.1 Chemometric thermal sensors

A common feature of chemical sensors is that the gas or odour O interacts with surface or bulk sites S on or in the sensing element. This basic process can often be described by the following chemical reaction, i.e.

$$O + S \underset{k_b}{\overset{k_f}{\Leftrightarrow}} OS \tag{1}$$

The forward and backward reaction rates, k_f and k_b, determine the equilibrium constant K that is related to the change in Gibbs free energy ΔG by

$$\Delta G = \Delta G_o + RT \ln(K) = 0 \tag{2}$$

The nature of this interaction depends upon the value of the standard free energy ΔG_o, the gas constant R and the temperature T. A high value of K relates to a highly sensitive detector, but unfortunately may lead to an irreversible reaction. Thus the equilibrium constant and the temperature limit the dynamic range and response of any chemosensitive element.

Thermal sensors measure the change in free energy of the reaction that produces a change in some other physical property such as temperature or conductance. High temperature sintered oxides have been used in thermistors for enzyme-catalysed reactions (Tran-minh and Vallin, 1978), but not yet for odour detection. However, there are a few reports of pyroelectric sensors that measure hydrogen, oxygen and nitrogen concentrations by detecting a change in the pyroelectric vector. By far the most common type of chemosensitive thermal device is the pellistor. The pellistor is basically a type of calori-

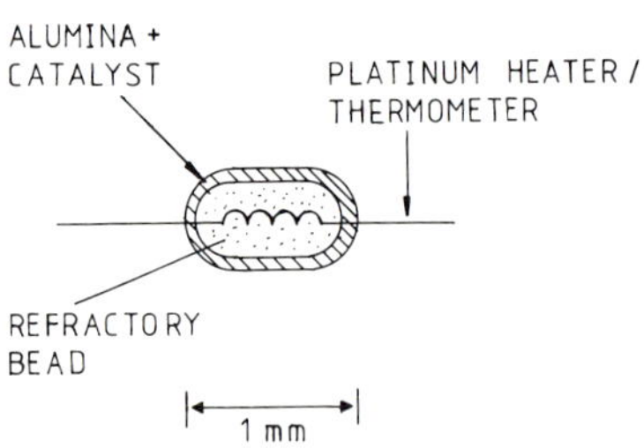

FIGURE 3
A chemosensitive thermal sensor: the pellistor gas detector

meter that measures the energy liberated following the oxidation of the gas and has been extensively reported by Gentry and Welsh (1987). Typically, a pellistor consists of a platinum coil coated with porous alumina that contains precious metal catalysts, see Figure 3. The gas diffuses through the porous coating and reacts changing the device temperature and the balance point of a Wheatstone bridge. Clearly, the sensor is non-specific, i.e. it responds to a range of chemical species. A feature that is shared by almost all the current sensors. The response time of thermistors is fast at 5—10 seconds, but unfortunately both thermistors and pellistors suffer from poisoning effects.

3.2 Chemoelectronic sensors

Since that important discovery of gas-sensitive semiconductor materials in 1953 by Brattain and Bardeen, the field of chemoelectronic sensors has grown into perhaps the largest. In chemoelectronic sensors, an interaction takes place between the measurand and sensing material that modifies either the charge generation or transfer mechanism. Chemoelectronic sensors may be conveniently classified according to their basic principle of operation into conductimetric (measurement of conductivity), potentiometric (measurement of potential or voltage), or amperometric (measurement of current). Electron or ion charge transfer mechanisms occur in chemoelectronic sensors, however the discussion here will be limited to those devices that are governed by electron transport.

In the mid 1960s the sensitivity of various chemoelectronic sensors to odours was under investigation. For example, Dravnieks studied the effect of several odorants upon the contact potential of coated materials in a potentiometric sensor (Dravnieks and Trotter, 1965). While Buck et al. (1965) were investigating the changes of the surface conductivity of metals and semiconductors, caused by an interaction between the charge carriers and odorants in a conductimetric device. By the late 1960s organic as well as inorganic materials were being studied for odour detection. One such study (Misra et al., 1968) discovered that single crystals of β-carotene responded conductimetrically to the adsorption of odours. A similar effect was also demonstrated in phosholipid layers by Cherry et al. (1970). A result that may be particularly relevant because the ciliary membranes of the olfactory receptor cells contain phospholipids.

Yet, the search for artificial odour detectors has undoubtedly been influenced by the rapid development of semiconductor gas sensors in the 1970s for the chemical industries. Some of these chemical sensors have been applied to odour sensing in a rather ad hoc manner. There are many forms of chemosensitive inorganic semiconductor materials, including single crystal, polycrystalline and amorphous oxides (for general information see: Fouletier, 1982, Wohltjen, 1984 or Moseley and Tofield, 1987). From the early 1970s onwards considerably research effort has been directed towards the chemosensitivity of sintered porous semiconductor materials. In particular, the properties of tin oxide and titanium oxide doped with high levels of heavy metal catalysts (eg. platinum, palladium, and tungsten) have been studied. The most successful commercial sensor is the Taguchi Gas Sensor (TGS) that was patented by Taguchi in 1971. Figure 4 shows the general arrangement of this conductimetric chemoelectronic sensor. A thick sintered layer of porous doped tin oxide surrounds an inert ceramic tube with an internal heater coil. In air the semiconducting oxide contains a significant number of adsorbed oxygen molecules bound to "surface" sites that act as electron acceptors. These surface states S lie within the oxide band gap, above the Fermi level, and produce a space-charge

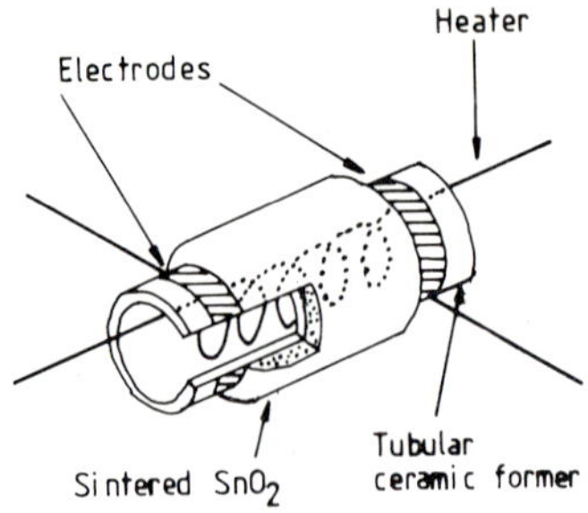

FIGURE 4
A conductimetric chemoelectronic sensor: the tin oxide gas detector (Taguchi)

region, or more precisely a depletion layer. In general terms the exposure of an n-type SnO_2 sensor to a gas causes a change in the width of this depletion layer as adsorbed oxygen is displaced. The resulting change in conductivity $\Delta\sigma$ is proportional to the change in carrier density Δn and given by,

$$\Delta\sigma = e\,\mu\,\Delta n \qquad (3)$$

Where the simplest reaction that describes an n-type oxide in oxygen is,

$$S + .5\,O_2 + n \Leftrightarrow O^- \qquad (4)$$

and following the introduction of the measurand R,

$$R + O^- \Leftrightarrow RO + n \qquad (5)$$

More detailed descriptions of the reaction kinetics may be found elsewhere which are extended to include the role of the catalysts (for further information see: Capehart, 1987, McAleer et al., 1987, Heiland and Luth, 1983). Semiconducting tin oxide sensors are particularly sensitive to the detection of low levels of alcohols (1—250 ppm) with saturation typically occurring at 1,000 ppm. The response time of porous tin oxide sensors is in the range of 10—30 seconds and is governed by a combination of reaction and diffusion kinetics (Gardner, 1989).

Recently, the use of oxide semiconductor gas sensors as odour detectors in a simple nose has been reported. For example, Shurmer et al. (1989a,b) have demonstrated the ability of SnO_2 sensors to discriminate alcohols, tobacco aromas and beers. Meanwhile Müller and Horner (1986) have reported on the sensor sensitivity to whisky aromas. Finally, the response to fish aromas has been investigated by Egashira et al. (1988) who show that ruthenium doped titanium oxide films can be used to monitor levels of trimethylamine in order to determine the freshness of fish.

The change in work function that accompanies the adsorption of a gas in potentiometric sensors has been studied by various workers. These include, tin oxide sensors by Mizsei and Hasanyi (1983), palladium-titanium oxide Schottky diodes by Yamamoto et al. (1980), palladium-silicon tunnel MIS structures by Murakami et al. (1988) and chemosensitive MOS structures by Lundstrom and Sodelberg (1981).

The majority of the chemosensitive devices discussed so far have several features in common, they have high operating temperatures, possess low specificity, high sensitivity to combustible gases and are insensitive to molecular shape. The property of low specificity is shared with their biological counterpart, however the artificial sensors use inorganic materials and operate at temperatures that would damage many fragile odours. The high sensitivity is also a shared feature with the biological process, however the response time of artificial devices is 10—30 seconds, much slower than olfactory receptor cells. Furthermore, odorants are typically small hydrophobic organic molecules containing simple polar groups and a mass range of 34—300 Da (Arctander, 1969). Yet these inorganic sensors take no direct account for molecular shape or mass which is believed to be important in odour sensing by Amoore (1970). Several "low temperature" organic crystalline and polymeric materials have been investigated for chemosensitivity. Vacuum sublimated phthyalocyanines are crystalised by annealing to produce thick film organic chemoelectronic sensors. Phthyalocyanines are particularly responsive to oxide gases (Miasik et al., 1986, Chadwick et al., 1986) as well as other electrophilic species such as HCl, SO_2 and H_2S. At present their application to odour sensing seems rather restricted, although their preparation by Langmuir-Blodgett techniques holds out more promise (Baker et al., 1983, Bott and Jones, 1984, Jones and Bott, 1986).

Recent interest has arisen in the use of polymers as the sensing element in conductimetric devices. The effect was first observed in chemically prepared samples of polypyrrole (Nylander et al., 1983), and later using electrochemically deposited polymers on interdigital electrodes (Hanawa et al., 1988, Bartlett et al., 1989, Bartlett and Ling-Chung, 1989a, b). A pyrrole-based polymer is most commonly used, and has been investigated as an ammonia chemiresistor by Nylander et al. (1983), and a hydrogen detector in suspended gate field effect transistors by Cassidy et al. (1986), and Josowicz & Janata (1988). The precise charge transfer mechanism in conducting polymers is still uncertain, although variable range hopping is often cited, yet they may offer considerable potential as odour detectors because of their low operating temperature, and the wide choice for their chemical structure (Pelosi and Persaud, 1989).

3.3 Chemosensitive mass sensors

When an odorant or gas is adsorbed onto the surface of a block of material it should result in a small increase in its mass. Chemosensitive mass sensors detect this change in mass and basically operate as microbalances. It might be expected that these sensors would suffer from a lack of sensitivity and poor stability. Yet, in practice mass sensors have been found to have a sensitivity in the ppm range and a low power consumption (cf oxide devices). One type of mass sensor uses a piezoelectric sensing element. A piezoelectric crystal can be forced to vibrate at its resonant frequency by the application of an external oscillatory electric field. Figure 5 shows the general arrangement of a piezoelectric crystal oscillator together with its mechanical lumped-system equivalent, consisting of a spring (with stiffness k), damper (with damping coefficient b) and a mass m in parallel. The frequency response of the piezoelectric sensor is well approximated by that of a lightly damped (i.e. $b/2\sqrt{km} \ll 1$) second order linear system, where its resonant oscillatory frequency f is approximately related to the mass m and spring constant k by,

$$f^2 = k/m \qquad (6)$$

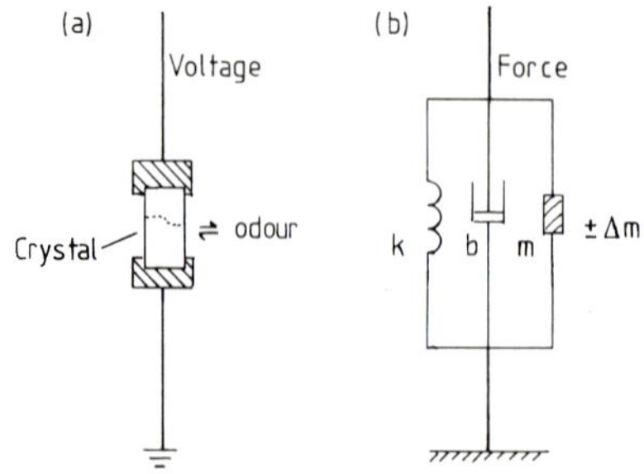

FIGURE 5
A chemosensitive mass sensor: (a): the piezoelectric crystal oscillator and (b): its mechanical model, a spring (spring coefficient k): damper (damping coefficient b): and mass m in parallel

The small change in mass Δm that accompanies the adsorption of an odour or gas causes a shift in the resonant frequency Δf according to a simplified version of the Saubery equation,

$$\Delta m/m = -2\,\Delta f/f \tag{7}$$

Bridge circuits are capable of detecting a change in frequency of up to one part per billion. Consequently, the change in mass that accompanies an odour concentration of only a few ppm can be detected. The application of piezoelectric chemoelectronic sensors for odour detection was first reported in the mid sixties by King (1964), and more recently by Guilbaut (1984) who used α-quartz to detect pesticides. Quartz piezoelectric oscillators typical run in the range of 1—50 MHz, and show a sensitivity of 0.01—0.1 Hz/

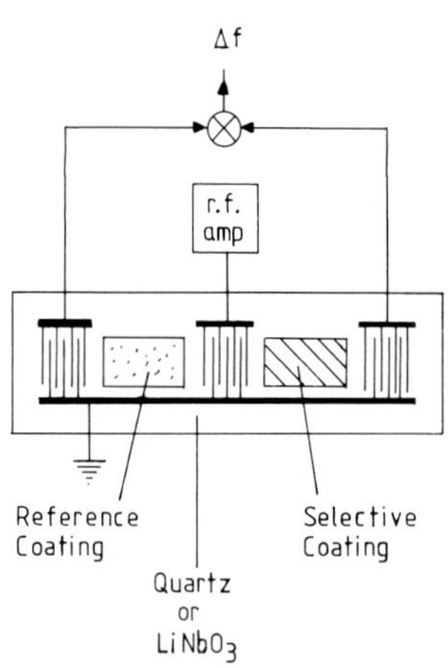

FIGURE 6
General arrangement of a SAW sensor comprising of selective and reference coatings

ppm. Piezoelectric sensors tend to suffer from several effects, namely temperature drift, moisture sensitivity and poor specificity. Carey et al. (1986, 1987) have reported an improved specificity by applying selective coatings. Employing pattern recognition techniques (see later), they found that an array of selectively coated piezoelectric sensors could classify a range of odours. Piezoelectric materials, commonly quartz and $LiNbO_3$, are also employed by Surface Acoustic Wave (SAW) sensors. Figure 6 shows a schematic diagram of a SAW sensor, where a chemically sensitive layer lies between a transmitter and receiver electrode. SAW devices have also been found to possess high sensitivity to the adsorbed mass (Wohltjen, 1984) that modifies the propagation frequency of the surface wave (through an increase in energy loss). Many organic vapours and odours have been detected by a 112 MHz SAW device with twelve selective coatings (Ballantine et al., 1986, Rose-Pehrsson et al., 1988). As in the work of Carey on piezoelectric crystals, it requires the application of pattern recognition techniques to discriminate the odours. An exciting possibility is the use of biological selective coatings as a more natural chemosensitive element in piezoelectric sensors for an artificial nose.

3.4 Other types of chemical sensors (including optical)

Adsorption spectroscopy is a technique that is well known to chemists for probing molecular structures. Its implementation as an odour sensor could, in principle, improve the specificity. Evanescent wave sensors have been reported by Giuliani et al. (1983) which show a response to ammonia. More recently Arnold (1985) has described an optical enzyme sensor that follows the equilibrium conditions of an enzyme reaction. Whereas Carpenter et al. (1985) report upon an optical technique to measure the gas-induced change of depletion layer width in a palladium/n-CdS diode. It relies on the principle that the width of the space-charge region modifies the intensity of photoluminescence. The recent advances in optoelectronic technology may enhance the exploitation of such sensors in the near future.

3.5 Comparison between artificial and biological sensing elements

Table 1 summarises some of the principal properties of chemosensitive devices and compares them to those of their biological counterpart: the olfactory receptor cells. The table shows the main types of artificial sensor elements, such as the pellistor, semiconducting oxide, piezoelectric and SAW. Their typical specificity, sensitivity, operating temperature, response time, stability and principle of operation are given. As can be seen artificial sensors share some common features with olfactory cells, such as a poor specificity — or broadband response — and high sensitivity. Yet, the response time of olfactory cells is much faster than that of the artificial sensors reported to date. Furthermore the basic principles upon which the artificial sensors work seems far removed from the biological system. In particular there is no deliberate use of stereography or molecular mass in order to improve their response, and they rely upon a modulation in signal amplitude rather than frequency. The utilisation of a frequency-based system for artificial chemosensitive sensors might lead to a reduction in baseline drift that limits their present application.

TABLE 1
Comparison between artificial and biological sensing elements

Type	Specificity	Sensitivity	Temp. (°C)	Rise Time*	Principle of Operation
Pellistor	Medium	ppm	250	5 s	Heat rise from combustion
Semiconducting Oxide	Low	ppm	400	30 s	Interaction with adsorbed oxygen
Phthyalocyanine	High	ppb-ppm	150	5 s	Action of oxidising gases
Piezoelectric/SAW	Low	ppm	50	60 s	Mass adsorption
Polymer	Low	ppt	20	5 m	Unknown
Receptor neurones	Low	ppm	37	< 1 s	Mass/shape

* Typical time to reach 90% of steady-state response at the. operating temperature shown.

Generally, artificial sensors monitor the change in equilibrium conditions in a reversible chemical reaction. Information is usually governed by amplitude rather than frequency (receptor cell): modulations.

4. SENSOR SIGNAL PROCESSING I: OLFACTORY BULB

The olfactory bulb processes the receptor cell output signals. It shows a neatly stratified structure with distinct layers: the olfactory nerve layer, the glomerular layer, the external plexiform layer, the mitral cell layer, the internal plexiform layer, and the granule cell layer. A recent review of the olfactory bulb's anatomy is given by Halasz (1990). Only some features, which may be important in this context, are outlined as follows: the outermost layer is the Olfactory Nerve Layer (ONL) where bundles of receptor cell axons distribute on the way to the glomerula. The next layer is the Glomerular Layer (GL). Glomerula are spherically shaped morphological entities in which the primary receptor fibres make synaptic contact with the secondary neurones, the Mitral Cells (MC). Mitral cells and many tufted cells are the only cells that send their output to higher brain structures. There is as yet no explanation for the fact that in most vertebrates the synaptic contacts between primary fibres and mitral cells form glomerula. However, in some lower vertebrates such as goldfish there are no typical glomerula and it seems that glomerula formation begins simultaneously with the appearance of interneurones which mediate information between glomerula. These interneurones, the periglomerular cells, are excited by primary fibres and act predominantly upon mitral cell dendrites, whereby this action is inhibitory in most of the periglomerular cells and excitatory in about 20% of the periglomerular cells. The information processing of periglomerular cells appears thus to be an important intermediate step in the information channel receptor neurones to mitral cells and it seems to require the evolutionary transition from an homogeneous, plexus-like glomerular layer as found in goldfish to separate glomerula as in all higher vertebrates.

Mitral cells in higher vertebrates form a narrow and sharp layer, the mitral cell layer (MCL). These cells have several dendrites only one of which, the long primary dendrite, reaches one glomerulum. The other secondary dendrites are oriented mainly orthogonally with respect to the primary dendrite. It is important to note the numbers of receptor neurones (N_r), glomerula (N_g), and mitral cells (N_m). Although these numbers differ from species to species, the ratios are approximately constant. The relation $N_r \gg N_m > N_g$ seems to hold in all studied species, where the ratio Nr:Ng:Nm is approximately given by (F*1000):1:F with the factor F in the range of 10 to 50.

Along the length of the primary Mitral Cell (MC) dendrites, there is a layer between the glomerula layer and the MC layer. In this layer, the external plexiform layer, information is conveyed vertically in the primary MC dendrite, from a glomerulum to a MC soma. The second type of relay neurones of the olfactory bulb, the tufted cells, are distributed in the EPL. Most of the rest of the information transfer in this layer consists in local lateral interactions between secondary dendrites of different MC, mainly through granule cell dendrites but also through other cell types such as short axon cells. Following the EPL and the MCL, there is the inner plexiform layer consisting mainly of numerous fibres,

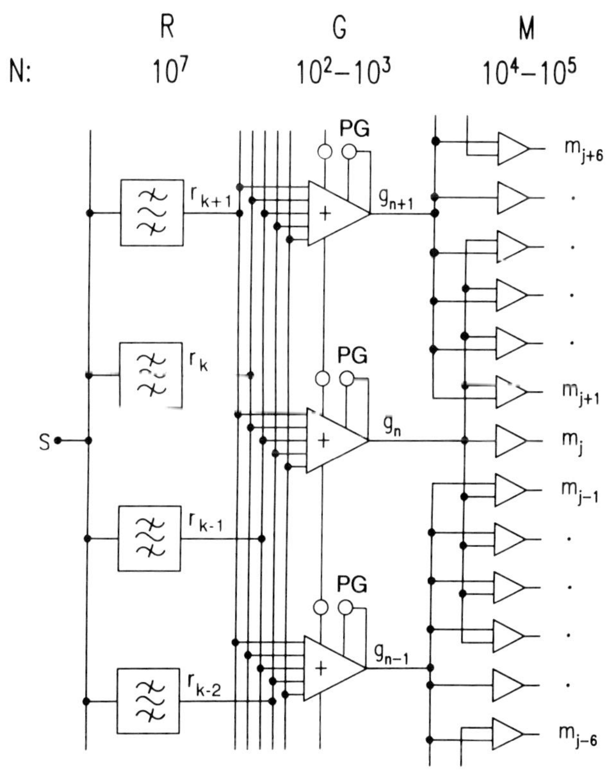

FIGURE 7

Input and output signals of glomerula. The stimulus S elicits responses r_i in receptor neurones shown as filters classes. Global data bus between filters and glomerula. Periglomerular cells (PG): mediate interactions within and between glomerula. Glomerular signals g_n project locally onto mitral cells.

particularly axon collaterals of mitral cells. The latter can also be observed in the MCL as well as in the following layer, the GCL, where the granule cells are distributed. These Gaba-ergic interneurones make reciprocal synapses onto mitral cell dendrites thereby mediating lateral inhibition between the MCs.

The most marked morphological structures in the olfactory system are the glomerula. A sketch of the input and output signals of glomerula is shown in Figure 7. The synaptic connections in the glomerula, i.e., the direct ones from olfactory axons to mitral cell dendrites and the indirect ones from primary nerves through periglomerular cells to mitral cell dendrites, accomplish a transform which is characteristic of the olfactory system. A crucial question is: according to what rule do the connections between receptor neurones and mitral cell dendrites establish? A simple assumption would be as follows (Schild, 1988): Call the receptor cell class activities r_i, the connection between class i and the j-th glomerulum c_{ij}, and the overall input to the j-th glomerulum $\underline{g}_j$. Then $\underline{g}_j = \Sigma c_{ij} r_i = \underline{c}_j\, \underline{r}$. This means that glomerula measure unnormalized correlations between input vectors $\underline{r}$ and connectivity patterns $\underline{c}_j$. In particular, if a stimulus S^k leads to a receptor response $\underline{r}^k$, then the stimulus response of all glomerula is the vector $\underline{g}^k = (g_1{}^k, g_2{}^k, \ldots)$. If the correlations $\underline{c}_j\, \underline{r}$ are to be meaningful, the components of $\underline{c}_j$ have to be adapted to the stimuli, for instance by a correlation learning rule,

$$c_{ij} \propto g_j r_i \tag{8}$$

and an appropriate normalization or saturation procedure.

Alternatively, instead of assuming one glomerular variable g_j, one could take into account the action of periglomerular cells. This would give a transform h_{ij} from certain receptor cell classes to some mitral cell inputs. Given the positive and negative feed-forward coupling of periglomerular cells, it can easily be imagined that glomerula perform a sort of orthogonal transform. Although it would seem possible and reasonable from a signal processing point of view that glomerular processing corresponds to correlation and factor analysis, this remains to be shown experimentally.

5. PROCESSING OF SENSOR SIGNALS II

5.1. Pattern recognition and artificial networks

The olfactory bulb processes the sensory signals that are generated by the olfactory receptor cells. In this respect its functional purpose is like that of a computer, but the architecture is rather different. Modern digital computers have 5 V DC signal levels that can switch states in a few nanoseconds, and employ a silicon based technology. In contrast, the nervous system has approximately 100 mV signals that have a pulse width of about one millisecond. The typical neurone dissipates about one picowatt of power compared to a microwatt for a silicon logic gate. Yet, the speed at which a single neurone can process information is nearly one million times slower than its artificial analog. The computer and olfactory bulb can both process signals, extract information and make a decision based on the output. Yet, the loss of a single logic gate can render a computer inoperable, while evidence suggests that a large part of the olfactory bulb can be damaged with no apparent loss of function. These could be the clues to understanding why our computers struggle to carry out pattern recognition. It should be noted that the visual system of a single person performs more signal processing than all the computers in use today! Many pattern recognition techniques have been developed in the past fifty years and implemented in software on computers. However, the efficiency of the processing is somewhat limited by the computer hardware. It is only now that we are moving away from the race for computational speed and realise, as nature has already worked out, that there is enormous benefit in the parallel distributed processing of signals. The transputer is under development in array processing and parallel programming languages, such as parallel Fortran and C, are constructed to handle the novel architecture. Yet, the processing power of a transputer far exceeds that of a single neurone, and it may not be until simpler elements like cellular automata (Barker, 1987) are developed that the gap between the artificial and biological system narrows. A plethora of pattern recognition techniques exist to solve the classification-type problems traditionally encountered in the chemical and physical sciences. In some respects these methods appear to be more of a statistical or chemometric tool, and are rather divorced from the biological processes in the olfactory bulb. Nonetheless they have been applied to odour detection with some success and we shall discuss them here, together with the more realistic technique of artificial neural networks that has its roots in neuroscience.

5.2 Classification by parametric techniques

Pattern recognition techniques may be conveniently subdivide into two basic types, namely parametric and non-parametric. Parametric in contrast to non-parametric techniques are based upon the use of a probability distribution of a parameter, calculated a priori, to characterise the system response. For example, for a system that consists of an array of conductimetric chemoelectronic sensors, the parameter is a vector G in which each of n elements may be associated with a change in the conductance g,

$$G = \{ g_1, g_2, \ldots\ldots g_n) \qquad (9)$$

The response vector G of an odour may now be compared to that of a standard vector G_s corresponding to a known or standard odour. For instance, Müller and Hornor 1986 report on the use of a correlation technique to classify gases, and later whisky aromas, from the output from an array of MOS sensing elements. The principle of the technique relies upon the calculation of a correlation coefficient r from

$$r = \frac{\Sigma_i(g_i - \langle g_{is} \rangle)}{\Sigma_i[(g_i - \langle g_i \rangle)^2 (g_{is} - \langle g_{is} \rangle)^2]^{1/2}} \qquad (10)$$

Where g_i is the response of sensor i to a test gas, and g_{is} the response to a standard gas or odour based upon the assumption that the sensor response is concentration-invariant ($\langle\ \rangle$ is the value averaged over n sensors). The value of the coefficient approaches unity when there is a match between the test and standard gas. A similar approach was used by Shurmer et al. (1989b) to classify alcohol vapours from the signals of three semiconductor oxide sensors. The normalisation procedure is important because it helps remove the concentration dependence of the response vector. The response of tin oxide sensors to alcohol vapour is linear with concentration below about 50 ppm (Gardner, 1989). Under these conditions the principle of linear superposition holds. Seven odours were analysed using a correlation method by Kaneyasu et al. (1987) who compared the output from an array of six semiconductor oxide sensors to the patterns of electrical activity in the rabbit's olfactory bulb.

When a mixture of odours or gases is analysed, the resulting classification is called multicomponent analysis. For a system containing several components or odours at different concentrations, then the response matrix is

$$G = X C \tag{11}$$

The columns of the response matrix G are the response vectors G, and X is a regression matrix that relates the odour concentrations in C to the response. Provided that there are more sensors than odours to detect, then the problem is over-determined and equation (11) is soluble. When the sensors only respond to one specific odour, all of the matrix X off-diagonal terms are zero. However the response of most types of chemical sensors is non-specific, so terms off the leading diagonal are non-zero. Under these conditions Multiple Linear Regression (MLR) and Partial Least Squares (PLS) are often used to solve the problem (see: Dobson, 1983). Carey et al. (1987) have used these techniques to classify the response of piezoelectric mass sensors to a range of organic vapours. They also discuss the relative merits of the two techniques.

5.3 Classification by non-parametric techniques

Feature weighting is a non-parametric pattern recognition technique that has been employed by Shurmer et al. (1989a) to discriminate between the response of tin oxide sensors to tobacco odours. Feature weighting is a preprocessing technique that allows the best display of the data for class discrimination. Figure 8 shows the original response

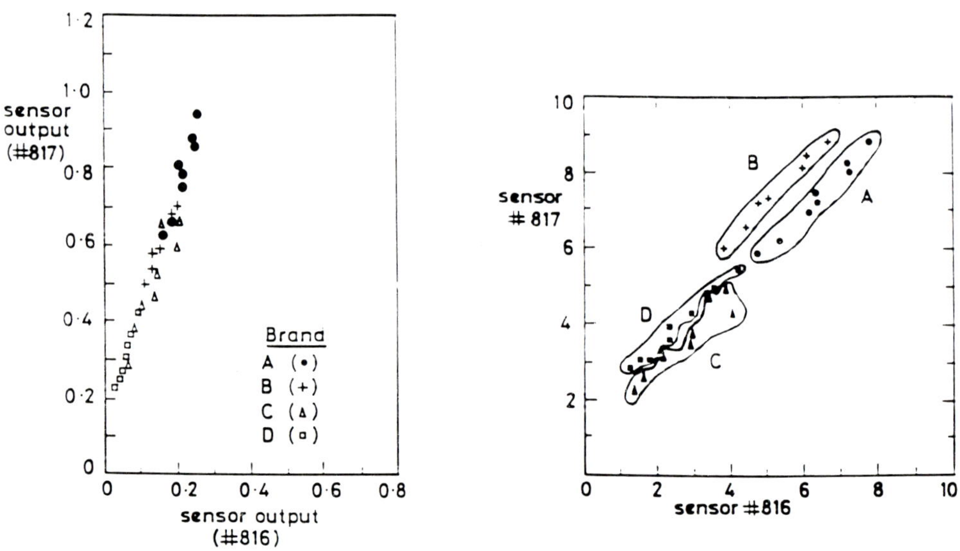

FIGURE 8
Scatter diagrams of (a): original and (b): transformed data plots of the conductance change (arbitrary units): of tin oxide gas sensors, TGS 816 & 817, to four tobacco blends (A—D)

data and the transformed data for each of four blends of tobacco. In a simple ratiometric transform the clusters in the response vectors are identifiable. Two non-parametric pattern recognition techniques are widely used in the chemical sciences, namely principal component and cluster analysis (see: Dobson, 1983). Principal Component Analysis (PCA) expresses the response matrix G in terms of a linear combination of orthogonal vectors, these are calculated to maximise the amount of information, or variance, in as few principal components as possible. PCA has been successfully applied to chemical sensing by Carey and Kowalski (1986) for arrays of selectively coated piezoelectric sensors. Similar work has been reported by Ballantine et al. (1986) and Rose-Pehrsson et al. (1988) to analyse the signals from an array of SAW devices. The technique could be used to detect such vapours as toluene, diethylsulphide and 1-butanol. Recent studies have been carried out by Abe et al. (1987,1988) into the sensing of odours by an array of eight tin oxide sensors. The odours were analysed and classified into five groups: Ethereal (E), Minty (M), Pungent (P), Ethereal-Minty (EM) and Ethereal-Pungent (EP). Figure 9 shows

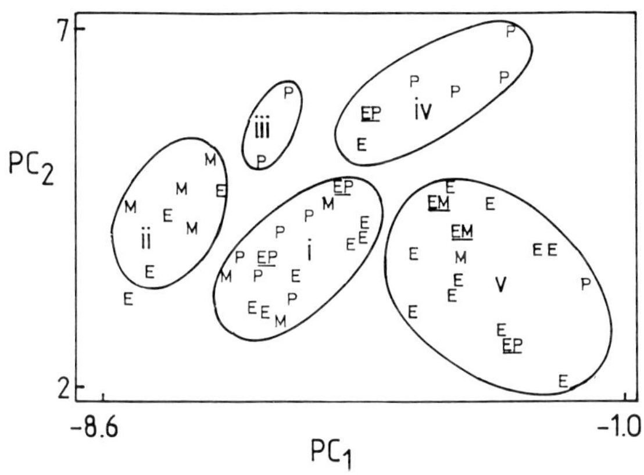

FIGURE 9
Results of a Principal Component analysis on 47 odours in which five odour classes are apparent, (i): Ethereal (ii): Minty (iii): Pungent (iv): Ethereal-pungent (v): Minty-ethereal (Abe et al., 1987)

the results of PCA on forty-seven odours that highlights these odour groups. The data was also analysed using cluster analysis that determines which responses lie near each other in n-dimensional vector space. Figure 10 shows the resulting dendrogram, i.e. a map of the similarity between the odours. The similarity decreases from left to right. Clearly, conventional pattern recognition techniques have been applied to odour sensing with a significant amount of success. Yet, these techniques bear little or no resemblance to the processing that occurs within the olfactory bulb and do not help us understand the underlying principles. In a radically different approach artificial neural networking techniques are now being developed that set out to mimic the biological neural system.

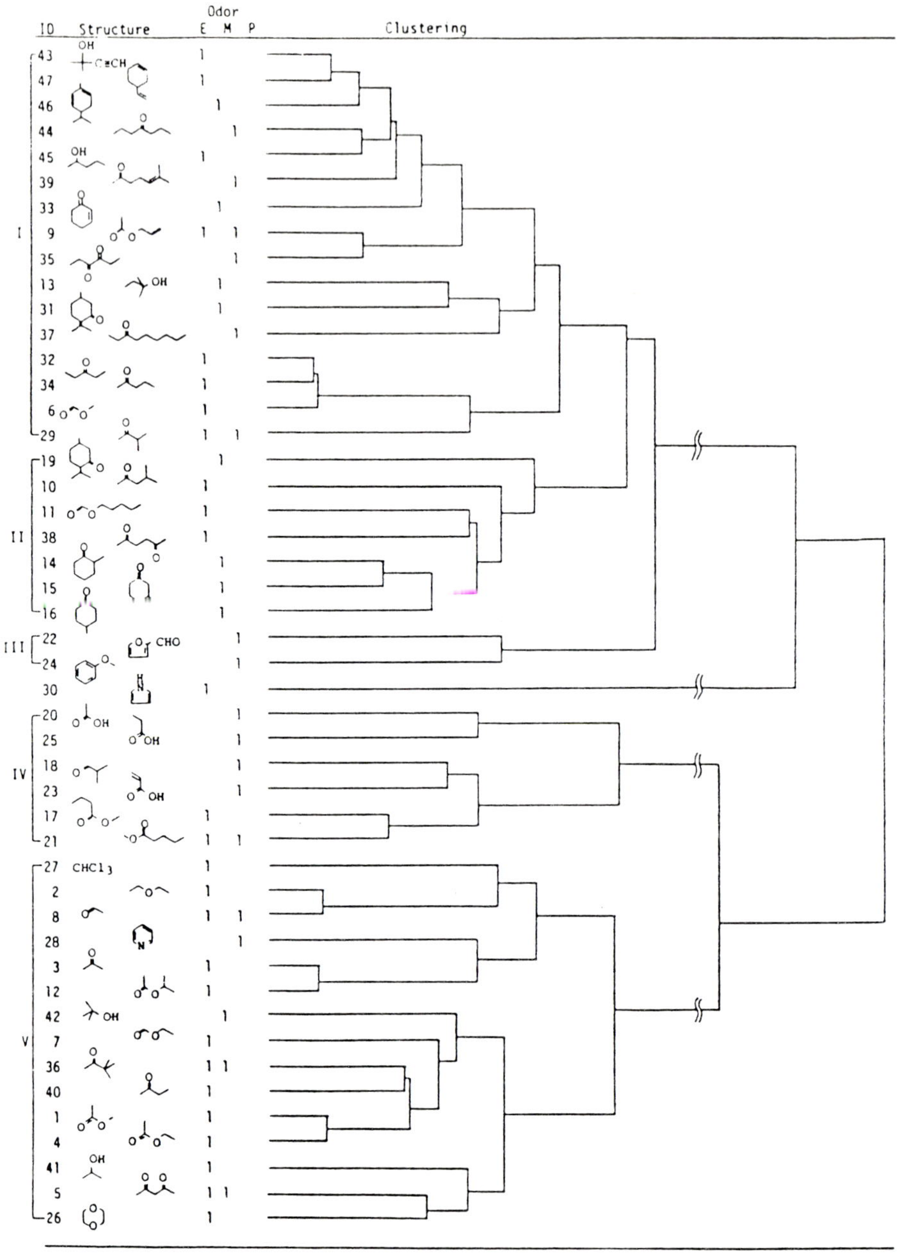

FIGURE 10

Results of a Cluster Analysis upon response of tin oxide sensors to odours. The classes found by principal component analysis are shown. (Abe et al., 1987)

5.4 Artificial neural networks

The architecture of the olfactory system has been discussed already in some detail. In essence there are about 10 million receptor cells that exhibit a low level of electrical activity (i.e. mean firing-rate) until an odour is adsorbed onto the distributed sites. These cells are synaptically linked into about a thousand glomerular nodes, followed by a third stage of around 10—100 thousand mitral cells within the olfactory bulb. Visual evidence of the temporo-spatial activity in the olfactory bulb shows that the process of odour detection is both non-specific and distributed (Kauer, 1990). The choice of a scheme for processing the odour signals is large. For example, Freeman and Davis (1990) have applied a state space approach to the coding of the neural activity. The use of chaos theory to model the firing of olfactory neurones is both intellectually appealing and is consistent with some of the experimental evidence. Yet the novel approach of artificial neural networks adheres closely to the biological system. Artificial Neural Networks (ANNs) consist of parallel interconnected processing elements with a hierarchical structure that is based upon the biological nervous system (Kohonen, 1988). The processing elements represent neurones (olfactory or otherwise) and their interconnections are the synaptic links. The architecture of the olfactory bulb suggests the use of a three layered network. Figure 11 shows the network adopted by Gardner and Hines (1990) to model the

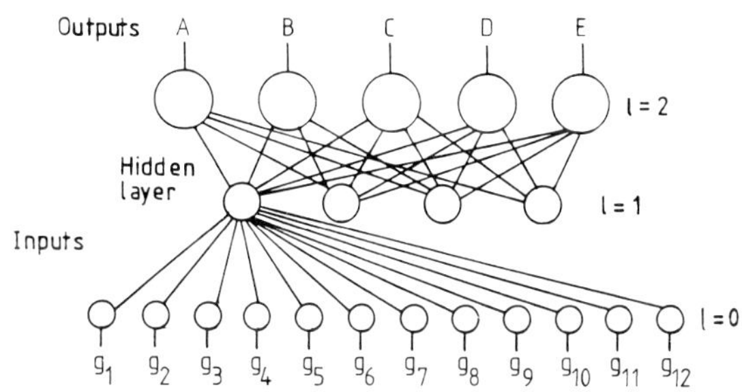

FIGURE 11
A three layered artificial neural network used to model the olfactory system (all elements in the hidden layer are connected to all elements in the input layer, but only one is shown)

olfactory system for use in the Warwick artificial nose (Gardner et al., 1990). The input layer (l=0) contains the elements that represent the olfactory cells or chemosensitive sensing elements; in the artificial system these are 12 different semiconductor oxide sensors. The signals presented to the processing element in the network need to lie between a value of [0,1], so the conductance change g_{ij} of each sensor i to odour j is normalised according to,

$$g_{ij}^{N} = g_{ij} / \Sigma_i \, g_{ij} \tag{12}$$

The next layer (l=1) is called the hidden layer because it does not interact with the outside world. The number of elements in this layer has been determined experimentally

TABLE 2
Results of analysing a three layer network for five alcohols: metanol, ethanol, propan-2-o!, 1-butanol and 2-methyl-1-butanol

Measured	Ideal output					Observed output					Inference
	A	B	C	D	E	A	B	C	D	E	
Methanol	1	0	0	0	0	.998	.005	.001	.000	.001	Methanol
Ethanol	0	1	0	0	0	.011	.840	.024	.014	.000	Ethanol
Propan-2-o1	0	0	1	0	0	.000	.021	.970	.005	.000	Propan-2-o1
1-Butanol	0	0	0	1	0	.042	.006	.000	.962	.000	1-Butanol
2-Methyl-1-butanol	0	0	0	0	1	.000	.000	.025	.000	.975	2-Methyl-1-butanol

and each element is like a glomerular node. The degree of convergence 3:1 is much lower than the presumed value for the biological system. Finally the outer layer (l=2) must possess a number of neurones or elements that is at least equal to the number of odours for detection. If a direct comparison to the mitral cells was applicable, then the biological system could detect 10—100,000 odours. However as we know that considerable redundancy exists in the olfactory bulb a lower value is more likely. The network shown in Figure 11 has been used to code five alcohols. Its performance is determined by the interconnectivity and learning rules that are applied to calculate the connection weightings. There are many algorithms that could be used to train this multi-layer network (or perceptron), but we have adopted the common back-up propagation techniques (Rumelhart and McClelland, 1986). First, the network is trained by presenting the normalised conductance signals of an odour to the input layer. In a feedforward learning process the output of the network is calculated from initially random weightings and a sigmoid activation function, and compared to the ideal output, i.e. 0 or 1. Second, the difference is used to modify the weightings in a lower layer systematically back through the network. The process is repeated until the weightings converge to stable values. Table 2 shows the results of training the network for 10,000 iterations and then attempting to recognise a new signal. The correct odour is calculated in each case applying the criterion that the value of the appropriate element exceeds 0.5 (Huang and Lippman, 1987). Similar results have been obtained by Japanese workers (Ema et al., 1989) who applied the back-propagation technique to the response of an array of piezoelectric sensors to several odours.

The development of artificial neural networks is still in its infancy. Yet the approach has enormous application potential due to features such as fault-tolerance, noise-tolerance, and adaptability. It should eventually be possible to construct real-time ANNs that dynamically analyse the response of chemical sensor arrays to odours, and then the gap between the artificial and biological nose will have narrowed.

REFERENCES

Abe, H., Yoshimura, T., Kanaya, S., Takahashi, Y., Miyashita, Y., Sasaki, S. I. (1987): Analytica Chimica Acta 194: 1.

Abe, H., Kanaya, S., Takahashi, Y., Sasaki, S. I. (1988): Analytica Chimica Acta 215: 155.

Amoore, J. E. (1970): Molecular basis of odour, Charles Thomas, Springfield, Ill. USA.

Arctander, S. (1969): Perfume and Flavour chemicals, published by author, USA.

Arnold, M. (1985): Anal Chem 57: 565.

Baker, S., Roberts, G. G., Petty, M. C. (1983): Proc IEE Part I 5: 260.

Ballantine, D. S., Rose, S. L., Grate, J. W., Wohltjen, H. (1986): Anal Chem 58: 3058.

Barker, J. R. (1987): Molecular Electronic Devices II, Dekker p. 639.

Bartlett, P. N., Archer, P. B. M., Ling-Chung, S. K. (1989): Sensors and Actuators 19: 125.

Bartlett, P. N., Ling-Chung, S. K. (1989a): Sensors and Actuators 19: 141.

Bartlett, P. N., Ling-Chung, S. K. (1989b): Sensors and Actuators 20: 287.

Bell, A. G. (1914): The National Geographic Magazine XXV: 649.

Bott, B., Jones, T. A. (1984): Sensors and Actuators 5: 43.

Brattain, W. H., Bardeen, J. (1953): Bell syst Tech J. 32: 1.

Buck, T. M., Allen, F. G., Dalton, M. (1965): Surface effects in. detection, Spartan Books Inc. USA; eds. T. Bregman and A. Dravnieks).

Capehart (1987): In: Solid State Sensors; eds. P. T. Moseley & B. C. Tofield; Adam-Hilger, Bristol.

Carey, W. P., Beebe, K. R., Kowalski, B. R. (1987): Anal. Chem. 59: 1529.

Carey, W. P., Beebe, K. R., Kowalski, B. R., Illmanm D. L., Hirshfield, T. (1986): Anal. Chem. 58: 149.

Carey, W. P., Kowalski, B. R. (1986): Anal. Chem. 58: 3077.

Carpenter, M. K., Van Ryswyk, H., Ellis, A. B. (1985): Langmuir 1: 605.

Cassidy, J., Foley, J., Pons, S., Janata, J. (1986): Anal. Chem. Symp. Ser. 25: 309.

Chadwick, A. V., Dunning, P. B. M., Wright, J. D. (1986): Mol. Cryst. Liq. Cryst. 134: 137.

Cherry, R. J., Dodd, G. H., Chapman, D. (1970): Biochem. Biophys. Acta 211: 409.

Dobson, A. J. (1983): An introduction to statistical modelling. Chapman & Hall, London.

Dravnieks, A., Trotter, P. J. (1965). J. Sci. Instruments 42: 624.

Egashira, M., Shimuzu, Y., Takao, Y. (1988): Jap. Chem. Letts 389.

Ema, K., Yokoyama, M., Nakamoto, T., Moriizumi, T. (1989): Sensors and Actuators 18: 289.

Farbman, A. I. (1986): Chem. Sens. 11: 1.

Fouletier, J. (1982/3): Sensors and Actuators 3: 295.

Freeman, W. J., Davis, G. W. (1990): Chemosensory Information Processing; ed. D. Schild. Springer, Berlin, p. 375.

Frings, S., Lindemann, B. (1988): J. Membrane Biol. 105: 233.

Gardner, J. W. (1989): Sensors and Actuators 18: 373.

Gardner, J. W., Bartlett, P. N., Dodd, G. H., Shurmer, H. V. (1990): Chemosensory Information Processing; ed. D. Schild. Springer, Berlin, p. 131.

Gardner, J. W., Hines, E. L. and Wilkinson, M. (1990): Meas. Sci. Techol 1:

Gentry, S. J., Walsh, P. T. (1987): Solid-state Sensors, Adam-Hilger, Bristol; eds. P. T. Moseley and B. C. Tofield, p. 33.

Getchell, T. V. (1986): Physiol. Rev. 66: 772.

Giuliani, J. F., Wohltjen, H., Jarvis, N. L. (1983): Opt. Lett. 8: 54.

Graziadei, P. P. C., Monti Graziadei, G. A. (1978): in Handbook of Sensory Physiology. Development of Sensory Systems, ed. Jacobson, M. Springer, New York, p. 55.

Guilbault, G. G. (1984): Methods and Phenomena, Vol. 7; eds. C. Lu and A. W. Czaderna, Elsevier, Amsterdam.

Halasz, N. (1990): in Chemosensory Information Processing; ed. Schild, D. Springer, Berlin, p. 175.

Hanawa, T., Susumu, K., Yoneyama, H. (1988): J. Chem. Soc. Faraday Trans. I 84: 1587.

Heiland, G., Luth, H. (1983): Chemical Physics of Solid Surfaces & Hetergeneous catalysts; eds. D. A. King, P. Woodruff. Elsevier, Amsterdam, p. 137.

Huang, W. Y., Lippman, R. P. (1987): INT Conf. on Neural Networks, San Diego, June 4, 1985.

Josowicz, M., Janata, J. (1988): Chemical Sensor Technology; ed. T. Seiyama, Elsevier, NY, p. 153.

Jones, T. A., Bott, B. (1986): Sensors and Actuators 9: 27.

Kaneyasu, M., Ikegami, A., Iwanaga, S. (1987): IEEE Trans on Components, Hybrids & Manu. Technol. 10: 267.

Kauer, J. S., Hamilton, K. A., Neff, S. R., Cinelli, A. R. (1990): Chemosensory Information Processing; ed. D. Schild, Springer, Berlin, p. 305.

King, W. H. (1964): Anal. Chem. 36: 1735.

Kohonen, T. (1988): Neural. Networks 1: 3.

Lundstrom, I., Sodelberg, D. (1981): Sensors and Actuators B2: 105.

McAleer, J. F., Moseley, P. T., Norris, J. O. W., Williamsm D. E. (1987): J. Chem. Soc. Faraday Trans. I 83: 1323.

Miazik, J. J., Hooper, A., Tofield, B. C. (1986): J. Chem. Soc. Faraday Trans I 82: 1117.

Mizsei, J., Hasanyi, J. (1983): Sensors and Actuators 4: 397.

Misra, T. N., Rosenburg, B., Switzer, R. (1968): J. Chem. Phys. 48: 2096.

Moseley, P. T., Tofield, B. C. (1987): eds. of Solid-state Sensors, Adam-Hilger, Bristol.

Moulton, D. G. (1974): Ann. NY Acad. Sci. 237: 52.

Moulton, D. G., Celebi, G., Fink, R. P. (1970): in Taste and Smell in Vertebrates; ed. Wolstenholme, G. E. W., Knight, J. J. & A. Churchill, London, p. 227.

Murakami, K., Ye, D., Yamamoto, T. (1988): Sensors and Actuators 13: 15.

Müller, R. S., Horner, G. (1986): Siemens Forsch.- u. Entwickl.-Ber. Springer 15: 95.

Nylander, C., Armgarth, M., Lundstrom, I. (1983): Anal. Chem. Symp. Series 17: 203.

Pelosi, P., Persaud, K. (1988): Nato ASI Ser. F, Sensors & Sensory systems for advanced robots; ed P. Dario. Springer-Verlag, Berlin, p. 361.

Rose-Pehrsson, S. L., Grate, J. W., Ballantine, D. S., Jurs, P. C. (1988): Anal. Chem. 60: 2801.

Rumelhart, D. E., McClelland, J. I. (1986): Parallel Distributed. Processing (Cambridge, MA: MIT press).

Schild, D. (1988): Biophys. J. 54: 1001.

Schild, D. (1989): Exp. Brain Res. 78: 223.

Schild, D. (ed.) (1990): Chemosensory Information Processing. Springer, Berlin.

Shurmer, H. V., Gardner, J. W., Chan, H. T. (1989a): Sensors and Actuators 18: 361.

Shurmer, H. V., Gardner, J. W., Corcoran (1989b): Sensors and Actuators B 1: 256.

Taguchi, N. (1971): British Patent No. 1257155.

Tran-Minh, C., Vallin, D. (1978): Anal. Chem. 50: 1874.

Vogt, R. G., Rybczynski, R., Lerner, M. R. (1990): in Chemosensory Information Processing; ed. Schild, D. Springer, Berlin.

Wohltjen, H. (1984): Anal. Chem. 56: 87.

Yamanoto, N., Tonomura, S., Matsuoka, T., Tsubomura, H. (1980): Surf. Sci. 92: 400.

MOLECULAR AND CELLULAR ASPECTS OF ERYTHROPOIETIN AND ERYTHROPOIESIS – FROM 1986 TO 1990

Ivan N. Rich

Department of Transfusion Medicine
of the University of Ulm
and the German Red Cross Blood Bank,
D-7900 Ulm/Donau, F.R.G.

KEYWORDS

ERYTHROPOIESIS, ERYTHROPOIETIN–MOLECULAR BIOLOGY, ERYTHROPOIETIN THERAPY, ERYTHROPOIETIN RECEPTORS, ERYTHROPOIETIN SENSITIVITY, ERYTHROPOIETIN PRODUCTION, KIDNEY, MACROPHAGE, OXYGEN SENSING, REVIEW.

Since the meeting on Molecular and Cellular Aspects of Erythropoietin and Erythropoiesis in September 1986 [1], we have witnessed several major developments in the study of red blood cell production. Several of these developments were discussed as theoretical considerations at that meeting. In the this report I would like to discuss some of the advances that have occurred over the past 4 years.

ERYTHROPOIETIC ORGANAZATION

All eight cell lineages of the hemopoietic and lymphopoietic system are derived from a self-renewing stem cell population compartment. This population we usually denote as "uncommitted"; that is, the decision has not been made into which lineage the stem cell is going to proceed on its way to become a functional blood cell. It is possible to envisage that at some time after fertilization, a population of totipotent stem cells are produced from the ectoderm, mesoderm and endoderm responsible for the development of different organs and tissues of the body. A population of totipotent stem cells could be specifically responsible for the production of blood cells. As such, these stem cells are in fact "committed" to become hemopoietic cells and nothing else.

The stem cell compartment comprises less that 0.1% of the total hemopoietic cells found in the bone marrow. Within this small population a cellular hierarchy exists. The majority of the stem cells are in a quiescent state and are only brought into cell cycle when required. The totipotent hemopoietic stem cell can be considered to possess the greatest proliferative potential; it is also the cell in the "deepest slumber". By mechanisms which are not understood, an influx from early to late stem cell populations occurs. With each step toward becoming committed to a particular hemopoietic cell lineage, the stem

NATO ASI Series, Vol. H 51
Cell to Cell Signals in Plants and Animals
Edited by V. Neuhoff and J. Friend

cell gradually looses its "stemness" by loosing its capacity of self-renewal and repopulating ability. In addition, commitment is accompanied by an increase in the proportion of cells transversing the cell cycle. In vitro stem cell populations designated high proliferative potential colony-forming cells or HPP-CFC [2] and multipotential stem cells also called CFC-Mix or CFC-GEMM [3, 4] are considered "early" stem cell populations, while the in vivo pluripotential stem cell or CFU-S [5] is a late stem cell perhaps on the verge of commitment. Incredibly, it is this small stem cell pool that maintains the production of blood cells throughout the life of the animal. With respect to erythropoiesis, this amounts to more than 2×10^6 erythrocytes per second throughout life.

Once a stem cell has made the irreversible decision to enter the erythroid cell line, the first detectable population is the cell that gives rise to macroscopic colonies of erythroid cells after 10 to 14 days in culture is designated the Burst-Forming Unit — Erythroid or BFU-E [6, 7]. By proliferation and differentiation, the BFU-E eventually gives rise to a population of cells that form small 8—32, sometimes 64 cells) colonies after 48 h *in vitro.* The cells that give rise to these colonies are called Colony-Forming Units — Erythroid or CFU-E [8, 9]. The latter may be considered similar to the first morphologically identifiable erythroblast, the pronormoblast [10]. In between the "early" BFU-E and "late" CFU-E are several erythropoietic populations. (We arbitrarily divide the hemopoietic system up into compartments for our own benefit based on functional assays. In actual fact, one differentiating population passes imperceptibly into the next producing a continuum of cells). If one observes the development of colonies, several colonies of different sizes are produced at different times. Indeed, it is the size of the colonies that gives away its age; the longer it takes for a colony to develop, the large it is and the more primitive the original cell was from which the colony was derived. So called 3-day erythroid burst colonies, derived from 3-day BFU-Es, have been defined as an intermediate population between the BFU-E and CFU-E. Similarly, culturing hemopoietic cells for 1 day gives rise to so-called erythroid clusters [11] which are derived from cells intermediate between the pronormoblast and basophilic erythroblast. Note, however, that all thes erythropoietic populations are defined under *in vitro* conditions; their *in vivo* counterparts still have to be identified. Furthermore, quantitation is dependent on many parameters which are extremely difficult to control. Several advances have been made however, especially those that take the physio-chemical parameters into consideration [12, 13]. Serum-free conditions have also been developed [14, 15], but since the primary cells require a certain amount of protein for growth, the latter is always present even if it is albumin. Unfortunately, whereas other areas of erythropoietic investigation have entered a new era, the analysis of erythropoietic organization has been left somewhat behind. Since the detection of the CFU-E and BFU-E populations, no real breakthrough has occurred to define erythropoietic precursor populations at a level other than their functional ability to produce colonies *in vitro.*

This is in contrast to our ever-increasing knowledge of the regulator molecule that controls erythropoiesis, erythropoietin (epo).

THE BIOCHEMISTRY AND MOLECULAR BIOLOGY OF ERYTHROPOIETIN

The DNA sequence for epo codes for 166 amino acids [16, 17] to which 3 carbohydrate chains are attached by N- and another by O-linkages [18, 19]. The glycosylated molecule has a total Mr of about 30,400; the polypeptide chain has a calculated Mr of 18,400 so that approximately 40% is attributable to the sugar moiety. The secondary structure of the molecule probably consists of two pairs of alpha-helixes parallel to each other. The chemistry of epo has recently been reviewed in detail by Lappin and Maxwell [20]. The gene is transcribed from 5 exons separated by 4 introns and contains a 27 amino acid leader sequence which is cleaved when the molecule is secreted. In addition, both human urinary and recombinant Chinese hamster ovary (CHO)-derived epo are truncated at the COOH terminus by removal of the arginine at position 166 [21]. In the mouse the epo gene resides on chromosome 5 [22], whereas in human cells it is found on chromosome 7 (q11—q22) [23]. There is about 80% conservation of the DNA sequence between human, monkey and mouse erythropoietins, with much of the untranslated 5' and 3' regions also being considerably conserved [24, 25].

In contrast to many cytokines, there is an absolute requirement for glycosylation of epo for its *in vivo* activity. That the carbohydrate is required for *in vivo* activity was demonstrated by Goldwasser and collaborators in 1974 [26]. Fully glycosylated, radiolabeled recombinant epo has a half-life of about 2 h in the rat, whereas the deglycosylated form is removed from the circulation within 10 min [27, 28]. Spivak and Hogen have shown that epo is found in the kidneys and bone marrow after 30 min after injection [28]. In man, the plasma half-life is between 6 and 9 h [28]. The specificity for the glycosylated molecule *in vivo* is not mirrored *in vitro* since deglycosylated epo can stimulate erythroid progenitor cells to form colonies in culture.

The molecular cloning of the epo gene leading to its large-scale production has had a considerable impact in several areas in the field of erythropoiesis.

ERYTHROPOIETIN AS A THERAPEUTIC AGENT

There is probably no doubt that the largest impact made by the accessibility of large quantities of recombinant epo has been its predictable action as a therapeutic agent. To understand this, one has also to understand the biology of epo.

Erythropoietin is a unique molecule. It is the only hemopoietic regulator molecule which, at present, has *in vivo* biological relevance. With the exception of M-CSF or CSF-1, it is the only hemopoietic regulator molecule found in the circulation under normal steady-state conditions. The *in vivo* production of epo is absolutely dependent on the prevailing partial oxygen tension (to be discussed later). Its activity in serum or plasma can be detected not only *in vitro* and by immunological means, but also by *in vivo* bioassay. Under normal conditions, a circulating concentration of about 12 mU/ml or 120 pg/ml (range, 8—24 mU/ml or 80 to 240 pg/ml) is present. This can be increased to more than 100 fold in aplastic anemia patients. On the other hand, it can be reduced to almost undetecable amounts in patients with polycythemia rubra vera. Its biology of action has been known since Carnot and Deflandre first postulated its presence in 1906 [29] and after its resurrection in the early 1950s, the *in vivo* experiments of many investigators proved its unequivocal specificity of action for the erythropoietic system. Only with the undauting efforts of Dr. Eugene Goldwasser and his associates at the University of Chicago in 1977 [30] to produce the first minute amounts of purified epo, was it possible 8 years later to publish the gene sequence [16, 17].

The effects of epo therapy in patients with anemia of end-stage renal failure were therefore predictable. The first clinical trials were preceded by several experiments in which was administered to large mammals [31]. The alleviation of the anemia of end-stage renal failure was first published by Eschbach and collegues followed shortly afterwards by Winearls and collegues in 1987 [32, 33]. These have now been followed by multicenter clinical trials in many countries of the world. There is absolutely no doubt that for this group of patients, in which the circulating concentration of epo does not correspond to the degree of anemia, the administration of recombinant epo significantly increases the quality of life.

The therapeutic administration of recombinant epo has also been performed in patients with other hemopoietic disorders, for example rheumatoid arthritis, anemia associated with malignancy, and myelodysplasia as well as for the use of autologous blood transfusion [34] and the anemia associated with AIDS. The use of epo in these and other groups of patients is questionable. There are few studies on circulating epo levels that clearly demonstrate or indicate that epo therapy would be useful. The question of therapy costs is also important. It is clear that a return for the investment by the companies involved plays another important role [35, 36], but biological and clinical cause and effect should be taken into account first.

ERYTHROPOIETIN ASSAY

To date, all main epo assays have been based on the WHO 2nd International Standard Preparation, a human urinary preparation with a specific activity of 1 unit per 0.2 mg protein [37]. With the cloning of the epo gene and its increased use in therapy and research, the requirement to base *in vivo, in vitro* and immunologic assays on a new WHO International recombinant standard is greater than ever. We can only hope that this will occur very soon.

Immunologic assays based on recombinant epo, as opposed to native material have now been developed and marketed. Egrie and associates [38] have shown that a radioimmunoassay based on recombinant reagents is similar to that based on native material. The epo assay palette has now been extended to enzyme-linked immunosorbant assays (ELISA). These are usually sandwich-based assays using monoclonal or polyclonal antibodies or a combination of the two. Wognum et.al. [39] have reported the use of an ELISA in which one monoclonal antibody is used for catching epo while the other is formed into a tetrametric complex consisting of anti-epo cross-linked to anti-alkaline peroxidase and rat anti-mouse monoclonal antibodies. The sensitivity is 5 to 10 mU/ml, which is similar to most RIAs. In 1986, Sawatzki reported the use of recombinant epo as a means of purifying, using automated FPLC, anti-epo derived from rabbits immunized with human recombinant epo [40]. This system has now been modified and expanded to produce highly purified monospecific polyclonal IgGs to recombinant epo. The antibodies have now been used to develop a sandwich ELISA in which the signal is amplified by the use of biotinylated anti-epo bound to anti-biotin conjugated to alkaline phosphatase. This ELISA has a sensitivity of about 0.5 mU/ml allowing both human serum and plasma as well as murine supernatants to be measured over an extremely wide range of epo concentrations [41, 42].

THE ERYTHROPOIENTIN RECEPTOR

With the cloning of the epo gene, it followed that the cloning of the epo receptor should not lag too far behind. In fact, 4 years elapsed for the DNA sequence of the epo receptor gene in uninduced murine erythroleukemic cells or MEL cells to be published. The sequencing and expression of the epo receptor gene was performed by D'Andrae and collegues [43]. The cDNA nucleotide sequence of 1773 bases codes for 507 amino acids with 167 bases in the 3' untranslated region ending with a poly(A) tail. The first initiator codon (ATG) is 43 bases from the 5' end. The receptor spans the membrane from the inside (carboxy terminus) to the outside (amino terminus) with 3 potential glycosylation sites. There is an extracellular region from amino acids 25—249 and an intracellular region from amino acids 273—507. This type of receptor is similar to the receptors of growth hormone, prolactin, the β-chain of IL-2, IL-3, IL-4, IL-6 and GM-CSF. However, there does not appear to be any sequence homology with a tyrosine kinase catalytic domain as seen with other growth factor receptors.

Although most investigators working in the field seem to agree that the epo receptor consists of two polypeptide chains, there is some discrepancy as to the actual sizes of the chains. By cross-linking studies, Sawyer and associates [44] found sizes of 85 and 100 kDa on mouse splenic erythroblasts, while Sasaki et.al. [45] detected sizes of 120 and 140 kDa on a Friend cell line and D'Andrae et.al. [43] found sizes of 65 and 105 kDa using MEL cells. These differences may be due to variations in glycosylation. Also in agreement is that the number of receptors is usually below 1000 per cell, although the actual number varies between laboratories and on the source of the cell population used. However, the number of receptors is divided between high (Kd range, 30 to 410 pM) and low affinity (Kd range, 210 to 600 pM) [43, 44, 46]. The low circulating epo concentration would imply that the high affinity epo receptors are physiologically important. In addition, when purified CFU-E from the mouse are treated with epo, a decrease (down-regulation) in the high affinity receptor occurs while the low affinity receptor number remains the same [47]. The latter results indicate that the high affinity receptor is internalized upon binding to epo. Indeed, the epo-receptor complex is rapidly internalized with epo being degraded by the lysosomes [44, 46].

What happens when epo binds to its receptor? The classic second messenger systems, cAMP and cGMP have been thoroughly investigated, but the evidence dose not support a role for either of these two compounds. In contrast, however, a clear and rapid (within 3 min) increase in Ca^{2+} has been observed in individual cells of human cord blood BFU-E colonies [48], but a similar finding was found with GM-CSF implying that this may not be specific. On the other hand, Ca^{2+}, and the presence of the calcium ionophores A23187 and Ro2-2985/1 can stimulate erythroid colony formation and that the effect of calcium can be blocked by EGTA chelation [49, 50]. However, no significant effect has been noticed on the inositol phosphate intermediates [51]. Therefore, although the role of Ca^{2+} appears to be important, there is still no clear indication as to the intracellular path of action of epo.

COMPETITION FOR STEM CELL RECEPTORS AND COMMITMENT TO A CELL LINEAGE

It is important to realize that the number of receptors per cell is usually an estimate of several cell populations. Nevertheless, one can in fact generalize since from receptor studies using other cytokines (with the exception of CSF-1), the number of receptors on normal myeloid precursors is similar to the number on epo precursors, namely less than 1000. This in turn implies that extremely low concentrations of regulator molecules are required in order to stimulate the responsive cells. As mentioned above, very low concentrations are found in the circulation. In addition, the interesting argument can be postulated that commitment to one or other hemopoietic cell lineages may be a matter of concentration of hemopoietic regulator molecules competing for receptors at the stem cell level.

The hypothesis of competition for commitment at the stem cell level is by no means new. Several investigations have implied such a mechanism [52—54] and some clinical studies using cytokines have also indicated that perhaps this mechanism may not be impossible. It remains to be seen whether receptors for the different hemopoietic regulator molecules are present on stem cell populations. For this, pure stem cell populations are required with steps in the right direction being made by several groups [55, 56].

THE SITE OF ACTION OF ERYTHROPOIETIN

The BFU-E has been reported to be sensitive to GM-CSF and IL-3 [57], while the CFU-E population is responsive to insulin and/or insulin-like growth factors [58, 59]. However, it was thought that erythropoiesis was regulated by two factors namely, epo and a factor originally called Burst Promoting Activity or BPA [60, 61]. The functional activity of BPA was shown to be indistinguishable from pure native IL-3 [62], and although the BPA gene has never been sequenced and cloned, we now assume that BPA is identical to IL-3. There were two main reasons for the "two-factor" hypothesis for the regulation of erythropoiesis. The first was that the BFU-E did not appear to be significantly affected by severe perturbation of the erythropoietic system [60]. The second was that the BFU-E required unphysiologically high concentrations of epo to be stimulated to form colonies [60]. As I have described previously [63], both of these characteristics are probably artifactual rather than real, since they are dependent on the sensitivity of the *in vitro* assay system.

I have argued that it is not the maximum or optimal epo concentration that is important for BFU-E stimulation, but the lowest epo concentration, since only 30% of the BFU-E are in the S-phase of the cell cycle under normal conditions [64]. Indeed, when BFU-E are cultured under physiologically relevant oxygen tensions, stimulation to form erythroid bursts occurs with epo concentrations as low as 12.5 mU/ml, which is in the same order of magnitude as the circulating epo concentration [13, unpublished results]. Furthermore, one may predict that if the BFU-E population bound epo, then this would occur in approximately 20—30%. This is because about 30% of the BFU-E are in cycle.

In short, there is good reason to believe that all BFU-E populations are indeed responsive to circulating concentrations of epo and not just the CFU-E. An inherent difference in epo sensitivity is advantageous, since normal concentrations would only allow the influx of a small (normal) proportion of BFU-E into the CFU-E and later compartments. At lower than normal circulating epo levels, the BFU-E would not be sensitive enough to feed into the CFU-E compartment, while at higher than normal levels, the reverse would be true. The resulting physiological manifestations are apparent. In fact, whereas the lowest stimulating concentration of epo for the early BFU-E is 12.5 mU/ml, that for the CFU-E population is more than one log lower (< 1 mU/ml) [65]. It would appear that epo is required throughout erythropoiesis, since if removed, the cells die. In other words, epo is required for both proliferation and maturity.

Whether epo is also a differentiation factor in terms of it being a requirement at the stage of determination into the erythroid lineage, still remains to be investigated.

THE SITE OF ERYTHROPOIETIN PRODUCTION

A. The Renal Source

With the use of molecular biology techniques, some facets of epo production have become clearer while others have provided more controversy. Let us first consider the kidney as an epo production site, if for no other reason than historical.

There is now absolutely no doubt that the kidney produces epo under hypoxic or severe anemic conditions. Difficult, however, is understanding its function under normal steady-state conditions. Here, even with molecular biology techniques, the results are by no means categorical. Using either poly A^+ mRNA in Northern blot analysis [65, 66] or RNase protection assays [67] have produced a signal which, according to the authors, is at the limit of detection. In a recent publication by Koury and associates, the number of cells expressing the epo gene per square centimeter of kidney cortex was determined using in situ hybridization and mRNA by Northern transfer by densitometry [66]. Groups of control mice had hematrocrits of 50% and 49% with serum epo concentrations of 26 and 63 mU/ml respectively. Unfortunately, in these two groups of animals, the renal epo mRNA was not determined, although a determination by radioactive in situ hybridization was performed producing about 5 epo-producing cells per cm^2. These inconsistent results do not help in deciding whether the normal kidney is producing epo or not. Similar problems may be faced if the polymerase chain reaction or PCR is used. With this extremely sensitive (perhaps too sensitive) technique, epo gene expression may be detected. The question then is whether the results correlate with the circulating concentration of epo, a particularly difficult question to answer since a quantitation by PCR is not possible. We have tried to follow a slightly different line by using non-isotopic in situ hybridization coupled with immunohistochemistry of intracellular epo [41, unpublished results]. In these experiments, neither a positive in situ hybridization nor a positive immunohistochemical reaction for the detection of intracellular epo in normal mouse kidney was observed. Therefore, we are still not certain as to whether the kidney can produce epo under normal steady-state conditions.

In contrast to the normal situation, epo has been detected by Northern transfer, RNase protection assay and in situ hybridization under anemic conditions [65, 66, 68—77]. It appears that epo mRNA can be detected in the kidney when the hematrocrit is between 30 to 35% [66]. There is also agreement that when epo production is observed, it occurs in the renal cortex [71 73, 75, unpublished results]. However, the cell type involved now causes problems. There are two schools of thought. One implicates the renal cortical interstitial or peritubular cell [71—73], while the other favors the tubular cell [76, 77, unpublished results]. The latter have used both in situ hybridization and immunohistochemistry using antibodies to epo and detect epo in the tubular cells. Unfortunately, the former school have not employed immunohisto-chemistry to detect intracellular epo. It could be argued that tubular cells may take up epo from the tubular lumen thereby producing a false-positive result. If this were the case, then no intracellular epo should be obtained under normal conditions, which is, in fact, the case (unpublished results). Therefore, this controversy remains to be clarified.

B. The Extrarenal Source

An interesting situation occurs in anephric animals or patients with end-stage renal failure. Although an anemia is present, the circulating concentration of epo does not correspond to the degree of anemia. That epo is detectable [78] indicates that an extrarenal source is present. This extrarenal source, at least in mice, has been considered to be the liver [79]. Here again, however, contradictory results have been obtained using Northern transfer analysis [65, 68]. Furthermore, the amount of epo that should be produced to alleviate the anemia is not generated.

The same situation is applicable to the bone marrow macrophage. The macrophage is an extremely interesting cell from many viewpoints. That the macrophage can release epo was first shown in 1980 [80], and the first in situ hybridization experiments to detect epo were described on cultured macrophages at the NATO workshop in 1986 [81]. Since that time, this work has been taken several steps forward by demonstrating not only that cultured bone marrow-derived resident macrophages express the epo gene and contain intracellular epo [75], but that the same is true for resident mouse bone marrow macrophages [75, 82]. Indeed, under normal steady-state conditions, about 0.2% of the total bone marrow macrophage population can simultaneously express the gene and contains intracellular immunoreactive epo [41, 82]. This proportion increases up to 5 fold when the hematocrit is reduced to about 17% [82], indicating, as in the kidney [66, unpublished results], that cells are recruited to produce more epo (unpublished results).

Of particular interest is the "erythropoietic functional unit" which can also be called the blood island. The latter consists of a central macrophage surrounded by developing erythroblasts [82]. The epo gene expression that we found in the normal mouse bone marrow was always seen in blood islands. We demonstrated that the central cell expressing the epo gene was also a macrophage by the presence of the mouse, macrophage-specific F4/80 antigen [75]. We have since shown that the central macrophage not only expresses the epo gene, but contains intracellular epo [82]. These results provide the first evidence for the blood island acting as a active microenvironmental niche for erythropoiesis to occur.

Since the bone marrow macrophage is also an extrarenal source of epo, one would assume that a compensatory effect by extrarenal epo sources would occur to alleviate the anemia observed under anephric or end-stage renal failure conditions. This does not occur. There could be many reasons for this. The presence of unspecific inhibitors is certainly a possibility since they could act by either being toxic to the cells or by destroying the cells. Physical parameters may also be a factor, for example a change in pH to more acidic conditions which may detrimental to the epo producing cells. Another likelihood is that the oxygen sensors in the extrarenal source are set at a different level than in the kidney. This now brings me to the final and most difficult topic of all; oxygen sensing and epo production.

OXYGEN SENSING AND ERYTHROPOIETIN PRODUCTION

There is no doubt that the oxygen sensing mechanism regardless of whether for the regulation of epo production or chemoreception, represents a "black box" (see [83] for review). Little if anything is known about the biological oxygen sensor and how it operates. Because, the oxygen sensing mechanism is coupled with gene expression, RNA synthesis and glycosylated protein synthesis and release, epo production is an ideal system in which to study this phenomenon. However, there are several problems involved.

Oxygen is brought into the vicinity of cells by its binding to hemoglobin. The oxygen is then released and, usually in a soluble form, diffuses to the outlying cells. Although all cells are "oxygen sensitive" with respect to their metabolism, we must assume that the function of specific cells in the body is acutely affected. Such is the case for epo-producing cells, regardless of whether they are macrophages in the bone marrow or in the liver or tubular or interstitial cells of the renal cortex (see [84, 85] for reviews). These cells in some way have the ability to sense and respond to changing oxygen tensions. The macrophage for example is extremely sensitive to changing partial oxygen tensions in the physiological range, that is 2% (15 mmHg) to 5% (40—45 mmHg), the oxygen tensions that occur in wounds and the blood capillaries respectively. The partial oxygen tension in the renal medulla has been measured to be about 10 mmHg [86]. In the tubules of the cortex, the situation is more difficult since the pO_2 changes from one tubule to another and it appears that large portions of the kidney exist on the brink of anoxia, this despite the high arterial and venous pO_2. However, in the cortex a range from 20—40 mmHg has been measured when the arterial pO_2 was 80—100 mmHg with a critical pO_2 for oxygen consumption between 6 and 28 mmHg [87]. In isolated tubules a range of 10—17 mmHg has been found [88]. It is obvious that the kidney is an extremely complex organ with respect to the study of oxygen sensing. One may also expect that because of the heterogeneity of pO_2 in the renal cortex, only specific areas may be involved in epo production. Thus, in contrast to the bone marrow macrophage which probably is subjected to a very small range of oxygen tensions, the renal cortical cells are probably not. Under hypoxic conditions, the situation in the kidney may be even more complicated. It is, therefore, certainly possible that the set points and/or range of responsiveness to pO_2 are different in the macrophage and kidney.

The oxygen sensing range is one side of the coin. The oxygen sensing mechanism is another. Several hypotheses have been put forward and range from changing H+ on concentrations, (pH), Ca^{2+} variations and enzyme oxygen affinities to the most recent involving heme proteins [89] and possible negative-acting DNA binding proteins [90, 91]. These mechanisms will be dealt with extensively in a future review on oxygen sensing. Suffice is to say that none satisfy the cellular, biochemical and molecular conditions necessary. A major problem is to discover the first important interaction. In this respect, a heme protein would satisfy many of the conditions. Indeed, one could even postulate that the heme protein itself could bind to specific sequences of the DNA and regulate transcription of the epo gene. Unfortunately, such a heme protein has not yet been found. Furthermore, such a molecular mechanism should not simply involve an on-off switch as appears to occur in the kidney. It would also have to involve steps or gradations of regulation to account for the fine control observed under normal steady-state conditions. A more appetizing hypothesis would be changing mRNA stability. Indeed, both the rate of transcription and stability of mRNA appear to be involved [92]. There are, therefore, indications that oxygen sensing may have an important molecular component.

QUESTIONS FOR THE FUTURE

I think it is quite clear that since the NATO Workshop in 1986, many of the topics discussed then have either become reality or are being investigated. There is still a great deal to learn. The promoter region of the epo gene is still an enigma, not to mention the properties of the 5' and 3' flanking regions. Knowledge of the 3-dimesional structure of epo should help us understand how it binds to its receptor. But we still want to know the processes that occur thereafter; we want to know how epo stimulates erythropoiesis at the molecular level, and we want to be able to identify the first epo sensitive cell. Another open question is how does the role of the microenvironment affect commitment of the hemopoietic stem cell into the erythropoietic cell lineage? The organisation of erythropoiesis during the early stages of differentiation must have its equal *in vivo,* not only *in vitro,* and techniques must be developed to detect these populations. Answers to these and many more questions will have considerable repercussions on how we understand the problems of biological differentiation and versatility.

ACKNOWLEDGEMENTS

This work was supported by the Deutsche Forschungsgemeinschaft, the Sonderforschungsbereich 322 and the German Red Cross.

REFERENCES

1) Rich, I.N. (1987): Molecular and Cellular Aspects of Erythropoietin and Erythropoiesis. NATO ASI Series, Vol. H8, Springer-Verlag, Heidelberg.

2) Hodgson, G.S., Bradley, T.R. and Radley, J.M. (1982): The organization of hemopoietic tissue as inferred from the effects of 5-fluorouracil. Exp. Hemat. 10: 26.

3) Johnson, G.R. and Metcalf, D. (1981): Pure and mixed erythroid colony formation in vitro stimulated by spleen conditioned medium with no detectable erythropoietin. Proc. Natl. Acad. Sci. USA 74: 3879.

4) Fausner, A.A. and Messner, H.A. (1979): Identification of megakaryocytes, macrophages and eosinophils in colonies of human bone marrow containing neutrophilic granulocytes and erythroblasts. BLOOD 53: 1023.

5) Till, J.E. and McCulloch, E.A. (1961): A direct measurement of the radiation sensitivity of normal bone marrow cells. Radiation Res. 14: 213.

6) Axelrad, A.A., McLeod, D.L., Shreeve, M.M .and Heath, D.S. (1973): Poperties of cells that produce erythrocytic colonies in vitro. In: Hemopoiesis in Culture, Robinson, W. (ed.) U.S. Gorvernment Printing Office, Washington D.C., p. 226.

7) Iscove, N.N. and Sieber, F. (1975): Erythroid progenitors in mouse bone marrow detected by macroscopic colony formation in culture. Exp. Hemat. 3: 32.

8) Stephenson, J.R., Axelrad, A.A., McLeod, D.L. and Shreeve, M.M. (1971): Induction of colonies of hemoglobin-synthesizing cells by erythropoietin in vitro. Proc. Natl. Acad. Sci. USA 44: 517.

9) Iscove, N.N., Sieber, F. and Winterhalter, K.H. (1974): Erythroid colony formation in cultures of mouse and human bone marrow; analysis of the requirement for erythropoietin by gel filtration and affinity chromatography on agarose-concanavalin A. J. Cell Physiol. 83: 309.

10) Rich, I.N. (1976): Erythroid colony formation (CFUe) in fetal liver and adult bone marrow and spleen from the mouse. Blut 33: 171.

11) Monette, F.C., Ouellette, P.L. and Faletra, P.P. (1981): Characterization of murine erythroid progenitors with high erythropoietin sensitivity in vitro. Exp. Hemat. 9: 249.

12) Rich, I.N. and Kubanek, B. (1982): The effect of reduced oxygen tension on colony formation of erythropoietic cells in vitro. Brit. J. Haemat. 52: 579.

13) Rich, I.N. (1986): A role for the macrophage in normal hemopoiesis. II. Effect of varying physiological oxygen tensions on the release of hemopoietic growth factors from bone-marrow-derived macrophages in vitro. Exp. Hemat. 14: 746.

14) Iscove, N.N., Guilbert, L.J. and Weyman, C. (1980): Complete replacement of serum in primary cultures of erythropoietin-dependent red cell precursors (CFU-E) by albumin, transferrin, iron, unsaturated fatty acid, lecithin and cholesterol. Exp. Cell Res. 126: 121.

15) Migliaccio, A.R., Bruno, M. and Migliaccio, G. (1987): Evidence for direct action of human biosynthetic (recombinant) GM-CSF on erythroid progenitors in serum-free culture. Blood 70: 1867.

16) Jacobs, K., Shoemaker, C., Rudersdorf, R., Neill, E.F., Kaufman, R.J., Mufson, A., Seehra, J., Jones, S.S., Hewick, R., Fritsch, E.F., Kawakita, M., Shimaza, T. and Miyake, T. (1985): Isolation and characterization of genomic and cDNA clones of human erythropoietin. Nature 313: 806.

17) Lin, F.K., Suggs, S., Lin, C.H., Browne, J.K., Smalling, R., Egrie, J.C., Chen, K.K., Fox, G.M., Martin, F., Stabinsky, Z., Badrawi, S.M., Lai, P.H. and Goldwasser, E. (1985): Cloning and expression of the human erythropoietin gene. Proc. Natl. Acad. Sci. USA 82: 7580.

18) Egrie, J.C., Strickland, T.W., Lane, J., Aoki, K., Cohen, A.M., Smalling, R., Trail, G., Lin, F.K., Browne, J.K. and Hinds, D.K. (1986): Characterization and biological effects of recombinant human erythropoietin. Immunobiology 172: 213.

19) Sasaki, H., Ochi, N., Dell, A. and Fukuda, M. (1988): Site-specific glycosylation of human recombinant erythropoietin: Analysis of glycopeptides or peptides at each glycosylation site by fast atom bombardment mass spectrometry. Biochem. 27: 8618.

20) Lappin, T.R.J. and Maxwell, A.P. (1989): Chemistry and assays of erythropoietin. In: Erythropoietin, Jelkmann, W. and Gross, A.J. (eds.). Springer-Verlag, Heidelberg, p. 7.

21) Recny, M.A., Scoble, H.A. and Kim, Y. (1987): Structural characterization of natural human urinary and recombinant DNA-derived erythropoietin. J. Biol. Chem. 262: 17156.

22) Lacombe, C., Tambourin, P., Mattei, M.G., Simon, D. and Guenet, J.L. (1989): The murine erythropoietin gene is localized on chromosome 5. Blood 72: 1440.

23) Law, M.L., Cai, G.Y., Lin, F.K., Wei, Q., Huang, S.Z., Hartz, J.H., Morse, H., Lin, C.H., Jones, C. and Kao, F.T. (1986): Chromosomal assaigment of the human erythropoietin gene and its DNA polymorphism. Proc. Natl. Acad. Sci. USA 83: 6920.

24) McDonald, J., Lin, F.-K. and Goldwasser, E. (1986): Cloning, sequencing, and evolutionary analysis of the mouse erythropoietin gene. Mol. Cell Biol. 6: 842.

25) Shoemaker, C.B. and Mitsock, L.D. (1986): Murine erythropoietin gene: Cloning, expression, and human gene homology. Mol. Cell Bio. 6: 849.

26) Goldwasser, E., Kung, C.K.-H. and Eliason, J.F. (1974): On the mechanism of erythropoietin-induced differentiation. XIII. The role of sialic acid in erythropoietin action. J. Biol. Chem. 249: 4202.

27) Fukuda, M.N., Sasaki, H., Lopez, L. and Fukuda, M. (1989): Survival of recombinant erythropoietin in the circulation: The role of carbohydrates. Blood 73: 84.

28) Spivak, J.L. and Hogans, B.B. (1989): The in vivo metabolism of recombinant human erythropoietin in the rat. Blood 73: 90.

29) Carnot, P. and Deflandre, C. (1906): Sur l'activité hémapoiétique des différents organes au cours de la régénération du sang. Compt. Rend. Acad. Sci. 143: 432.

30) Miyake, T., Kung, C.K.-H. and Goldwasser, E. (1979): Purification of human erythropoietin. J. Biol. Chem. 252: 5558.

31) Eschbach, J.W., Mladenovic, J., Garcia, J.F., Wahl, P.W. and Adamson, J.W. (1985): The anemia of chronic renal failure in sheep. Response to erythropoietin-rich plasma in vivo. J. Clin. Invest. 74: 434.

32) Eschbach, J.W., Egrie, J.C., Downing, M.R. and Adamson, J.W. (1987): Correction of the anemia of end-stage renal disease with recombinant human erythropoietin. New Engl. J. Med. 316: 73.

33) Winearls, J.A., Oliver, D.O., Pippard, M.J., Reid, C., Downing, M.R. and Cotes, P.M. (1986): Effect of human erythropoietin derived from recombinant DNA on the anemia of patients maintained by chronic haemodialysis. Lancet 8517: 1175.

34) Graber, S.E. and Krantz, S.B. (1989): Erythropoietin: Biology and clinical use. In: Hematology/Oncology Clinics of North America. Hematopoietic Growth Factors, Golde, D.W. (ed.). W.B. Saunders Company, Philadelphia, USA, p. 369.

35) Growing pains for Amgen as epoetin wins US approval. Nature (1989) 339: 493.

36) Court battle ends at the start. Nature (1989) 342: 846.

37) Annable, L., Cotes, P.M. and Mussett, M.V. (1972): The second international reference preparation of erythropoietin, human, urinary, for bioassay. Bull Wld. Hlth. Org. 47: 99.

38) Egrie, J.C .Cotes, P.M. Lane, J. Gaines Das, R.E., Tam, R.C. (1987): Development of radioimmunoassays for human erythropoietin using recombinant erythropoietin. J. Immunol. Meth. 99: 235.

39) Wognum, A.W., Lansdorp, P.M., Eaves, A.C. and Krystal, G. (1989): An enzyme-linked immunosorbant assay for erythropoietin using monoclonal antibodies, tetrameric immune complexes, and substrate amplification. Blood 74: 622.

40) Sawatzki, G. (1987): Isolation of anti-erythropoietin by automated immuno-affinity FPLC using recombinant EPO. In: Molecular and Cellular Aspects of Erythropoietin and Erythropoiesis. NATO ASI Series, Vol. H8, Rich, . I.N. (ed.). Springer-Verlag, Heidelberg, p. 409.

41) Vogt, Ch., Noe, G. and Rich, I.N. (1990): Normal steady-state hemopoiesis. Assay of erythropoietin by ELISA, localization of erythropoietin in the kidney and expression of hemopoietic regulator molecules in the bone marrow. In: In: Proceeding of the 5th Annual Symposium on Molecular Biology of Hematopoiesis, Konwalinka, G. and Abraham, N.G. (eds.). Plenum Press (in Press).

42) Rich, I.N. and Noé, G. (1989): A sensitive sandwich ELISA for erythropoietin using highly purified polyclonal anti-erythropoietin IgG. Blood 74 Suppl. 1: 331a (Abstract).

43) D'Andrae, A.D., Lodish, H.F. and Wong, G.G. (1989): Expression cloning of the murine erythropoietin receptor. Cell 57: 277.

44) Sawyer, S.T., Krantz, S.B. and Goldwasser, E. (1987): Binding and receptor-mediated endocytosis of erythropoietin in Friend Virus-infected erythroid cells. J. Biol. Chem. 262: 5554.

45) Sasaki, R., Yanagawa, S.I., Hitomi, K. and Chiba, H. (1987): Characterization of erythropoietin receptor of murine erythroid cells. Eur. J. Biochem. 168: 43.

46) Fukamachi, H., Tojo, A., Saito, T., Kitamura, T., Nakata, M., Urabe, A. and Takaku, F. (1987): Internalization of radioiodinated erythropoietin and the ligand-induced modulation of its receptor in murine erythroleukemia cells. Intl. J. Cell Cloning 5: 209.

47) Landschultz, K. and Boyer, S. (1988): Natural history of erythropoietin (EPO)-binding during erythropoiesis. Blood 72: 92a.

48) Miller, B.A., Scaduto, R.C., Tillotson, D.L., Botti, J.J. and Cheung, J.V. (1988): Erythropoietin stimulates a rise in intracellular free calcium concentration in single early human erythroid precursors. J. Clin. Invest. 82: 309.

49) Misti, J. and Spivak, J.L. (1979): Erythropoiesis in vitro: Role of calcium. J. Clin. Invest. 64: 1573.

50) Sawyer, S.T. and Krantz, S.B. (1984): Erythropoietin stimulates $^{45}Ca^{2+}$ uptake in friend virus-infected erythroid cells. J. Biol. Chem. 259: 2769.

51) Thompson, L.P., Sawyer, S.T. and Blackmore, P.F. (1988): A search for the second messenger of erythropoietin. FASEB J 2: A813.

52) Hellman, S., Grate, H.E., Chaffey, J.T. and Carmel, R. (1970): Hematopoietic stem cell compartment: patterns of differentiation following radiation or cyclophosphamide. In: Hemopoietic Cellular Proliferation, Stohlman, F. Jr. (ed.). Grune & Stratton, New York.

53) Kubanek, B., Bock, O., Heit, W., Bock, E. and Harriss, E.B. (1973): Size and proliferation of stem cell compartments in mice after depression of erythropoiesis. In: Haemopoietic Stem Cells, (eds.). Ciba Foundation Symposium. Elsevier., Amsterdam, p. 243.

54) Van Zant, G., Goldwasser, E. and Baron, J.M. (1976): Study of the haemopoietic microenvironment in vitro. Nature 260: 609.

55) Spangrude, G.J., Heimfeld, S. and Weissman, I.L. (1988): Purification and characterization of mouse hematopoietic stem cells. Science 241: 58.

56) Ploemacher, R.E. and Brons, R.H.C. (1989): Separation of CFU-S from primitive cells responsible for reconstitution of the bone marrow hemopoietic stem cell compartment following irradiation: evidence for a pre-CFU-S cell. Exp. Hemat. 17: 263.

57) Sonoda, Y., Yang, Y.C., Wong, G.G. and et al. (1988): Analysis in serum-free culture of the targets of recombinant human hemopoietic growth factors: Interleukin 3 and granulocyte/macrophage-colony-stimulating factor are specific for early developmental stages. Proc. Natl. Acad. Sci. USA 85: 4360.

58) Kurtz, A., Haertl, W., Jelkmann, W., Zapf, J. and Bauer, C. (1984): Activity in fetal bovine serum that stimulates erythroid colony formation in fetal mouse livers is insulinlike growth factor 1. J. Clin. Invest. 76: 1643.

59) Sawada, K., Krantz, S.B. and Dessypris, E.N. (1989): Human colony-forming units—erythroid do not require accessory cells, but do require direct interaction with insulin-like growth factor 1 and/or insulin for erythroid development. J. Clin. Invest. 83: 1701.

60) Iscove, N.N. (1977): The role of erythropoietin in regulation of population size and cell cycling of early and late erythroid precursors in mouse bone marrow. Cell Tissue Kinet 10: 323.

61) Iscove, N.N., Roitsch, C.A., Williams, N. and Guilbert, L.J. (1982): Molecules stimulating early red cell, granulocyte, macrophage, and megakaryocyte precursors in culture: Similarity in size, hydrophobicity, and charge. J. Cell Physiol. Suppl. 1: 65.

62) Goldwasser, E., Ihle, J.N., Prystowsky, M.B., Rich, I.N. and Van Zant, G. (1983): The effect of interleukin-3 on hemopoietic precursor cells. In: Normal nad Neoplastic Hematopoiesis, (eds.) Alan R. Liss, Inc., New York, p. 301.

63) Rich, I.N. (1984): Haemopoietic regulation in vitro: In vivo significance of functionally similar multiactive potentiating factors. Ann. Immunol. 135C: 280.

64) Rich, I.N. (1988): The macrophage as a production site for hematopoietic regulator molecules: Sensing and responding to normal and pathophysiological signals. Anticancer Res. 8: 1015.

65) Bondurant, M.C., Koury, M.J. (1986): Anemia induces accumulation of erythropoietin mRNA in the kidney and liver. Mol. Cell Bio. 6: 2731.

66) Koury, S.T., Koury, M.J., Bondurant, M.C., Caro, J. and Graber, S.E. (1989): Quantitation of erythropoietin-producing cells in kidneys of mice by in situ hybridization: Correlation with hematocrit, renal erythropoietin mRNA, and serum erythropoietin concentration. Blood 74: 645.

67) Koury, M.J., Bondurant, M.C., Graber, S.E. and Sawyer, S.T. (1988): Erythropoietin messenger RNA levels in developing mice and transfer of 125I-erythropoietin by the placenta. J. Clin. Invest. 82: 154.

68) Goldwasser, E., McDonald, J. and Beru, N. (1987): The molecular biology of erythropoietin and the expression of its gene. In: Molecular and Cellular Aspects of Erythropoietin and Erythropoiesis, Rich, I.N. (ed.). Springer-Verlag, Heidelberg, p. 11.

69) Caro, J., , Schuster, S., Besarab, A. and Erslev, A.J. (1987): Renal bio-genesis of erythropoietin. In: Molecular and Cellular Aspects of Erythropoietin and Erythropoiesis, Rich, I.N. (ed.). Springer-Verlag, Heidelberg, p. 329.

70) Schuster, S.J., Badiavas, E.V., Costa-Giomi, P., Weinmann, R., Erslev, A.J. and Caro, J. (1989): Stimulation of erythropoietin gene transcription during hypoxia and cobalt exposure. Blood 73: 13.

71) Koury, S.T., Bondurant, M.C. and Koury, M.J. (1988): Localization of erythropoietin synthesizing cells in murine kidneys by in situ hybridizati. Blood 71: 524.

72) Bruneval, P., Da Silva, J.L., Lacombe, C., Salzmann, J.L., Tambourin, P., Varet, B., Camilleri, J.P. and Bariety, J. (1989): Erythropoietin synthesis in the anemic mouse kidney as observed by morphological techniques. In: Erythropoietin, Jelkmann, W. and Gross, A.J. (eds). Springer-Verlag, Heidelberg, p. 26.

73) Lacombe, C., Da Silva, J.-L., Bruneval, P., Fournier, J.-G., Wendling, F., Casadevall, N., Camilleri, J.-P., Bariety, J., Varet, B. and Tambourin, P. (1988): Peritubular cells are the site of erythropoietin synthesis in the murine hypoxic kidney. J. Clin. Invest. 81: 620.

74) Schuster, S.J., Wilson, J.H., Erslev, A.J. and Caro, J. (1987): Physiologic regulation and tissue localization of renal erythropoietin messenger RNA. Blood 70: 316.

75) Vogt, Ch., Pentz, S. and Rich, I.N. (1989): A role for the macrophage in normal hemaopoiesis. III. In vitro and in vivo erythropoietin gene expression in macrophages detected by in situ hybridization. Exp. Hemat. 17: 391.

76) Maxwell, A.P., Lappin, T.R.J., Johnston, C.F. and Bridges, J.M. (1989): Erythropoietin production by kidney tubular cells — immunohisto-chemical and in situ hybridization studies. Exp. Hemat. 17: 223. (Abstract)

77) Maxwell, A.P., Lappin, T.R.J., Johnston, C.F., Bridges, J.M. and McGeown, M.G. (1990): Erythropoietin production in kidney tubular cells. Brit. J. Haemat. (in Press).

78) Nathan, D.G., Schupak, E. and Stohlman, F. (1964): Erythropoiesis in anephric man. J. Clin. Invest. 43: 2158.

79) Fisher, J.W. (1979): Extrarenal erythropoietin production. J. Lab. Clin. Med. 93: 695.

80) Rich, I.N., Heit, W. and Kubanek, B. (1980): An erythropoietic stimulating factor similar to erythropoietin released by macrophages after silica treatment. Blut 40: 297.

81) Rich, I.N. (1987): Erythropoietin production by macrophages: Cellular response to physiological oxygen tensions and detection of erythropoietin gene expression by in situ hybridization. In: Molecular and Cellular Aspects of Erythropoietin and Erythropoiesis. NATO ASI Series, Vol. H8, Rich, I.N. (ed.). Springer-Verlag, Berlin Heidelberg, p. 291.

82) Vogt, Ch., Noé, G. and Rich, I.N. (1990): The role of the blood island during normal and 5-fluorouracil-perturbated hemopoiesis. Blood Cells (in Press).

84) Acker, H. (1989): PO_2 chemoreception in arterial chemoreceptors. Ann. Rev. Physiol. 51: 835.

85) Rich, I.N. (1988): Oxygen tension and erythropoietin production: The role of the macrophage in regulating erythropoiesis. In: Oxygen Sensing in Tissues, Acker, H. (ed.). Springer-Verlag, Berlin Heidelberg, p. 113.

86) Bauer, C. (1988): Metabolic events that may activate erythropoietin production in the hypoxic kidney. In: Oxygen Sensing in Tissues, Acker, H. (ed.). Springer-Verlag, Heidelberg, p. 93.

87) Epstein, F.H. (1985): Hypoxia of the renal medulla. Quarterly J. of Med. 224: 807.

88) Jones, D.P. (1986): Renal metabolism during normoxia, hypoxia, and ischemic injury. Ann. Rev. Physiol. 48: 33.

89) Balaban, R.S., Soltoff, S.P., Storey, J.M. and Mandel, L.J. (1980): Improved renal cortical tubule suspension: Spectrophotometric study of O2 delivery. Am. J. Physiol. 241: F50.

90) Goldberg, M.A., Dunning, S.P. and Bunn, H.F. (1988): Regulation of the erythropoietin gene: Evidence that the oxygen sensor is a heme protein. Science 242: 1412.

91) Beru, N., Smith, D.H. and Goldwasser, E. (1989): Evidence for a negative factor regulating erythropoietin gene expression. Leukemia Res. 13, Suppl. 1: 3 (Abstract).

92) Beru, N., Smith, D. and Goldwasser, E. (1989): Evidence for a negative regulator of the erythropoietin gene. Blood 74, Suppl 1: 192a (Abstract).

93) Goldberg, M.A., Gaut, C.C. and Bunn, H.F. (1989): Erythropoietin mRNA levels are regulated by both transcriptional events and by changes in RNA stability. Blood 74, Suppl 1: 191a (Abstract).

CELLULAR AND MOLECULAR BASIS OF SYNAPTIC TRANSMISSION

Herbert Zimmermann

AK Neurochemie,
Zoologisches Institut der J.W. Goethe-Universität,
Siesmayerstraße 70, D-6000 Frankfurt am Main,
Federal Republic of Germany

This commentary is based on the proceedings of a NATO workshop held in Göttingen from September 9—13, 1987 on the cellular and molecular basis of synaptic transmission. The reader is mainly referred to the literature quoted in the articles. Due to the limited space available complementary literature and literature which appeared since the meeting can only be cited to a very limited extent. While knowledge regarding both presynaptic and postsynaptic signalling mechanisms and their regulative control is evolving rapidly the article aims to highlight recent developments as well as open questions and future perspectives. Consensus mechanisms are emphasized but the structural, functional and also molecular diversity of the various transmission systems is not ignored.

EVOLUTION OF NERVOUS SYSTEMS AND OF NEURONAL SIGNALLING

The most sophisticated form of chemical signalling

Synaptic transmission can be regarded as the most advanced and specialized form of cell to cell signalling. The development of a nervous system in animal evolution (from Coelenterates onwards) represents a tremendous gain in individual fitness. It is not surprising that there was a high selection pressure on further improving its efficiency: This did involve not only an increase in the number of nerve cells and in the complexity of neuronal interconnections. It had a great influence also on the efficiency of the communication mechanisms between nerve cells and on the diversity of mechanisms involved.

Chemical synaptic transmission refers to the direct communication between nerve cells or from a nerve cell to an effector cell by means of chemical messengers. However, it is easily overlooked that this general definition categorizes a wide variety of communication mechanisms which all represent adaptions to the various functional needs of neuronal information processing. Chemical signalling may be very rapid and focal and involved in the direct and fast communication from and to the periphery or between specified nerve cells in a given network. On the other hand, synaptic transmission may be

NATO ASI Series, Vol. H 51
Cell to Cell Signals in Plants and Animals
Edited by V. Neuhoff and J. Friend

less focal or even non-focal, not directed to only a single and structurally defined target. Its function may rather be in the tuning of other signalling systems than in immediate signal transfer. In this case synaptic transmission and hormonal regulation of neuronal function can no longer clearly be distinguished.

The evolution of chemical transmission produced a very wide variety of different chemical messengers. Whereas it was thought until recently that a minimal degree of simplicity was guaranteed by the "one neuron — one transmitter" doctrin, the observation of the simultaneous presence of several chemical messengers (in varying combinations) in a single neuron and their release has rendered our understanding of the physiology of synaptic transmission even more difficult.

Diversity and consensus

Thus, the study of synaptic transmission will focus on the one hand on the analysis of the diverse transmission systems: The various structural specializations involved in the transmission process, the biochemistry of the individual neurotransmitter substances and of their receptor systems, and their cytoarchitecture in the nervous systems. In addition, the structural and biochemical variability of the synaptic transmission process throughout the animal kingdom is taken into account. On the other hand, from a cell biological point of view, the general and unifying mechanisms common to all or at least a large proportion of transmitting systems will come into focus. This involves e.g. general mechanisms of regulation of neurotransmitter synthesis and release and of their inactivation or of transmitter receptor interactions, and receptor mediated intracellular signalling. Parallels can also be drawn to non-synaptic secretory and signalling systems, and their receptor actions.

STRUCTURAL BASIS OF SYNAPTIC TRANSMISSION

Variations of a theme

Signal transmission by means of chemical messengers generally is restricted to a structurally specialized region referred to as synapse. The structural detail of synapses varies with the chemical and functional heterogeneity. Typically, a synapse is comprised of a presynaptic element which contains and releases the neurotransmitter and a postsynaptic target element equipped with receptor molecules. The presynaptic element is normally represented by a terminal axon ramification whereas — in vertebrates — dendrites, soma and axon of a target neuron can function as postsynaptic element.

The structure of synapses in the peripheral and central nervous system has been investigated in detail using conventional electron-microscopical techniques, immunocytochemistry, and freeze fracture and deep etching techniques (1, 2, 3, 4). In spite of considerable differences in neuronal configuration the principal structural elements of the chemical synapse have been preserved from annelids to mammals. A general structural feature of the presynaptic element is the accumulation of (synaptic) vesicles which contain the messenger molecules. Furthermore, the presynaptic plasma membrane in both the central and peripheral nervous system contains regions spezialized for exocytotic fusion of synaptic vesicles and transmitter release (3, 4). As influx of extracellular Ca^{++}-ions is the major trigger for transmitter release such "release sites" are likely to be closely connected to Ca^{++} channels of the plasma membrane (5). The subsynaptic part of the postsynaptic membrane is rich in transmitter receptors. It may contain structural specializations like intracellularly attached electron-dense material or infoldings enriched in neurotransmitter receptors. In the peripheral nervous system the presynaptically released neurotransmitter has to pass through a basal lamina before it can reach the receptors at the postsynaptic membrane.

In the case of non-focal release an immediate postsynaptic element may be missing. There, neurotransmitter can be released along extensive axon ramifications which carry varicosities and pass through the nervous or target tissue. Such varicosities are rich in synaptic vesicles and carry the presynaptic release apparatus. This type of release is reminiscent of the release of hormones from terminal axon ramifications of neurosecretory neurons (6, 7).

Still many unknowns

There are a variety of synaptic specializations which are not ubiquitous and whose function is not always understood. For example, the subsynaptic cytoplasm at mammalian dendritic spines may contain a spine apparatus, a membrane stack possibly involved in Ca^{++} sequestration. In addition, the subsynaptic cytoplasm of both dendritic spines and of neuromuscular junctions may contain electron-dense bodies or plates connected to filamentous material closely connected to the postsynaptic membrane and the endoplasmic reticulum (2). It may be envisaged that such specializations are involved in the targeting of receptors to their specific locations in the postsynaptic membrane or in receptor turnover.

STORAGE ORGANELLES FOR NEUROTRANSMITTERS AND NEUROHORMONES

Life cycle: from where the vesicle comes and where it goes

Storage organelles for neurotransmitters and neurohormones originate from the trans-Golgi apparatus and are transferred to the axon terminals via fast axonal transport (8). Within the axon terminals or axon varicosities synaptic vesicles are accumulated presumably by specific trapping within the cytonet (9). In the nerve terminal the vesicle may go through several cycles of exo- and endocytosis resulting in vesicular heterogeneity (6, 10, 11, 12). At present we do not understand which cellular signal will eventually guide the synaptic vesicle membrane compartment from the nerve terminal back into the axon and to the perikaryon after it has fulfilled its function. It appears, however, that the vesicle will soon lose its identity on retrograde transport through the axon by melting into a pool of endosomal organelles including multivesicular bodies or tubular membrane structures (8). It is also an open question to what extent synaptic vesicle membranes are being recycled in the pericaryon for immediate reuse and what factors are involved in the regulation of synaptic vesicle formation.

Two or more types of vesicles and the consequences

One can differentiate at least two morphological species of storage organelles: the small electron-lucent synaptic vesicles (ca. 45 nm) and larger vesicles with electron-dense core (ca. 100 nm). Most neurons contain both types of vesicles with the electron-lucent type being in the majority. The small electron-lucent vesicles contain low molecular weight neurotransmitters (13) whereas the electron-dense vesicles (also referred to as granules) contain physiologically active peptides, proteins and possibly also low molecular weight transmitter (12, 14). There are apparent differences in the release characteristics of low and high molecular weight messengers regarding stimulation conditions and Ca^{++}-dependency (15), and also in the cellular site of exocytotic fusion of the two vesicle types (16): The small electron-lucent vesicles make use of morphologically specified release sites at the presynaptic membrane. The larger peptide-loaded granules release their contents distal to the synaptic cleft (parasynaptic release, 16).

A major question is at present that of the homology of vesicle types in the various neuronal and neurosecretory systems. Do all electron-dense vesicles (granules) in the various types of neurons belong to one category and the electron-lucent ones to another? Can the membrane of electron-dense vesicles after retrieval following exocytosis form a smaller vesicle (12, 17)? Is this membrane compartment of the recycled and previously large electron-dense vesicle identical with a typical electron-lucent vesicle or does it represent yet another vesicle compartment?

It is possible that the recycled membrane compartment of a granule may constitute a new small synaptic vesicle which becomes reloaded locally with the small molecular weight components only (11, 12, 17). This would also explain the presence of electron-lucent vesicles in e.g. the hormone granule loaded varicosities of neurosecretory neurons. A supportive argument for this possibility would be the ubiquitous presence of membrane proteins between the two vesicle types in a given transmitter system: There are a number of synaptic vesicle membrane proteins which are common to vesicles in both, neurons and endocrine cells suggestive of their wide distribution (18, 19, 20, 21). But the experimental evidence for a common localization of a number of such membrane proteins in both small and large vesicles of the same type of neuron is still debated. Immunocytochemical observations (negative) stand at present against biochemical findings (positive) (12, 18, 22).

Biochemical properties: diversity and homology

In general, the biochemical properties of synaptic vesicles are as diverse as the different transmitter systems. Some store catecholamines like noradrenaline and dopamine, others serotonin, others amino acids, like glutamate, GABA or glycine and yet others acetylcholine and ATP. Amongst peptides stored in synaptic vesicles (electron-dense type) are substance P, neuropeptide Y, enkephalins or vasoactive intestinal polypeptide (VIP). Each type of vesicle has its own set of membrane proteins that mediates the uptake of its specific non-peptide transmitter substance(s) or of the precursor, followed by intravesicular synthesis (only noradrenaline). The mechanisms of vesicular uptake of acetylcholine and catecholamines and recently also of amino acid transmitters have been characterized in detail (23, 24, 25, 27, 28). This also settles the long debate about the vesicular storage (and release) of amino acid transmitters.

In spite of certain diversity, the homology of the transmitter storage organelles in the various transmitter systems is exemplified by a number of common molecular characteristics. A major common constituent of small vesicles and also large granules is an ATPase of the vacuolar type (29) which in all systems investigated is involved in the production of an electrochemical gradient over the vesicle membrane. It provides the driving force for transmitter uptake. A recently described second ATPase which forms a phosphointermediate (P-type) may also be a general constituent of vesicles and granules (30). The vesicle-membrane associated phosphoproteins synapsin I and synapsin II are associated ubiquitously with vesicles of the small, electron-lucent type but are absent from the larger electron-dense granules (19, 31). Quite the contrary, the soluble core proteins of the chromogranin-family (32) have a wide distribution between electron-dense neuronal vesicles and granules of secretory cells (11, 12, 19). But they are missing from small electron-lucent vesicles. Proteoglycans of varying types may be contained in electron-lucent and in electron-dense vesicles (11, 12, 33).

Primary structures and the search for a function

Presently, we observe the elucidation of the primary structure of an increasing number of synaptic vesicle proteins (19, 20, 21) but their functional characterization is only slowly evolving. Often of proteins whose function is well established protein biochemical information is very partial. Among the best characterized are synapsin Ia and Ib and synapsin IIa and IIb which comprise a family of four homologous synaptic vesicle coating proteins and whose messenger RNAs are generated by differential splicing of transscripts from two genes (31). Both types of synapsins are phosphorylated on excitation of the nerve terminal. They serve as substrate for Ca^{++}, calmodulin-dependent as well as for cyclic adenosine monophosphate-dependent protein kinases. It has been suggested that synapsin I can link synaptic vesicles to cytoskeletal elements like actin, tubulin and spectrin. Thus, synapsin I may be the central molecular mediator which fixes the synaptic vesicle in the nerve terminal cytonet (9). As phosphorylation decreases the binding affinity of synapsin I to synaptic vesicles, synaptic activation may uncoat synapsin I from vesicles and thus liberate vesicles from the cytonet for exocytosis (34). An example of a well characterized hydrophobic integral membrane protein is synaptophysin (18, 35, 36). This Ca^{++}-binding glycoprotein of apparent Mr 38 kDa is present in nerve terminals and also in neuroendocrine tissues. Six identical subunits can form a rosette-like homooligomeric structure which closely resembles transmembrane channels. From analogies in amino acid sequence and in hydropathy profiles as well as from the structural configuration a relationship to connexins which form the channels of gap junctions has been derived. Indeed, the potential for translocation of ions through synaptophysin reconstituted in planar lipid bilayers has been documented. At present neither the physiological gating mechanism nor the function of the synaptophysin channel are known. There is also physiological evidence for ion channels (37) in synaptic vesicles. Another speculation would be that synaptophysin serves in the coupling of the synaptic vesicle to a similar channel protein of the presynaptic membrane during the onset of vesicle exocytosis.

THE CONCEPT OF COTRANSMISSION

The neuron speaks with many tongues

The concept of cotransmission was developed along various lines: More than one physiologically active substance was found in a single neuron in histochemical investigations and to be released when a nerve of "defined" transmitter type was activated. Isolated synaptic vesicles of known transmitter type contained in their lumen more messenger substances than a single "classical" neurotransmitter.

Cholinergic synaptic vesicles of the electron-lucent type costore high concentrations of ATP and Ca^{++}-ions in addition to acetylcholine (38). Large cholinergic dense-cored vesicles may co-store acetylcholine and the peptides VIP, calcitonin gene-related peptide (CGRP) or galanin (15), and the large dense-cored vesicles of adrenergic nerves may costore noradrenaline and ATP and also peptides like enkephalin and neuropeptide Y (NPY), and furthermore proteins of the chromogranin family (11, 12, 14, 32). Substance P can be costored with serotonin, and acetylcholine and GABA and possibly also glutamate may be released from the same nerve cell. The isolation of subpopulations of synaptosomes of defined transmitter type provides further insight into the multiple possibilities of costorage and corelease of individual transmitters (39). Support for the costorge and corelease of "classical" transmitters comes also from the demonstration of the simultaneous presence of the carriers for the uptake of e.g. GABA and choline on the same synaptosomes (40). It has to be noted that it is impossible by biochemical means to demonstrate that two or more substances really reside in one and the same vesicle (rather than in different vesicles contained in the same sample). But immunocytochemical results at the electron-microscopical level are now also supportive.

How can it all be understood?

This has complicated our picture of the synaptic transmission process in a fascinating way (41). The exocytotic release of a cocktail of soluble messengers from one single vesicle will result in the formation of a mixed quantum. The targeting of the coreleased messengers can be quite complicated. The messengers may act at the same or at different sites. They may signal back via autoreceptors to their own nerve terminal and thus mutually affect their own release. One may act directly at the postsynaptic membrane and the other one may find its target at further distance and at different cellular elements of the tissue. In the sweat gland, acetylcholine acts to activate the postsynaptic gland cells whereas the coreleased VIP may cause vasodilation and thus an improved blood supply of the activated gland (42). The frequency dependence of the release of the two substances is different: At lower stimulation frequencies mainly acetylcholine is released whereas on increased activation VIP enters the game acting at the blood vessels.

In sympathetic transmission which has been studied in great detail (43) we now know that it is the coreleased ATP which triggers the fast depolarization of target cells whereas the "classical" transmitter noradrenaline and also coreleased NPY may often act without depolarization. There are, however, differences between tissues regarding the relative preponderance in signal transfer. Furthermore, both classical transmitters and cotransmitters may also signal back via autoreceptors to their own nerve terminal and thus modulate their own release. Coreleased ATP may even modulate neurotransmitter release by phosphorylation of membrane proteins via surface located ectoprotein kinases (44). A physiological role of ATP in the corelease with acetylcholine is also emerging (38, 45, 46).

Whereas the small molecular weight classical transmitter substances enter the vesicles via specific carrier systems in the nerve terminal, peptides are synthesiszed and packaged into vesicles in the perikaryon and are generally processed by limited proteolysis on their way to the synapse (17). Thus, peptides are presumably not recycled after exocytosis and not reloaded within the nerve terminal into the synaptic vesicle membrane compartment — as is the case for classical transmitter substances. Recycled membranes of peptide (and low molecular weight messenger) storing vesicles may thus be further used exclusively for the reuptake and release of low molecular weight messengers: as might be the case for the small dense-cored vesicles in adrenergic neurons (11, 12, 14).

Differential elimination of the chemical signal

If the phasic character of the synaptic signal is to be maintained the target site has to be cleared from the released messengers. For the classical transmitter substances mechanisms of inactivating the released substance like extracellular hydrolyis (acetylcholine) or reuptake into nerve cells or other cells (catecholamines, amino acids) have been described. Synaptically released peptides can be hydrolyzed and inactivated extracellularly by membrane bound surface located ectopeptidases. Two of these (endopeptidase 24.11 and angiotensin converting enzyme) are capable of hydrolyzing enkephalins and have thus received particular attention. Since there is limited substrate specificity of peptidases, specificity for the hydrolysis of a particular peptide may be achieved by the availability of the enzyme at the target site (47, 48, 49). Primary sequence information on a number of these ectoenzymes (e.g. aminopeptidase 24.11 or the various molecular forms of acetylcholinesterase) has revealed a detailed understanding of their membrane topology and the mechanisms of membrane anchorage (50). Similarly, released ATP can be hydrolyzed extracellularly to adenosine by synaptically located ectophosphohydrolases involving at least an ATPase and an AMPase (5'-nucleotidase) (38, 45, 51) and possibly also an ATP:AMP-phosphotransferase activity (52). The adenosine formed may be taken up again by a specifc high affinity uptake system. At the same time adenosine may act to inhibit transmitter release as a neuromodulatory substance in its own right. The speed of the inactivation process and thus the resulting period of messenger activity may differ for each individual coreleased messenger.

PRESYNAPTIC ION CHANNELS

Neuronal versus presynaptic channels

Regulation of ion flow across cell membranes through water filled pores, or channels is a widespread phenomenon in practically all cells. In nerve cells it is difficult to determine what types of ion channels are exactly localized at the terminal axon compartment specialized in neurotransmitter release. A variety of drugs and toxins interfering with specified channel functions interfere with induced transmitter release. This may serve as a first indication for a presynaptic location of the channels involved. In addition, the use of isolated synaptosomes (13, 53) has proven very helpful in both the localization and identification of presynaptic ion channels.

Multiplicity of types and functions

Neurons share with muscle fibres the voltage-dependent rapidly activating Na^+ channels which mediate the action potential. On the basis of their sensitivity to a variety of drugs and toxins (e.g. veratridine, batrachotoxin, tetrodotoxin and saxitoxin) these Na^+ channels have also been identified in isolated synaptosomes (6, 53). It is therefore likely that the action potential can directly invade and thus depolarize the terminal ramifications of axons.

The transmembrane concentration of Ca^{++} ions is about 10^4 times higher than that for Na^+ or K^+ ions. Whereas flux of Na^+ and K^+ through voltage sensitive ion channels is the prerequisite for axonal conduction of action potentials the influx of Ca^{++} ions through voltage gated Ca^{++} channels at the nerve terminal is the major trigger for transmitter release. As revealed from studies on neurotransmitter release from intact neurons and also from isolated synaptosomes the terminal axon compartment may possess at least two types of voltage gated Ca^{++} channels (5, 6, 54, 55, 56, 57). At present it is not entirely clear whether the nerve terminal calcium channels fall directly into the three main categories of voltage dependent Ca^{++} channels (T-, N-, and L-type) which are of widespread occurance and which are categorized on the basis of physiological and pharmacological properties (e.g. activation properties, single channel conductance, sensitivity to dihydropyridine drugs or Ω-conotoxin, the peptide toxin of the marine snail *Conus geographus* (58, 59). The inhibition of either transmitter release or Ca^{++} influx into isolated synaptosomes by Ω-conotoxin in the absence of an inhibitory effect of dihydropyridines suggests the presence of an N-type like channel in nerve terminals (55). Other studies support the existence in axon terminals also of L-type (dihydropyridine sensitive) or of both N-type and L-type channels (6, 60, 61). Interestingly, the release of peptides and exocytosis of large dense-cored vesicles has been related to the activation of L-type channels whereas release from small vesicles and exocytosis of classical transmitter may be mediated by the N-type channels (6, 15). This is correlated with the preferential activation of these channels by high and low frequency stimulation respectively. Although little is known concerning the topography of the various Ca^{++} channel types it is possible that the N- or L-type channels are selectively clustered at the axon terminals yielding specific release sites and release conditions for the two different vesicle types.

In contrast to Ca^{++} channels which are voltage gated and directly initiate synaptic signal transfer, presynaptic K^+ channels are rather involved in counteracting depolarizing effects and stabilizing cell voltages. In this respect they can effect or regulate e.g. the degree of voltage-dependent Ca^{++}-influx through the axonal plasma membrane. K^+ channels represent the most diverse group of ion channels and representatives of various subclasses have also been described for isolated synaptosomes (43, 53, 56). In ^{86}Rb efflux studies five different K^+ channels have been identified in synaptosomes isolated from rat forebrain. Amongst these are two voltage-gated K^+ channels, a Ca^{++}-activated, and an opiate-activated K^+ channel, as well as a K^+ channel responsible for the resting K^+-conductance. Although it is not clear at present to what extent the different types of K^+ channels are colocalized on individual nerve terminals and how they could be functionally related, their presence points to multiple mechanisms for the modulation of voltage-dependent phenomena at the nerve terminal (62). Thus, cyclic AMP-dependent presynaptic K^+ channels are involved in an elementary form of learning in the sea snail *Aplysia*. Their phosphorylation causes channel blockade and thus prolonged depolarization and opening of Ca^{++} channels. This in turn results in augmented transmitter release and facilitation of behavioural response (63).

A superfamily, which one was the first?

The analysis and comparison of the primary structures of the voltage-activated Na^+ channel, the dihydropyridine-sensitive Ca^{++} channel (L-type) and of an A-current K^+ channel (that mediates a rapidly inactivating K^+ current) from *Drosophila* has revealed illuminating information concerning membrane topology, functional properties and also evolutionary origin (64, 65). The principal subunits of all these voltage-sensitive channels are homologous members of a gene family. The α-subunits of the Na^+ and the Ca^{++} channels form the ion conducting channels and contain both four repeated units of strong sequence homology. Each of the repeats consists of six hydrophobic (α-helical) and membrane spanning segments. While the *Drosophila* K^+ channel differs from this scheme in that it is composed of several comparatively short subunits, the hydropathy profiles of their sequences display the same characteristic pattern as each of the repeats of the larger channel proteins. From the sequence data and from the occurence of the various channel types from protists to higher eucaryots it has been proposed that a K^+ channel was the original voltage sensitive channel whereas Ca^{++} and later Na^+ channels arose from it during evolution.

MECHANISM OF TRANSMITTER RELEASE

Long debated, the cellular source of transmitter release

In analogy to secretory release from a large variety of cellular systems, secretion of neurotransmitter from nerve terminals is thought to occur by exocytosis. This involves the release of soluble synaptic vesicle contents after controlled fusion with the presynaptic plasma membrane. Support for this view comes from ultrastructural analysis of nerve terminals on induced transmitter release, uptake of extracellular markers into synaptic vesicles, immunocytochemical analysis, the comparison of synaptic vesicle contents and products released, biochemical studies of synaptic vesicle life cycle and synaptic vesicle heterogeneity, and the quantal nature of the postsynaptic signal (1, 6, 10, 11, 13, 66, 67). A major challenge is still the detail of the molecular cascade leading from Ca^{++} influx into the axon terminal to synaptic vesicle exocytosis.

Using rapid freezing techniques it could be shown at the frog neuromuscular junction that exocytotic fusion of synaptic vesicles occurs about 0.5 msec before the onset of the postsynaptic signal (1). Under conditions of intense synaptic activation or conditions which impair synaptic vesicle recycling a net reduction in synaptic vesicle numbers has been observed and the counts of postsynaptically recorded quanta could be directly related to a loss in synaptic vesicle numbers at the presynaptic terminal (1). If synaptic vesicle recycling is blocked the incorporation of the synaptic vesicle membrane compartment into the presynaptic plasma membrane can be directly visualized by immunocytochemical methods using antibodies against a synaptic vesicle membrane protein (1). However, an immediate correlation between the loss of synaptic vesicle numbers and a decrease in the amplitude of the synaptic signal is not observed under all experimental conditions (51, 68). The observation that the classical quantal event may in fact consist of subquanta (subminiature endplate potentials, 69, 70) has raised a debate as to whether one postsynaptic quantum corresponds to the release of the contents of a single synaptic vesicle (1).

At least for the low molecular weight transmitter substances there appears in addition to quantal release a non-quantal transmitter leakage (71). The mechanism of this leakage is not quite clear. It is thought to result from the transient incorporation of the vesicular acetylcholine transporter into the presynaptic plasma membrane (10, 72). Whereas molecular leakage predominates under resting conditions, on evoked secretion it is quantal release. For the cholinergic system much of the earlier confusion concerning the source of transmitter released could be resolved using vesamicol (AH5183), a drug which selectively blocks uptake of ACh into synaptic vesicles (73). The experiments show that vesicular storage of newly synthesized acetylcholine is a precondition for its release. The demonstration of vesicular storage of amino acids, the Ca^{++} dependence and the pharmacological characteristics of their release all suggest that these transmitter substances are also released by exocytosis (74, 75). It must not be overlooked, however, that the experimental evidence available leaves room for additional interpretations and hypotheses. It has been suggested that the neurotransmitter acetylcholine is not released by exocytosis of synaptic vesicles but through pores in the presynaptic membrane which are formed by the association of membrane proteins referred to as mediatophore. The Ca^{++} activated opening of a mediatophore complex would give rise to a quantum and the function of synaptic vesicles would be solely in transmitter storage (76).

Structural and physiological diversity of the releasing axon segment

Much of the discussion related to exocytotic release and the quantal nature of the postsynaptic signal has been related to the cholinergic neuromuscular junction in which the small electron-lucent synaptic vesicle type predominates and which has a relatively simple synaptic geometry. It should be noted, however, that the geometry of nerve terminal ramifications competent for transmitter release can be extremely variable. The terminal ramifications of postganglionic sympathetic axons for example contain numerous vesicle filled varicosities and release transmitter over an extensive axon segment. Interestingly, on activation of the axon synaptic messengers are released not from all but only from individual varicosities. The release is subject to various local receptor mediated control mechanisms (43). The situation may be similar at neurosecretory neurons which also release hormones from varicosities (7). Furthermore — as discussed above — the release of small and large dense-cored synaptic vesicles may be regulated differentially and occur at different sites.

The big challenge, the molecular cascade underlying exocytosis

The number of molecular events which are implicated in the mechanism of synaptic vesicle exocytosis is expanding rapidly. They include an increase of the cytosolic Ca^{++} concentration to the micromolar range (77, 78), the activation of protein kinases, in particular of protein kinase C (79), synaptic vesicle associated proteins like synapsin I and II (31) or a chromaffin granule-binding protein (80), the phosphoproteins B-50 (GAP-43) (81) and P92 (possibly the MARCKS protein) (66), cytoskeletal proteins like actin (82) and fodrin (83), G proteins (84) and even others (85, 86). It is, however, not clear how these various molecular components interact and how their interaction is regulated. It appears obvious that in action potential-evoked transmitter release the elevation of the intracellular calcium level following influx through voltage gated calcium channels generally is the first step of the cascade (77, 78). A receptor-mediated synaptic phosphoinositide turnover with the formation of inositol 1,4,5-trisphosphate as a second messenger for mobilizing calcium from internal stores may be regarded as a modulatory synaptic mechanism (87). It is a major disadvantage that the immediate Ca^{++} target is not known. An early step will be the mobilization of synaptic vesicles from their association with the cytonet (9, 66) and possibly the removal of a cytoskeletal axolemmal undercoating at the presynaptic release site. This may explain the involvement of synapsin I and various cytoskeletal proteins in the release process. Ca^{++} and protein phosphorylation presumably play a major role in the control of these processes (31). Furthermore, there is convincing evidence that activation of protein kinase C potentiates transmitter release. Some of its presynaptic targets like B 50 (89) or MARCKS protein (90) are by now well defined. The endogenous activator of the enzyme is thought to be diacylglycerol which is formed together with the inositol trisphosphate on phospholipase C mediated breakdown of phosphatidylinositol 4,5-bisphosphate (91). It is not clear at the moment whether a net increase in diacylglycerol needs to be involved in action potential-evoked transmitter release or whether protein kinase C may become sufficiently activated by a rise in cytosolic Ca^{++} concentrations with diacylglycerol concentrations remaining constant. Furthermore, we do not know how the substrate proteins for protein kinase C interpose between synaptic vesicles and the presynaptic plasma membrane. These proteins are fatty-acetylated and in this form would be membrane-anchored. If the fatty acid was removed by protein kinase C dependent phosphorylation, the membrane link would be lost (92). How this could be related to docking of synaptic vesicles to tho presynaptic membrane — which is protein kinase C dependent (66) — is unclear. When docking of synaptic vesicles has been achieved this step is followed by a yet undefined mechanism of membrane fusion and fission. It is thought that the interaction of the two membrane compartments is initiated by a channel-like structure possibly involving corresponding protein channels in the synaptic vesicle membrane and in the presynaptic plasma membrane (synaptophysin, 35, 36). Eventually soluble messengers may be released through a short lived "fusion pore" (86) and the synaptic vesicle membrane will be immediately recycled for reuse. It has to be seen to what extent the small electron-lucent and the large dense-cored synaptic vesicles obey the same molecular releasing mechanism.

Modulatory mechanisms become overwhelming

Synaptic transmission modulates and is subject to modulation. This is of great importance e.g. for synaptic information processing and modulation processes involved in behavioural plasticity. Releasing nerve terminals may possess receptors for other transmitter substances but also receptors for the synaptic messengers which they themselves release. This modulation may involve receptor mediated modifications of e.g. K^+ or Ca^{++} channels and thus control the availability of extracellular Ca^{++} for transmitter release. External messengers may, however, also act via receptor-activated G proteins controlling phospholipase C and thus the molecular cascade involved in vesicle exocytosis (84). Among messengers which act via presynaptic receptors are practically all neurotransmitter substances like acetylcholine, catecholamines, ATP (adenosine), opioids or peptides like substance P and somatostatin or amino acids (6, 40, 41, 45, 93, 94). Some substances are active only after extracellular modification. It could be shown that at least part of the adenosine acting on presynaptic neurotransmitter release is derived from released ATP (45). It is possible that some of the presynaptic modulators may display their activity after they have been taken up into the nerve terminals via specific carriers (40). The multiple modulatory mechanisms at both the presynaptic and postsynaptic side underline the superiority of chemical transmission in neuronal information processing over electrical transmission which operates via direct electrical coupling of cells.

NEUROTRANSMITTER RECEPTORS

It all falls into shape

Synaptic messengers represent a broad spectrum of substances without immediate chemical relationship, and until recently this was also the impression concerning the variety of transmitter receptors. Not only is there a separate receptor for each synaptic messenger. Most receptors also display a variety of subtypes which differ in their pharmacological properties and in tissue distribution. From the analysis of the primary structures together with a comparison of hydropathy profiles and general functional properties, however, a new system for receptor grouping has emerged which does not bear obvious relations to the chemical nature of the messenger.

From a functional point of view one may differentiate one group of receptors which represent ligand gated ion channels. They meet the requirement of fast (microsecond range) conductance change. For the second group of receptors the immediate consequence of activation by a synaptic messenger is to promote the binding and the activation of a GTP-binding protein (G protein). The G protein acts to regulate enzymes which alter the concentration of diffusible intracellular messengers like cyclic nucleotides or inositol trisphosphate. These ultimately may act in the conversion of a channel protein. They operate on a slower time scale. In some cases the G protein may regulate ion channels directly. Interestingly, members of these two groups each fall into a single structural category (family) which is based on a common evolutionary origin. The receptors representing ligand gated ion channels consist of several subunits a combination of which must be inserted into the membrane to achieve ligand gated conductance. Subunits between different receptors display a high degree of homology. Amongst these are the nicotinic acetylcholine receptor, the glycine receptor and the $GABA_A$ receptor. The second group of G protein-linked receptors includes receptors for peptides like substance P and substance K, the muscarinic acetylcholine receptor, the α- and β-adrenergic receptors, the 5-hydroxytryptamin receptor, and rhodopsin, the "receptor" for quanta of light in the retinal photoreceptor cells. These receptors are formed by a single protein and have in common seven potentially membrane spanning sequences. They also display high amino acid sequence homology (95, 96, 97).

Ligand gating, fast and successful

The hitherto most intensely studied nicotinic acetylcholine receptor of *Torpedo* electric organ and mammalian muscle has a pentameric structure composed of 4 subunits (α_2,β,Γorε,δ). The subunits are encoded by different but phylogenetically closely related genes and display a high degree of amino acid homology (98, 99, 100, 101). Each subunit contains four putative transmembrane segments part of which are involved in the formation of a central water filled pore (channel) which allows the gated passage of cations (65). Neuronal (brain) nicotinic acetylcholine receptors differ pharmacologically from their counterparts in muscle and they may fall into different subtypes (99, 102). Although sequence data suggest a close similarity with the muscle nicotinic receptor the subunit stoichiometry of the neuronal nicotinic acetylcholine receptors is not clear at the moment (103).

The glycine receptor forms a similar pentameric channel but functions as a chloride channel. It is probably composed of three α and two β subunits. These subunits again are encoded by different but closely related genes and with their transmembrane topology (four hydrophobic segments) they closely resemble the nicotinic acetylcholine receptor (36, 104). The brain $GABA_A$ (benzodiazepine) receptor which also represents a gated chloride channel similarly consists of two different subunits encoded by two related genes and both subunits also display the typical four putative transmembrane regions (36, 105). To date the genes for several isoforms of the α subunit have been cloned. This wealth of information on protein primary structure gives rise to many further questions. What is the exact membrane topography of each subunit (9)? Which segments of the individual subunits form the ion channel and how are they structurally arranged? Which element of the receptor is responsible for the structural transition on gating? Regarding the close homology of the different receptors types — what differences in molecular structure account for the differences in ligand and also in ion channel specificity?

Interestingly, different functional receptor subtypes can be accomplished by alterations in subunit composition. The nicotinic acetylcholine receptor in mammalian skeletal muscle contains a Γ subunit whereas the receptor in the adult muscle instead contains a (homologous) ε subunit. Receptors containing the Γ subunit also appear in large number in denervated muscle and are mostly incorporated into extrasynaptic regions (80, 101). These Γ channels possess a lower conductance and longer open time than the adult ε channel — an impressive example for a transcriptional regulation of receptor types during development. A similar developmental heterogeneity has also been infered for glycine receptors (106).

A family of channels meets a family of G protein

The G protein-linked channels also show diversity regarding messenger specificity and the formation of subtypes. In all cases investigated so far, subtypes were found not to be the result of modification or differences in membrane environment but of the expression of different (though closely related) genes. Analogous to the differences in channel specificity displayed by the ligand-gated receptor channels the G protein-linked receptors can mediate variable intracellular signals. The type of message transferred is determined by the G protein to which the receptor is coupled. These G proteins themselves form a large family of differing functional specificity (107). Amongst these are stimulation of phospholipase C and phosphoinositide breakdown, inhibition or stimulation of adenylate cyclase activity or direct regulation of ion channels.

Although all G protein-linked receptors belong to a large gene superfamily and carry the typical seven clusters of hydrophobic amino acids, differences in primary structure would explain the different associations with G proteins. Indeed sequence homologies are more stringent in the presumptive membrane spanning regions which appear to embody functions common to all of the receptors. Interestingly, the atrial muscarinic acetylcholine receptor has 38% overall amino acid homology with the cerebral muscarinic receptor. But it possesses a completely non-homologous large cytoplasmic domain which could play a role in differential effector coupling (108). For comparison, the β_1 and β_2 adrenergic receptors which are functionally very similar (both lead to stimulation of adenylate cyclase activity) have very similar amino acid sequences (109). Additional cloning information together with the identification of the type of G protein link for each receptor type may soon reveal concensus sequences responsible for G protein binding.

A peculiarity of the above mentioned scheme certainly is the fact that acetylcholine rides on both receptor families. It thus broadens range and variability of its message. There is no evidence for a phylogenetic link of the two superfamilies. The use by acetylcholine of both muscarinic and nicotinic receptors obviously has originated independently and thus has to be regarded as "molecular" convergence. The finding that both receptors nevertheless share consensus sequence elements which may represent acetylcholine binding sites does not come unexpected (110).

More receptors — more principal receptor types?

Future determination of primary structures will reveal whether additional receptor families exist. Glutamate (and aspartate) exite a multiplicity of receptors in brain which differ both with regard to their pharmacological specificity as well as to the mechanism of action. One may differentiate NMDA receptors activated by N-methyl-D-aspartate and non-NMDA receptors (111). Whereas the non-NMDA receptor class produces fast synaptic potentials by increasing conductance of Na^+ and K^+ ions the NMDA receptors produce slowly rising and long-lasting (200—300 ms) depolarizations and are permeable to Ca^{++} as well as to Na^+ and K^+ ions. Furthermore, NMDA receptors share with voltage gated ion channels the principal property that their activation is voltage dependent. The current is regenerative and increases with depolarization similar to the Na^+ current in the action potential. Thus, the NMDA receptor needs to be "gated" by the action of other excitatory messengers (possibly by activation of non-NMDA receptors). As neurons may posses non-NMDA and NMDA receptors, and in addition voltage gated Ca^{++} channels this opens a broad spectrum for affecting the activity of neurons and neuronal circuits and has much attracted the interest of physiologists investigating neuronal plasticity phenomena. In fact these receptor systems have been implicated in a variety of activity dependent processes like temporal integration, rhythmic firing and long term potentiation (112). Yet, another class of glutamate receptors with high activity during synaptogenesis transfers the signal by stimulating the inositol phosphate/diacylglycerol pathway (113).

Although ATP and adenosine are ubiquitous in all tissues and cells they act via specific receptors as neuronal modulators and presumably also transmitters (114). Recent studies have strengthened the position for ATP as a neurotransmitter acting via P_2 receptors either as a cotransmitter or as a genuine transmitter (15, 114). In smooth muscle ATP is released from sympathetic axons and gives rise to postsynaptic quanta (6, 43). The conductance activated by ATP is cation selective (115) but does not appear to discriminate strongly between Na^+, K^+ and Mg^{++} (116). P_2 receptor activation increases inositol trisphosphate production in a variety of cell types including myocardial cells (117) and vertebrate skeletal muscle (46). Furthermore, extracellular ATP induces synaptic potentiation in hippocampal slices (118). The analysis of the molecular biology of this somewhat neglected receptor appears to be a promising goal.

Among the peptide receptors the opiate receptors have been intensely characterized but a more detailed understanding of their molecular configuration and action has to await sequence data (119). Their location appears to be mainly presynaptic acting in the regulation of the release of other messengers. In pharmacological terms four major types of opioid receptors may be differentiated. Regarding the mechanism of action the formation of cyclic AMP as second messenger, but also inhibition of adenylate cyclase and the opening of synaptic K^+ channels or Ca^{++} channels have been implied for the various receptor subtypes.

The hen or the egg?

Neurotransmitter receptors are only part of the ubiquitous and highly specialized chemical communication system in higher organisms. Sensitivity for substances which serve as synaptic messengers in higher animals can be detected almost in parallel with the onset of the evolution of nervous systems and thus reach back 500 million years and more (120). This raises the question what existed first, the messenger or the receptor, or the hen or the egg. In fact, there is evidence that messengers and their synthesizing enzymes existed a substantial period before their respective receptors. The various families of receptors may have evolved at different times and independently, broadening the range of accepted messengers and molecular mechanisms of action during evolution by gene duplication. Neither do we know which receptor was first nor which was the first ligand. At present, much of the molecular analysis of transmitter receptors is restricted to mammalian systems. Further functional analysis and comparison of primary structures throughout the animal kingdom may yield additional insight into the genesis of receptors. Ligand and ion channel specificities in invertebrates may be quite different from those observed in higher vertebrates. Opposing effects of two neurotransmitters may even be achieved by their differential effect on the same type of receptor (121). There are even close relations to receptors in plants. Thus, the putative mating factor of yeast, a peptide pheromone, possesses seven putative transmembrane segments and shares many structural features with one class of mammalian cell surface receptors (122). The archetyp of a receptor may be envisaged as a membrane spanning protein eventually aquiring the competence of ligand binding and transmembrane signalling.

REFERENCES

Note: All references marked with * are contained in the proceedings of the workshop and refer to "In: Zimmermann, H. (ed.), Cellular and Molecular Basis of Synaptic Transmission, NATO ASI Series H: Cell Biology, Vol. 21, Springer. Berlin, Heidelberg, New York, London, Paris, Tokyo, 1988."

1) Ceccarelli, B., Valtorta, F., Hurlbut, W.P. (1988) New evidence supporting the vesicle hypothesis for quantal secretion at the neuromuscular junction. *, 51—72.

2) Couteaux, R., Spacek, J. (1988): Specializations of subsynaptic cytoplasms. Comparison of axospinous synapses and neuromuscular junctions. *, 25—50.

3) Hirokawa, N. (1983): Membrane specializations and cytoskeletal structures in the synapse and axon revealed by the quick-freeze, deep-etch method. In: Chang, D.C., Tasaki, I., Adelmanm, W.J., Leuchtag, H.R. (1983): Structure and Function of Excitable Cells. Plenum Press. New York, London. pp. 113—141.

4) Pfenninger, K., Akert, K., Moor, H., Sandri, C. (1972): J. Neurocytol 1: 129—149.

5) Smith, S.J., Augustine, G.J. (1988): Calcium ions, active zones and synaptic transmitter release. TINS 11: 458—464.

6) Stjärne, L. (1989): Basic mechanisms and local modulation of nerve impulse-induced secretion of neurotransmitters from individual sympathetic nerve varicosities. Rev. Physiol. Biochem. Pharmacol. 112: 1—137.

7) Nordmann, J.J., Dayanithi, G. (1988): Release of neuropeptides does not only occur at nerve terminals. Biosci. Rep. 8: 471—483.

8) Janetzko, A., Zimmermann, H., Volknandt, W. (1989): Intraneuronal distribution of a synaptic vesicle membrane protein: antibody binding sites at axonal membrane compartments and *trans*-Golgi network and accumulation at nodes of Ranvier. Neuroscience 32: 65—77.

9) Hirokawa, N., Sobue, K., Kanda, K., Harada, A., Yorifuji, H. (1989): The cytoskeletal architecture of the presynaptic terminal and molecular structure of synapsin 1. J. Cell Biol. 108: 111—126.

10) Zimmermann, H. (1988): Cholinergic synaptic vesicles. Handbook Exptl. Pharmacol. 86: 349—382.

11) Winkler, H. (1988): Occurence and mechanism of exocytosis in adrenal medulla and sympathetic nerve. Handb, Exptl, Pharmacol, 90/I: 43—118.

12) Winkler, H., Fischer-Colbrie, R., Obendorf, D., Schwarzenbrunner, U. (1988): Adrenergic and cholinergic vesicles: Are there common antigens and common properties? *, 305—314.

13) Whittaker, V.P. (1988): The cellular basis of synaptic transmission: An overview. *, 1—23.

14) Klein, R.L., Duncan, R.W., Selva, T.J., Kong, J.-Y., Clayton, W.E., Liaw, Y.-L., Rezk, N.F., Thureson-Klein, A. (1988): Sympatho-adrenal costorage, release and synthesis of enkephalins and catecholamines induced by acute CNS ischemia in pig. *, 377—393.

15) Agoston, D.V. (1988): Cholinergic co-transmitters. Handb. Eptl. Pharmacol. 86: 479—533.

16) Thureson-Klein, A.K., Klein, R.L., Zhu, P.C., Kong, J.-Y. (1988): Differential release of transmitters and neuropeptides co-stored in central and peripheral neurons. *, 137—151.

17) Boarder, M.R. (1989): Presynaptic aspects of cotransmission: relationship between vesicles and neurotransmitters. J. Neurochem. 53: 1—11.

18) Wiedenmann, B., Franke, W.W. (1988): Synaptophysin: A major cell type-specific protein of neuro-endocrine cells. *, 351—356.

19) Volknandt, W., Henkel, A., Zimmermann, H. (1988): Synaptic vesicle proteins. *, 337—350.

20) Volknandt, W., Zimmermann, H. (1990): Identical properties of transmembrane synaptic vesicle protein Mr 100,000 in *Torpedo* and Mr 86,000 in bovine brain. Neurochem. Int.: in press.

21) Südhof, T.C., Baumert, M., Perin, M.S., Jahn, R. (1989): A synaptic vesicle membrane protein is conserved from mammals to Drosophila. Neuron. 2: 1475—1481.

22) Trifaro, J.-M., Fournier, S., Novas, M.L. (1989): The p65 protein is a calmodulin-binding protein present in several types of secretory vesicles. Neuroscience 29: 1—8.

23) Parsons, S.M., Noremberg, K., Rogers, G.A., Gracz, L.M., Kornreich, W.D., Bahr, B.A., Kaufman, R. (1988): Complexity and regulation in the acetylcholine storage system of synaptic vesicles. *, 325—335.

24) Fonnum, F., Fykse, E.M., Paulsen, R. (1988): Excitatory amino acids; physiology, anatomy and biochemistry. *, 172—183.

25) Fykse, E.M., Christensen, H., Fonnum, F. (1989): Comparison of the properties of Γ-aminobutyric acid and L-glutamate uptake into synaptic vesicles isolated from rat brain. J. Neurochem. 52: 946—951.

26) Isambert, M.-F., Gasnier, B., Laduron, P.M., Henry, J.-P. (1989): Photoaffinity labelling of the monoamine transporter of bovine chromaffin granules and other monoamine storage vesicles using 7-azido-8-(^{125}I)iodoketanserin. Biochemistry 28: 2265—2270.

27) Hell, J.W., Maycox, P.R., Stadler, H., Jahn, R. (1988): Uptake of GABA by rat synaptic vesicles isolated by a new procedure. EMBO J. 7: 3023—3029.

28) Kish, P.E., Fischer-Bovenkerk, C., Ueda, T. (1989): Active transport of Γ-aminobutyric acid and glycine into synaptic vesicles. Proc. Natl. Acad. Sci. USA 86: 3877—3881.

29) Cidon, S., Sihra, T.S. (1989): Characterization of a H^+-ATPase in rat brain synaptic vesicles. Coupling to L-glutamate transport. J. Biol. Chem. 264: 8281—8288.

30) Yamagata, S.K., Noremberg, K., Parsons, S.M. (1989): Purification and subunit composition of a cholinergic synaptic vesicle glycoprotein phosphointermediate-forming ATPase. J. Neurochem. 53: 1345—1353.

31) Südhof, T.C., Czernik, A.J., Kao, H.-T., Takei, K., Johnston, P.A., Horiuchi, A., Kanazir, D., Wagner, M.A., Perin, M.S., De Camilli, P., Greengard, P. (1989): Synapsins: Mosaics of shared and individual domains in a family of synaptic vesicle phosphoproteins. Science 245: 1474—1480.

32) Simon, J.-P., Aunis, D. (1988): Biochemistry of the chromogranin A protein family. Biochem. J. 262: 1—13.

33) Stadler, H., Borroni, E., Ploghöft, M., Kiene, M.-L. (1988): Structure and function of synaptic vesicles: new aspects. *, 315—324.

34) Sihra, T.S., Wang, J.K.T., Gorelick, F.S., Greengard, P. (1989): Translocation of synapsin I in response to depolarization of isolated nerve terminals. Proc. Natl. Acad. Sci. USA 86: 8108—8112.

35) Thomas, L., Hartung, K., Langosch, D., Rehm, H., Bamberg, E., Franke, W.W., Betz, H. (1988): Identification of synaptophysin as a hexameric channel protein of the synaptic vesicle membrane. Science 242. 1050—1053.

36) Betz, H. (1990): Homology and analogy in transmembrane channel design: lessons from synaptic membrane proteins. Biochemistry, in press.

37) Rahamimoff, R., De Riemer, S.A., Sakmann, B., Stadler, H., Yakir, N. (1988): Ion channels in synaptic vesicles from Torpedo electric organ. Proc. Natl. Acad. Sci. USA 85: 5310—5314.

38) Grondal, E.J.M., Zimmermann, H. (1988): Extracellular metabolism of ATP at the cholinergic electromotor synapse: characterization and location of ectonucleotidases. *, 395—410.

39) Bradford, H.F., Dochtery, M., Joh, T.H., Wu, Y.-Y. (1988): The detection, isolation and properties of subpopulations of mammalian brain synaptosomes of defined transmitter type. *, 153—159.

40) Raiteri, M., Bonanno, G., Pittaluga, A., Versace, P. (1988): Heterologous carriers located on rat CNS nerve endings and their possible role in modulation of transmitter release. *, 199—213.

41) Bartfai, T., Iverfeldt, K., Brodin, E., Ögren, S.-O. (1986): Functional consequences of coexistence of classical and peptide neurotransmitters. Prog. Brain. Res. 68: 321—330.

42) Hökfelt, T., Johansson, O., Ljungdahl, A., Lundberg, J.M,. Schultzberg, M. (1980): Nature 284: 515—521.

43) Stjärne, L. (1988): The new paradigm: sympathetic neurotransmission by lateral interaction between single mixed quanta acting in two different biophases. *, 357—375.

44) Ehrlich, Y.H., Snider, R.M., Kornecki, E., Garfield, M.G., Lenox, R.H. (1988): Modulation of neuronal signal transduction systems by extracellular ATP. J. Neurochem. 50, 295—301.

45) Brown, S.J., Luzio, J.P., Richardson, P.J. (1988): The application of immunoaffinity techniques to the study of cholinergic neurobiology. *, 411—424.

46) Heilbronn, E., Häggblad, J. (1988): Transmitter/modulator induced events related to excitation-contraction coupling in skeletal muscle of vertebrates. Action of ATP. *, 425—441.

47) Dowdall, M.J., Downie, D.L. (1988): Identity and functional role of neuropeptides in *Torpedo* electric organ. *, 443—452.

48) Bauer, K. (1985): Degradation and biological inactivation of neuropeptides. In: Hamprecht, B., Neuhoff, V. (eds.): Neurobiochemistry, 36. Colloquium — Mosbach 1985, Springer. Berlin, Heidelberg, New York, Toronto, pp. 43—54.

49) Kenny, A.J. (1986): Endopeptidase-24.11: An ectoenzyme capable of hydrolyzing regulatory peptides at the surface of many different cell types. In: Kreutzberg, G.W., Reddington, M., Zimmermann, H. (eds.): Cellular Biology of Ectoenzymes. Springer. Berlin, Heidelberg, New York, Tokyo, pp. 257—271.

50) Kenny, A.J., Turner, A.J. (1987): Ectoenzymes: an update. In: Kenny, A.J., Turner, A.J. (eds.): Mammalian Ectoenzymes. Elsevier. Amsterdam, New York, Oxford, pp. 329—351.

51) Pappas, G.D., Kriho, V., Becker, R,B. (1988): Fine structural correlates of calcium dynamics in the presynaptic terminal. *, 121—135.

52) Nagy, A.K., Shuster, T.A., Delgado-Escueta, V. (1989): Rat brain synaptosomal ATP:AMP-phosphotransferase activity. J. Neurochem. 53: 1166—1172.

53) Blaustein, P., Bartschat, D.K,. Benishin, C.G., Brown, W.E., Colby, K.A., Krueger, B.K., Schneider, M.J., Sorensen, R.G. (1988): Potassium channels in rat brain synaptosomes: pharmacology and toxicology. *, 241—261.

54) Carvalho, A.P., Santos, M.S., Henriques, A.O., Tavares, P., Carvalho, C.M. (1988): Calcium channels and Na^{+}/Ca^{2+} exchange in synaptosomes. *, 263—284.

55) Suszkiw, J.B. (1988): Properties of presynaptic voltage-sensitive calcium channels in rat synaptosomes. *, 285—291.

56) Schauf, C.L. (1987): Ion channel diversity: a revolution in biology? Sci. Prog. Oxford 71: 459—478.

57. Hosey, M.M., Lazdunski, M. (1988): Calcium channels: molecular pharmacology, structure and regulation. J. Membr. Biol. 104: 81—105.

58) Lemos, J.R., Nordmann, J.J., Cooke, I.M., Stuenkel, E.L. (1986): Single channels and ionic currents in peptidergic nerve terminals. Nature 319: 410—412.

59) Tsien, R.W., Hess, P., McCleskey, E.W., Rosenberg, R.L. (1987): Calcium channels: mechanisms of selectivity, permeation, and block. Ann. Rev. Biophys. Biophys. Chem. 16: 265—290.

60) Dunn, S.M.J. (1988): Multiple calcium channels in synaptosomes: voltage dependence of 1,4 dihydropyridine binding and effects of function. Biochemistry 27: 5275—5281.

61) Lemos, J.R., Nowycky, M.C. (1989): Two types of calcium channls coexist in peptide-releasing vertebrate nerve terminals. Neuron. 2: 1419—1426.

62) Krnjevic, K., Leblond, J. (1988): Mechanisms of failure of synaptic transmission during anoxia. * 293—304.

63) Kandel, E.R., Schwartz, J.H. (1982): Molecular biology of learning: Modulation of transmitter release. Science 218: 433—443.

64) Catterall, W.A. (1988): Structure and function of voltage-sensitive ion channels. Science 242: 50—81.

65) Maelicke, A. (1988): Structural similarities between ion channel proteins. TIBS 13: 199—202.

66) Zimmermann, H., Volknandt, W., Henkel, A., Bonzelius, F., Janetzko, A., Kanaseki, T. (1990): The synaptic vesicle membrane: Origin, axonal distribution, protein components, exocytosis and recycling. Cell Biol. Int. Rep.: in press.

67) Whittaker, V.P. (1988): Model cholinergic systems: an overview. Handb. Exptl. Pharmacol. 86: 3—22.

68) Kadota, T., Kadota, K., Tatsuoka, H., Mizote, M., Yamaai, T. (1988): Poststimulation increase of synaptic vesicle number in the preganglionic nerve terminals of the cat sympathetic ganglion *in vivo*. *, 73—82 .

69) Kriebel, M.E., Fox, G.Q., Kötting, D. (1988): Effect of nerve stimulation, K^+ saline and hypertonic saline on classes of quanta, quantal content and synaptic vesicle size distribution of *Torpedo* electric organ. *, 97—120.

70) Fox, G.Q., Kriebel, M.E., Kötting, D. (1988): Synaptic vesicle classes in Torpedo and skate electric organ and muscle. *, 83—95.

71) Vizi, S.E. (1984): Physiological role of cytoplasmic and non-synaptic release of transmitter. Neurochem. Int. 6: 435—440.

72) Licht, R., Michaelson, D.M. (1988): Spontaneous release of acetylcholine from *Torpedo* electromotor neurons. *, 215—227.

73) Marshall, I.G., Parsons, S.M. (1987): The vesicular acetylcholine transport system. TINS 10, 174—177.

74) Nicholls, D., Sanchez-Prieto, J. (1988): Synaptosomal bioenergetics and the mechanism of amino acid transmitter release. *, 185—198.

75) Nicholls, D.G. (1989): Release of glutamate, aspartate, and Γ-aminobutyric acid from isolated nerve terminals. J. Neurochem. 52: 331—341.

76) Israel, M. (1988): The mechanism of acetylcholine release: its essential component, the mediatophore. *, 229—239.

77) Knight, D.E., Scrutton, M.C. (1986): Gaining access to the cytosol: the technique and some applications of electropermeabilization. Biochem. J. 234: 497—506.

78) Augustine, G.J., Charlton, M.P., Smith, S. (1987): Calcium action in synaptic transmitter release. Ann. Rev. Neurosci. 10: 633—693.

79) Dekker, L.V., De Graan, P.N.E., Versteeg, D.H.G., Oestreicher, A.B., Gispen, W.H. (1988): Phosphorylation of B-50 (GAP43): is correlated with neurotransmitter release in rat hippocampal slices. J. Neurochem. 52: 24—30.

80) Schweizer, F.E., Schäfer, T., Tapparelli, C., Grob, M., Karli, U.O., Heumann, R., Thoenen, H., Bookman, J., Burger, M. (1989): Inhibition of exocytosis by intracellularly applied antibodies against chromaffin granule-binding protein. Nature 339: 709—712.

81) Dekker, L.V., De Graan, P.N., Oestreicher, A.B., Versteeg, D.H.G., Gispen, W.H. (1989): Inhibition of noradrenaline release by antibodies to B 50 (GAP-43). Nature 342, 74—76.

82) Burgoyne, R.B., Cheek, T.R. (1987): Reorganization of peripheral actin filaments as a prelude to exocytosis. Biosci. Rep. 7: 281—288.

83) Perrin, D., Langley, O.K., Aunis, D. (1987): Anti-α-fodrin inhibits secretion from permeabilized chromaffin cells. Nature 326: 498—501.

84) Burgoyne, R.D. (1987): G proteins. Control of exocytosis. Nature 328: 112—113.

85) Burgoyne, R.D. (1988): Calpactin in exocytosis? Nature 331: 20.

86) Penner, R., Neher, E. (1989): The patch-clamp technique in the study of secretion. TINS 12: 159—163.

87) Gill, D.L. (1989): Receptor kinships revealed. Nature 342: 16—18.

88) Berridge, M.J., Irvine, R.F. (1989): Inositol phosphates and cell signalling. Nature 341: 197—205.

89) Benowitz, L.T., Routtenberg, A.: TINS 10: 527—532.

90) Stumpo, D.J., Graff, J.M., Albert, K.A., Greengard, P., Blackshear, P.J. (1989): Molecular cloning, characterization, and expression of a cDNA encoding the "80- to 87-kDa" myristoylated alanine-rich C kinase substrate: A major cellular substrate for protein kinase C. Proc. Natl. Acad. Sci. USA 86: 4012—4016.

91) Nishizuka, Y. (1986): Studies and perspectives of protein kinase C. Science 233: 305—312.

92) Aderem, A.A., Albert, K.A., Keum, M.M., Wang, J.K.T,. Greengard, P., Cohn, Z.A. (1988): Stimulus-dependent myristoylation of a major substrate for protein kinase C. Nature 332: 362—364.

93) Gray, D.B., Pilar, G.R., Ford, M.J. (1989): Opiate and peptide inhibition of transmitter release in parasympathetic nerve terminals. J. Neurosci. 9: 1683—1692.

94) Kilbinger, H. (1988): The autonomic cholinergic neuroeffector junction. Handbook Exptl. Pharmacol. 86: 581—614.

95) Barnard, E.A. (1988): Separating receptor subtypes from their shadows. Nature 355: 301—302.

96) Nakanishi, S., Ohkubo, H., Masu, Y., Takumi, T., Tamaki, H., Yokota, Y. (1989): Molecular characterization of neuropeptide receptors and of a novel membrane protein involved in K^+ permeation. In: Maelicke, A. (ed.): Molecular Biology of Ion Channels, NATO ASI Series H: Cell Biology, Vol. 32, Springer. Berlin, Heidelberg, New York, London, Paris, Tokyo, pp. 269—277.

97) O'Dowd, B.F., Lefkowitz, R.J., Caron, M.G. (1989): Structure of the adrenergic and related receptors. Ann. Rev. Neurosci. 12: 67—83.

98) Witzemann, V., Sakmann, B. (1988): Mechanisms of regulating the expression and function of acetylcholine receptor. *, 453—461.

99) Darlison, M.G., Hicks, A.A., Cockcroft, V.B., Squire, M.D., Barnard, E.A. (1988): Brain α-neurotoxin-binding proteins and nicotinic acetylcholine receptors. *, 475—492.

100) Maelicke, A., Fels, G., Plümer-Wilk, R. (1988): Antibodies as probes of the structure of the nicotinic acetylcholine receptor. *, 463—473.

101) Steinbach, J.H., Ifune, C. (1989): How many kinds of nicotinic acetylcholine receptors are there? TINS 12: 3—6.

102) Sher, E., Gotti, C., Fornasari, D., Chini, B., Ogando, A.E., Clementi, F. (1988): Human neuroblastoma cells: an in vitro model for the study of mammalian nicotinic receptors. *, 493—506.

103) Steinbach, J.H. (1989): Structural and functional diversity in vertebrate skeletal muscle nicotinic acetylcholine receptors. Ann. Rev. Physiol. 51: 353—365.

104) Zopf, D., Grenningloh, G., Hermans-Borgmeyer, I., Rienitz, A., Becker, C.-M., Schmitt, B., Gundelfinger, E.D., Betz, H. (1988): Molecular biology approaches to the function and development of CNS synapses.*, 507—525.

105) Barnard, E.A., Darlison, M.G., Seeburg, P. (1987): Molecular biology of the $GABA_A$ receptor: the receptor/channel superfamily. TINS 10: 502—509.

106) Akagi, H., Miledi, R. (1988): Heterogeneity of glycine receptors and their messenger RNAs in rat brain and spinal cord. Science 242: 270—273.

107) Gilman, A.F. (1987): G proteins: transducers of receptor-generated signals. Ann. Rev. Biochem. 56: 615—649.

108) Peralta, G., Winslow, J.W., Peterson, G.L., Smith, D.H., Ashkenazi, A., Ramachandran, J., Schimerlik, M.I., Capon, D.J. (1987): Primary structure and biochemical properties of an M_2 muscarinic receptor. Science 236: 600—605.

109) Frielle, T., Kobilka, B., Lefkowitz, R.J., Caron, M.G. (1988): Human β_1- and β_2-adrenergic receptors: structurally and functionally related receptors derived from distinct genes. TINS 11: 321—324.

110) Peterson, G.L. (1989): Consensus residues at the acetylcholine binding site of cholinergic proteins. J. Neurosci. Res. 22: 488—503.

111) MacDermott, A.B., Dale, N. (1987): Receptors, ion channels and synaptic potentials underlying the integrative actions of excitatory amino acids. TINS 10: 280—284.

112) Collingridge, L., Bliss, T.V.P. (1987): NMDA receptors-their role in long-term potentiation. TINS 10: 288—293.

113) Sladeczek, F., Recasens, M., Bockaert, J. (1988): A new mechanism for glutamate receptor action: phosphoinositide hydrolysis. TINS 11: 545—549.

114) Gordon, J.L. (1986): Extracellular ATP: effects, sources and fate. Biochem. J. 233: 309—319.

115) Benham, C.D., Bolton, T.B., Byrne, N.G., Large, W.A. (1987): Action of externally applied adenosine triphosphate on single smooth muscle cells dispersed from rabbit ear artery. J. Physiol. (Lond) 387: 473—488.

116) Friel, D.D. (1988): An ATP-sensitive conductance in single smooth muscle cells from rat vas deferens. J. Physiol. (Lond) 401: 361—380.

117) Legssyer, A., Poggiolo, J., Renard, D., Vassort, G. (1988): ATP and other adenine compounds increase mechanical activity and inositol trisphosphate production in rat heart. J. Physiol. (Lond) 401: 185—199.

118) Wierasko, A., Seyfried, T.N. (1989): ATP-induced synaptic potentiation in hippocampal cells. Brain Res. 491: 365—359.

119) Simon, E.J. (1988): Recent studies on opioid receptors. *, 525—539.

120) Venter, J.C,. Di Porzio, U., Robinson, D.A., Shreeve, S.M., Lai, J., Kerlavage, A.R., Fracek, S.P., Lentes, K.-U., Fraser, C.M. (1988): Evolution of neurotransmitter receptor systems. Prog. Neurobiol. 30: 105—169.

121) Pfeiffer-Linn, C., Glantz, R.M. (1989): Acetylcholine and GABA mediate opposing actions on neuronal chloride channels in crayfish. Science 245: 1249—1251.

122) Nakayama, N., Miyajima, N., Arai, K. (1985): Nucleotide sequences STE2 and STE3, cell type-specific sterile genes from Saccharomyces cerevisiae. EMBO J. 4: 2643—2648.

CELLULAR AND MOLECULAR BIOLOGY OF MYELINATION

G. Jeserich[1], H.H. Althaus[2], T.V. Waehneldt[2]

[1]Abteilung Zoophysiologie, Universität Osnabrück, D-4500 Osnabrück, FRG
[2]Max-Planck-Institut für experimentelle Medizin, D-3400 Göttingen, FRG

The process of myelination is one of the key events during nervous system development and represents the ultimate step in cellular differentiation by which proper neuronal function is attained. Knowledge of the biochemistry and cell biology of the myelin-forming glia has advanced rapidly in recent years. This progress was to a large extent conveyed by the concerted application of new experimental tools, including improved cell culture systems availability of specific immunological probes and the powerful development of recombinant DNA technology and cell transfection.

The multidisciplinary aspects of glial cell biology and myelination were comprehensively discussed by leading neuroscientists from Europe, Israel, USA and Canada at a Nato Advanced Research Workshop held at Monastery Ohrbeck, near Osnabrück, August 28 — September 2, 1989.

The meeting concentrated on the following major topics:

STEPS IN GLIAL CELL DIFFERENTIATION AND MYELINOGENESIS

An immunological characterization of the different steps of oligodendroglial differentiation as observed in cell culture from embryonic brain was given. Using a panel of monoclonal antibodies against cell surface glycolipids (gangliosides, sulfatide, galactocerebroside) the sequence of stages in oligodendrocyte lineage progression was defined as follows: the earliest stage was the O—2A bipotential progenitor, which is a bipolar, proliferative and migratory cell capable to differentiate either into myelinating oligodendrocytes or type II astrocytes, depending on environmental conditions (1, 2). The next step is characterized by the appearence of the O4 antigen to produce "proligodendrocytes" (3), which in vivo still have a significant capacity for proliferation. Finally the cells develop into terminally differentiated oligodendrocytes indicated first by the occurrence of galactocerebroside (GalC) on the cell surface and followed by the sequential expression of myelin proteins, including 2', 3'cyclic 3'-nucleotide phosphodiesterase (CNP), myelin basic protein (MBP) and proteolipid protein (PLP). To study the role of stage specific cell

NATO ASI Series, Vol. H 51
Cell to Cell Signals in Plants and Animals
Edited by V. Neuhoff and J. Friend

surface antigens in oligodendrocyte differentiation a series of antibody perturbation experiments were performed: it was found that treatment of proligodendrocytes with monoclonal antibody O4 (4) induced the formation of cellular aggregates and as a consequence stimulated the expression of specific myelinogenic properties, including occurrence of myelin basic protein, MBP messenger RNA and CNP activity (5). Antibodies against galactocerebroside (6), on the other hand, blocked the developmental progression from the $O4^+$ proligodendrocyte to the $GalC^+$ phenotype (7). The identity of the antigen responsible for the perturbation could not be firmly established, since the antibodies crossreacted with both galactocerebroside and sulfatide as well as an unidentified antigen "Y".

SIGNAL TRANSDUCTION AND REGULATORY EVENTS IN MYELIN-FORMING CELLS

Experimental studies with immunoisolated O—2A progenitors revealed the importance of several polypeptide growth factors for normal glial cell development. Platelet-derived growth factor (PDGF) and fibroblast growth factor (FGF) are both mitogenic for the O—2A progenitor cells (8, 9). PDGF is a disulfide-linked dimer of A and B chains, with the structure AA, AB or BB depending on its source. O—2A progenitor cells seem to possess only type A PDGF receptors and respond better to PDGF-AA than to PDGF-BB (10, 11). In rat optic nerve cell cultures type I astrocytes were identified as the major source of this growth factor. Growth stimulation of O—2A cells by PDGF leads to transcriptional activation of several genes including the protooncogenes c-fos and c-myc (12). O—2A progenitors in culture do not proliferate indefinitely when maintained in the presence of PDGF, but stop responding to this factor after a number of mitotic cycles and spontaneously differentiate into oligodendrocytes. PDGF therefore seems to be important not only for expanding the pool of progenitor cells, but also for controlling the time and rate of cell differentiation (13). The loss of PDGF responsiveness is not elicited at the level of the receptors, but seems to lie at some point downstream the mitotic signalling pathway, since newly-produced oligodendrocytes possess PDGF-receptors, which are indistinguisable from those of dividing O—2A progenitors concerning their ability to stimulate an increase in cytosolic Ca^{2+} and to activate the expression of Fos und Myc proteins.

Fibroblast growth factors are even more strongly mitogenic for progenitor cells and seem to stimulate the expression of PDGF-A-receptors on the surface of these cells while concomitantly blocking the expression of myelin genes (14). Transforming growth factor β (TGF-β), on the other side, which is synthesized by type I astrocytes, seems to inhibit PDGF-induced mitosis (15). Thus it appears that a subtle interplay between several growth factors is necessary to regulate the switch between mitotic growth and differentiation of O—2A lineage cells.

Another important growth and differentiation factor for O—2A lineage cells is insulin-like growth factor I (IGF-I), which promotes proliferation of both O—2A progenitors as well as newly differentiated oligodendrocytes expressing CNP (16). IGF-I furthermore induces bipotential precursors to differentiate into oligodendrocytes (17) and stimulates the synthesis and accumulation of myelin in aggregating brain cell cultures. Experiments with transgenic mice, that over-expressed IGF-I and accordingly developed twice as much myelin as their non-transgenic littermates, suggest that IGF-I acts as an inducer of myelination in vivo as well (18).

Attempts to identify potential growth factors for human adult oligodendrocytes, which eventually could be used to enhance oligodendroglial proliferation after injury or in diseased states (e.g. multiple sclerosis) failed, however, since none of a variety of known factors tested were able to induce cell division in cultured adult human oligodendrocytes (19, 20). Fetal human Schwann cells, in contrast, were effectively stimulated to divide by a number of mitogens, including glial growth factor from bovine pituitary, nerve growth factor and PDGF.

On the intracellular side of the signal transduction pathway cyclic AMP was identified as an important link in the regulatory cascade of oligodendrocyte development and myelin gene expression. Treatment of embryonic oligodendroglial cultures with membrane permeant cyclic AMP analogs (8-bromo-cAMP, dibutyryl-cAMP) accelerated oligodendrocyte differentiation without affecting cell lineage decisions (21). This became apparant by an accelerated synthesis of CNP (22) which was preceded by a two-to three-fold increase of CNP mRNA, indicating that cyclic AMP may be acting at the transcriptional level. Choleratoxin and forskolin which both activate adenylate cyclase activity increased the level of CNP expression in a similar manner as exogenous cyclic AMP analogs.

Furthermore evidence was gained that cyclic AMP is involved as a second messenger in oligodendroglial process formation during regeneration, since stimulation of proteinkinase A with forskolin or dibutyryl cyclic AMP initiated the production of flattened membrane vesicles by cultured oligodendrocytes isolated from adult pig brain. Activation of proteinkinase C by treatment of the cultures with phorbolesters, on the other hand, resulted in the formation of long fiber pocesses (23). Hence it was concluded that fiber formation and the production of flattened membrane sheets are under the control of different kinases and that a balanced interaction of both kinases is very important for oligodendroglial remyelination.

Immunocytochemical localization of proteinkinase C in white matter tracts of adult rat brain (24) confirmed the presence of this enzyme in cells with an oligodendrocyte morphology, whereas myelin itself was not significantly labeled. This further supports a role for proteinkinase C in the regulation of oligodendroglial cell functions. Investigations on ontogenic changes of endogenous substrate proteins for proteinkinase C in rat white matter tracts (25) revealed the presence of a major component of about 66, 000 dalton MW in the soluble fraction, which was most highly phosphorylated in the neonate and progressively decreased with age. In the particle fraction of white matter a number of substrate proteins were detected in MW ranging from 87, 000 to 46, 000 daltons.

Oligodendrocytes do not only receive biochemical signals from their environment but also actively secrete molecules into their growth medium. A discrete number of sulfated and fucosylated glycoproteins (26) as well as chondroitin sulfate proteoglycans (27) were constitutively secreted into the culture medium. Secretion of these compounds was in part dependent on adhesion of the cells to a substrate and temporally modulated (28). A glycoprotein of 45, 000 dalton MW (GP45) , e.g., was induced fourfold between the second and third week in culture. Though as yet no definite functional role could be ascribed to any of the various secretory products, GP45 was suggested to have a trophic or tropic function either on oligodendrocytes themselves or on other cells.

As an experimental alternative to the classical adhesion culture systems rotation-mediated aggregating fetal rat brain cultures were presented which proved particularly suitable to study myelination and demyelination under *in vitro* conditions (29). The principal advantage of this system lies in the formation of three-dimensional structures allowing optimal cell differentation (30). In these cultures compact myelin is formed around axons which shows a biochemical composition similar to those isolated from adult rat brain (31). The histotypic structure of the culture, including myelinated axons is maintained during several months in serum-free culture conditions. The addition of triiodothyronine (T_3) to these cultures increases the activity of two enzymes involved in myelinogenesis: galactosylceramide sulfotransferase (a myelin lipid synthesizing enzyme) and CNP (a myelin-oligodendroglial marker) (32). Furthermore the continuous presence of T_3 is necessary to maintain normal levels of MBP. It was shown that T_3 thereby does not stimulate MBP transcription but significantly influences the stability of MBP mRNAs.

DEMYELINATION, REMYELINATION AND GLIAL CELL TRANSPLANTATION

The mechanisms underlying inflammatory demyelination in demyelinating diseases such as multiple sclerosis and the animal model experimental allergic encephalomyelitis (EAE) were evaluated in detail (33). It was proposed that activated macrophages are largely responsible for the tissue damage in the lesions, being initiated by secretion products of these cells. In particular secretion of plasminogen activator, which catalyzes the formation of plasmin, a trypsin-like neutral protease, could degrade myelin proteins (34, 35). The specificity of the mechanism for myelin could lie in the extreme sensitivity of myelin basic protein to proteolysis. Experimental inhibition of plasminogen, plasmin and other neutral protases in EAE sensitized rats, accordingly, gave a significant protection against the clinical expression of the disease. (36)

Moreover the behaviour of oligodendrocytes in response to axotomy and Wallerian degeneration was reevaluated in detail. It was documented that oligodendrocytes from mature animals are able to survive the loss of their axons for long periods of time and respond to Wallerian degeneration with a small degree of cell proliferation. Limited phagocytosis does occur altough the major part of it is carried out by microglia and macrophages.

Demyelination induced by dietary elemental tellurium was presented as a new nonimmunological animal model for Schwann cell demyelination (37), which appeared uniquely suited for biochemical studies of primary demyelination in the peripheral nervous system (PNS) (38). In this model remyelination is initiated by one week after starting the experiment even in the continued presence of tellurium in the diet. The active component was more exactly defined as the oxidized form, tellurite, which is probably formed after ingestion in the gastrointestinal tract. Its biochemical mode of action lies in the interference with the biosynthesis of cholesterol by blocking the conversion of squalene to squalene epoxide. Though this metabolic block is induced in all tissues, the most prominent site of pathology is the myelinating Schwann cell, which could be explained by the extremly high rate of membrane synthesis, i.e. the relatively high demand for cholesterol, in actively myelinating Schwann cells.

The remyelination capabilities of myelin deficient mutant animals was studied using the shiverer mouse mutation, the myelin of which is completely deprived of myelin basic protein (39) and on the ultrastructural level where myelin appears abnormally thin and uncompacted (40). Following lysolecithin-induced demyelination a spontaneous remyelination of fibers occurs, the chronological process of remyelination being comparable in shiverer and normal mouse spinal cord (41). A striking difference, however, was that in the shiverer mutation from the first step of remyelination ensheathment proceeded by folding rather than by spiralling (42). The newly formed myelin was similar to the surrounding myelin, showing great irregularities in the number and organization of lamellae and exhibiting no clear relationships between the axonal diameter and the myelin thickness. These abnormalities were particularly frequent and severe in the remyelinated areas. By contrast when a graft containing normal oligodendrocyte precursor cells was placed at some distance from the lesioned site in the shiverer spinal cord, cells were able to migrate from the grafted site to the lesion and to participate in the remyelination process (43) In this case, remyelination was mixed, i.e. previously demyelinated shiverer axons were ensheathed either by normal or by shiverer myelin. Since transplanted oligodendrocytes formed normal myelin whereas the myelin produced by host (shiverer) oligodendrocytes showed the typical abnormalities it was concluded that the process of myelin formation in the lesion was completely dependent on the origin of the oligodendrocytes and not on specific axonal properties.

The ability of transplanted glial cells (oligodendrocytes, astrocytes, Schwann cells) to influence axonal remyelination was further investigated using the ethidium bromide lesion model, where glial cells are killed by intracerebral injection of ethidium bromide and the demyelinated axons lie in a glial-free environment (44). Spontaneous remyelination occurs in this system, too, but it is carried out mainly by Schwann cells invading from peripheral sites (45). When mixed glial cell cultures containing a high proportion of oligodendrocytes were transplanted into the lesion, there was an extensive oligodendrocytic remyelination instead. Surprisingly the extent of Schwann cell repair was restored when the proportion of type I astrocytes was depleted in the transplant, whereas high amounts of type I astrocytes increased the extent of oligodendrocyte content of the transplant. This indicates that astrocytes, rather than oligodendrocytes, are the critical factor which determines the distribution of peripheral- and central-type remyelination in this system (46). It was suggested that astrocytes probably prevented the invasion of Schwann cells and additionally potentiated the recruitment of host oligodendrocytes by the production of soluble migratogenic and/or mitogenic factors.

TRANSFECTED CELLS AS A TOOL IN MYELIN RESEARCH

New avenues of glial cell biology were pursued by studies with transfected cells. HELA cells were transfected with cDNA encoding the major myelin glycoprotein of the peripheral nervous system, P_0. One day after transfection P_0 could be immunocytochemically detected in the perinuclear region, consistent with its site of synthesis on the rough endoplasmic reticulum and Golgi apparatus (47, 48). Staining on the cell surface was rather weak, in general. Defined regions of membrane contact between adjacent cells, however, exhibited enhanced expression of P_0 suggesting a role for this protein in membrane adhesion. HELA cells transfected with the 14 Kd rat MBP cDNA unexpectedly exhibited a similar perinuclear staining, though MBP is not an integral membrane protein and in oligodendrocytes is synthesized on free polyribosomes (49). This could be explained by an association of MBP with the cytoplasmic side of the perinuclear membranes, which is consistent with the strong affinity that MBP has for intracellular membranes under physiological conditions. By contrast, CNP_I, which like MBP is synthesized on free polyribosomes in oligodendrocytes (50), largely behaved like a soluble protein in the transfected HELA cells, seeming to fill the cytoplasm, even into the long microvilli on these cells.

To faciliate studies on the molecular and cellular mechanisms controlling the growth of oligodendrocytes and its progenitors in vitro, cells from the rat optic nerve were immortalized by infection with a retroviral vector expressing a temperature sensitive transforming protein (SV 40 antigen). A clonal cell line expressing the A2B5 antigen at the permissive temperature of 33°C was selected. Epidermal growth factor, fibroblast growth factor and platelet-derived growth factor increased the proliferation of these cells at 33°C. Addition of dibutyryl cyclic AMP elevated the number of galactocerebroside positive cells twofold confirming a role for cyclic AMP in the differentiation of oligodendrocytes (22). By raising the ambient temperature to 39°C (the non-permissive temperature of the oncogene) the cells could be induced to differentiate morphologically and immunocytochemically to express some of the features characteristic of oligodendrocytes, including myelin basic protein and proteolipid protein.

Furthermore it was tried to introduce the LacZ gene of *Escherichia coli* (coding for the enzyme β-galactosidase) into mature oligodendrocytes or bipotential progenitor cells to use it as a marker for O—2A lineage studies *in vivo* and for grafting experiments with non-mutant mice. Application of conventional methods for cell transfection (51), however, lead to either low yields of transfected cells or to drastic cell death. More promising results were obtained by retroviral infection (52) of cultured glial precursor cells with a recombinant Moloney murine leukemia virus in which viral structural genes were replaced by the *E. coli* LacZ gene. Up to 7% of the progenitor cells showed expression of the β-galactosidase gene under these conditions and no evidence for gene dilution was obtained.

PHYLOGENETIC ASPECTS OF MYELINATION

The interesting potential of myelin proteins as tools to delineate phylogenetic pathways of vertebrate evolution was demonstrated (53). By a comparative electrophoretic and immunoblot analysis of myelin proteins in the CNS of the various vertebrate classes it was found that in bony and cartilaginous fishes P_0-like glycoproteins occur as major myelin constituents functionally replacing PLP (54, 55). Myelin in the CNS of bony fishes was additionally characterized by a novel component of about 36 Kd MW, termed 36K (56), which is not glycosylated. The myelin protein composition of lungfish spinalcord, which contained a glycosylated PLP instead of a P_0-like component and no 36K was of particular interest in a transitional phylogenetic sense (57). CNS myelin of the coelacanth *Latimeria* which is considered as a living fossil and a common ancestor of land-living vertebrates surprisingly contained unglycosylated PLP, immunologically related to the PLP of tetrapod myelin. On the basis of the immunochemical data a cladogram was constructed illustrating the proposed phylogenetic relationships among the various vertebrate classes, demonstrating closer links of coelocanths to tetrapods than lungfish.

The observation that myelin produced by glial cells in the CNS of fish exhibited striking biochemical similarities with those in the peripheral nervous system of higher vertebrates, prompted the question as to the nature of the myelin-forming cells in the brain of fish. Since mammalian oligodendrocytes and Schwann cells are known to basically differ in their dependence from axonal signals to maintain their myelin-related phenotype (58), the interest was focused on the regulation of myelinogenic expression by fish glial cells. When cultured in the absence of neurons the myelin-forming glia of trout CNS soon acquired a morphology resembling those of mammalian oligodendrocytes. In terms of phenotypic expression, however, interesting parallels to the *in vitro* behavior of Schwann cells were revealed (59): galactocerebroside, which in mammalian oligodendrocytes is persistently expressed over long periods of time *in vitro* in the absence of axonal contacts, rapidly disappeared from the surface of cultured trout oligodendrocytes. In contrast, the fish CNS myelin glycoprotein IP2, which is immunologically related to the major myelin product of Schwann cells P_0 (60), was continuously expressed over weeks in culture in the entire absence of neurons.

In a further phylogenetic approach the ultrastructure of myelin was comparatively investigated by X-ray diffraction and immunoblotting and correlated with available amino acid sequence data to get insight into the molecular basis of myelin membrane adhesion (61). Thus it was shown that the extracellular apposition of myelin lamellae may be mediated either by P_0 or PLP and that a large extracellular spacing is present in P_0-containing myelin. A proposed homophilic interaction between P_0 molecules may mediate the extracellular apposition and is likely to be hydrophobic. A comparison between shark and rodent sequence data for myelin proteins (62) reveals that P_0 has a similar β-propensity curve as the immunoglobulins.

COMPONENTS AND STRUCTURES OF MYELIN

Most recent developments in the field of myelin membrane biochemistry were elucidated. Two novel protein components were described which appeared particulary promising for studies on myelin formation and maintenance: a quantitatively minor polypeptide was isolated from bovine myelin with an apparant molecular weight of 55 Kd as determined by SDS-polyacrylamidegel electrophoresis (63). This protein, termed MOG (for myelin-oligodendrocyte glycoprotein) tended to associate with PLP and its isoform DM-20 during isolation and seemed to play a role in the immunopathology of antibody mediated demyelination in vivo. Using mouse monoclonal antibodies against MOG the protein was localized to the external surface of myelin and oligodendrocytes in actively myelinating rats (64, 65). Developmental studies *in vitro* showed that MOG is an antigenic marker for mature oligodendrocytes, which is expressed after the major structural myelin proteins (66).

Furthermore a mannose-binding protein was detected in rat white matter, which corresponded to a molecule previously isolated from the developing rat cerebellum and therefore was termed "cerebellar soluble lectin" (CSL) (67). Active CSL molecules occured as macromolecular complexes consisting of about 40 identical monomers formed by either a 33 Kd or a 31, 5 Kd component. The lectin was immunohistochemically localized in CNS and PNS myelin, and in oligodendrocytes as well as in Schwann cells (68). Oligodendrocytes allowed to adhere to CSL coated surfaces were immediately and significantly stimulated to divide (69). Two major ligands for CSL were found in cultured oligodendrocytes with apparent molecular weights of 100 Kd and 16 Kd, respectively. In isolated myelin membranes a series of additional ligands were identified, including glycoproteins of 50, 31, 29, 24 and 19 Kd MW. The 31 Kd component was identified as an axonal glycoprotein present on the whole surface of young axons which disappeared almost completely at the period of synaptogenesis (70). A hypothetical model was presented suggesting a role for CSL in cell and membrane adhesion functionally related to myelination.

Evidence was obtained for the presence of muscarinic cholinergic receptors in rat CNS myelin membranes and the interaction of these receptors with phospholipase C and the adenylate cyclase system (71, 72). In addition the occurrence of guanine nucleotide regulatory proteins (G-proteins) was revealed and efforts were made to isolate them. A polypeptide of 40 Kd MW, which was ADP-ribosylated if myelin was treated with pertussis toxin, was tentatively identified as the α-subunit of the G-protein linking the muscarinic receptor to phospholipase C. Furthermore a series of GTP-binding proteins of relatively low molecular weight (19—22 Kd) were found in myelin as well as in crude membrane fractions of oligodendrocytes (73). It was suggested that these small GTP-binding proteins, which are generally thought to mediate intracellular membrane traffic leading to secretion (74), eventually could be involved in the process of myelin assembly.

MOLECULAR BIOLOGY OF GENES CODING FOR MYELIN PROTEINS

Another topic of major interest was related to the molecular structure and regulation of genes coding for myelin proteins. A detailed description of the amino acid sequence and the gene structure of the proteolipid protein was given. Based on biochemical and immunocytochemical analyses a membrane integration model was presented for this myelin compound (75, 76), embracing three α-helices which span the hydrophobic core of the myelin bilayer and two cis-membranal α-helices. According to this model disulfide bonds on the extracellular side connect the N-terminus with the penultimate hydrophobic sequence at the C-terminus, leading to a clustering of the trans and cis-membranal helices. The majority of the SH-groups and disulfide bonds within the hydrophobic domains are suggested to be oriented toward the extracytosolic surface, the others within the first and third intermembranal domain. Five disulfide bonds lead to a compact arrangement of the protein. The acylation at threonin_{208} fostering hydrophobic interactions with the apposed next membrane layers and the surplus positive charges carried by the outer surface allowing ionic interaction with anionic polar head groups of membrane lipids further add to the compacting properties of PLP. The biochemical data on the primary structure of the protein were confirmed by nucleotide sequence analyses of rat and human PLP cDNA (77, 78). Analyses of the PLP gene structure revealed the presence of seven exons spread over 17 kilobases (79). The first intron, separating exons 1 and 2, is 7 Kb in length. Exon 1 contains only the 5' untranslated region, a methionine codon, and a guanosine nucleotide for the first codon triplet. Multiple mRNA transcription start sites within exon 1 were identified by S1 nuclease protection assays, allowing to assign putative cap sites and a TATA box (80). Moreover the precise relationship of PLP and the related DM-20 protein was determined (81). It was found that DM-20 mRNA represents a subpopulation of PLP gene transcripts, which due to an internal deletion at the 3' end of exon 3 contains only exon 3a and all other exons, while PLP transcripts utilize exons 3a and 3b. This internal deletion occurred when an alternative donor splice signal within exon 3 was used, removing the 3' half of exon 3 as intron material. By Southern blot hybridization analysis using a suitable Eco RI fragment of the human PLP gene and restricted DNA from human-mouse hybrid cell lines the PLP gene was allocated to chromosome $X_{q12-q22}$ (82).

Furthermore the regulation of expression of the MBP gene during brain development was studied. Transcriptional activation of this gene was revealed by measuring the rate of MBP gene transcription in isolated nuclei. Evidence for the region of the MBP gene required to respond to developmental signals was gained from experiments with the shiverer dysmyelination mutant, which shows correct developmental regulation of transcription of the undeleted 5' portion of the gene. Hence it was concluded that the required sequence elements reside in the 5' half of the gene. At later stages of development (three weeks postnatally) a significant decrease in MBP gene transcription occurs in normal mouse brain, whereas in the shiverer mouse, which cannot synthesize MBP, transcription of the undeleted 5' end of the MBP gene continues at the maximal rate seen at three weeks (83, 84). Hence the accumulation of mature MBP, or the synthesis of mature, compact myelin seem to be essential for the normal downregulation of MBP gene expression at later stages in myelination (regulatory feed back loop). In addition a strategy for direct cloning of MBP transcription-inducing factors was developed: it involves the creation of a "primed" cell line which carries a selectable marker gene (e.g. neomycin resistance gene) under control of the MBP promoter but which does not express its endogenous MBP gene. The primed cell line is then transfected with human genomic DNA and selective pressure is applied to grow only those cells which now express the selectable marker gene. The human gene responsible is then cloned based on proximity to Alu sequences, or marking with cotransfected prokaryotic DNA.

Furthermore the regulation of MBP gene expression was studied in the myelin deficient (mld) mouse mutant, to get insight into the molecular basis of the genetic defect. A drastic reduction in the concentration of translatable MBP mRNA was found in the mutants (85), in spite of normal rates of MBP transcription measured in the brain nuclei. This was due to the occurrence of antisense RNA in the mld nuclei which was not transported to the cytoplasm and thus lead to an inhibition of MBP gene expression (86). Much of the residual sense RNA was arrested in the nuclear compartment as well and hence was not translated into protein either. The MBP_{mld} allele additionally gave rise to pleiotropic effects on the expression of other myelin protein genes as well, being most pronounced in the case of myelin-associated glycoprotein (MAG). MAG concentration in mld myelin was greatly reduced (87), though a large amount of MAG was processed in the endoplasmic reticulum of mld oligodendrocytes. It was concluded that the MBP_{mld} allele influences MAG expression in two ways, i.e. directly by increasing MAG gene transcription and indirectly through destabilization of the myelin sheath, which stimulates MAG degradation.

Finally the cascade of gene expression that leads to myelination in the peripheral nervous system was analysed using purified Schwann cell cultures from neonatal rat sciatic nerve (88). This cascade requires about 24 hours for completion and culminates in the expression of genes coding for myelin-specific structural proteins. Both for initial induction and for maintained expression of these genes Schwann cells require an ongoing axonal contact. Hence in permanently transected sciatic nerves *in vivo* as well as in cell cultures lacking neurons Schwann cells rapidly downregulate mRNAs for major myelin proteins to nearly background levels (89). The necessity for axonal contacts can be *in vitro* substituted by agents elevating intracellular cyclic AMP, like forskolin, cholera toxin or membrane permeant cyclic AMP analogs. Since upregulation of myelin mRNA levels following elevation of intracellular cyclic AMP takes 18—24h and is beyond maintained at high levels, it is assumed that cyclic AMP induction acts via a cascade of intermediate regulatory steps.

A cyclic AMP inducible protein, termed SCIP, was detected, which is expressed at particularly high levels in the rapidly myelinating optic and sciatic nerves and which is absent in non-neuronal tissues (90). In its putative DNA binding domain this protein is stucturally related to classical transcription factors (91). Analysis of the kinetics of cyclic AMP induction of the SCIP gene relative to the induction of major myelin genes revealed that upregulation of SCIP mRNA precedes those of myelin gene mRNAs by approximately 12 hours and remains stable as long as cyclic AMP is elevated. Moreover evidence was obtained that the SCIP gene is under the control of a transcriptional repressor such as c-jun; accordingly, inhibition of the synthesis of this repressor by pretreatment of the cultures with cycloheximide markedly potentiated the subsequent induction of SCIP mRNA by cyclic AMP.

REFERENCES

1) Raff, M.C., Miller, R.H. and Noble, M.D. (1983): A glial progenitor cell that develops in vitro into an astrocyte or an oligodendrocyte depending on the culture medium. Nature 303: 390—396.

2) Levi, G., Aloisi, F., Wilkin, G.P. (1987): Differentiation of cerebellar bipotential glial precursors into oligodendrocytes in primary culture: Developmental profile of surface antigens and mitotic activity. J. Neurosci. Res. 18: 407—417.

3) Gard, A.L., Pfeiffer, S.E. (1989): Oligodendrocyte progenitors isolated directly from developing telencephalon at a specific phenotypic stage: Myelinogenic potential in a defined environment. Development 106: 119—132.

4) Sommer, I., Schachner, M. (1981): Monoclonal antibodies (O1 to O4): to oligodendrocyte cell surfaces: An immunocytological study in the central nervous system. Dev. Biol. 83: 311—327.

5) Bansal, R., Gard, A.L., Pfeiffer, S.E. (1988): Stimulation of oligodendrocyte differentiation in culture by growth in the presence of a monoclonal antibody to sulfated glycolipid. J. Neurosci. Res. 21: 260—267.

6) Ranscht, B., Clapshaw, P.A., Noble, M., Seifert, W. (1982): Development of oligodendrocytes and Schwann cells studied with a monoclonal antibody against galactocerebroside. Proc. Natl. Acad. Sci. USA 79: 2709—2713.

7) Bansal, R., Pfeiffer, S.E. (1989): Reversible inhibition of oligodendrocyte progenitor differentiation by a monoclonal antibody against surface galactolipids. Pro. Natl. Acad. Sci. USA 86: 6181—6185.

8) Besnard, F., Perraud, F., Sensenbrenner, M., Labourdette, G. (1989): Effects of acidic and basic fibroblast growth factors on proliferation and maturation of cultured rat oligodendrocytes. Int. J. Develop. Neurosci.; in press.

9) Richardson, W.D., Pringle, N., Mosley, J.D., Westermark, B., Dubois-Dalcq, M. (1988): A role for platelet-derived growth factor in normal gliogenesis in the central nervous system. Cell 53: 303—319.

10) Hart, C.E., Forstrom, J.W., Kelly, J.D., Seifert, R.A., Smith, R.A., Ross, R., Murray, M.J., Bowen-Pope, D.F. (1988): Two classes of PDGF receptor recognize different isoforms of PDGF. Science 240: 1529—1534.

11) Hart, I.K., Richardson, W.D., Heldin, C.-H., Westermark, B., Raff, M.C. (1989): PDGF receptors on cells of the oligodendrocyte-type-2 astrocyte (O—2A): cell lineage. Development 105: 595—603.

12) Cohen, D.R., Curran, T. (1988): *fra*-1: a serum-inducible cellular immediate-early gene that encodes a Fos-related antigen. Mol. Cell. Biol. 8: 2063—2069.

13) Noble, M., Murray, K., Stroobant, P., Waterlield, M.D., Riddle, P. (1988): Platelet-derived growth factor promotes division and motility and inhibits premature differentiation of the oligodendrocyte/type-2 astrocyte progenitor cell. Nature 333: 560—562.

14) Mc Kinnon, R.D., Matsui, T.A., Aaronson, S., Dubois-Dalcq, M. (1989): FGF inhibits myelin gene expression an induces the PDGF-a receptor in differentiating 0—2A glial progenitor cells. American Society for Cell Biology, 29th Annual Meeting.

15) Van-Obberghen-Schilling, E., Behar, T., Sporn, M.B., Dubois-Dalcq, M. (1987): Signalling between type 1 astrocytes and their glial 0—2A progenitors: Modulation by transforming growth factor-beta (TGFβ). J. Cell Biol. 105 (4): (Part 2): 318a.

16) McMorris, F.A., Smith, T.M., DeSalvo, S., Furlanetto, R.W. (1986): Insulin-like growth factor I/somatomedin C: a potent inducer of oligodendrocyte development. Proc. Natl. Acad. Sci. USA 83: 822—826.

17) McMorris, F.A., Dubois-Dalcq, M. (1988): Insulin-like growth factor I promotes cell proliferation and oligodendroglial commitment in rat glial progenitor cells developing in vitro. J. Neurosci. Res. 21: 199—209.

18) Carson, M.J., Behringer, R.R., Mathews, L.S., Palmiter, R.D., Brinster, R.L., McMorris, F.A. (1988): Myelin content increased in transgenic mice producing elevated levels of insulin-like growth factor-I (IGF-I). Neurosci. Absts. 14: 119.

19) Yong, V.W., Kim, S.U., Pleasure, D.E. (1988): Growth factors for fetal and adult human astrocytes in culture. Brain Res. 444: 59—66.

20) Yong, V.W., Kim, S.U., Kim, M.W., Shin, D.H. (1988): Growth factors for human glial cells in culture. Glia 1: 113—123.

21) Raible, D.W., McMorris, F.A. (1989): Cyclic AMP regulates the rate of differentiation of oligodendrocytes without changing the lineage commitment of their progenitors. Dev. Biol. 133: 437—446.

22) McMorris, F.A. (1983): Cyclic AMP induction of the myelin enzyme 2', 3'-cyclic nucleotide 3'-phosphohydrolase in rat oligodendrocytes. J. Neurochem. 41: 506—515.

23) Schröter, J., Althaus, H.H. (1987): The phorbolester TPA dramatically accelerates oligodendroglial process regeneration. Naturwissenschaften 74: 393—394.

24) Girard, P.R., Mazzei, G.J., Wood, J.G., Kuo, J.F. (1985): Polyclonal antibodies to phospholipid/Ca^{2+}-dependent protein kinase and immunocytochemical localization of the enzyme in rat brain. Proc. Natl. Acad. Sci. USA 82: 3030—3034.

25) Turner, R.S., Raynor, R.L., Mazzei, G.J., Girard, P.R., Kuo, J.F. (1984): Developmental studies of phospholipid-sensitive Ca^{2+}-dependent protein kinase and its substrates and phosphoprotein phosphatases in rat brain. Proc. Natl. Acad. Sci. USA 81: 3143—3147.

26) Yim, S.H., Szuchot, S. (1989): Oligodendrocyte products of secretion: I. Isolation and initial characterization. Submitted.

27) Yim, S.H., Szuchet, S. (1988): Proteoglycans associated with oligodendrocyte membrane. Trans. Am. Soc. Neurochem. 19: 260.

28) Szuchet, S., Yim, S.H. (1989): Oligodendrocyte products of secretion: II. Temporal modulation. Submitted.

29) Honegger, P., Matthieu, J.-M., Lassmann, H. (1989): Demyelination in brain cell aggregate cultures, induced by a monoclonal antibody against the myelin/oligodendrocyte glycoprotein (MOG). Schw. Arch. Neurol. Psy. 140: 1013.

30) Seeds, N.W. (1973): Differentiation of aggregating brain cell cultures. In: Sato, G. (ed.): Tissue Culture of the Nervous System. Plenum Press, New York, p. 35—53.

31) Honegger, P. (1987): Oligodendrocyte development and myelination in serum-free aggregating brain cell cultures. In: Serlupi-Crescenzi, G. (ed.): A Multidisziplinary Approach to Myelin Disease. NATO-ASI Series. Plenum Publishing Corp, New York, p. 161—169.

32) Almazan, G., Honegger, P., Matthieu, J.-M. (1985): Triiodothyronine stimulation of oligodendroglial differentiation and myelination. A developmental study. Dev. Neurosci. 7: 45—54.

33) Norton, W.T., Cammer, W. (1984): Chemical pathology of diseases involving myelin. In: Morell, P. (ed.): Myelin, 2nd edn. Plenum Press, New York, p. 369.

34) Brosnan, C.F., Cammer, W., Bloom, B.R., Norton, W.T. (1980): Initiation of primary demyelination in vivo by a plasminogen activator (urokinase). J. Neuropathol. Exper. Neurol. 39: 344.

35) Cammer, W., Brosnan, C.F., Basile, C., Bloom, B.R., Norton, W.T. (1986): Complement potentiates the degradation of myelin proteins by plasmin: Implications for a mechanism of inflammatory demyelination. Brain Res. 364: 91—101.

36) Brosnan, C.F., Cammer, W., Norton, W.T. (1980): Proteinase inhibitors suppress the development of experimental allergic encephalomyelitis. Nature 285: 235—237.

37) Hammang, J.P., Duncan, I.D., Gilmore, S.A. (1986): Degenerative changes in rat intraspinal Schwann cells following tellurium intoxication. Neuropathol. Appl. Neurobiol. 12: 359—370.

38) Harry, G.J., Goodrum, J.F., Bouldin, T.W., Wagner-Recio, M.W., Toews, A.D., Morell, P. (1988): Tellurium-induced neuropathy: Metabolic alterations associated with demyelination and remyelination in rat sciatic nerve. J. Neurochem. 52: 938—945.

39) Dupouey, P., Jacque, C., Bourre, J.M., Cesselin, F., Baumann, N. (1979): Immunochemical studies of myelin basic protein in shiverer mouse devoid of major dense line of myelin. Neurosci. Lett. 12: 113—118.

40) Privat, A., Jacque, C., Bourre, J.M., Dupouey, P., Baumann, N. (1979): Absence of the major dense line in the mutant mouse shiverer. Neurosci. Lett. 12: 107—112.

41) Gumpel, M., Gout, O., Gansmuller, A. (1989): Spontaneous remyelination and intracerebral grafting of myelinating cells in the mammals. In “Neuronal Grafting and Alzheimer Disease”, Gage, F., Privat, A., Chrisren, Y., ed. Springer Verlag: 43—53.

42) Gumpel, M., Gout, O., Lubetzki, C., Gansmuller, A., Baumann, N. (1989): Myelination and remyelination in the central nervous system by transplanted oligodendrocytes using the shiverer model. Discussion on the remyelinating cell population in adult mammals. Dev. Neurosci. 11: 132—139.

43) Gout, O., Gansmuller, A., Baumann, N., Gumpel, M. (1988): Remyelination by transplantated oligodendrocytes of a demyelinated lesion in the spinal cord of the adult shiverer mouse. Neurosci. Lett. 87: 195—199.

44) Graca, D.L., Blakemore, W.F. (1986): Delayed remyelination in the rat spinal cord following ethidium bromide injection. Neuropathol. Appl. Neurobiol. 12: 593—605.

45) Blakemore, D.L. (1983): Remyelination of demyelinated spinal cord axons by Schwann cells. In: Kao, C.C., Bunge, R.P., Reier, P.J., eds. Spinal cord reconstrunction. Ravan Press, New York: 281—293.

46) Blakemore, W.F., Crang, A.J. (1989): The relationship between type-1 astrocytes, Schwann cells and oligodendrocytes following transplantation of glial cell cultures into demyelinating lesions in the adult rat spinal cord. J. Neurocytol. (in press).

47) Poduslo, J., Dyck, P.J., Berg, C.T. (1985): Regulation of myelination: Schwann cell transition from a myelin-maintaining state to a quiescent state after permanent nerve transfection. J. Neurochem. 44: 388—400.

48) Trapp, B.D., Itoyama, Y., Sternberger, N.H., Quarles, R.H., Webster, H. de (1981): Immunocytochemical localization of P_0 protein in Golgi membranes and myelin of developing rat Schwann cells. J. Cell. Biol. 90: 1—6.

49) Colman, D.R., Kreiblich, G., Frey, A.B., Sabatini, D.D. (1982): Synthesis and incorporation of myelin polypeptides into CNS myelin. J. Cell. Biol. 95: 598—608.

50) Trapp, B.D., Bernier, L., Andrews, S.B., Colman, D.R. (1988): Cellular and subcellular distribution of 2'3-cyclic nucleotide 3'-phosphodiesterase and its mRNA in the rat central nervous system. Neurochem. 51: 859—868.

51) Chen, C., Okayama, H. (1987): High efficiency transformation of mammalian cells by plasmid DNA. Mol and Cell. Biology 7: 2745—2752.

52) Price, J., Turner, D., Cepko, C. (1987): Lineage analysis in the vertebrate nervous system by retrovirus mediated gene transfer. Proc. Natl. Acad. Sci. 84: 156—160.

53) Waehneldt, T.V., Matthieu, J.-M., Jeserich, G. (1986a): Appearance of myelin proteins during vertebrate evolution. Neurochem. Int. 9: 463—474.

54) Waehneldt, T.V., Kiene, M.-L., Malotka, J., Kiecke, C., Neuhoff, V. (1984): Nervous system myelin in the electric ray, *Torpedo marmorata*: morphological characterization of the membrane and biochemical analysis of its protein components. Neurochem. Int. 6: 223—235.

55) Jeserich, G., Waehneldt, T.V. (1986): Bony fish myelin: evidence for common major structural glycoproteins in central and peripheral myelin of trout. J. Neurochem. 46: 525—533.

56) Jeserich, G. (1983): Protein analysis of myelin isolated from the CNS of fish: Developmental and species comparison. Neurochem. Res. 8: 957—969.

57) Waehneldt, T.V., Matthieu, J.-M., Jeserich, G. (1986b): Major central nervous system myelin glycoprotein of the African lungfish (*Protopterus dolloi*): crossreacts with myelin proteolipid protein antibodies indicating a close phylogenetic relationship with amphibians. J. Neurochem. 46: 1387—1391.

58) Mirsky, E., Winter, J., Abney, E.R., Pruss, R.M., Gavrilovic, J., Raff, M.C. (1980): Myelin specific proteins and glycolipids in rat Schwann cells and oligodendrocytes in cultures. J. Cell. Biol. 84: 483—393.

59) Jeserich, G., Rauen, T. (1990): Cell cultures enriched in oligodendrocytes from the CNS of trout in terms of phenotypic expression exhibit paralells with cultured rat Schwann cells. Glia 2: (in press).

60) Jeserich, G., Waehneldt, T.V. (1987): Antigenic sites common to major fish myelin glycoproteins (IP): and to major tetrapod PNS myelin glycoprotein (P_0): reside in the amino acid chains. Neurochem. Res. 12: 821—825.

61) Kirschner, D.A., Inouye, H., Ganser, A.L., Mann, V. (1989): Myelin membrane structure and composition correlated: a phylogenetic study. J. Neurochem., in press.

62) Saavedra, R.A., Fors, L., Aebersold, R.H., Arden, B., Horvath, S., Sanders, J., Hood, L. (1989): The myelin proteins of the shark brain are similar to the myelin proteins of the mammalian peripheral nervous system. J. Mol. Evol. 29: 149—156.

63) Linington, C., Webb, M., Woodhams, P.L. (1984): A novel myelin-associated glycoprotein defined by a mouse monoclonal antibody. J. Neuro. Immunol. 6: 387.

64) Linington, C., Bradl, M., Lassmann, H., Brunner, C., Vass, K. (1988): Augmentation of demylination in rat acute allergic encephalomyelitis by circulating mouse monoclonal antibodies directed against a myelin/oligodendrocyte glycoprotein (MOG). Am. J. Pathol. 130: 443.

65) Brunner, C., Lassmann, H., Waehneldt, T.V., Matthieu, J., Linington, C. (1989): Differential ultrastructural localization of myelin basic protein, myelin/oligodendroglia glycoprotein, and 2', 3'-cyclic nucleotide 3'-phosphodiesterase in the CNS of adult rats. J. Neurochem. 52: 296.

66) Scolding, N.J., Frith, S., Linington, C., Morgan, B.P., Campell, A.K., Compston, D.A.S. (1989): Myolinoligodendrocyte glycoprotein (MOG): is a surface marker of oligodendrocyte maturation. J. Neuroimmunol. 22: 169.

67) Zanetta, J.P., Meyer, A., Kuchler, S., Vinsedon, G. (1987): Isolation and immunochemical study of a soluble cerebellar lectin delineating its structure and function. J. Neuochem. 49: 1250—1257.

68) Kuchler, S., Fressinaud, C., Saelieve, L.L., Vincedon, G., Zanetta, J.P. (1988): Cerebellar soluble lectin is responsible for cell adhesion and participates in myelin compaction in cultured rat oligodendrocytes. Dev. Neurosci. 10: 199—212.

69) Fressinaud, C., Kuchler, S., Sarlieve, G., Vincendon, G., Zanetta, J.P. (1988): Adhesion on a matrix made of the nervous lectin CSL induces increased proliferation of oligodendrocytes. C. R. Acad. Sci. Paris 307: 863—868.

70) Kuchler, S., Herbein, G., Sarlieve, L.L., Vincedon, G., Zanetta, J.P. (1989): An endogenous lectin CSL interacts with major glycoprotein components in peripheral nervous system myelin. Cell Molec. Biol. (in press).

71) Larocca, J.N., Ledeen, R.W., Dvorkin, B., Makman, M.H. (1987a): Muscarinic receptor binding and muscarinic receptor mediated inhibition of adenylate cyclase in rat brain myelin. J. Neurosci. 7: 3863—3876.

72) Larocca, J.N., Cervone, A., Ledeen, R.W . (1987b): Stimulation of phosphoinositide hydrolysis in myelin by muscarinic agonist and potassium. Brian Res. 436: 357—362.

73) Bernier, L., Horvath, E., Braun, P. (1989): GTP binding proteins in CNS myelin. Trans. Amer. Soc. Neurochem. 20: 254.

74) Bourne, H.R. (1988): Do GTPases direct membrane traffic in secretion? Cell 53: 669—671.

75) Stoffel, W., Hillen, H., Giersiefen, H. (1984): Structure and molecular arrangement of proteolipid protein of central nervous system myelin. Proc. Natl. Acad. Sci. USA 81: 5012—5016.

76) Stoffel, W., Subkowski, T., Jander, S. (1989): Topology of proteolipid protein in the myelin membrane of central nervous system. Biol. Chem. Hoppe Seyler 370: 165—176.

77) Schaich, M., Budzinski, R.M., Stoffel, W. (1986): Cloned proteolipid protein and myelin basic protein cDNA. Biol. Chem. Hoppe-Seyler 367: 825—834.

78) Milner, R.J., Lai, C., Nave, K.-A., Lenoir, D., Ogata, J., Sutcliffe, J.G. (1985): Nucleotide sequences of two mRNAs for rat brain myelin proteolipid protein. Cell 42: 931—939.

79) Diehl, H.J., Schaich, M., Budzinski, R.M, . Stoffel, W. (1986): Individual exons encode the integral membrane domains of human myelin proteolipid protein. Proc. Natl. Acad. Sci. USA 83: 9807—9811.

80) Macklin, W.B., Campagnoni, C.W., Deininger, P.L., Gardinier, M.V. (1987): Structure and expression of the mouse myelin proteolipid protein gene. J. Neurosci. Res. 18: 383—394.

81) Nave, K.-A., Lai, C., Bloom, F.E., Milner, R.J. (1987): Splice side selection in the proteolipid protein (PLP): gene transcript and primary structure of the DM-20 protein of CNS myelin. Proc. Natl. Acad. Sci. USA 84: 5665—5669.

82) Mattei, M.G., Alliel, P.M., Dautigny, A., Passage, E., Pham-Dinh, D., Mattei, J.F., Jolles, P. (1986): The gene encoding for the major brain proteolipid (PLP): maps on the q-22 band of the human X chromosome.

83) Roach, A., Takahashi, N., Pravtcheva, D., Ruddle, F., Hood, L. (1985): Chromosomal mapping of mouse myelin basic protein gene and structure and transcription of the partially deleted gene in shiverer mutant mice. Cell 42: 149—155.

84) Molineaux, S.M, . Engh, H., de Ferra, F., Hudson, L., Lazzarini, R.A (1.986): Recombination within the myelin basic protein gene created the dysmelinating shiverer mouse mutant. Proc. Natl. Acad. Sci. USA 83: 7542—7546.

85) Roch, J.M, . Braun-Luedi, M., Cooper, B.J., Matthieu, J.M. (1986): Mice heterozygous for the mld mutation have intermediate levels of myelin basic protein mRNA and its translational products. Mol. Brain Res. 1: 137—144.

86) van der Krol, A.R., Mol, J.N.M., Stuitje, A.R. (1988): Modulation of eukaryotic gene expression by complementary RNA or DNA sequences. BioTechniques 6: 958—975.

87) Matthieu, J.M., Roch, J.M., Omlin, F.X., Rambaldi, I., Almazan, G., Braun, P. (1986): Myelin instability and oligodendrocyte metabolism in myelin-deficient mutant mice. J. Cell Biol. 103: 2673—2682.

88) Lemke, G. (1988): Unwrapping the genes of myelin. Neuron 1: 535—543.

89) Lemke, G., Chao, M. (1988): Axon regulate Schwann cell expression of the major myelin and NGF receptor genes.

90) Monuki, E.S., Weinmaster, G., Kuhn, R., Lemke, G. (1989): SCIP: a glial POU domain gene regulated by cyclic AMP. Neuron 3, in press.

91) Sturm, R.A., Das, G., Herr, W. (1988): The ubiquitous octamer-binding protein Oct-1 contains a POU domain with a homeobox subdomain. Genes. Dev. 2: 1582—1599.

PLASTICITY
IN CELL TO CELL COMMUNICATION

Exploring Nature's Roulette Wheel: Chaos in Biological Systems

Arun V. Holden and Lars Folke Olsen

Centre for Nonlinear Studies, The University,
Leeds, LS2 9JT, U.K. and Department of Biochemistry,
Odense University, Odense, Denmark

1. INTRODUCTION

Chaos is irregular, unpredictable behaviour generated by a simple, nonlinear deterministic system; the paradoxical combination of unpredictability (in the long term) with simple determinism results from sensitivity to initial conditions. Turbulent irregularity dominates macroscopic physical phenomena, from laboratory to galactic physics, and the current enthusiasm for chaotic dynamics [14, 44] follows from the discovery of universality in chaos [7]. Although irregularity is characteristic of biological signals, it has usually been approached using the theory of stochastic processes; for example, the

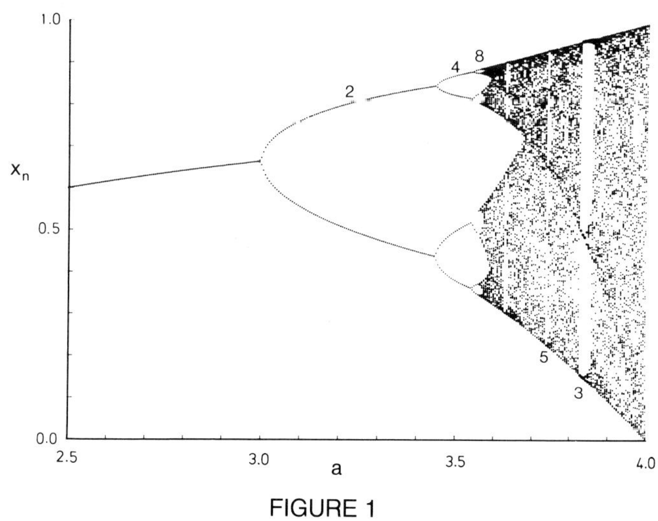

FIGURE 1
Bifurcation diagram for the logistic equation (1) plots points in the attracting set against the bifurcation parameter a.

NATO ASI Series, Vol. H 51
Cell to Cell Signals in Plants and Animals
Edited by V. Neuhoff and J. Friend

analysis of miniature endplate potentials, or of single channel currents, is based on the Poisson point process. Interest in chaos in biology dates from Bob May's review article [27], which considered the simple quadratic map:

$$x_{n+1} = ax_n(1 - x_n), \qquad 0 < a \leqslant 4. \tag{1}$$

x_n, $0 < x \leqslant 1$, may be considered to represent the relative number of individuals in a population at generation n, where the successive generations do not overlap in time. a is a parameter that controls the steepness of the single-humped curve of x_{n+1} against x_n.

If $a < 3$, whatever the initial value x_1, successive iterations will settle down to a fixed value that depends on a; there is a stable, fixed point attractor. At $a = 3$ there is a qualitative change in behaviour, the fixed point becomes unstable and a stable, period-2 attractor emerges: there is a bifurcation at the value of the bifurcation parameter $a = 3$. This is a period-doubling bifurcation; further increases in the parameter a give a cascade of period-doubling bifurcations until at $a \simeq 3.58$ the period is infinite, or the successive iterations do not form a periodic sequence: they are apparently random but since they have been generated by a simple, deterministic process they are an example of *chaos*. The values a_k at which period-doubling occurs converge in a geometric progression, $a_k \simeq a_\infty - c\,\mathcal{F}^{k}$, where c is a constant (2.6327 for the quadratic map) and $\mathcal{F}$ is Feigenbaum's constant with a value of 4.6692.. For $3.58 \leqslant a \leqslant 4$ the band of chaos is interrupted by windows of periodicity, that occur in a peculiar and intriguing order. This is the Sharkovskii sequence: if the integers are ordered, first the odd numbers, and their even powers, in ascending order; and then the powers of 2 in descending order we will obtain all the integers above 1 in a shuffled order starting with 3 and ending with 2:

$3 \rightarrow 5 \rightarrow 7 \rightarrow \ldots 6 \rightarrow 10 \rightarrow 14 \ldots 12 \rightarrow 20 \rightarrow 28 \ldots 3.2^n \rightarrow 5.2^n \rightarrow 7.2^n \rightarrow 2^m \rightarrow 2^{m-1} \rightarrow \ldots 16 \rightarrow 8 \rightarrow 4 \rightarrow 2$.

For the quadratic map, if at a value a^* of the bifurcation parameter a a period p cycle occurs, then all cycles of period q such that $p \rightarrow q$ in the Sharkovskii sequence must also have occurred. Hence "period 3 implies chaos" [25]. A further feature of the bifurcation diagram is its *self-similarity;* within the periodic windows the same pattern of a period-doubling cascade into chaotic bands, with windows of periodicity, repeats itself.

These numerical peculiarities may be interesting, but they do not appear to be very relevant to cell-to cell signalling. Their importance follows from universality; the same bifurcation patterns are seen in other maps, and for the same class of maps the bifurcation patterns are quantitatively as well as qualitatively similar — Feigenbaum's constant $\mathcal{F}$ is universal, and may be computed or estimated from actual experimental data. Further, the same patterns are seen for nonlinear ordinary differential equations, and, more importantly, for experimental observations from a wide range of chemical, physical and biological systems. The observation of chaotic behaviour, period-doubling bifurcations into chaos, and other routes into chaos in biological experiments means that one can understand some of the complicated behaviours seen in biology using very simple nonlinear models, which can be quantitatively accurate even though they may not be mechanistically precise.

2. BIOCHEMICAL SYSTEMS AND MODELS

Until recently only few biochemists believed that chaos and other types of complex dynamics might play any role in the functioning of the living cell. There are several reasons, apart from technical difficulties, why this is so. One reason should be sought for in the word "homeostasis". Homeostasis, which occurs frequently in biochemical and physiological textbooks, implies the maintainance of steady states in an organism by regulatory processes. Even today many biochemists assume that cell metabolism sits in some stable steady state. Consequently experimental observations of aperiodic fluctuations are usually regarded as sampling errors or as due to random perturbations from without. Another reason is that despite the enormous amount of information now available about the kinetics and regulatory properties of biochemical pathways relatively few attempts have been made to simulate their long-term behaviour. Most studies of biochemical models have been restricted to searches for stable steady states.

However, there is a growing experimental and theoretical literature demonstrating that biochemical reactions are not likely to be in a steady state. A review by Rapp lists about 140 cellular oscillators [35] and this number is still increasing. Many of these oscillators are involved in inter- and intracellular signal transduction and cellular differentiation. Some of them may also function as biological clocks. Furthermore, it is common knowledge that biochemical systems have evolved to optimize their thermodynamic efficiency. As pointed out by Ross and Schell [37] oscillating reactions show greater efficiency than reactions proceeding at a constant rate. The same holds for chaotic processes. Feedback regulations are important factors in adaptations to changes in environment. Adaptation requires a certain amount of variability and complex dynamics can provide the cell with such variability in contrast to steady state behaviour which cannot (Conrad [6]). Finally, steady state and simple periodic behaviours are often very sensitive to random perturbations in contrast to chaotic dynamics which are usually relatively robust to such perturbations (Schaffer et al. [41], Olsen [29])

Chaos is a phenomenon associated with nonlinear dynamical systems. No linear system can show such behaviour. Biological systems in general and biochemical systems in particular are governed by nonlinear rate laws. For example, the simplest equation that can describe a biochemical reaction, the Michaelis-Menten equation, is nonlinear and most enzymes show kinetics which are far more complex than this equation. Furthermore, many enzymes are regulated by their substrates or products or by other intermediates in the same or another metabolic pathway. Such feedback regulations add to the inherent nonlinearities. Finally cell metabolism is compartmentalised through the existence of cellular and intracellular membranes. The diffusive exchanges of low molecular weight metabolites through these membranes are highly regulated (Sporns et al. [43]). Consequently cell metabolism contains all the ingredients that are necessary for complex dynamics to show up. This idea has received further suppport from studies of long-term behaviours of simple biochemical models. Most of the models showing chaos and other types of complex dynamics involve two feedback regulated enzymes arranged in series or one feedback regulated enzyme with a periodically varying substrate (Decroly; Sinha and Ramaswamy; Moran and Goldbeter, all in [9]). Since a single feedback regulated reaction can generate simple periodic oscillations, the coupling of two feedback regulated enzymes is equivalent to the coupling of two periodic oscillators. It is well known that such a coupling often results in chaotic dynamics (Tomita [46]). Another more realistic biochemical model is the model of cyclic AMP (cAMP) signalling in the slime mold *Dictyostelium discoideum* (Goldbeter and Martiel in [9]. Upon starvation these amoebae aggregate in a wave-like manner around centres in response to cAMP signals. The model of the c-AMP signalling system is based on activation of intracellular cAMP synthesis as a response to binding of cAMP to the cell surface receptor, and on desensitisation of the receptor on incubation with cAMP. The model shows periodic and chaotic oscillations of intracellular cAMP. This may explain that while the wild type of *Dictyostelium* amoebae aggregate in a periodic fashion a mutant Fr17 shows aperiodic aggregation (Durston, [11])

In addition there are experimental observations of chaos in biochemical systems. In gylcolysis glucose is converted to ethanol and carbon dioxide, with pyruvate and other low molecular weight subatances as intermediates. The function of glycolysis is to generate chemical energy in the form of ATP and intermediates for the biosynthesis of amino acids, fat and membrane lipids. Phosphofructokinase is a key enzyme in the regulation of the flow of metabolites through this pathway (Boiteau and Hess, [3]). In yeast extracts, which are rich in glycolytic enzymes, the intermediates of glycolysis oscillate when glucose is supplied at a constant rate. The regulatory properties of phosphofructokinase are responsible for these oscillations. Chaotic oscillations are observed if the supply of glucose is periodic (Markus et al., [26]; Hess and Markus [18]). The peroxidase-oxidase reaction is another enzyme reaction showing chaotic dynamics. In the reaction NADH is oxidized with molecular oxygen as the electron acceptor. The reaction is catalysed by a single enzyme, peroxidase. When both substrates are supplied continuously, the reaction exhibits simple and complex oscillations and chaos. (Olsen and Degn, [30, 31]). The reaction mechanism is not yet understood in detail. However, it is known that it involves autocatalysis (Degn and Mayer, [10]). A simple model with two autocatalytic reactions arranged in series can simulate the complex periodic oscillations and chaos seen in the experimental system (Olsen, [29]; Aguda and Larter, [1]).

To summarise, there seems to be good indirect evidence that chaos exists in cell metabolism. At this point it is not possible to assign a precise role for such dynamics in the functioning of the cell. However, it is interesting to note that chaotic dynamics may be asociated with a healthy condition, whereas periodic oscillations and steady states may signal a pathological condition (Goldberger and West, in [9]; Pool, [34]).

3. CELLULAR SYSTEMS AND MODELS

The Hodgkin-Huxley equations [19], that describe membrane excitation in the squid giant axon, are the best example in biology of a quantitatively accurate system of nonlinear ordinary differential equations, that have been intensively studied using both numerical and mathematical techniques [20]. The squid giant axon forms part of an escape reflex system, and so it is either at rest, or conducting an action potential. However, by experimental manipulations (constant current injection, change in extracellular fluid composition, action of potassium channel blockers) the membrane can sustain repetitive activity; this corresponds to periodic solutions of the membrane excitation equations, and these periodic solutions emerge via Hopf bifurcation. Labouriau in [9] has applied singularity theory techniques to the Hodgkin-Huxley membrane equations, to examine the change from periodic solutions to single action potentials (homoclinic solutions). Although the Hodgkin-Huxley equations have 4 kinetic variables, are strongly nonlinear, and have a large number of parameters that can act as bifurcation parameters, periodic-to-periodic bifurcations, and chaos, are not seen: in the absence of forcing, the Hodgkin-Huxley membrane equations behaves as if they were a two variable differential system, capable only of periodic and equilibrium solutions. Cardiac muscle is normally rythmically active, and so differential equations that describe the electrical activity of sino-atrial nodal tissue are to be expected to have periodic solutions; for example the Irisawa-Noma [23] model, or its three variable simplification studied by Guevara in [9]. Once again, although the bifurcations into periodicity are well understood, no chaotic behaviour is found, even though for nonlinear differential systems with three or more variables chaos is generic and so to be expected. Chaotic behaviour in the electrical activity of the normal heart has been inferred from the 1/f Fourier spectrum of the EEG (Goldberger and West in [9]).

However, many biological systems, although they may be described by nonlinear differential equations, do not exhibit chaotic solutions, or only exhibit chaotic solutions in peculiar, or very restricted, conditions. This does not mean that chaos is abnormal, or a symptom of a pathology; the normal mode of behaviour of glucose-stimulated pancreatic β-cells in some mice strains is irregular bursting. Lebrun in [9] suggests that the difference between the regular and irregular bursting seen in different strains in mice is a result of dietary produced changes; this would correspond to a bifurcation between periodicity and chaos, with the bifurcation parameter influenced by diet. The differential equations, obtained from voltage clamp experiments, that represent pancreatic β-cells, are similar to those that represent somatic membrane of snail neurones, and the high order system of ordinary differential equations can be reduced to a third order system that has been numerically explored by Chay in [9]. This system shows simple periodic, patterned periodic and chaotic solutions that simulate the background discharge patterns of central neurones, suggesting that the different behaviours may be due to different parameter values for their membrane conductances. The mechanism for burst production is sketched in Figure 2. The variables of the Chay system are membrane potential *V,* potassium activation *n,* and intracellular Ca^{++} activity *C;* the fast variables of the inward current system are so fast that they are replaced by their steady state values as a function

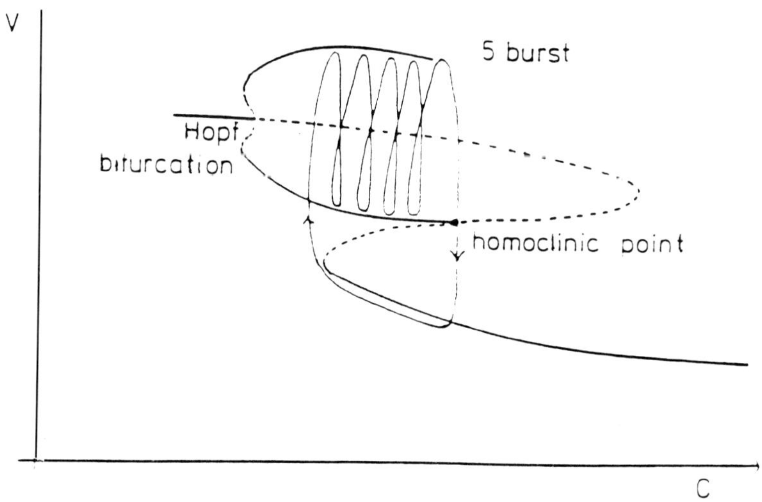

FIGURE 2
Dynamic mechanism of burst production. C acts as a slow variable that modulates the dynamics of the V-n subsystem, which has oscillatory solutions arising at a Hopf bifurcation and which terminate by coallescing with a saddle point, giving a homoclinic orbit.

of membrane potential. During a burst each action potential gives an increase in intracellular Ca^{++}, and when the Ca^{++} exceeds a threshold a Ca^{++}-activated K^{+} selective conductance is activated, which terminates the burst.

The excitation systems that describe the behaviour of nerve and muscle membrane are high order, and contain large number of parameters, and so using numerical integrations to map the possible behaviours that can be obtained with different combinations of parameters would be computationally excessive. A more realistic approach is to use path-following algorithms [4] to obtain bifurcation diagrams. These algorithms follow bifurcations in parameter space, and so the regions where interesting behaviours are to be found may be readily located.

A periodically forced nonlinear oscillator, with an appropriate amplitude and frequency of forcing, can show phase-locked and entrained periodic behaviour and quasi-periodic and chaotic behaviour. Chaos can readily be produced in sinusoidally forced squid membrane, and in numerical solutions of the Hodgkin-Huxley equations; this occurs at frequencies of approximately 200 Hz and has no biological significance. What is noteworthy is the close correspondence seen between the experimental records of Aihara and Matsumoto in [9], and the numerical solutions of the Hodgkin-Huxley equations. Periodically forced squid membrane may not be a biologically relevant preparation, but it is an excellent laboratory testbed for evaluating methods for identifying and quantifying chaos in experimental time series, and has been used by Everson in [9] for assessing methods for the reconstruction of attractors.

4. CELLULAR NETWORKS

Glass and Mackey [13] introduced the idea of a dynamical disease, in which the control parameters of a physiological system are outside their normal range of values, and as a result of bifurcation there is abnormal behaviour. Belair in [9] described the regulation of blood platelet production by a functional differential equation with time delays: the system shows oscillations in platelet concentrations that simulate those seen in thrombocytopenia. Jensen in [9] modelled renal glomerular filtration by a system of eight ordinary differential equations that represent a single nephron and its coupling to four neighbours, and simulated periodic and chaotic fluctuations in proximal intratubular pressure. In both these cases a complicated, structured physiological system is represented by a much simpler caricature, in which both the physiology and the mathematical representation has been grossly simplified. Simplification can introduce artefactual dynamics — for example, the chaotic behaviour of the logistic map is not seen in the differential system it approximates. Although these models do simulate observed regular and irregular oscillations a model may provide a good, quantitatively accurate, simulation of a physiological signal while misrepresenting the dynamics of the system; for a dynamic model to be meaningful the bifurcation structure of the model should describe, and explain, the effects of changes in parameters. The promise of nonlinear dynamics is that it may be possible to describe complicated behaviours in complicated systems by simple, nonlinear models with relatively few variables.

A single neurone can have a regular, or irregular, background discharge, corresponding to periodic or chaotic behaviour. Within the nervous system, neurones are part of an interconnected, interacting network, and the behaviour of a neurone is strongly determined by its connectivity. Labos in [9] has approached the problem of chaotic and patterned behaviour in neural networks by examining the behaviour of small networks of very simple model neurones: threshold devices, piecewise linear interval maps and polynomial recursions. Such networks can provide remarkably convincing simulations of real neuronal discharge patterns, and in general it is not possible to predict the behaviour of the network from a knowledge of its components and connectivity, without actually simulating the behaviour. Thus, within an interacting system like the central nervous system there is the possibility of chaotic behaviour in the sense introduced by Wolfram [48], in that the only way to predict the long term result of a process is to actually run it.

5. POPULATION DYNAMICS

Some of the first mathematical models to exhibit chaos were ecological (May, [27, 28]; Rogers [36]). In fact, to avoid chaos in an ecological model is often a difficult task (Schaffer; Markus et al. in [9]). Nevertheless, most ecologists look upon chaos with a great deal of scepticism, which is to some extent justified as there are few if any real ecological data sets in which the "fieldmarks" of chaos can be seen. The data sets that are available today are either corrupted by too much environmental noise or they are too short to give a real clue about the underlying dynamics (May, [28]; Pool, [33]). The time scales over which natural populations develop are very long and in order to make a case for chaos one would often need records which span hundreds of years. Such records are very rare. One of the few data sets that meets these requirements is the classical Hudson Bay Company fur trade records for the Canadian lynx from 1735 to 1940. The number of furs received by trappers in a given year is assumed to be an indirect measure of the size of the lynx population in that year. If this is correct the lynx population oscillates with a period of about 10 years (Elton and Nicholson, [12]). However, the amplitude of these oscillations vary considerably from one cycle to the next. Schaffer [38] analyzed the lynx records by extracting return maps from the data. These maps, which are graphs that express the population size at one excursion against the population size at the preceding excursion, indicate that the dynamics may be chaotic. However, the lynx data also contains some noise which may prevents an unambiguous diagnosis of the dynamics. Most other population data are considerably worse and it will probably never be possible to present solid evidence for chaos in any of the existing data. One could even suspect that all the existing long-term data are those with only regular dynamics. It is tempting to believe that researchers collecting long-term data are more comfortable with stable populations than populations that jump about in a seemingly random fashion.

One further obstacle to establishing chaos in ecology comes from the study of seasonally breeding insects under natural and laboratory conditions (Hassel et al., [17]). Using short term demographic data, not actual time series of species abundance, these authors analysed 28 insect populations in terms of a simple mathematical model and estimated the model parameters. For each parameter set they determined the model's dynamics; and only one of these populations had parameters that would generate chaos. Although it is generally acknowledged that mathematical models are usually too simple to reproduce faithfully the dynamics of natural populations, this study convinced the majority of ecologists that chaos is probably nothing more than a mathematical curiosity.

Thus in order to find evidence for chaotic dynamics in ecological systems within the foreseeable future, researchers have gone back to the laboratory to look at microbial predator-prey systems in chemostats to which nutrients are continuously supplied. Such experiments are inspired by analyses of predator-prey models which show that for a constant supply of nutrients the populations of the predator and the prey converge either to a steady state or to a simple limit cycle oscillation. However, if the supply of nutrient is seasonally dependent the result is often chaotic dynamics (Kot, [24]). If the same can be realised in the laboratory there is additional evidence that natural populations may sometimes exhibit chaotic dynamics.

Support for these observations comes from a field very much related to ecology, namely human epidemiology. Humans infected by bacteria or a virus are analogous to a predator-prey system. Since the late 18th century the monthly or weekly cases of a number of infectious diseases have been recorded in western industrialised countries. These data sometimes show a seasonal variation throughout the year and on top of this tremendous year-to-year fluctuations (Schaffer in [9]). Traditionally, such fluctuations were regarded as due to environmental noise. However, recently Schaffer and Kot [40] analysed two classical data sets of measles infections in New York and Baltimore by computing return maps. These maps revealed a well defined relationship between infection rates in successive years and hence suggested that the dynamics of measles infections are chaotic. Since then other measles data sets have been tested for chaos with positive results (Olsen in [9]). Not all epidemiologists agree that measles epidemics are chaotic. Some believe that the data are better explained by a high/low 2-year cycle which is perturbed by environmental noise (Pool, [32]). In order to settle the debate Schaffer et al. [42] have computed invariant measures such as Lyapunov exponents and fractal dimensions (see section 6) of the measles data from 8 large cities. If the dynamics are chaotic the Lyapunov exponent is a positive number and the fractal dimension is non-integer. However, if the dynamics correspond to a periodic oscillation the Lyapunov exponent is zero and the fractal dimension is one. For the eight measles data sets the Lyapunov exponents all fell around 0.5 bits/year and the fractal dimension around 2.5. Furthermore, measles dynamics can be simulated by a very simple mathematical model, the Susceptible-Exposed-Infective-Recovered (SEIR) model. If the contact rate, i.e the number of contacts between an infective and susceptibles, is seasonally dependent the dynamics of the SEIR model, with parameters corresponding to measles infections, will change from a simple oscillation to chaos through a series of period-doubling bifurcations as seasonality increases. (Aron and Schwartz, [2]; Schaffer, [40]). It is interesting that for parameters giving rise to chaos the Lyapunov exponent of the SEIR simulations is 0.5 bits/year whereas the fractal dimension is around 2.5 (Schaffer et al. [42]). This agreement between experimental observations and theoretical data leaves little doubt that measles infections in large homogenous populations are chaotic.

6. DETECTING AND QUANTIFYING CHAOS

How does one recognise chaos in a fluctuating time series? Although chaotic systems generate dynamics which at first glance appear to be random thay also enclose signatures in the time series which identify their deterministic origin. Within the last decade various methods have been developed which probe these signatures and help to distinguish between chaos and other types of dynamics. These methods include reconstruction of attractors, construction of Poincare sections and return maps and compuation of Lyapunov exponents and fractal dimensions.

The experimental biologist is often confronted with the fact that only one variable in a multivariable system can be measured. To visualise and characterise a chaotic attractor one would in principle need to know all the variables. However, Takens [45] has shown that certain properties of an attractor can be reconstructed from a single time series *x(t)*, where *x(t)* is one variable in an *n*-dimensional system $\{x(t), y(t), z(t), ...\}$. The reconstruction is made by chosing a time delay, τ, and forming the vector $\{x(t), x(t+\tau), x(t+2\tau), ..., x(t+[m-1]t)\}$. Provided that m is greater or equal to $2n+1$ this new vector will have the same properties as the original *n*-dimensional vector. To illustrate the method we consider a simple sinusoidal oscillation. It is fairly easy to see that if τ is chosen to be smaller than the period a two dimensional reconstruction will be a closed loop corresponding to a limit cycle. The reconstructed attractors can be analysed using the same methods as one would apply to the original attractor.

If we intersect a chaotic attractor with a plane normal to its direction and record the intersection points we end up with a Poincare section. Often this section is a flattened, al-

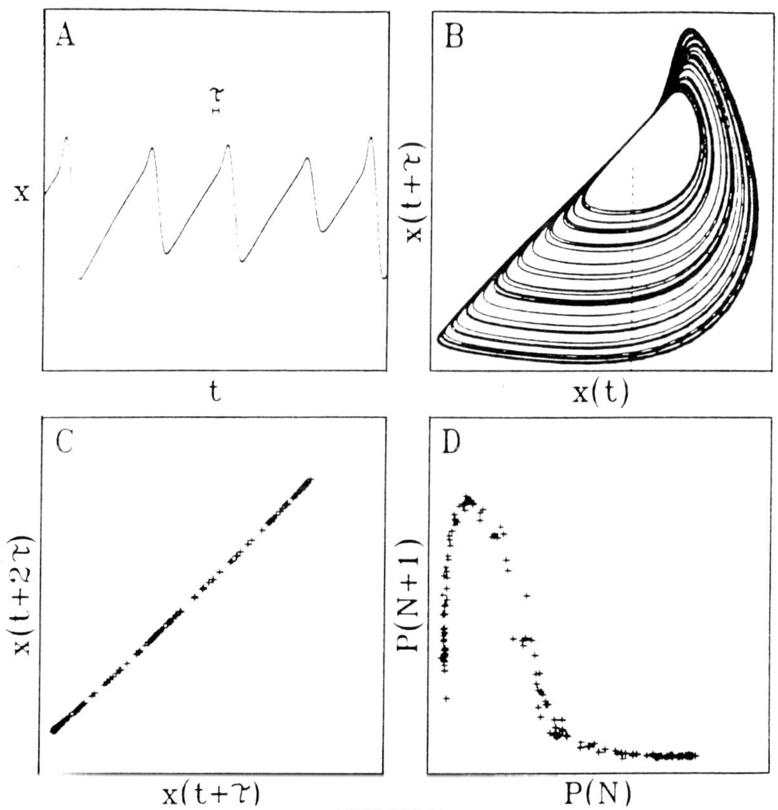

FIGURE 3

Steps in graphical identification of chaos (a) time series, (b) view of reconstructed attractor obtained by plotting x(t + τ) against x(t), (c) Poincare section and (d) return map. Data for Belousov-Zhabotinsky reaction.

most one dimensional curve. If we then plot the distance from the origin of each point on the Poincare section against the distance for the preceeding point we obtain a return map. Such return maps sometimes define a unimodal map as illustrated in Figure 3. However, usually the maps are not unimodal, but appear as a thick bundel of curves, but still with a well defined and recognisable structure. If we apply the same procedure to a noisey signal, e.g. a noise perturbed limit cycle oscillation, we may still obtain a reasonably flat Poincare section. However, the return map will be a random scatter of points with a dense covering. Construction of Poincare sections and return maps is often the first, graphical, approach to detecting chaos in a time series (Aihara and Matsumoto; Hayashi et al.; Schaffer; Olsen, all in [9])

Two other, numerical, properties that can be used to distinguish chaotic attractors from other attractors are their Lyapunov exponents and fractal dimensions. A dynamical system with n independent variables has n Lyapunov exponents. The exponents are generalised eigenvalues and measure the rates of orbital divergence or convergence on the attractor. The exponents are ordered according to their magnitude. The largest Lyapunov exponent is negative if the system corresponds to a stable steady state, zero for a limit cycle oscillation and positive for a chaotic attractor. An algorithm for computing the maximal Lyapunov exponent of experimental time series was devised by Wolf et al. [47]. The algorithm makes use of the reconstruction scheme mentioned above and has successfully been applied, sometimes in a modified form, to experimental time series from biological systems (Everson; Olsen in [9])

Many chaotic attrcators have a fractal geometry; their trajectories in state space form thick bands which on magnification disclose themselves as composed of sub-bands which again are composed of sub-sub-bands... . Such fractal, or "self-similar" structures which are only connected in one dimension take non-integer dimensions as opposed to simple attractors (steady states and limit cycles) that have integer dimension. There are several methods by which the dimension of an attractor can be computed. Unforetunately, most of these methods are computationally intensive and require larger amounts of data than are normally accesible to the experimental biologist. However, a realtively simple algorithm devised by Grassberger and Procaccia [16] computes an approximate value for the dimension of chaotic attractors and requires relatively few experimental data points. The algorithm has one further advantage that it often separates the inevitable noise in experimental data from their deterministic part. The application of this method to biological signals is illustrated by Albano et al. in [9] in their analysis of the EEG.

The compuations of Lyapunov exponents and fractal dimensions are often non-trivial procedures. In many cases the algorithms return the answer Lyapunov exponent (or dimension) "unresolved". Usually this due to too much noise in the data. Reducing the noise by smoothing the data can be dangerous; for example, Grassberger [15] showed that smoothing a time series produced by a random number generator can result in spurious estimates of the fractal dimension. A more sophisticated noise reduction technique is singular value decomposition. Here the time series is initially reconstructed in a high dimensional state space and then projected down to a few significant directions. The effects of singular value decomposition can be quite dramatic as exemplified by Albano et al. and Everson in [9] for neural signals.

The above discussion was focussed on detecting and quantifying chaos in biological data. However, not all fluctuating data are chaotic. An example of such non-chaotic fluctuations is a weakly damped oscillation perturbed by a large amount of noise. The nonlinear methods discussed above would classify such dynamics as stochastic, but they do not detect the underlying structure. Demongeot et al. in [9] discuss socalled confiners which are the stochastic counterparts to deterministic attractors, and introduces some new methods for their study. A combination of these methods with the above nonlinear techniques results in a powerful toolkit for analysing the time series of biological data.

7. TOWARDS SPATIO-TEMPORAL IRREGULARITY AND PATTERNING

Chaotic dynamics may be used to account for some of the irregular time series that are obtained in experimental biology. However, many biological processes have a spatial as well as temporal aspect, and so one is interested in spatio-temporal patterning (or coherence) and irregularity. Structure and chaos are the two faces of nonlinearity — see for example the papers in [5], and coherent behaviour may be more important than chaos in biology. Solitons — nonlinear waves that behave as particles, in that, unlike the nerve impulse, they are not annihilated on collision, have been proposed to mediate energy transport in macromolecules [8], and are stable at biological temperatures (Cruzeiro et al., in [9])

A key problem in biology is morphogenesis, the formation during development of differentiated structure from a system of identical cells. Reaction diffusion systems provide illustrations of morphogenesis, and so there is interest in understanding the simplest reaction diffusion models that generate structure (Parisi et al. in [9]), even though these may be far from biological reality. Two-dimensional chemical excitable media can generate concentric and spiral waves, similar to the waves seen during aggregation in slime molds. The Belousov-Zhabotinsky reaction is the best studied chemical reaction-diffusion, excitable system, and its structures can be perturbed by convection, so providing a test-bed for developing measures of spatial structure and irregularity (Markus et al in [9]). A NATO ARW on nonlinear wave phenomena in excitable media, with particular emphasis on biological systems, will be held in 1989 [22].

REFERENCES

1) Aguda, B. and Larter, R. J. Chem. Phys. (1989).

2) Aron, J.L. and Schwartz, I.B. J. Theor. Biol. 112: 403 (1985).

3) Boiteux, A. and Hess, B. Phil. trans. Royal Soc. B293: 5 (1981).

4) Brindley, J., Kaas-Petersen, C. and Spence, A. Physica-D 34: 456 (1989).

5) Christiansen, P.L. and Parmentier, R.D. Structure, coherence and chaos in dynamical systems. Manchester University Press 1989.

6) Conrad, M. in [9].

7) Cvitanovic, P. (ed.) Universality in Chaos Adam Hilger, Bristol 1985.

8) Davydov, A.S. Biology and quantum mechanics. Pergamon, Oxford (1982).

9) Degn, H., Holden, A.V. and Olsen, L.F. (eds.) Chaos in Biological Systems, Plenum, New York, 1988.

10) Degn, H. and Mayer, D. Biochem Biophys. Acta 180: 291 (1969).

11) Durston, A.J. Devel. Biol. 38: 308 (1974).

12) Elton, C. and Nicholson, M.J. J. Animal Ecol. 11: 96 (1942).

13) Glass, L. and Mackey, M.C. Annals N.Y. Acad. Sci. 316: 214 (1979).

14) Gleick, J. Chaos: making of a new science Heinemann, London 1988.

15) Grassberger, P. Nature (London) 323: 609 (1986).

16) Grassberger, P. and Procaccia, I. Physica-D 9: 189 (1983).

17) Hassel, M.P., Lawton, J.H. and May, R.M. J. Animal Ecol. 45: 471 (1976).

18) Hess, B. and Markus, M. Trends Biochem. Sci., 12: 45 (1987).

19) Hodgkin, A.L. and Huxley, A.F. J. Physiol (London) 117: 500 (1952).

20) Holden, A.V. The mathematics of excitation. In Biomathematics in 1980, ed. L.M. Ricciardi and A.C. Scott. North-Holland, Amsterdam (1982).

21) Holden, A.V. (ed.). Chaos. Manchester University Press, Manchester, 1986; Princeton University Press, Princeton, New Jersey, 1986.

22) Holden, A.V. , Markus, M. and Othmer, H. (eds.) Nonlinear wave processes in excitable media. Plenum, New York, 1990.

23) Irisawa, H. and Noma, A. In Cardiac Rate and Rhythm ed. L.N. Bouman and H.J. Jongsma. Martinus Nijoff, The Hague p. 35 (1982).

24) Kot, M. (in preparation).

25) Li, Y.-T. and Yorke, J. Am. Math. Monthly, 82: 985 (1975).

26) Markus, M., Kuschmitz, D. and Hess, B. FRBS Lett. 172: 235 (1984).

27) May, R.M. Nature (London) 261: 459 (1976).

28) May, R.M. Proc. Royal Soc. Lond. A413: 27 (1987).

29) Olsen, L.F. Phys. Lett 94A: 454 (1983).

30) Olsen L.F. and Degn, H. Nature 267: 177 (1977).

31) Olsen, L.F. and Degn, H. Q. Rev. Biophys. 18 : 165 (1985).

32) Pool, R. Science 243: 25 (1989).

33) Pool, R. Science 243: 310 (1989).

34) Pool, R. Science 243: 604 (1989).

35) Rapp, P. J. Exp. Biol. 81: 281 (1979).

36) Rogers, T.D. Prog. Theor. Biol. 6: 91 (1981).

37) Ross, J. and Schell, M. Ann. Rev. Biophys. Biophys. Chem. 16: 401 (1987).

38) Schaffer, W.M. Am. Nat. 124: 798 (1984).

39) Schaffer, W.M. IMA J. Math. Appl. Med. & Biol. 2: 221 (1985).

40) Schaffer, W, M, and Kot, M. J. Theor. Biol 112: 403 (1985).

41) Schaffer, W.M., Ellner, S.E. and Kot, M. J. Math. Biol. 24: 479 (1986).

42) Schaffer, W.M., Olsen, L.F., Truty, G.L. and Fulmer, S.L. Proc. AAAS Annual meeting, San Francisco 1989, in press.

43) Sporns, O., Roth, S. and Seeliog, F.F. Physica-D 26: 215 (1987).

44) Stewart, I. Does God Play dice? The mathematics of chaos. Blackwell, Oxford 1989.

45) Takens, F. Lect. Notes Math. 898: 366 (1981).

46) Tomita, K. Phys. Rep. 86: 113 (1982).

47) Wolf, A., Swift, J.B., Swinney, H.L. and Vastano, J.A. Physica-D 16: 285 (1985).

48) Wolfram, S Physica-D 10: 1 (1984).

PHYTOCHROME PROPERTIES AND BIOLOGICAL ACTION

Brian Thomas

Institute of Horticultural Research
Littlehampton, West Sussex BN17 6LP, UK

INTRODUCTION

Aims of this chapter

Unlike the majority of chapters in this book, this is being written in advance of the relevant NATO workshop. This chapter cannot therefore be a distillation of the collective wisdom of the participants. Instead it represents the author's personal view of the aspects of phytochrome research which have made the most dramatic progress during the life of the NATO Cell Signalling Programme. The mechanism by which phytochrome brings about its effects is still not fully understood. However, recent rapid advances in our understanding of the phytochrome molecule and how it modifies development, especially through the regulation of gene expression can be traced back to three developments. These are the isolation and characterisation of undegraded phytochrome (1, 2), the isolation of the phytochrome gene (3) and the application of the techniques of genetic engineering to phytochrome-regulated gene expression. Reflecting these, the chapter deals primarily with molecular and biochemical progress towards understanding the cellular signals generated by phytochrome in response to light. There are a number of sources for information on more general or specific aspects of phytochrome to which the reader wishing for more information is directed (4, 5, 6).

Photomorphogenesis

A major difference between plants and animals is that plants harvest energy from the range of the electromagnetic spectrum which we call light. This energy supports all life on the planet and the ability of plants to monitor and respond to changes in the light environment is vital to their, and ultimately our, survival. Plants have two distinct types of mechanism for absorbing and utilising light. Photosynthesis is a mechanism for capturing light and converting it into chemical energy. It is primarily a process in which the quantity of light absorbed and the utilisation of the captured energy is maximised, or at least optimized. In parallel with this plants posess independent mechanisms for detecting light and converting its energy to information used to modify patterns of growth and development.

NATO ASI Series, Vol. H 51
Cell to Cell Signals in Plants and Animals
Edited by V. Neuhoff and J. Friend

This process is called photomorphogenesis and requires the involvement of photoreceptor pigments. These absorb light and provide the initial biochemical signal to trigger the plant's photoresponse.

In terms of cell signalling photomorphogenesis provides a context peculiar to plants in which photoreceptors act as primary signals. Transduction of the photoreceptor signals along signalling pathways, on the other hand, almost certainly does not involve unique mechanisms. The identification and characterization of photoreceptors and elucidation of their mechanism of action is a major goal of plant biochemists and an essential part of the study of cell signalling in plants.

Phytochrome

A large number of photomorphogenetic responses are under the control of the chromoprotein photoreceptor phytochrome. Phytochrome is found in all green plants including algae, mosses and ferns. The protein exists as two photoisomers. One form (Pr) has a characteristic absorbance spectrum with a major absorbance maximum (λ_{max}) at about 660nm and a secondary maximum at about 380nm. The other form (Pfr) has a λ_{max} at about 730nm and a secondary peak at about 400nm . When either form absorbs light a series of photochemical and protein conformational changes is initiated leading within milliseconds to the formation of the other photoisomer. For many responses red light at about 660nm is the most effective part of the spectrum and the effect of red can be prevented by a subsequent far-red irradiation at about 703nm. In such cases red, far-red photoreversibility is frequently repeatable over several cycles of irradiation. This is understandable in terms of the photoreversible interconversions between Pr and Pfr as long as it is assumed that Pfr is biologically active and Pr inactive. The following represents a simple model for the phytochrome system.

$$\mathrm{Pr} \underset{\text{Far-red light}}{\overset{\text{Red light}}{\rightleftharpoons}} \mathrm{Pfr} \longrightarrow \text{Response} \qquad (h\nu)$$

Under continuous illumination a dynamic photoequilibrium is established between Pr and Pfr, but because Pr and Pfr have different absorbance characteristics the relative proportions of the two forms varies depending on the spectral distribution of incident light. Red light at 660nm establishes the maximum proportion of phytochrome as Pfr (80—90%) whereas far-red light at 730nm will result in only about 3% as Pfr.

The net result of these photochemical properties is that phytochrome is uniquely placed to serve as a molecular monitor of the light environment. To give two examples: phytochrome is synthesised as Pr and in seedlings penetrating the soil the formation of Pfr as light first impinges upon the plant acts as an early warning of microenvironmental change. The plant responds by modifying its strategy from searching for light to maximising its absorbtion. Developmental emphasis shifts from shoot elongation to leaf expansion coupled with switching on the synthesis of photosynthetic proteins and protective pigments. This is achieved through the co-ordinated activation and deactivation of specific plant genes.

Secondly, variation in the relative amounts of light energy in the red and far-red parts of the spectrum is caused by vegetational shading. Consequential changes in the Pfr/P total of shaded plants enable them to detect this shading. How the plant responds will depend partly on the ecological strategy it uses for survival.

PHYTOCHROME CHARACTERISATION

Phytochrome genes

The cloning of Avena phytochrome cDNA clones by Peter Quail's group (7) represents something of a watershed in phytochrome research. This enabled the sequence of the protein to be deduced and resolved several outstanding controversies regarding its properties. It also made possible the regulation and autoregulation of its synthesis to be characterised in detail.

Phytochrome cDNA clones were first isolated and characterised from *Avena sativa* (3, 7). Restriction endonuclease site polymorphism between clones indicated that phytochrome exists as a small gene family in this species. There are at least four members expressed. No clone was obtained with the complete coding region and analysis was based on partial sequences. DNA sequence analysis of three of the classes showed a 98% squence similarity in the coding regions and 5' untranslated regions. Complete sequences for two members of the gene family shows them to both have an open reading frame of 3384bp. This equates to proteins 1128 amino acid residues long with molecular masses of 124, 870 and 124, 949 Da. The attachment site for the tetrapyrrole chromophore was located at Cysteine-321.

Based on restriction endonuclease analysis, two classes of genomic clone were isolated from Avena and one of these has been sequenced (8). This gene, designated type 3, is approximately 5.9kbp long and contains six exons and 5 introns. The mRNA coded for by this gene has a total length of 3781 nucleotides consisting of 142 nucleotides of untranslated 5' sequence, 249 nucleotides of untranslated 3' sequence and 3390 nucleotides of amino acid coding sequence, including initiator and stop codons.

Subsequently cDNAs and genomic clones have been isolated from a number of species. At the present time details have been published for phytochrome genes from *Cucurbita* (9), *Pisum* (10) and *Oryza* (11). No overall pattern has yet emerged regarding the organisation of phytochrome genes. It is claimed that only one copy of the phytochrome gene is present in *Pisum* and *Oryza,* compared to four in *Avena* and at least two in *Cucurbita.* Multiple transcripts have, however been detected in *Avena, Cucurbita* and *Pisum.* In the latter case multiple transcripts are detectable by primer extension analysis indicating that the differences lie in the length of the 5' untranslated sequence (10). Resolution of these apparent anomalies requires further work.

Comparison of phytochrome sequences from different species reveals about 65% nucleotide and amino acid sequence similarity between monocot and dicot phytochromes. Dicot phytochromes have a higher level of similarity at about 77% (12). Sequence conservation is not evenly distributed. Most of the amino-terminal two thirds of the aligned polypeptide chains show localised regions with a high degree of similarity. A very high degree of conservation is seen at the chromophore attachment site. A stretch of 29 amino acids is shown to be entirely the same in *Avena* and *Cucurbita* at this part of the molecule. This presumably reflects the highly constrained nature of the protein chromophore interactions which contribute to the unique photochemical properties of phytochrome. Sequence similarity is by contrast low at the amino and carboxy-termini of the molecule.

Properties of the Protein

Phytochrome is a large soluble chromoprotein of 1127 amino acids in *Avena,* normally occuring as a dimer. The molecular weight of the monomer is 124 kDa in *Avena* and *Secale,* 127 kDA in *Zea,* 121 kDa in *Pisum,* and 120 kDa in *Cucurbita* (13). Cross-reactivity of monoclonal antibodies to phytochrome from a range of higher plants indicates that at least some of the sequence is highly conserved (14, 15). This is in agreement with the comparison of nucleic acid and deduced amino acid sequences. Microsequencing of a peptide generated by subtilisin digestion of *Avena* phytochrome revealed that the sequence was heterogeneous (16), demonstrating that several of the genes for phytochrome are expressed.

Phytochrome is considered to be composed of three domains, based on the high susceptibility of two regions of the molecule to proteases. An N-terminal 10 kDa domain is cleaved more readily from the Pr form than from the Pfr form (17) and is thought to shield the chromophore from the aqueous environment. The chromophore is contained within a relatively protease-resistant 64 kDa core domain, which will undergo phototransformation, although λ_{max} for Pfr and Pr in the absence of the N-terminal domain differ from that of the undegraded protein (17). The function of the remaining 55 kDa C-terminal domain is more obscure, although it appears that the site of phytochrome dimerisation is located in this region (18). The site of chromophore attachment is bounded on either side by regions of hydrophobic residues which could be expected to form a hydrophobic cleft (19, 20).

CD measurements indicate 52% α-helical structure, 21% β-turn and 27% random coil for the Pr form of phytochrome, with little or no β-sheet (20). These agree only loosely with Chou-Fasman predictions. Postranslational modifications other than chromophore attachment include a single phosphate group at an undefined site per monomer chain (21) and the N-terminus blocked with N-acetyl serine (22).

The chromophore is a single, linear tetrapyrrole linked to cysteine 321 via a thioether linkage. The chromophore in Pr is thought to be protonated and in an extended conformation (23). Spectral differences between Pr and Pfr arise, in part, from a cis-trans isomerization at the C15 bridge between rings C and D with the Pr chromophore being the cis isomer. However, this is not sufficient in itself to cause the wavelength shift to 730 nm; chromophore-protein interactions are also required. Loss of the N-terminal 6—10 kDa of the protein results in a shift of λ_{max} for Pfr from 730 to 720 nm. This N-terminal domain appears to interact with the chromophore and take part in a conformational change on phototransformation.

Types of phytochrome

Almost all the information on the phytochrome molecule is based on the properties of phytochrome extracted from dark-grown tissues. However, phytochrome extracted from light-grown plants differs in several respects from that in etiolated plants. Tokuhisa et al.

(24) purified phytochrome from light-grown *Avena* and found a photoreversible species with altered difference spectra, molecular weight, peptide mapping and immunoreactivity from that isolated from etiolated tissue. Similar results were obtained by Shimazaki & Pratt (25) in both green and norflurazon-bleached *Avena* and by Abe et al. in *Pisum.*

The phytochrome present in etiolated plants at high levels, designated Type I phytochrome, is destroyed rapidly on transfer to light leaving a residual pool of Type II phytochrome which is more stable as Pfr. Types I and II phytochrome have also been detected in germinating embryos of *Avena* (27, 28) and in embryo tissue from *Pisum* (29).

Type II phytochrome is characterised by poor antigenicity with antibodies raised against Type I phytochrome (30). Other differences are a shift in λ_{max} from 666 nm to 652 nm for Pfr, and a molecular weight of 118 kDa (31). It has recently been suggested that Type II phytochrome is itself made up of two immunochemically distinct species (32) and this is consistent with the organisation of phytochrome genes in the *Arabidopsis* genome (P. Quail, personal communication). The final picture of the phytochrome system of higher plants is therefore likely to include three different photoreversible proteins.

Phytochrome synthesis and destruction

Phytochrome is synthesised as Pr which normally accumulates in darkness, because destruction from this form is slow. Pfr, on the other hand apparently has a much shorter half life and following photoconversion to Pfr phytochrome is destroyed rapidly. Pfr destruction in dark grown seedlings which receive a brief irradiation is rapid with a half-life of 1—2 hours. This is accountable by the loss of Type I phytochrome. Type II phytochrome is more stable with a Pfr half-life of 7—8 hours (33). Destruction occurs predominantly from the Pfr form although Pr destruction is enhanced if phytochrome is cycled through the Pfr form (34).

Phytochrome destruction is a complex process involving biochemical modification and subcellular relocation. Phytochrome detectable by immunocytochemistry becomes redistributed in the cell following a red light treatment. From being cytoplasmically dispersed phytochrome rapidly becomes aggregated in discrete areas of the cell, probably small vacuoles (35). This is followed by loss of immunocytochemical staining as phytochrome is broken down. The biochemical mechanism involved in phytochrome destruction apparently involves covalent binding of ubiquitin to the molecule. This has been investigated by Vierstra's group who have shown the appearance of phytochrome-ubiquitin conjugates in cell extracts following irradiation of dark-grown plants (36).

Chromophore biosynthesis and holoprotein assembly

During the last few years Lagarias and his colleagues, in particular, have made considerable progress in characterisation chromophore synthesis. Much of the progress results from the discovery that gabaculine, an inhibitor of 5-aminolevulinic acid synthesis, inhibited the accumulation of the photoreversible product and that phytochrome apoprotein accumulated in these tissues (37). The synthesis of the chromphore and apoprotein is not, therefore, tightly regulated. Elich and Lagarias (38) added a range of linear tetrapyrroles to gabaculine treated tissues and found that with the exception of biliverdin IIIa, a photoreversible product with a covalently attached chromophore was formed. From these studies they proposed the following biosynthetic pathway

Heme → biliverdin IXa → phytochromobilin + apoprotein → Pr.

The final step combining the bilin and apoprotein requires a bilin lyase. Recent evidence suggests that this is an autocatalytic process, with phytochrome itself acting as the necessary enzyme (39). An extrapolation from this work is that it may now be feasible to make photoreversible phytochrome synthetically for e.g crystallisation and structural studies in bacterial or eukaryotic expression systems.

SIGNAL TRANSDUCTION

The mechanisms by which phytochrome becomes activated by light and by which this information is transduced within the cell to bring about the photomorphogenetic response is still not well understood. Responses are diverse, ranging from short term changes in membrane properties to the activation or repression of specific genes. Although a cohesive picture remains to be assembled, the last few years has produced much information which will ultimately prove relevant to the problem. A number of approaches have been taken.

Differences between Pr and Pfr

Identifying differences between the properties of Pr and Pfr is a prerequisite for understanding how the light signal is transduced because Pfr acts as the first element in the chain of cellular events but Pr cannot. Several lines suggest that there are significant changes in protein conformation when photoconversion takes place.

When phytochrome is proteolytically digested as Pr or Pfr the products of proteolysis are different, presumably because conformational changes lead to exposure of different stretches of amino acids. By sequencing the N-terminus of the proteolytic fragments produced, and comparing these with the known amino acid sequence for the entire protein, the cleavage sites can be identified. In this way specific sequences differentially exposed on phototransformation have been determined. Grimm et al. (40) showed that there are several cleavage sites in the first 60 amino acids at the N-terminus when Pr is digested. On the other hand, cleavage at glutamate-354 and lysine-753 occured only in Pfr suggesting that in this form specific sites become exposed on the chromophore and carboxyterminus domains.

A complementary approach has been to compare the reaction of Pr and Pfr with protein-modifying chemicals. For example Pfr contains reactive sulphydryls which are inaccesible in Pr (41, 42), one of which is located in the chromophore domain. Conformational changes in the N-terminus and chromphore-bearing domains are also suggested by hydrogen/tritium exchange experiments. These, coupled with studies on the chromophore using oxidising and hydrophobic compounds suggest that a cleft in the core domain of the molecule, containing the chromophore, is shielded by the N-terminal domain in Pfr but not in Pr (19, 20).

Phosphorylation of phytochrome by protein kinases has been used as a probe of conformational change. (43). In this work a site near the N-terminus was preferentially phosphorylated in Pr. In Pfr loss of phosphorylation at this site was accompanied by a new phosphorylation site in the C-terminal domain.

Conformational changes have also been characterised using monoclonal antibodies and site-directed antibodies raised to synthetic peptides corresponding to a known part of the phytochrome sequence. Antibodies have been reported from a several laboratories which bind with higher affinity to either Pr or Pfr (see 30 for a recent review). Generally speaking, antibodies which bind to the N-terminal domain (type 1 antibodies) (44) exhibit higher affinities for the Pr than the Pfr form. Several N-terminal antibodies modify the spectral properties of Pfr and increase the rate of dark reversion of Pfr to Pr. (45). Cordonnier et al. (46) also found that type 1 antibodies can protect the N-terminal region from proteolysis.

Type 2 antibodies are defined as those which bind to the chromophore domain (44). Several of these exhibit preferential binding to Pfr (30), in agreement with proteolysis and protein modification studies. Type 2 antibodies do not alter the spectral properties of phytochrome. Type 3 antibodies bind to the C-terminal domain but show no consistent pattern in Pr or Pfr preference; both having been described (30). The C-terminal domain has one epitope which is highly conserved in phytochrome from algae, a moss and both monocotyledonous and dicotyledonous plants although its functional significance is unknown (14).

Taken together the different approaches indicate that phototransformation involves conformational changes across the whole of the phytochrome molecule. In particular, the N-terminal domain undergoes significant alteration and is certainly required for correct spectral transformation. As yet there is no evidence as to which part of the molecule is involved in generating the initial cell signal leading to developmental changes.

Phytochrome and membranes

Hendricks and Borthwick proposed in 1967 that the primary action of phytochrome was mediated by modification of membrane properties (47). After attracting much attention for the following ten years or so the interaction of phytochrome with membranes then received relatively little study up until the present. Progress has however continued to be made, albeit slowly. From analysis of the amino acid sequence it can be deduced that there is no membrane spanning domain on the molecule, thus ruling out the possibility that phytochrome is an intrinsic membrane protein. Overall the properties of the molecule are characteristic of cytosolic proteins although there are several localised hydrophobic regions which could allow phytochrome to interact with membranes or membrane proteins.

Attempts to demonstrate phytochrome interaction with cell membranes by immunolocalisation techniques have been unsuccessful or produced ambiguous data (48). Membrane preparations from dark-grown plants have however frequently been shown to contain photoreversible phytochrome. Recent developments have come from the use of immunochemical techniques which have shown that such phytochrome is not proteolytically degraded (49). A major problem with interpretation of these and similar data is assessing whether the association is an artifact of homogenisation arising out of the "sticky" nature of the protein. Against this is that two laboratories have shown that the phytochrome associated with plasma membranes prepared by phase partition does not change in parallel with the total cytosolic pool (50, 51) suggesting a level of specificity in the association.

Undoubtedly, progress in this area has been hampered by a lack of suitable *in vitro* model systems. There have been a number of reports of phytochrome modulation of enzyme activity on isolated membranes or organelles but as yet none have been sufficiently robust to be demonstrated independently by another laboratory. A semi-*in vitro* response was found by Blakeley et al. (52) who showed that the osmotic swelling of isolated protoplasts from etiolated wheat leaves was under phytochrome control. This was confirmed also for *Avena* by Kim et al. (53) and subsequently further investigated by Bossen et al. (54). The latter workers were able to demonstrate that the response was Ca^{2+} dependent and they also provided evidence for the involvement of G-proteins, protein kinases and inositol triphosphate turnover (56). Work on unravelling the components of transduction chains is in its infancy. Isolated reports of effects of light on protein kinases, inositol triphosphate turnover, GTP-binding proteins and also on Ca^{2+} transport indicate that transduction mechanisms comparable to those in animal and other plant regulatory systems are involved. The next few years will see rapid progress in this area, with protoplast and isolated plasma-membrane systems being used extensively to study the early steps in the phytochrome signal transduction chain.

Phytochrome as an Enzyme

An attractive hypothesis for its mechanism of action is that phytochrome is an enzyme with Pfr being the active configuration. Preliminary indications that phytochrome might be a protein kinase based on apparent autophosphorylation by highly purified preparations has subsequently been shown to be the result of co-purification of a kinase (56, 57). As yet the only enzyme activity associated with phytochrome appears to be the bilin lyase activity of the apoprotein which catalyses chromophore attachment.

PHYTOCHROME AND GENE EXPRESSION

Phytochrome autoregulation

In addition to being regulated by light at the level of protein stability through the lability of Pfr Duke et al. (58) observed that light reduced the ability of corn seedlings to synthesise photodetectable phytochrome. *In vitro* translation subsequently confirmed that in dark-grown *Avena* seedlings translatable phytochrome mRNA was dramatically reduced after exposure to light (59—61). The availability of phytochrome cDNA enabled the amount of mRNA sequence to be direcly measured following light treatments. In *Avena* phytochrome mRNA abundance dropped rapidly after a saturating red pulse or transfer to white light. Levels were reduced to 10% five hours after a red pulse and the half-life of phytochrome mRNA was estimated at about one hour. The mRNA abundance response showed a lag of about 15 minutes (62). Assay of transcriptional activity by "run-on" experiments in this system, however, indicated repression of *phy* transcription within 5 minutes (63). The response was unimpaired by treatments which inhibited protein synthesis by over 90%, indicating that all the components of the transduction chain are present in the cell prior to irradiation. The response is very sensitive to small quantities of Pfr and a far-red light treatment can often generate enough Pfr to inhibit synthesis. Red — far red photoreversibility can be demonstrated if appropriate protocols are used, showing that phytochrome is the photoreceptor for its own regulation (62). In germinating *Avena* seeds photoregulation of phytochrome synthesis is by a stable Pfr population (64, 65). Thus Type II phytochrome may be capable of regulating type I phytochrome synthesis. Down regulation of the phytochrome gene is not a universal feature of phytochrome in plants. *Lycopersicon esculentum* shows little effect of light on phytochrome mRNA levels and several species have a partial response to light (12, 66). Whether this variation has biological significance is not known.

Other photoregulated genes

The expression of a number of genes has been shown to be up or down regulated by phytochrome at the transcriptional or post-transcriptional level (67). Genes are often classified as very low fluence (VLF) or low fluence (LF) genes depending upon the amount of light required for their activation. VLF responses are induced by far-red alone and therefore do not show red / far red reversibility. For the majority of genes, reversibility can be demonstrated, placing them in the LF group. Examples of VLF genes are *Cab* and *phy* with ferredoxin and *rbcs* being genes which fall into the LF category (68). The response to light shows complexity in a number of other features (69). Genes show differences in the time courses of their response to light. Where mRNA abundance for a number of genes has been measured following a single light pulse a complex pattern has been obtained, with almost every gene having a distinct pattern of response (70, 71). Complexity is also exhibited in variation in the time course of escape from reversibilty by far red in LF genes, by dependence on the developmental state of the tissues or interaction with other photoreceptors (72). Photoregulated genes may also be regulated by an endogenous circadian rhythm (69).

Transcriptional control of gene activation involves the binding of activators or repressors to regulatory elements of the gene in question. Elucidation of these interactions typically has used transformation with foreign genes containing light promoter or enhancer sequences and analysis of the response of the transformants. For example the wheat Cab-1 gene is phytochrome-regulated in transgenic *Nicotiana* plants and pea *rbcs* genes in transgenic *Petunia* (73, 74). A number of regulatory sequences have been implicated in conferring light sensitivity. In both *Cab* and *rbcS,* 5' upstream sequences are sufficient for phytochrome mediated light induction. The region from −169 to −112 in the *rbcS*-3A contains three sequence elements or "boxes" which are conserved in all pea *rbcS* genes (68). Boxes II and III have homologues repeated further upstream and can function as both positive and negative light-responsive elements (LREs). They share homology with enhancer elements in viral and constitutive mammalian genes. The cab-I gene also contains an upstream enhancer element which confers phytochrome-mediated expression (75). There is no obvious homology between this and pea LREs and as yet no evidence for universal sequences for photoregulation.

In addition to *cis*-acting DNA sequence elements in the upstream region of the genes, protein factors which bind to these regulatory elements have been identified. A protein factor which binds to boxes II and III in the pea *rbcS* gene has been designated GT-1 (68). These genes are positively regulated by light, yet GT-1 also binds to the upstream sequence of the negatively photoregulated phytochrome genes of *Avena* and *Oryza* (11). Recently it has been reported that another DNA-binding protein designated AT-1 binds to conserved AT-rich elements, (the AT-1 box) within the promoters of certain pea *rbcS* and *cab* genes (76). The binding activity of AT-1 is reversibly modified by phosphorylation thus providing a potential mechanism by which transcription might be regulated.

The characterisation of *cis*- and *trans*-acting factors represents major progress towards understanding the regulatory mechanisms involved in the regulation of gene transcription by light. The number of regulatory sequences and DNA binding proteins identified so far is relatively small. Even with these, however, the potential interactions possible when sequence elements and protein factors are in combination are numerous. Many of these interactions are probably subtle and this may be a major factor in establishing the complexity of expression patterns in response to light.

GENETIC APPROACHES TO PHYTOCHROME ACTION

Photoreceptor mutants

Defined genetic mutants can prove valuable tools in unravelling complex developmental mechanisms. A major advance in the last few years has been the identification of plant mutants with altered photomorphogenetic responses. Koorneef, Kendrick and their colleagues have identified a range of mutants in *Arabidopsis, Cucumis* and *Lycopersicon* characterised in the most cases by an insensitivity in hypocotyl extension to light (77). Of these long hypocotyl mutants, *hy-1* and *hy-2* mutants of *Arabidopsis* (78) and *au* and *yg-2* mutants of *Lycopersicon* (79) were identified as possible phytochrome mutants based on their spectral sensitivity and a reduced expression of several phytochrome-induced responses. These include seed germination, hypocotyl inhibition and anthocyanin formation. Phytochrome could not be detected spectrophotometrically in seeds or hypocotyls of *hy-1* and *hy-2* mutants of *Arabidopsis* or *au* mutants of *Lycopersicon.* The *au* mutant was analysed by Quail's group who confirmed, using immunoblot analysis, that the mutant contained no more than 5% of the wild type level of phytochrome. However, further analysis by Northern blotting and *in vitro* translation, revealed that the mutant had a comparable level of phytochrome mRNA to the wild type (66). As the mutant apparently has the same ability to synthesise the apoprotein as the wild type the reduced phytochrome content suggests a high rate of turnover of the phytochrome polypeptide.

Mutants are potentially useful in distinguishing the functional difference, if any, between Types I and II phytochrome. A cucmber long hypocotyl mutant *(lh)* contains the same level of phytochrome as its wild type in dark-grown seedlings (80). It also shows a striking absence of phytochrome control in de-etiolated seedlings, leading to the proposal that the mutation might be in Type II phytochrome, in contrast to the *au* mutant which lacks Type I phytochrome. Immunochemical studies with Type II specific antibodies surprisingly have failed to confirm this (79).

A number of groups are now attempting to use other strategies in producing photoresponse mutants. One of the most interesting of these is by Chory et al. (81) who have characterised a mutant designated det1 in *Arabidopsis* which develops as a light grown plant in darkness. They have proposed that the primary role of light in controlling gene expression is mediated by the activation of leaf development. Other promising strategies are the use of "suicide genes" to select mutants in the phytochrome transduction pathway leading to transcriptional changes in *Arabidopsis* (82), and the use of haploid *Nico-*

tiana pumbaginifolia to select phytochrome-response mutants (83). Undoubtedly this is an area where there will be vigorous efforts and rapid advances in the next few years.

Transgenic plants

A complementary genetic approach to the use of genetic lines with defined alterations in photoresponse is to introduce phytochrome genes into plants by transformation and evaluate their effect. At the time of writing this chapter the first reports of such transgenic plants are appearing. This is an area of high activity and the first results are dramatic. Keller et al. (84) expressed genes from *Avena sativa* in *Nicotiana tabacum* and *N. plumbaginifolia.* The genes were under the transcriptional control of either a light-regulated *Avena* phytochrome promoter or the constitutively active cauliflower mosaic virus 35S promoter. The phytochrome promoter was ineffective but RNA was produced with the 35S constructs. Full length phytochrome polypeptide with photoreversible spectral activity was produced. The phytochrome content of the transformed plants was markedly higher than the untransformed controls. Overexpression of phytochrome genes was associated with a number of phenotypic changes which were not observed in kanamycin resistant transformants not expressing *Avena* phytochrome. There was increased green pigmentation, increased tillering, reduced rooting capacity in the presence of kanamycin, reduced apical dominance and shorter internodes. Kay et al. (85) made constructs of the *Oryza* phytochrome gene and also transformed *Nicotiana.* In these transformed plants the period of the free running circadian rhythm in *Cab* gene expression was extended. This was also taken as an indication of the biological activity of the rice gene in the transformed plants.

Work with transgenic plants has in each case involved using Type I phytochrome from a monocot and overexpressing it in light grown dicots. The appearance of spectrally active phytochrome indicates a well preserved mechanism for chromophore attachment in species which are not closely related. This might be in part explained by the ability of the phytochrome molecule to catalyse chromophore attachment to itself. However the availability of precursor requires correct intracellular localisation or targetting of the apoprotein. It is not yet known whether other posttranslational modifications occur correctly in the transgenic tissues although presumably any essential for phytochrome function must take place.

Taken with the pattern of response observed with mutants it is clear that the quantity of phytochrome in plant tissues has a significant effect on plant phenotype. The reduced response to light when phytochrome levels are reduced in mutants such as *au* in *Lycopersicon* compared the opposite with elevated levels of phytochrome in transgenic plants argues strongly that photomorphogenesis is a positive consequence of Pfr formation, rather than a negative reaction to the removal of Pr. In addition it is difficult to sustain the idea that the ratio of Pr to Pfr is the decisive factor in phytochrome responses, as has been proposed (86), rather the amount of Pfr.

The *au* mutants and transgenic plants both have aberrant expression of Type I phytochrome genes and both show dramatic alteration in phenotype, albeit in opposite ways, when growing in the light. This argues strongly that the function of Type I phytochrome is not only to act as the photoreceptor for de-etiolation and greening, but also to regulate development in light-grown plants. Physiological experiments show High Irradiance Responses, usually assumed to be attributable to Type I phytochrome, in light-grown plants (87). By using quantitative immunoassay these can be correlated with changes in the quantity of Type I phytochrome apoprotein.

CONCLUSIONS

Our knowledge and understanding of how light regulates plant development has increased dramatically over the last few years. this has undoubtedly arisen through the application of molecular techniques to the problems of photomorphogenesis. We now know a lot about the phytochrome molecule, on one hand, and specific genes which it regulates, on the other. The steps in between, however, are still obscure. Characterising the signal transduction pathways between phytochrome and cellular response now presents a major challenge to innovative scientists. There are many exciting discoveries yet to be made.

REFERENCES

1) Vierstra, R.D. and Quail, P.H. (1983). Purification and initial characterization of 124 kilodalton phytochrome from *Avena.* Biochemistry 22, 2498—2505.

2) Litts, J.C., Kelly, J.M. and Lagarias, J.C. (1983). Structure function studies on phytochrome. Preliminary characterization of highly purified phytochrome from *Avena sativa* enriched in the 124 kilodalton species. J. Biol. Chem. 258, 11025—11031.

3) Hershey, H.P., Barker, R.F., Idler, K.B., Lissemore, J.L. and Quail, P.H. (1985). Analysis of cloned cDNA and genomic sequences for phytochrome: complete amino acid sequences for two gene products expressed in etiolated *Avena.* Nucleic. Acids Res. 13, 8543—8558.

4) Kendrick, R.E. and Kronenberg, G.H.M. (1986). Photomorphogenesis in plants. Martinus Nijhoff, Dordrecht, The Netherlands.

5) Jordan, B.R., Partis, M.D. and Thomas, B. (1986). The biology and molecular biology of phytochrome. Oxford surveys of Plant Molecular Biology Vol 3, 315—362.

6) Furuya, M. (1987). Phytochrome and photoregulation in plants. Academic Press, Tokyo.

7) Hershey, H.P., Colbert, J.T., Lissemore, J.L., Barker, R.F. and Quail, P.H. (1984). Molecular cloning of cDNA for *Avena* phytochrome. Proc. Nat. Acad. Sci. U.S.A. 81, 2332—2336.

8) Hershey, H.P., Barker, R.F., Idler, K.B., Murray, M.G. and Quail, P.H. (1987). Nucleotide sequence and characterization of a gene encoding the phytochrome polypeptide from *Avena.* Gene 61, 339—348.

9) Lissemore, J.L., Colbert, J.T. and Quail, P.H. (1987). Cloning of cDNA for phytochrome from etiolated *Cucurbita:* coordinate photoregulation of the abundance of two distinct phytochrome trancripts. Plant Mol. Biol. 8, 845—896.

10) Sato, N. (1988). Nucleotide sequence and expression of the phytochrome gene in *Pisum* sativum: Differential regulation by light of multiple transcripts. Plant Mol. Biol. 11, 697—710.

11) Kay, S.A., Keith, B., Shinozaki, K., Chye, M.-L. and Chua, N.-H. (1989). The rice phytochrome gene: structure, autoregulated expression, and binding of GT-1 to a conserved site in the 5' upstream region. The Plant Cell 1, 351—360.

12) Colbert, J.T. (1988). Molecular biology of phytochrome. Plant Cell Environ. 11, 305—318.

13) Vierstra, R.D., Cordonnier, M.-M., Pratt, L.H. and Quail, P.H. (1984). Native phytochrome: immunoblot analysis of relative molecular mass and *in vitro* proteolytic degradation for several plant species. Planta 160, 521—528.

14) Cordonnier, M.-M., Greppin, H. and Pratt, L.H. (1986). Identification of a highly conserved domain on phytochrome from angiosperms to algae. Plant Physiol. 80, 982—987.

15) Schneider-Poetsch, H.A., Schwartz, H., Grimm, R. and Ru diger, W. (1988). Cross-reactivity of monoclonal antibodies against phytochrome from Zea and *Avena.* Planta 173, 61—72.

16) Grimm, R., Lottspeich, F. and Rudiger, W. (1987). Hetrogeneity of the amino acid sequence of phytochrome from etiolated oat seedlings. FEBS Lett. 225, 215—217.

17) Vierstra, R.D. and Quail, P.H. (1982). Proteolysis alters the spectral properties of 124 kdalton phytochrome from *Avena.* Planta 156, 158—165.

18) Jones, A.M. and Quail, P.H. (1986). Quaternery structure of 124-kilodalton phytochrome from *Avena sativa* L. Biochemistry 25, 2987—2995.

19) Vierstra, R.D. and Quail, P.H. (1986). The Protein. in Photomorphogenesis in Plants (Kendrick, R.E. & Kronenberg, G.H.M. eds.) Nihoff/Junk, Dordrecht, The Netherlands. pp. 35—60.

20) Song, P.-S. (1988). The molecular topography of phytochrome: chromophore and apoprotein. J. Photochem. Photobiol. 2, 43—57.

21) Hunt, R.E. and Pratt, L.H. (1980). Partial characterization of undegraded oat phytochrome. Biochemistry 19, 390—394.

22) Grimm, R., Kellerman, J., Schafer, W. and Rudiger, W. (1988). The amino-terminal structure of oat phytochrome. FEBS Lett. 234, 497—499.

23) Rudiger, W. (1986). The Chromophore. in Photomorphogenesis in Plants (Kendrick, R.R & Kronenberg, G.H.M. eds.) Nihoff/Junk, Dordrecht, The Netherlands. pp. 17—34.

24) Tokuhisa, J.G., Daniels, S.M. and Quail, P.H. (1985). Phytochrome in green tissue: spectral and immunochemical evidence for two distinct molecular species of phytochrome in light-grown *Avena sativa.* Planta 64, 321—332.

25) Shimazaki, Y. and Pratt, L.H. (1985). Immunochemical detection with rabbit polyclonal and mouse monoclonal antibodies of different pools of phytochrome from etiolated and green *Avena* shoots. Planta 164, 333—344.

26) Abe, H., Yamamoto, K.T., Nagatani, A and Furuya, M. (1985). Characterization of green tissue-specific phytochrome isolated immunochemically from pea seedlings. Plant Cell Physiol. 26, 1387—1399.

27) Hilton, J.R. and Thomas, B. (1985). A comparison of seed and seedling phytochrome in *Avena sativa* using monoclonal antibodies. J. Exp. Bot. 36, 1937—1946.

28) Tokuhisa, J.G. and Quail, P.H. (1987). The levels of two distinct species of phytochrome are regulated differently during germination in *Avena sativa* L. Planta 172, 371—377.

29) Konomi, K., Abe, H. and Furuya, M. (1987). Changes in the content of phytochrome I and II apoproteins in embryonic axes of pea seeds during imbibition. Plant Cell Physiol. 28, 1443—1451.

30) Cordonnier, M.-M. (1989). Monoclonal antibodies: molecular probes for the study of phytochrome. Photochem. Photobiol. 49, 821—831.

31) Tokuhisa, J.G. and Quail, P.H. (1989)Phytochrome in green tissue: Partial purification of the 118-kilodalton phytochrome species from light-grown *Avena sativa* L. Photochem. Photobiol. 50, 143—152.

32) Cordonnier, M.-M. and Pratt, L.H. 1989 Phytochrome from green *Avena* characterized with monoclonal antibodies directed to it. in Photobiology Proceedings, 10th International Congress on Photobiology (Riklis, E. ed.). Plenum Press, New York, in press

33) Jabben, M. and Holmes, M.G. (1983). Phytochrome in light-grown plants. in Photomorphogenesis. (Shropshire, W. Jr. and Mohr, H. eds.) Encyclopaedia of Plant Physiol., New Series 16, pp. 704—722. Springer-Verlag, Heidelberg.

34) Dooskin, R.H. and Mancinelli, A.L. (1968). Phytochrome decay and coleoptile elongation in *Avena* following various light treatments. Bull. Torrey Bot. Club. 95, 474—487.

35) McCurdy, D.W. and Pratt, L.H. (1986). Kinetics of intracellular redistribution of phytochrome in *Avena* coleoptiles after its photoconversion to the active, far-red absorbing form. Planta 167, 330—336.

36) Jabben, M., Shanklin, J. and Vierstra, R.D. (1989). Ubiquitin-phytochrome conjugates: pool dynamics during *in vivo* phytochrome degradation. J. Biol. Chem. 264, 4998—5005.

37) Jones, A.M., Allen, C.D., Gardner, G. and Quail, P.H. (1986). Synthesis of phytochrome apoprotein and chromophore are not coupled obligatorily. Plant Physiol. 81, 1014—1016.

38) Elich, T.D. McDonagh, A.F., Palma, L.A. and Lagarias, J.C. (1989). Phytochrome chromophore biosynthesis. Treatment of tetrapyrrole-deficient *Avena* explants with natural and non-natural bilatrienes leads to formation of spectrally active holoproteins. J. Biol. Chem. 264, 183—189.

39) Lagarias, J.C. and Lagarias, D.M. (1989). Self-assembly of synthetic phytochrome holoprotein *in vitro.* Proc. Nat. Acad. Sci. U.S.A. 86, 5778—5780.

40) Grimm, R., Lottspeich, F., Schneider, H.A.W. and Rudiger, W. (1986). Investigation of the peptide chain of 124kDa phytochrome: localization of proteolytic fragments and epitopes for monoclonal antibodies. Z. Naturforsch. 41c, 993—100.

41) Smith, W.O. and Cyr, K.L. (1988). Modification of sulfhydryl groups on phytochrome and their influence on physicochemical differences between the red- and far-red absorbing forms. Plant Physiol. 87, 195—200.

42) Eilfeld, P., Widerer, G., Malinowski, H., Rudiger, W. and Eilfeld, P.G. (1987). Topography of the phytochrome molecule as determined from chemical modifications of S-H groups. Z. Naturforsh. 43c, 63—73.

43) Wong, Y.-S., Cheng, H.-C., Walsh, D.A. and Lagarias, J.C. (1986). Phosphorylation of *Avena* phytochrome *in vitro* as a probe of light-induced conformational changes. J. Biol. Chem. 261, 12089—12097.

44) Daniels, S.M. and Quail, P.H. (1984). Monoclonal antibodies to three separate domains on 124 kilodalton phytochrome from *Avena.* Plant Physiol. 76, 622—626.

45) Lumsden, P.J., Yamamoto, K.T., Nagatani, A. and Furuya, M. (1985). Effect of monoclonal antibodies on the *in vitro* Pfr dark reversion of pea phytochrome. Plant Cell Physiol. 26, 1313—1322.

46) Cordonnier, M.-M., Greppin, H. and Pratt, L.H. (1985). Monoclonal antibodies with differing affinities to the red-absorbing and far-red-absorbing forms of phytochrome. Biochemistry 24, 3246—3253.

47) Hendricks, S.B. and Borthwick, H.A (1967). The function of phytochrome in regulation of plant growth. Proc. Nat. Acad. Sci. U.S.A. 58, 2125—2130.

48) Pratt, L.H. (1986). Localization within the plant. in Photomorphogenesis in Plants (Kendrick, R.E. & Kronenberg, G.H.M. eds.) Nihoff/Junk, Dordrecht, The Netherlands. pp. 61 82.

49) Jordan, B.R., Partis, M.D. and Thomas, B. (1984). a study of phytochrome-membrane association using an enzyme-linked immunosorbent assay and western blotting. Physiol. Plant. 60, 416—421.

50) Terry, M.J., Thomas, B. and Hall, J.L. (1989). Analysis of the association of phytochrome with wheat leaf plasma membranes by quantitative immunoassay and western blotting. Abstracts of the European Symposium on Photomorphogensis in Plants, 1989. Freiburg. No 22.

51) Eisinger, W.R., Short, T.W. and Briggs, W.R. (1989). Light regulation of calcium fluxes in isolated membrane vesicles from pea (*Pisum sativum* L.) seedlings. Abstracts of the European Symposium on Photomorphogensis in Plants, 1989. Freiburg. No 33.

52) Blakeley, S.D., Thomas, B., Hall, J.L. and Vince-Prue, D. (1983). Regulation of swelling of etiolated wheat-leaf protoplasts by phytochrome and gibberellic acid. Planta 158, 416—421.

53) Kim, Y.-S., Moon, D.-K., Goodin, J.R. and Song, P.-S. (1986). Swelling of etiolated oat protoplasts induced by cAMP and red light. Plant Cell Physiol. 27, 193—197.

54) Bossen, M.E., Dassen, H.A., Kendrick, R.E. and Vredenberg, W.J. (1988). The role of calcium ions in phytochrome-controlled swelling of etiolated wheat (Triticum aestivum) protoplasts. Planta 174, 94—100.

55) Bossen, M.E., de Win, A.H.N., Kendrick, R.E. and Vredenberg, W.J. (1989). a GTP-binding protein is invoved in phytochrome-regulated changes in plasma membrane properties of etiolated wheat protoplasts. Abstracts of the European Symposium on Photomorphogensis in Plants, 1989. Freiburg. No S24.

56) Grimm, R., Gast, D. and Rudiger, W. (1989). Characterization of a protein-kinase activity associated with phytochrome from etiolated oat (*Avena sativa* L.) seedlings. Planta 174, 199—206.

57) Kim, I.-S., Bai, U. and Song, P.-S. (1989). A purified 124-kDa Oat phytochrome does not possess a protein kinase activity. Photochem. Photobiol. 49, 319—323.

58) Duke, S.O., Naylor, A.W. and wickliff, J.L. (1977). Phytochrome control of longitudinal growth and phytochrome synthesis in maize seedlings. Physiol. Plant. 40, 59—68.

59) Gottman, K. and Schafer, E. (1982). *In vitro* synthesis of phytochrome apoprotein directed by mRNA from light and dark-grown *Avena* seedlings. Photochem. Photobiol. 35, 521—525.

60) Gottman, K. and Schafer, E. (1983). Analysis of phytochrome kinetics in light-grown *Avena* seedlings. Planta 157, 392—400.

61) Colbert, J.T., Hershey, H.P. and Quail, P.H. (1983). Autoregulatory control of translatable phytochrome mRNA levels. Proc. Nat. Acad. Sci U.S.A. 80, 2248—2252.

62) Colbert, J.T., Hershey, H.P. and Quail, P.H. (1985). Phytochrome regulation of phytochrome mRNA abundance. Plant Mol. Biol. 5, 91—101.

63) Lissemore, J.L. and Quail, P.H. (1988). Rapid transcriptional regulation by phytochrome of the genes for phytochrome and chlorophyll a/b-binding protein in *Avena sativa.* Mol. Cell. Biol. 8, 4840—4850.

64) Hilton, J.R. and Thomas, B. (1987). Photoregulation of phytochrome synthesis in germinating embryos of *Avena sativa L.* J. Expt. Bot. 38, 1704—1712.

65) Thomas, B., Penn, S.E. and Jordan, B.R. (1989). Factors affecting phytochrome transcripts and apoprotein synthesis in germinating embryos of *Avena sativa* L. J. Ept. Bot. 40, 1299—1304.

66) Sharrock, R.A., Parks, B.M., Koorneef, M. and Quail, P.H. (1988). Molecular analysis of the phytochrome deficiency in an aurea mutant of tomato. Molec. Gen. Genet. 213, 9—14.

67) Tobin, E.M. and Silverthorne, J. (1985). Light regulation of gene expression in higher plants. Ann. Rev. Plant Phys. 36, 569—593.

68) Nagy, F., Kay, S.A. and Chua, N.-H. (1988). Gene regulation by phytochrome. Trends Genet 4, 37—42.

69) Schafer, E. and Briggs, W.R. (1986). Photomorphogenesis from signal perception to gene expression. Photobichem. Photobiophys. 12, 305—320.

70) Kaufman, L., Briggs, W.R. and Thompson, W. (1985). Phytochrome control of specific mRNA levels in pea buds: the presence of both very low and low fluence responses. Plant Physiol. 78, 388—393.

71) Kaufman, L., Roberts, L.L., Briggs, W.R. and Thompson, W. (1986). Phytochrome control of specific mRNA levels in developing pea buds: kinetics of accumulation, reciprocity and and escape kinetics of the low fluence response. Plant Physiol. 81, 1033—1038.

72) Jenkins, G.I. (1988). Photoregulation of gene expression in plants. Photochem. Photobiol. 48, 821—832.

73) Lamppa, G., Nagy, F. and Chua, N.-H. (1985). Light-regulated and organ-specific expression of a wheat *Cab* gene in transgenic tobacco. Nature, 316, 750—752.

74) Fluhr, R. and Chua, N.-H. (1986). Developmental regulation of two genes encoding ribulose-bisphosphate carboxylase small subunit in pea and transgenic petunia plants: phytochrome response and blue light induction. Proc. Nat. Acad. Sci. U.S.A. 83, 2358—2362.

75) Nagy, F., Boutry, M., Hsu, M.-Y. and Chua, N.-H. (1987). The 5' proximal region of the wheat cab-1 gene contains a 286bp enhancer-like sequence for phytochrome response. EMBO J. 6, 2537—2542.

76) Datta, N. and Cashmore, A. (1989). Binding of a pea nuclear protein to promoters of certain photoregulated genes is modulated by phosphorylation. The Plant Cell 1, 1069—1077.

77) Adamse, P., Kendrick, R.E. and Koorneef, M. (1988). Photomorphogenetic mutants of higher plants. Photochem. Photobiol. 48, 833—841.

78) Koorneef, M., Rolff, E. and Spruit, C.J.P. (1980). Genetic control of light-inhibited hypocotyl elongation in *Arabidopsis thaliana* L. Heynh. Z. Pflanzenphysiol. 100, 147—160.

79) Koorneef, M., Cone, J.W., Dekens, R.G., O'Herne-Robers, E.G., Spruit, C.J.P. and Kendrick, R.E. (1985). Photomorphogenetic responses of long-hypocotyl mutants of tomato. J. Plant Physiol. 120, 153—165.

80) Adamse, P., Jaspers, P.A.P.M., Bakker, J.A., Kendrick, R.E. and Koorneef, M. (1987). Photomorphogenetic responses of a long-hypocotyl mutant of *Cucumis sativus* L. J. Plant Physiol. 127, 481—491.

81) Chory, J., Peto, C., Feinbaum, R., Pratt, L.H. and Ausubel, F. (1989). *Arabidopsis thaliana* mutant that develops as a light-grown plant in the absence of light. Cell 58, 991—999.

82) Karlin-Neumann, G., Sun, L., Brusslan, J. and Tobin, E.M. (1989). A strategy to use a "suicide gene" to select mutants in the phytochrome transduction pathway leading to transcriptional changes in *Arabidopsis thaliana.* Abstracts of the European Symposium on Photomorphogensis in Plants, 1989. Freiburg. No 26.

83) Kaye, C., Pratt, L.H. and Malmberg, R.L. (1989). Development of a positive selection system for phytochrome-response mutants in haploid *Nicotiana plumbaginifolia.* Abstracts of the European Symposium on Photomorphogensis in Plants, 1989. Freiburg. No 27.

84) Keller, J.M., Shanklin, J., Vierstra, R.D. and Hershey, H.P. (1989). expression of a fuctional monocotyledonous phytochrome in transgenic tobacco. EMBO J. 8, 1005—1012.

85) Kay, S.A., Nagatani, A., Keith, B., Deak, M., Furuya, M. and Chua, N.-H. (1989). Rice phytochrome is biologically active in transgenic tobacco. The Plant Cell 1, 775—782.

86) Smith, H. (1981). Evidence that Pfr is not the active form of phytochrome in light-grown plants. Nature, 293, 163—165.

87) Carr-Smith, Thomas, B. and Johnson, C.B. (1989). An action spectrum for the effects of continuous light on flowering in wheat. Planta in press.

MODULATION OF SYNAPTIC TRANSMISSION AND PLASTICITY IN NERVOUS SYSTEMS

Hanns-Christof Spatz

Institut für Biologie III,
Universität Freiburg i. Br.
W.-Germany

SUMMARY

In recent years the study of learning and memory proceeded to an analysis on the cellular and molecular level. In some cases conditioned behaviour could be traced down to the modulation of synaptic transmission between nerve cells. Different approaches to elucidate the basic elements of the molecular mechanisms seem to converge to a unifying picture.

The genetic information for an organism must fit into every one of its cells. It is, therefore, too limited to endow an active animal with a sufficiently large repertoire of reactions to respond to a changing environment. A second system of information processing, the nervous system was developed, refined and enlarged to enormous complexity during evolution.

The brain may indeed be considered as an organ, specialized to collect, process and store information about the status of the body and about the environment and to formulate commands for appropriate interactions. The huge dimension of the task is represented in the stupendous number of nerve cells in the brain. It is, however, a simplification to see complexity only as a system property. Part of the information processing occurs already at the level of the individual nerve cell (1). While this seems to be hard to accept for some "neural network" theoreticians, it is obvious to the student of microscopic neuroanatomy who observes the diversity of size and shape of the nerve cells and yet realizes the high degree of order in the system.

Whenever I consider the transmission of electrochemical signals in a nerve cell as sketched in Fig. 1, I cannot escape the analogy to the flow of information in a modern democracy. The dendritic tree receives a multitude of signals from receptor cells or other nerve cells giving rise to excitory postsynaptic potentials (EPSPs) or inhibitory postsynaptic potentials (IPSPs), the Yes or No votes. These signals are conducted along the branches of the dendritic tree and attenuated along the way. There are even losses of this information by shunting inhibitions. Signals which have to travel a long way to the cell body are attenuated more than signals originating from synapses near the cell body. Thus votes can carry different weights. Depending on the geometry of the dendritic tree and the distribution and properties of its ionic channels, the incoming information is processed and channelled to the cell body, in our analogy the decision making body. If the

NATO ASI Series, Vol. H 51
Cell to Cell Signals in Plants and Animals
Edited by V. Neuhoff and J. Friend

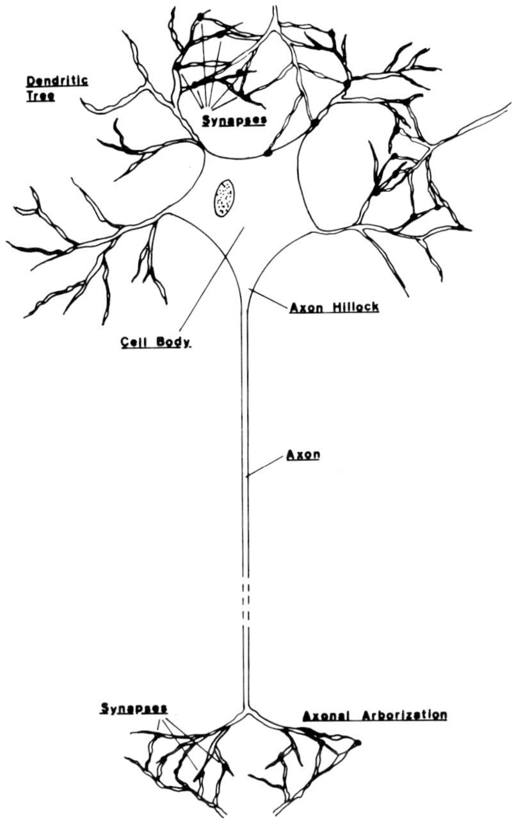

FIGURE 1
Schematic drawing of a nerve cell.

Yes votes sufficiently exceeded the No votes and the signal reaches a certain threshold, an action potential originates at the axon hillock. It travels down the axon without alteration in its form all the way to the executive, the axonal arborizations and the synapses at their ends. Executives have their own ways. In the synapse the stereotype action potential is translated into a chemical signal. This, however, is by no means a stereotype process. A considerable degree of modulation is observed in synaptic transmission.

The term plasticity is used to describe instances, where the response of the nervous system depends not only on current variables but also on past "experience". This includes all forms of learning. It can be realized in two entirely different ways:

a) Structural plasticity, i.e. growth or destruction of nerve cells or formation and loss of connections between nerve cells.
b) Changes of the efficacy of existing synaptic connections.

This article will focus on modulation of synaptic transmission and its role for plasticity in nervous systems. It is not intended as a review and will only try to highlight some of the issues raised during a NATO workshop in Il Ciocco, Italy, in September 1987 and to describe the developments since, as for instance discussed in a follow-up workshop in Freiburg, September 1989.

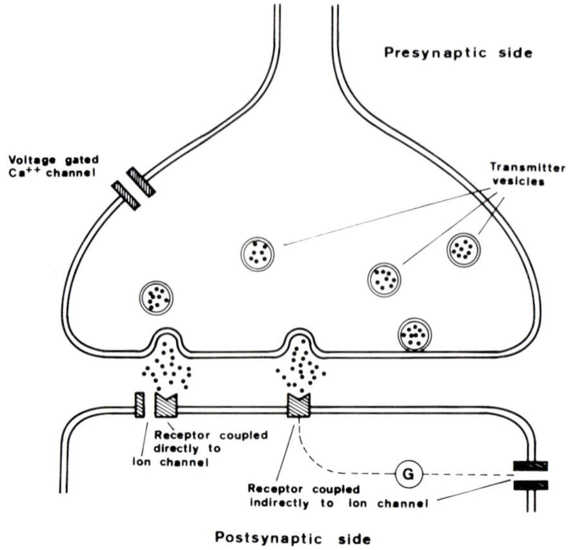

FIGURE 2
Schematic drawing of a synapse.

Communication between nerve cells is considered to occur predominantly by means of chemical transmitters in specialized structures called 'synapses' (Fig. 2). In suit of an action potential neurotransmitter is released from the presynaptic site, diffuses through the synaptic cleft and interacts with receptors on the membrane of the postsynaptic side. This interaction can directly lead to opening of ionic channels or it gives rise to another chemical signal which through a multistep cascade of biochemical reactions may act upon opening and closing of ionic channels. While the presynapse can be understood as an interpreter of electrical to chemical signals, chemically gated ionic channels translate the biochemical signal back to an electrical signal: the EPSP or IPSP. Since direct electrical synapses can also be found in nervous systems, the rather time consuming mode of communication by chemical synapses must have important advantages. One of these lies in the multitude of chemical signals. Even more important may be the degree of fine tuning and of modulation that a cascade of biochemical reactions allows.

"The old, oversimplified, rigid model of neurotransmission has collapsed. The new one now emerging has to account for an almost unbelievable diversification of neurotransmission in different junctions: The choice is not between chemical *or* electrical transmission — there is room for both. The rapidly growing list of 'putative messengers' now includes several monoamines, a nucleotide (presumably ATP), several amino acids, and at least 50 neuropeptides. Many neurons use several transmitters, stored in different combinations in different transmitter vesicles. The nerve impulse pattern, and input from 'presynaptic' receptors, promote or restrict secretion of the contents of one or the other kind of transmitter vesicles, and hence alter the composition of the secreted 'transmitter cocktail'. Thus nerves do not 'talk' to other cells in a simple binary language: + (excitatory) or — (inhibitory); 'chemical coding' of the message allows subtiler communication. Nerve impulses release transmitters in multimolecular packets ('quanta') of a single transmitter, or a mixture of different cotransmitters. For many signal substances there exist multiple receptors, coupled to multiple transducers, which act via multiple second messenger systems and mediate multiple actions in effector cells. Nerve terminals possess receptors to numerous chemical signal substances: Their own messengers, other neurotransmitters, substances released from effector cells and hormones reaching them by diffusion from the blood stream. They adjust the amount and combination of transmitters they secrete in response to the next impulse according to the integrated information conveyed by the nerve impulse pattern, as well as by the 'presynaptic' receptors which they use to monitor the chemical composition of their microenvironment. Neural messengers mediate transmission both synaptically, i.e., in well-defined close contact junctions, and para- and non-synaptically, diffusing over considerable distances and exerting excitatory, inhibitory or modulatory pre- and/or postjunctional effects on distant targets, via specific and selectively distributed receptors" (cited from (2)).

As expressed here, the multitude of transmitters is not only mirror imaged on the receptor side. For a number of transmitter substances pharmacologically different receptor types and subtypes are known. In the case of dopamine the D1 receptors are known to stimulate adenylate cyclase and at least some of the D2 receptors inhibit cyclase activity (3).

"... If a chemical neurotransmitter is to change the electrical activity of a recipient neurone, its biochemical action on the receptor has to be transduced into a change in ionic current flow. Functionally, the transduction step from receptor activation to ionic current flow may be classified into two forms — DIRECT and INDIRECT. This functional division has a physical basis in the structure of the receptor.

Direct transduction — here the 'receptor' — that is, the binding site for the transmitter — is an integral component of a macromolecular complex which also incorporates a potential ion channel. As a result, binding of the transmitter induced a very rapid (< 100 μsec) conformational change in the protein complex allowing the transient opening of the ion channel. This form of transmission, with its short-latency, is ideally suited for rapid point-to-point communication, the transmitter effectively 'closing the switch' in a hard-wired circuit. The number of chemical transmitter substances fulfilling this function appears rather limited: principal excitatory (i.e. cation-channel generating) transmitters in vertebrates are acetylcholine in the peripheral nervous system (acting on nicotinic receptors) and glutamate in the central nervous system, while glycine and GABA (γ aminobutyric acid) fulfill inhibitory transmitter functions, the channel being anionselective. A close homology between the structure of the receptors for acetylcholine, GABA and glycine has recently been shown supporting this functional analogy.

Indirect transduction — In this case, the receptor protein does not contain a potential ion channel; instead, the receptor and ion channel comprise quite separate molecules. Thus, transduction of the signal received by the receptor (in the form of transmitter binding) into a change in ionic channel function involves a number of intermediary steps. Several characteristics distinguish this form of transmission from direct transmission.

a) The intermediary steps in the transduction process confer a finite lag of some tens of milliseconds (or more) in the response to the transmitter. Thus, this form of transmission is ill-suited to rapid information processing but instead provides a form of fine-tuning, controlling neuronal excitability over periods of seconds or minutes.

b) There is considerable flexibility in the response to a given transmitter, both in terms of which ion channels are affected, and in terms of the change in ion channel function — i.e. whether they are opened or closed or whether their gating properties are modified. Though certain constraints are emerging, depending on the structure of the individual receptors, coupling (see below) and ion channels, at present these are insufficient to permit fully predictive rules.

c) This form of transmission is operated by a remarkably wide variety of transmitters, such as acetylcholine (via muscarinic receptors), and peptides, all acting on physically discrete receptors. However available information suggests that these receptors may all conform to a generally homologous structure, distinct from the direct-transmitting ionophore-receptors, and constituting a second 'super-family' of receptors.

d) In most systems so far investigated, the immediate post-receptor transduction step comprises coupling with one or a family of GTP-binding proteins (G-proteins), resulting in dissociation of the G-protein into its constituent α and β_γ components and activation of its GTPase function" (cited from (4)).

G-proteins seem to be part of nature's general toolbox. They carry specific functions as links between a receptor and an enzyme like adenylate cyclase or phosphodiesterase. But from a general point of view they can be regarded as timers. Upon activation by the receptor the α subunit of a G-protein exchanges bound GDP with GTP and this way becomes free to act as a special courier. The α subunit possesses GTPase activity. Within a certain time the GTP molecule will be hydrolyzed to GDP and the courier returns home, to reassociate with the β_γ subunits.

"Activation of the G-protein may then lead to ion channel modulation in one or more of the following ways:

i) Activated G-protein subunits may couple directly to an ion channel, to open or close it.
ii) The G-protein may activate or inhibit one of several intracellular enzymes, such as adenylate cyclase, cyclic nucleotide phosphodiesterase or phospholipase C, to modify the levels of soluble second messengers such as cyclic adenosine monophosphate, cyclic guanosine monophosphate, inositol phosphate or diacylglycerol.
iii) The soluble messengers may interact directly with the channel, or, in turn, activate kinase enzymes, modifying the state of channel protein phosphorylation" (cited from (4)).

It is suggestive, that the capacity for tuning the chemical synapse offers a mechanistic explanation for some of the processes which endow the nervous system with its remarkable possibility for plastic responses. Through learning individuals can adapt to the changing living conditions, an important evolutionary advantage in addition to the wasteful genetic adaptation. It is not surprising that a substantial degree of plasticity is already utilized by animals at a low level of organisation (5). For a number of so called "Simple Nervous Systems" the studies of learning processes have been carried all the way to the molecular level. This is most notably the case for the marine snails *Aplysia* and *Hermissenda,* for the honeybee and for the fruitfly *Drosophila*.

Work on the molecular biology of learning in invertebrate systems was greatly stimulated by the pathbreaking work of Kandel and Schwartz (6), who could trace sensitization of a simple reflex in *Aplysia* to a complete chain of biochemical reactions, in an identified neuron. In essence the model can account for short-term and intermediary-term memory, both in *Drosophila* and in *Aplysia,* while the mechanism of persistent memory is not yet elucidated.

"A fundamental problem in neurobiology is to understand the basic processes that endow the nervous system with the ability to store and retrieve long-term memory. A related problem is to understand the relationship between the mechanisms for the induction, storage and retrieval of short-term memory and those for long-term memory. While behavioral studies of vertebrates indicate that short-term memory and long-term memory are distinct entities, there must be some mechanistic link between the two types, since short-term memories can be consolidated into long-term ones.

In order to examine the mechanisms of long-term memory and their relationship to short-term memory, we have used a relatively simple identified neural circuit in the nervous system of *Aplysia* that controls defensive withdrawal of the tail and siphon. This preparation allows for the analysis of long-term cellular changes that occur in sensory neurons as a result of sensitization, a form of non-associative learning. In addition, with this preparation it is becoming possible to elucidate some of the subcellular processes that are responsible for converting short-term memory into long-term memory.

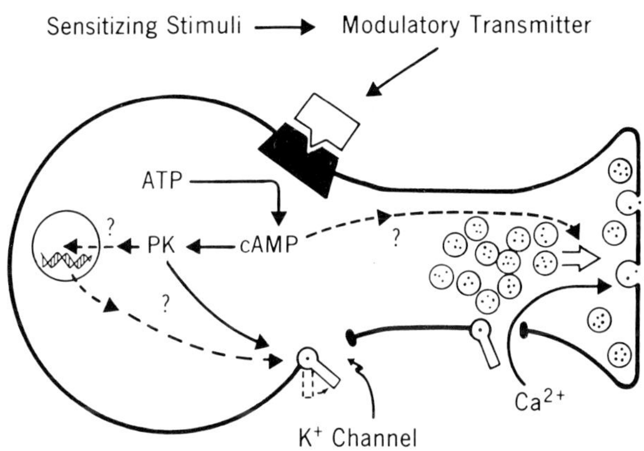

FIGURE 3
Model of heterosynaptic facilitation that contributes to sensitization. (See text for details, ATP = adenosine triphosphate, cAMP = cyclic adenosine monophosphate, PK = Protein kinase).

A model that summarizes some of the neural and molecular mechanisms contributing to short- and long-term sensitization is illustrated in Fig. 3. Sensitizing stimuli lead to the activation of facilitatory neurons that release a neurotransmitter with properties similar to, or the same as, serotonin. This transmitter acts on the sensory neurons to increase the levels of cAMP, which in turn leads to closure of membrane K^+ channels through protein phosphorylation. Closure of these K^+ channels leads to membrane depolarization and an enhancement of the excitability of the sensory neurons. An additional consequence of reduced K^+ channel activity is a reduction of the repolarizing current during an action potential. Thus, action potentials initiated in the sensory neurons are broader. This may serve to enhance Ca^{2+} influx in the presynaptic terminal and increase transmitter release from the sensory neurons.

The mechanisms for the long-term changes are not yet fully understood. However, it appears that the same intracellular messenger that leads to the expression of the short-term changes, also triggers cellular processes that lead to the long-term changes. One possible mechanism for this action is through the regulation of protein synthesis by cAMP. Our results with protein analysis using 2D-PAGE, as well as experiments using protein synthesis inhibitors support a role for this mechanism. It is also possible that the long-term changes in membrane currents could be due at least in part to changes in the ratio of subunits of the cAMP-dependent protein kinase (7).

In addition to the long-term biochemical and biophysical changes that we have described, morphological changes have been observed in the synaptic terminals of the sensory neurons as a consequence of long-term sensitization. This may contribute to the synaptic facilitation that has been observed under similar conditions. Thus, both short- and long-term sensitization are associated with a number of cellular alterations that appear to act in concert to increase the excitability of the sensory neurons and enhance transmitter release, thereby producing an enhanced withdrawal response" (cited from (8)).

In *Drosophila* most nerve cells are so small that intracellular recordings or injection of biochemical reagents into identified cells are impossible. However, the analysis of the biochemical defects of learning mutants (9) as well as the behavioural analysis of flies treated with specific drugs has provided evidence for a key role of the cAMP metabolism in learning processes. A question, which puzzled students of *Aplysia* (7) as well as students of *Drosophila,* is, how a *short stimulus* can be translated into a *long lasting change* in the efficacy of synaptic transmission. It has become clear lately, that a solution of the problem can be found if the kinetics of the reactions in the cAMP cascade are considered.

A single short stimulus from an interneuron leads to release of a modulatory transmitter (Fig. 3). The receptor activates an adenylate cyclase via a G-protein. This leads to a transient increase of cAMP levels in the cell, transient because phosphodiesterases rapidly convert cAMP into AMP, which has no second messenger function. cAMP leads to the dissociation of catalytic and regulatory subunits of the cAMP-dependent protein kinase. Only the dissociated catalytic subunit is active to phosphorylate other proteins. Upon return to normal cAMP levels the two subunits will reassociate. The rate of reassociation, however, depends on the state of phosphorylation of the regulatory subunit, being itself a substrate of the catalytic subunit and its oponent a phosphatase. Phosphorylation of the regulatory subunit extends the rate of reassociation and thus the activity of the protein kinase into the range of minutes after cAMP decay.

Only under special circumstances will protein phosphorylation be a quasi irreversible process. In general phosphatases will rapidly remove a phosphate group from a previously phosphorylated substrate. If a long term change of phosphorylation is to be achieved, the balance between kinase and phosphatase action has to be shifted. In *Drosophila* this can occur through proteolytic modification of the regulatory subunit of the protein kinase (10). The dissociated regulatory subunit can be truncated by a Ca^{++}-dependent protease to a modified regulatory subunit. This changes its affinity to the catalytic subunit in such a way that the activation curve of the enzyme is shifted to lower cAMP concentrations. Thus at normal maintained cAMP concentrations in the cell a considerable fraction of the kinase remains dissociated and therefore active for periods of time rating in hours, until protein turnover restores the original state.

Figure 4 gives a schematic representation of this chain of signal extension reactions. Learning in *Drosophila* is strongly correlated with the degree of modulation permissible in every one of these steps:

a) The learning mutant *rutabaga* is known to have lost the Ca^{++}/Calmodulin dependent fine tuning of the adenylate cyclase.

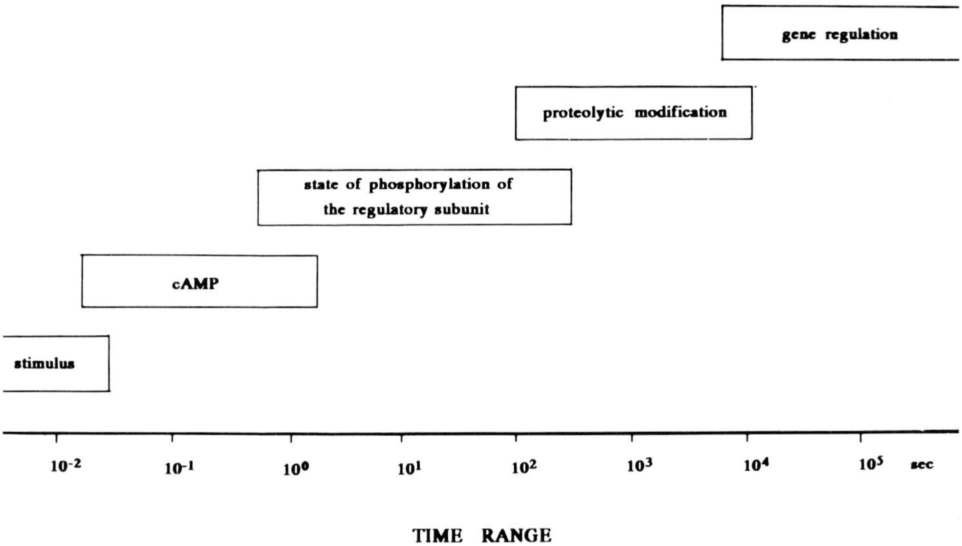

FIGURE 4
Schematic representation of a chain of signal extension reactions. A logarithmic time scale is used to indicate the kinetics of the individual steps.

b) Deviation from the normal degree of phosphorylation of the free regulatory subunit leads to a diminished ability to learn.

c) Concentrations of the proteolytically modified regulatory subunit twice as high as normal characterize bad learners.

Most remarkable is the observation, that it is the protein kinase, which is not only the key enzyme for information processing but also the site of information storage.

Although in *Drosophila* the bridge between kinase activity and transmitter release is not yet built, phosphorylation and proteolytic modification may serve as biochemical correlates of short term and intermediary term memory. Processes extending memory into the range of days or longer are still unknown, except that altered gene expression must be involved. Intensive research is devoted to this problem.

The cAMP cascade is a special example of a chain of signal extension reactions. Quite surprisingly, however, in most systems where learning has been studied at the molecular level, protein kinases seem to play a key role. The Ca^{++}/Calmodulin dependent protein kinase for instance is involved in a nonassociative form of learning (11). In mammalian systems much attention has been given to the role of protein kinase C in processes associated with classical conditioning of the eyeblink response of the rabbit.

Impressively advanced are the studies of "Long Term Potentiation" (12). This phenomenon first observed in the archicortex of the mammalian brain, the hippocampus, resembles learning processes in several respects.

"There is increasing evidence to suggest that at least some types of learning are encoded in the mammalian brain by a form of synaptic plasticity that may be broadly described as Hebbian, in that underlying changes in synaptic strength occur as the result of coincident pre- and post-synaptic activity. A widely-studied example of this sort of plasticity is the enduring enhancement of synaptic transmission known as long-term potentiation (LTP), which is produced with beguiling ease by the application of brief trains of high-frequency stimulation to any of the main excitatory pathways of the hippocampal formation. In the last few years a consensus has developed regarding the cellular events leading to the initiation of LTP; in this view, LTP develops, in susceptible synapses, whenever there is a conjunction of synaptic activity and strong postsynaptic depolarization, conditions which allow the opening of the voltage-dependent ion channel associated with the NMDA subtype of glutamate receptor. A similar consensus on how the potentiated state is maintained has yet to emerge, with evidence for persistent changes in transmitter release and in synaptic morphology, possibly reflecting changes in protein phosphorylation, redistribution of calcium stores, and changes in protein synthesis" (cited from (12)).

Although induction of LTP seems to be a postsynaptic event, maintenance is characterized by an increased release of the putative transmitter glutamate. It has therefore been proposed that a retrograde messenger is released from the postsynaptic site of induction, which acts presynaptically to increase transmitter release. This may be a lipoxygenase metabolite of arachidonic acid (13), which would be in concert with the observation that protein kinase C causes a prolonged potentiation of transmission in pyramidal cells of the hippocampus. In addition, postsynaptic activation of the Ca^{++}/Calmodulin protein kinase plays an important role in the maintenance of LTP (14).

A major breakthrough in our understanding of the molecular mechanism of LTP may come from the elucidation of the function of B-50, a synaptosomal protein and specific substrate of protein kinase C. The increase of neurotransmitter release in the hippocampus is correlated with the degree of phosphorylation of B-50 (15). Even more directly: chemical depolarization of synaptosomes induces transmitter release and protein kinase C mediated B-50 phosphorylation. Antibodies against B-50, which specifically interfere with its phosphorylation inhibit noradrenaline release (16). Since B-50 is present throughout the brain and since the protein kinase C system plays a role in the release of a number of transmitters, B-50 phosphorylation may be an essential part of such a mechanism.

We begin to see parts of the bridge between protein phosphorylation and modulation of synaptic transmission. Other parts are still in the haze but in contrast to my statement in the preface of the proceedings of the workshop (17), a unifying picture of basic elements of the molecular mechanism underlying learning processes seems to emerge, in part due to the impact of the "Il Ciocco" workshop in September 1987.

REFERENCES

1) Bullock, Th.H., R. Orkand, and A. Grinnell: "Introduction to Nervous Systems". W.H. Freeman & Co., 1977.

2) Stjärne, L.: "New Paradigm: Neurotransmission by Single Quanta of Multiple Messengers Released from Multiple Varicosities and Acting via Multiple Receptors". Abstract for the NATO Advanced Research Workshop, Il Ciocco, 1987.

3) Kelly, E., and S.R. Nahorski: "Attempts to Characterise Dopamine Receptor-Effector Mechanisms in the Brain". NATO ASI Series, Vol. H19, Springer Verlag, 1988.

4) Brown, D.A., H. Higashida, P.R. Adams, and N.V. Marrion: "Postsynaptic Signal Transduction in Neuroblastoma and Ganglion Cells: Receptor-Mediated Control of K-Currents". NATO ASI Series, Vol. H19, Springer Verlag, 1988.

5) Dudai, Y.: "The Neurobiology of Memory". Oxford University Press, 1989.

6) Kandel, E.R., and Schwartz, J.H.: "Molecular Biology of Learning: Modulation of Transmitter Release". Science *218,* 433—443, 1982.

7) Greenberg, S.M., V.F. Castellucci, H. Bayley, and H. Schwartz: "A Molecular Mechanism for Long-Term Sensitization in Aplysia". Nature, *329,* 62—65, 1987.

8) Byrne, J.H., A. Eskin, and K.P. Scholz: "Neural and Molecular Mechanisms of Short- and Long-Term Sensitization in *Aplysia*". NATO ASI Series, Vol. H19, Springer Verlag, 1988.

9) Tully, T.: "On the Road to a Better Understanding of Learning and Memory in *Drosophila melanogaster*". NATO ASI Series, Vol. H19, Springer Verlag, 1988.

10) Müller, U., and H.-CH. Spatz: "Ca^{2+}-Dependent Proteolytic Modification of the cAMP-Dependent Protein Kinase in Drosophila Wild-Type and Dunce Memory Mutants". J. Neurogenetics, *6,* 95—114, 1989.

11) Mitschulat, H., and R. Willmund: "Molecular Aspects of Plasticity in Phototaxis". NATO ASI Series, Vol. H19, Springer Verlag, 1988.

12) Bliss, T.V.P., M.L. Errington, and M.A. Lynch: "Induction and Maintenance of Long-Term Potentiation in the Hippocampus". NATO ASI Series Vol. H19, Springer Verlag, 1988.

13) Lynch, M.A., M.P. Clements, K.L. Voss, M.L. Errington, J.H. Williams, and T.V.P. Bliss: "Role of Arachidonic Acid in Long-Term Potentiation: Is it a Retrograde Messenger?" Abstract for the Research Workshop "Cellular Signals and Plasticity in Nervous Systems", Freiburg i. Br., September 1989.

14) Malenka, R.C., J.A. Kauer, D.J. Perkel, and R.A. Nicoll: "The Impact of Postsynaptic Calcium on Synaptic Transmission — Its Role in Long-Term Potentiation". TINS, *12,* 444—450, 1989.

15) Dekker, L.V., P.N.E. De Graan, F.M.J. Heemskerk, L.H. Schrama. A.B. Oestreicher, P. Schotman, and W.H. Gispen: "The Role of Protein Phosphorylation in Long-Term Potentiation". NATO ASI Series Vol. H19, Springer Verlag, 1988.

16) Dekker, L.V., P.N.E. De Graan, A.B. Oestreicher, D.H.G. Versteeg, and W.H. Gispen: "Inhibition of Noradrenaline Release by Antibodies to B-50 (GAP-43)". Nature *342,* 74—76, 1989.

17) Hertting, G., and H.-CH. Spatz: "Modulation of Synaptic Transmission and Plasticity in Nervous Systems". NATO ASI Series, Vol. 19, Springer Verlag, 1988.

GANGLIOSIDES AND MODULATION OF NEURONAL FUNCTIONS

H. Rahmann* and H. Wiegandt**

* Institute of Zoology,
University of Stuttgart-Hohenheim,
D-7000 Stuttgart 70, FRG,
and
** Physiol. Chem. Institute,
University of Marburg,
D-3550 Marburg an der Lahn, FRG

INTRODUCTION

Gangliosides (= amphiphilic sialic acid containing glycosphingolipids) are components of all plasma membranes and as such ubiquitous in vertebrate tissues. They are especially abundant and complex in neuronal membranes, where they are anchored in the outer leaflet of the membrane bilayer by their hydrophobic ceramide moiety, while exposing their negatively charged hydrophilic and spacious sialo-oligosaccharide residue into the extracellular space (Fig. 1). Together with neutral glycolipids, glycoproteins and glycosaminoglycans, gangliosides comprise the glycocalix of the cell surface.

Despite intensive research on gangliosides ever since Ernst Klenk discovered them in 1935 (for review comp. Yamakawa, 1), the biological function of these "sphinx-like" glycosphingolipids remains elusive. Over the past few years, several thousands of articles on gangliosides have been published demonstrating their very special properties and their possible involvement in diverse biological phenomena. As far as we know up to the present time, these properties may be due to particular physico-chemical attributes of these compounds, especially in combination with calcium.

The results of research on gangliosides obtained so far have shed light not only on possible functional roles of these glycolipids with regard to participation in cell recognition, cell-to-cell contact formation, receptor binding and modulation, immunological phenomena, bio-signal transduction etc., but have also unexpectedly opened up a whole new field of possible therapeutic applications of gangliosides in case of various neuropathies of various etiology.

However, since there still exists a considerable diversity of opinion among the scientists on the subject and partial contradictions among the various scientific and clinical data, these discrepancies are urgently in need of clarification. With regard to this, experts working with various aims and techniques in the broad field of ganglioside research were invited to participate in the NATO Advanced Workshop on "Gangliosides and Modulations of Neuronal Functions" held at the University of Stuttgart-Hohenheim, Stuttgart, FRG, October 20—24, 1986. A critical synopsis of the present state of knowledge with

NATO ASI Series, Vol. H 51
Cell to Cell Signals in Plants and Animals
Edited by V. Neuhoff and J. Friend

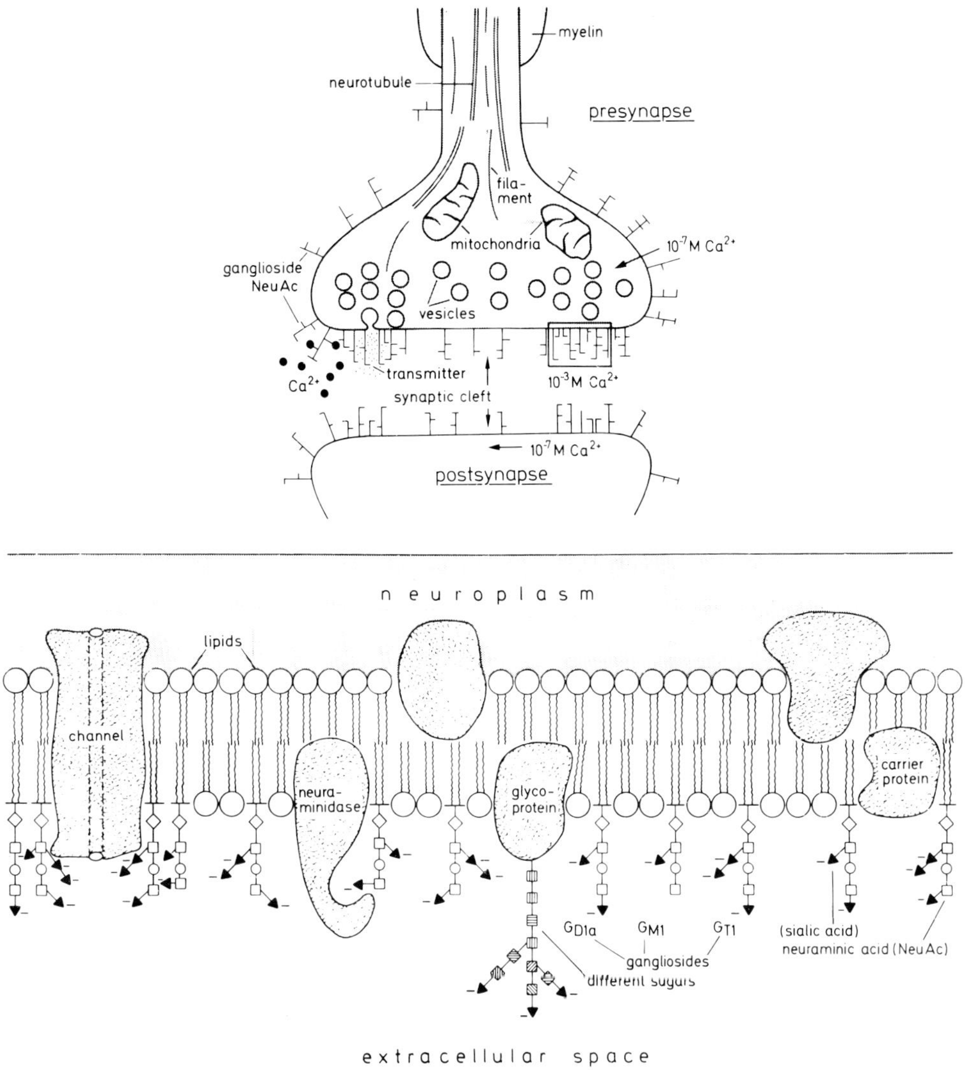

FIGURE 1

Scheme of a synaptic terminal with localization of ganglioside clusters in the outer membrane surface, and molecular membrane including differently polar gangliosides, integral glycoproteins, carrier- and channel proteins and the enzyme neuraminidase (= sialidase), clearing gangliosides

respect to brain ganglioside research, and of the relevance and possibilities of their therapeutic applications was elaborated.

The purpose of the workshop and the resulting volume (NATO ASI Series H7) was to summarize and assess in the form of survey lectures the disturbingly huge flood of publications in a wide range of topics on the following ganglioside themes: I. Fundamentals for research on the function of gangliosides; II. Brain ganglioside metabolism; III. Cell specific distribution and differentiation-related expression of brain gangliosides; IV. Gangliosides and neuronal plasticity; and V. Bio-medical potential of exogenous ganglioside application on neuronal function.

Therefore, in the following *overviews* are presented on the current state of knowledge in the *fields of ganglioside research*, that were addressed at the meeting. *Recent advances*, in this area made since the workshop took place are included, emphasizing some *new proposals for future research activities*.

RESULTS

I. Fundamentals for Research on the Function of Gangliosides

As was reviewed by Wiegandt (2), the gangliosides are those glycosphingolipids that carry one or more sialic acid molecules attached to their neutral sugar portion (Fig. 2). Gangliosides, therefore, occur typically only in such animals that synthesize sialic acids, i.e. species of the *Deuterstomia* including the mammals. In contrast, in *Protostomia*, animals without sialic acid, it appears as though sialic acid may be replaced by other sugar acids, e.g. by glucoronic acid in the arthrosides of insects (comp. chapter III).

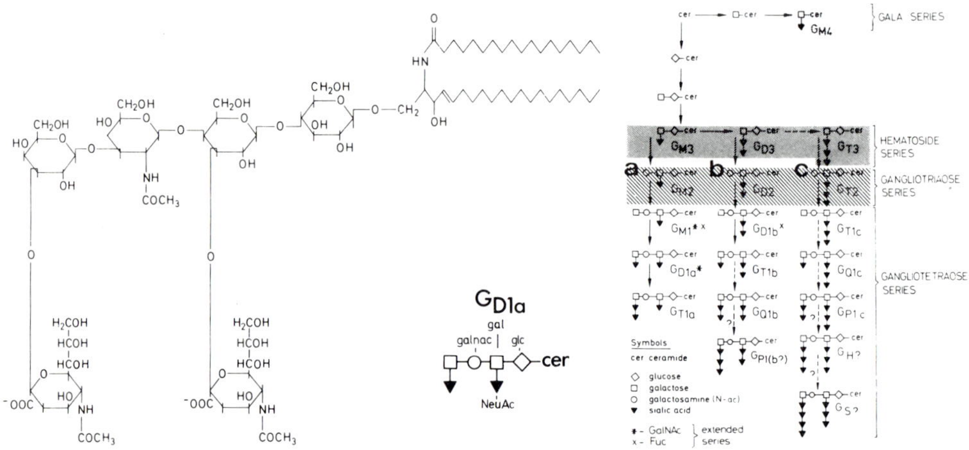

FIGURE 2
Molecular structure of the di-sialoganglioside GD_{1a} and major pathways of ganglioside biosynthesis

The neutral carbohydrate backbone of the gangliosides may be derived of either of several carbohydrate "series". These glycosphingolipid-oligosaccaride series are, within an animal species specific repertoire, characteristic for a given cell type. The carbohydrate series that are known in the Deuterostomia comprize the Lacto- (Lc), Muco- (Mc), iso-Muco- (iMc), Globo- (Gb), iso-Globo- (iGb), Ganglio- (Gg) and the Sialo-ganglio- (Neu-Gg) series (Fig. 3).

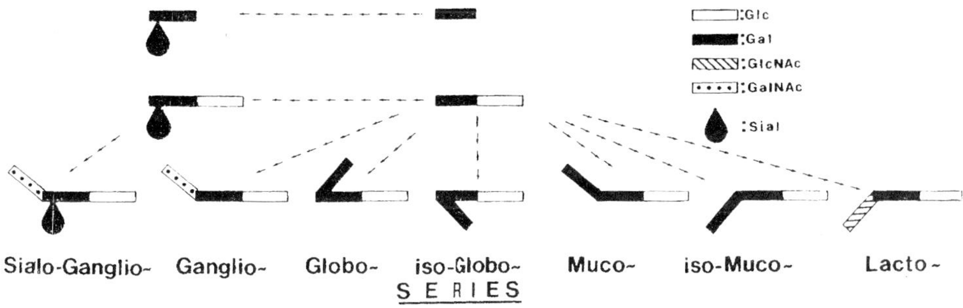

FIGURE 3
Scheme of oligosaccharide series of glycosphingolipids

For several reasons, including their particulary high concentration and therefore the possibility of their likely involvement in special functions of the central nervous system, particular attention has always been directed to the gangliosides of the brain. With the exception of the glial ganglioside NeuAc-GalCer, these gangliosides belong exclusively to the sialo-ganglio-series.

An up-to-date review of the chromatography of the gangliosides was provided by Ando et al. (3) centering on new methods of the high pressure liquid chromatographic separation of the underivatized glycolipids.

Fundamental to modern methods for the structural elucidation of gangliosides and other glycosphingolipids in general is the application of NMR- and MS-techniques, as reviewed by Yu (4) and Egge and Peter-Katalinic (5), respectively. It is the proton magnetic resonance spectroscopy, in particular various two-dimensional methods that can provide information not only on the primary, but also the secondary structure of the gangliosides, i.e. their complete structural analysis. Introduction of "softly ionizing" fast atom bombardment mass spectrometry (FAB-MS) has also contributed significantly to the further development of ganglioside analysis. The method allows, e.g. the analysis of highly sialylated gangliosides without any previous derivatisation. FAB-MS furnishes data not only on the sugar portion of the gangliosides but also on the composition of their ceramide moiety.

Physico-chemical properties of the gangliosides in solution were reviewed by Corti et al. (6) centering mostly on their suprastructural behaviour in aqueous systems as micelles or, as in the case of the gangliosides NeuAc-GalCer and II^3NeuAc-LacCer, lamellae. The all important physico-chemical behaviour of gangliosides in membranes was thoroughly reviewed by Gant (7). Among other agents, it is Ca^{2+}-ions that may directly modulate the ganglioside distribution in a biomembrane. In the light of this observation, Rahmann's data (8—10) on an involvement of gangliosides in synaptic transmission appear all the more significant. More recently, Rahmann and coworkers (11—13) could show by special electronmicroscopic observation that Ca^{2+}-stores at the synaptic contact region of nerve terminals are colocalized with gangliosides and a high affinity Ca^{2+}-ATPase. This adds to the notion that Ca^{2+} plays a central role in nerve related biological functions of the gangliosides (comp. chapter IV).

From the laboratory of Sandhoff (14, 15), routes for the chemical modification of gangliosides were reported. Such ganglioside derivatives are instrumentally useful in the elucidation of physico-chemical as well as biological roles of this important class of sialo-glycoconjugates.

II. Brain Ganglioside Metabolism

a. *Pathways for de novo Biosynthesis of Gangliosides*

As reviewed by Tettamanti et al. (16) and Sandhoff et al. (15) the ganglioside biosynthesis, in brain as well as in other tissues, is the result of sequential additions of monosaccharide units to a ceramide. Each reaction is catalyzed by a glycosyltransferase, the saccharide unit being transferred from the corresponding sugar-nucleotide to the receptor (Fig. 4). Some glycosyltransferases have a strict specificity and constitute the key points for the formation of different ganglioside series. The sequential order of glycosylations governing ganglioside biosynthesis is assumed to be guaranteed by the assembly of the involved glycosyltransferases in a multienzyme system located in the Golgi apparatus. From here the transport of gangliosides to the plasma membrane is likely mediated by vesicles (15, 16).

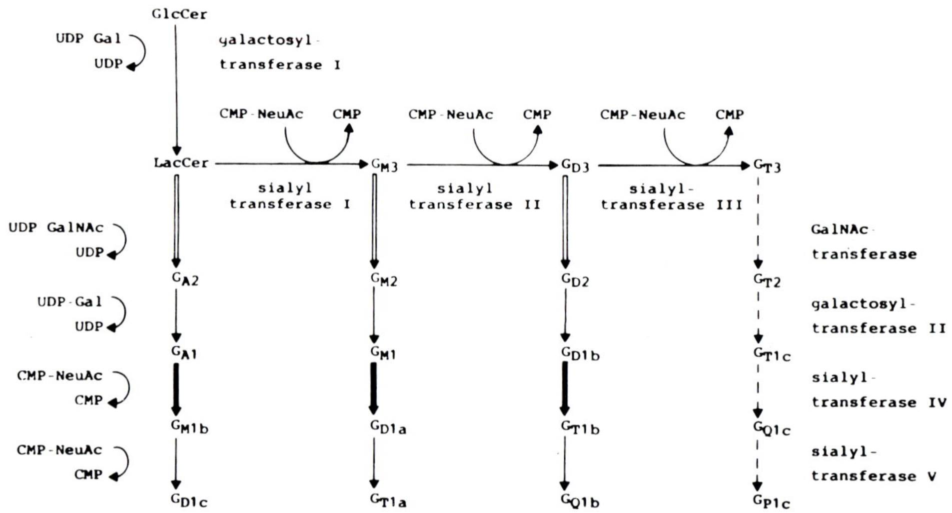

FIGURE 4

Proposed model for ganglioside biosynthesis: Successive transfer of N-acetylgalactosamine (GalNAc), galactose, and sialic acid to Lac Cer, GM_3, and GD_3 (and possibly GT_3) leading to the asialo, a- and b-series (and possibly c-series) gangliosides is catalyzed by the same set of glycosyltransferases (According to 15)

Lipidic and saccharide precursors for ganglioside biosynthesis can be recruited by de novo synthesis as well as by recycling of catabolic lysosomal by-products. The latter process consists in the use for biosynthetic purposes of compounds like sphingosine, ceramide, sialic acid originating from lysosomal degradation of gangliosides. In addition to this, the possibility exists that a ganglioside, derived from the plasma membrane by-passes the lysosomal apparatus and reaches the Golgi-apparatus, where it is further di-

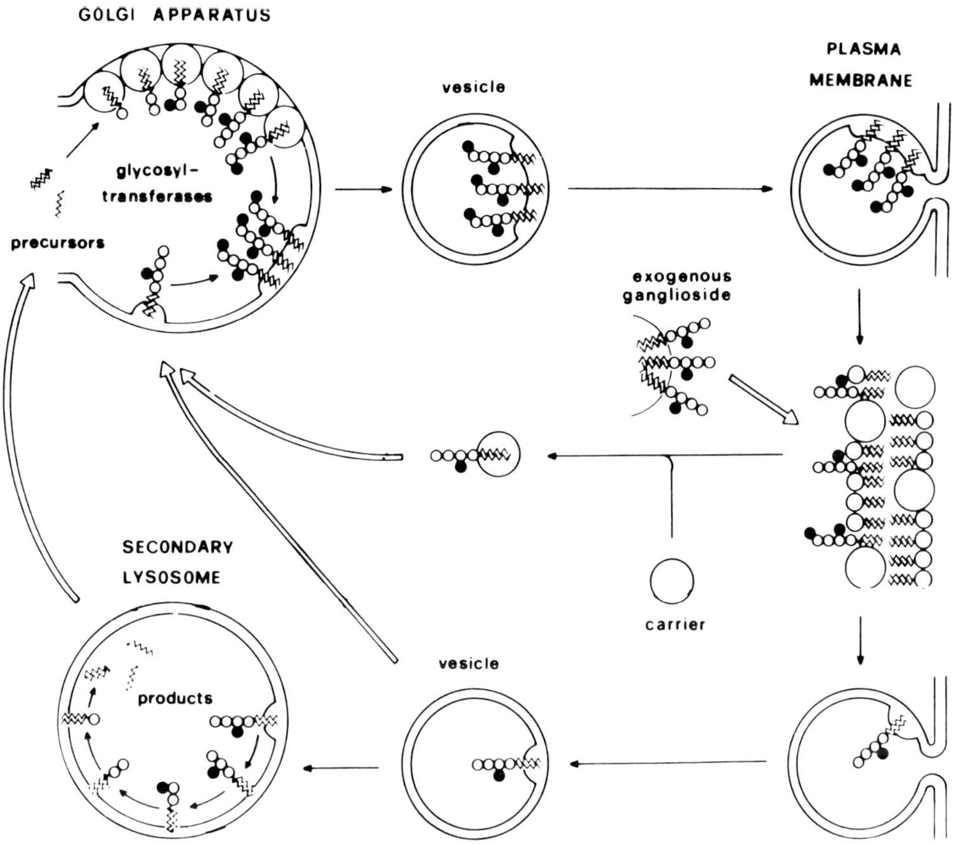

FIGURE 5
Functional coordination of subcellular compartment in ganglioside biosynthesis. De novo biosynthesis, recycling of catabolic by-products, and direct glycosylation of plasma membrane-derived gangliosides are represented (According to 16)

rectly glycosylated (16, Fig. 5). Most probably, by exocytotic vesicles, the final products are delivered from the Golgi-cisternae to the plasma membrane. Turnover rates of gangliosides are rather low with half lifes of several weeks.

Exogenous gangliosides can be taken up as well *as in vitro* by cultured cells from the medium (17, 18) or *in vivo* following direct application of these lipids to neurons (e.g.) retina ganglion cells, 19).They are slowly inserted with their ceramide portion into the bilayer of the plasma membrane from where they reach by vesicle-bound movements either the lysosomal compartment for degradation or the Golgi cisternae for further biosynthesis of higher gangliosides.

Enzymes involved in ganglioside biosynthesis have not been well purified and characterized so far. Sugar nucleotides (CMP-NeuAc, UDP-Glc, UD-Gal, UDP-GalNAc) needed for biosynthesis are provided by the cytosol and enter the Golgi apparatus through carrier proteins. Regulation of ganglioside biosynthesis most probably takes place at the level of transcription of synthesizing enzymes, at the level of the individual enzymes (feed-back regulation), and at the level of passage of precursors from one compartment to the other.

Gene transfection of cells with adenovirus and retrovirus cause dramatic changes in the ganglioside biosynthesis (20). A gene controlling biosynthesis of GM_1 (in the liver of inbred mice) has been mapped already at the locus centromeric to H-2K of chromosome 17 (21).

b. *Neuronal Transport of Gangliosides*

As reviewed by Ledeen et al. (22) in dendrites and axons of neurons, gangliosides participate as well in the anterograde as in the retrograde vesicle-bound fast neuronal transport, thus supporting the trophic feed-back system of the neurons between the nerve cell body and the synaptic terminals (9, 10, 23). As had been reported before, the bulk biosynthesis of gangliosides takes place in the Golgi apparatus, being located in the perikaryon of the neuron, whereas the nerve fiber terminals are assumed to be devoid of biosynthesis and catabolism of gangliosides (23). However, in vitro studies, on the one side, allow for a conversion of disialo to monosialogangliosides at nerve endings by a membrane-bound sialidase (15). On the other side, also a peripheral biosynthesis in nerve endings has to be considered, since in an *in vivo* model (optic system of chicken) it had been found that tritiated GM_1 following its uptake by retina ganglion cell bodies and after anterograde transport to the nerve terminals in the optic tectum is partially converted into tritiated GD_{1a} (19).

In the same optic system of chicken, Caputto and collaborators (24) identified a functional regulation of ganglioside biosynthesis: An increased labelling of gangliosides occurs during light perception, and a decreased labelling during urethane anesthesia. In addition to this, studies with isolated membranes from retina showed an inhibition of GM_2- and GD_3-synthases by end product gangliosides.

c. *Pathway of Ganglioside Degradation*

According to Sandhoff et al. (15) gangliosides are degraded by the sequential action of exohydrolases, most of which are localized in lysosomes. In addition, ganglioside cleaving sialidases have been localized in plasma membranes, especially in those of neurons and myelin. Some of the degrading hydrolases have been purified and characterized (Fig. 6), and different modes of glycolipid-enzyme interactions were observed. A water soluble hydrolase may recognize either directly the lipid substrate or through mediation of an activator protein, which facilitates the interaction. A membrane-bound hydrolase, such as sialidase, may interact with its substrate inserted into the same membrane by lateral diffusion.

Regulation of ganglioside degradation and the subcellular localization of ganglioside desialylation (lysosomal and/or membrane-bound sialidase) still remain unclear. Nevertheless, genetically fixed disturbances of the pathway of sphingolipid degradation, especially of ganglioside degradation, on different levels, were found to be responsible for metabolic blocks of known mental diseases (comp. Fig. 6).

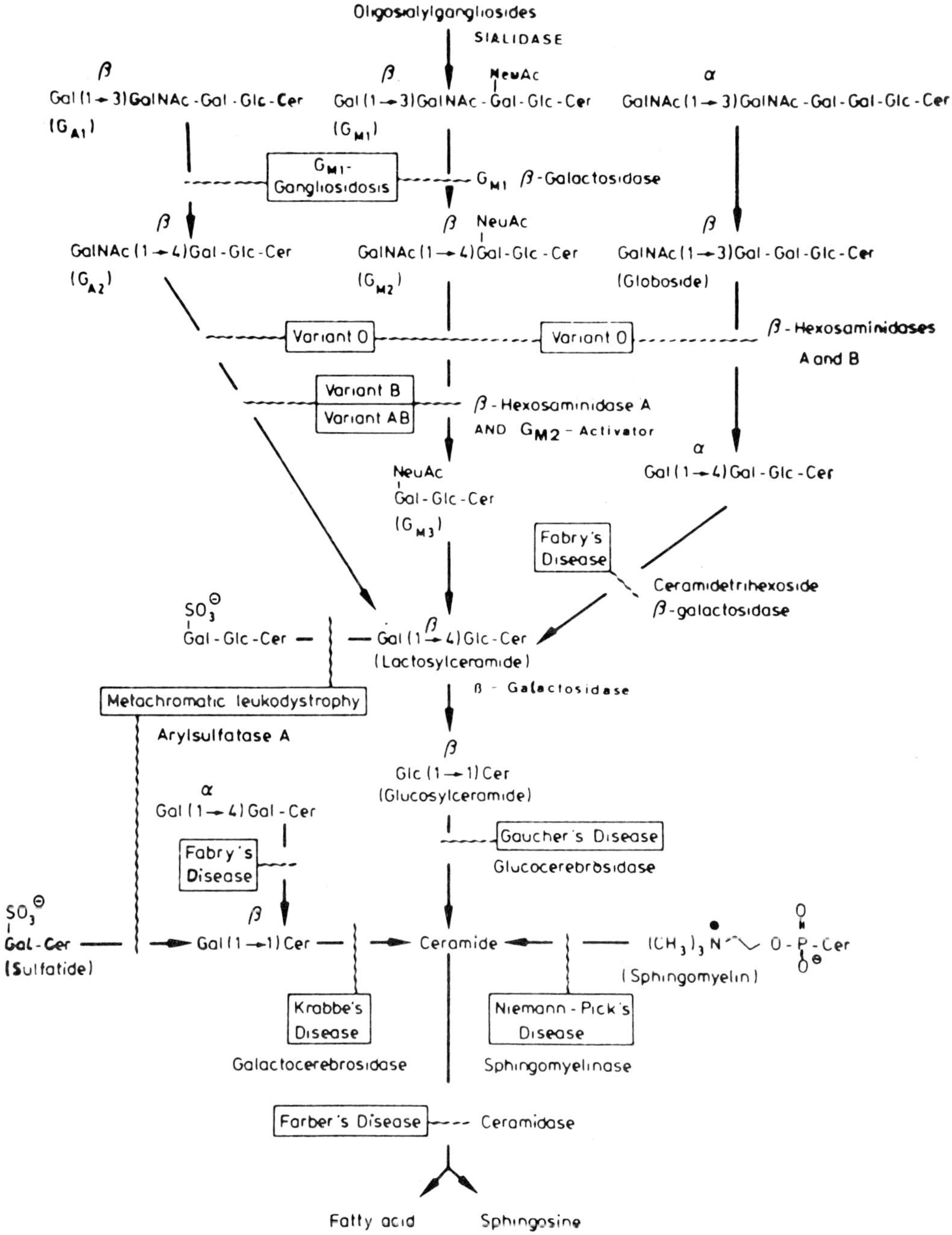

FIGURE 6

Proposed degradation scheme of sphingolipids denoting metabolic blocks of known diseases (lipidosis) (According to 15)

III. Cell Specific Distribution and Differentiation-related Expression of Brain Gangliosides

For an understanding of the physiological function of brain gangliosides it may prove important to have detailled knowledge about the phylogenetic occurrence of these glycosphingolipids among the animal kingdom as well, as the ontogenetic development including genetic mutants. In addition, it is also of great importance to focus on gangliosides as autoantigens and the possibility of the formation of antibodies against gangliosides and their functional consequences.

a. *Phylogenetic Aspects*

Negatively charged glycosphingolipids, such as the gangliosides are widely distributed throughout the animal kingdom. Some of these compounds exist as zwitterions. Besides sialic acid, they may carry glucoronic acid, sulfate, phosphate or phosphonate. With regard to phylogenetic aspects, the animal kingdom is subdivided in the *Protostomia* (= lower invertebrates like worms, arthropods, molluscs, etc.) and the *Deuterostomia* (= echinoderms and vertebrates). *Sulpho-glycosphingolipids and sialo-glycosphingolipids* (= gangliosides) have been found only amongst the Deuterostomia, (for review compare Hilbig and Rahmann, 25). In the Protostomia sialic acid is absent. Until now, *glucuronic acid containing glycosphingolipids* have been isolated from both deuterostomian tissue (human peripheral nerves) and protostomian tissue (spermatozoa of the fresh water mussel Hyriopsis and the flies Calliphora vicina (26) and Lucilia caesar. These diptera insects also contain phosphate diester groups, i.e. 2-aminoethyl-phosphate linked to N-ametylglucosamine. Finally, *aminophosphonic acid containing glycosphingolipids* were detected in the marine snail Aplysia kurodai and in the snail shellfish.

Concerning sialoglycosphingolipids, as reviewed by Hilbig and Rahmann (25), during evolution of vertebrates, an increase in *concentration of gangliosides* occurred with phylogenetic progress of nervous organization and function: In lower ectothermic vertebrates (fish, amphibians, reptiles) the content varies between 100 and 800 μg sialic acid/g fresh wt. of brains; in endothermic, warm-blooded birds and mammals it ranges between 400 and 1500 μg. (Concomittantly a significant decrease in the content of neuronal sialoglycoproteins is observed). Comparing the ganglioside content from single brain structures (e.g. forebrain, midbrain, brain stem, cerebellum, spinal cord), a general trend can be followed indicating a decrease of the concentration from the phylogenetically younger brain regions (forebrain) to the older structures (spinal bulb = medulla).

Concerning the *classification of brain gangliosides* as reviewed by Wiegandt (2), about 50 different types of gangliosides were found, in vertebrates, differing as well in their sialosyl-oligosaccharide headgroup (number of sialic acid residues, N-, or N,O-acetylneuraminic or N-glycolylneuraminic acids) as in their ceramide portion (differences in the length of fatty acid chain, sphingenine or eicosaphingenine).

More in detail, the brain ganglioside composition in *cold-blooded lower vertebrates* consists mainly of a larger number of polar fraction components based on lactose-, gangliotriaose- or gangliotetraose-carbohydrate core, whilst in the higher evolved warm-blooded *birds and mammals* the brain ganglioside pattern is composed mainly of a smaller number of less polar fractions (GM_1, GD_{1b}, GD_{1a} and GT_{1b}) mostly from the gangliotetraose family.

In phylogenetic younger vertebrates (rodents and primates including man among the mammals) a trend to a topographic diversity of the brain ganglioside pattern, combined with a reduction in the number of fraction can be observed.

b. Ontogenetic Aspects

During ontogenetic development, dramatic changes in the formation of brain gangliosides occur, which are correlated with the stage of neuronal differentiation (= neurogenesis). As reported by Felding-Habermann and Wiegandt (27) with the chick embryo as a model system, it was found that even at a *pre-organ stage of early ontogenetic development* glycolipids of all carbohydrate series that are typical for the adult animal are being synthesized. These data speak in favour of the socalled restriction hypothesis of early development indicating that a later tissue-specific glycolipid distribution is achieved by cell-type-specific restrictions in glycolipid biosynthesis during differentiation.

Concerning the analysis of the *developmental profile of brain ganglioside biosynthesis throughout life span*, Rösner and Rahmann (28) reviewed the corresponding results obtained so far from various vertebrates. Gangliosides can serve as marker compounds for the level of neuronal differentiation insofar as the expression of individual ganglioside fractions can be correlated with characteristic stages of nerve cell development (Fig. 7):

- *The phase of proliferation* of neuronal and precursor cells is characterized by a relative simple set of gangliosides, predominantly of GD_3 (and GD_2), and by the absence of more complex ganglioside species.
- *The period of cell migration and differentiation* is correlated with a drastic decrease in GD_3 but a major biosynthesis of polar polysialogangliosides belonging predominantly to the c-pathway of biosynthesis (GT_{1c}, GQ_{1c}, GP_{1c}, G"H", G"S").

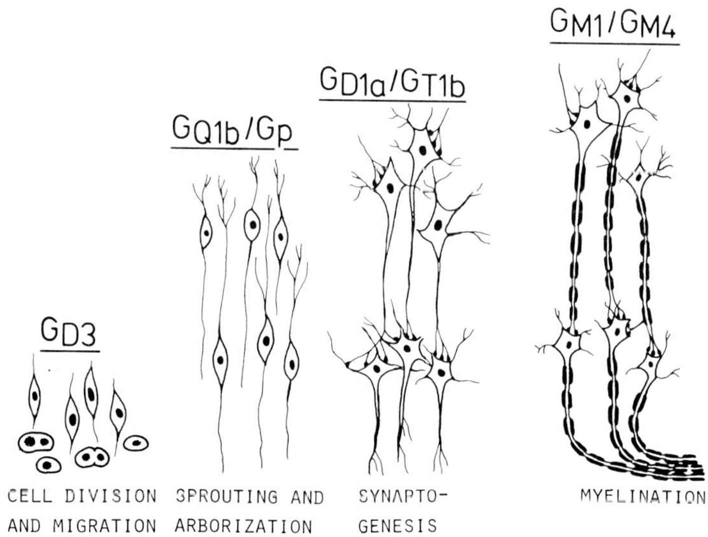

FIGURE 7
Scheme of the marker-function of brain gangliosides for the determination of different developmental stages of neurogenesis (for nomenclature of gangliosides comp. Fig. 2)

- *The phase of fiber growth and synapse formation*, during which the total ganglioside biosynthesis is enhanced several-fold, is characterized by a sharp rise of GD_{1a} and GT_{1b}; at the same time the expression of c-pathway polysialogangliosides decreases.
- *The onset of myelination* induces an increased accumulation of GM_1 and GM_4. This last developmental phase, which continues to adulthood, is preceeded by a second, smaller rise of GD_3, probably due to oligodendroglial cell proliferation.
- Finally, during *the phase of senescence*, a continuous decrease of GD_{1a}, and by this a relative higher proportion of polysialogangliosides, may contribute to a change in the polarity (= rigidity) of the neuronal membranes.

c. *Neurological Abnormalities of Brain Gangliosides*

Many neurological mutations affect specific stages of nervous system development. As Baumann et al. (29) reviewed, attempts had been made to correlate alterations of differentiation of nerve cells during neurogenesis to abnormalities of gangliosides. Concerning this aspect laboratory rodents, especially mice, are useful models for such kind of investigations:

In Jimpy mice mutant, being affected in myelin formation, the content of ganglioside-bound sialic acid is reduced. In the Quaking mutant, which is characterized by a major defect in myelin compaction, a pool of GM_1 is particularly linked to myelin. The Trembler mutant affects mainly the peripheral nerves; it is associated with abnormal Schwann cell proliferation. The biosynthesis of GD_3, which in normal cases is indicative for cell proliferation, in this mutant has not been evidenced. GT_{1a} has been found to be significantly reduced in mice mutations resulting in Purkinje cell degenerations in the cerebellum.

Similar as in humans, also in mice (and cat), lysosomal storage diseases, resembling in some cases Niemann-Pick disease were detected, in which abnormal accumulations of GM_2, GM_1 or GM_2 took place (comp. Fig 6, denoting corresponding metabolic blocks of known lipidosis).

Using monoclonal antibodies to glioma-associated gangliosides Fredman (30) has shown that human glioma tissue is characterized by a ganglioside pattern different from normal brain tissue. The large variations, which occurred in eleven glioma cell lines, were interpreted to be a reflection of the genotypic and phenotypic heterogeneity of gliomas.

d. *Monoclonal Antibodies Against Gangliosides*

A very important tool for the elucidation of the functional role of different ganglioside species are antibodies directed against individual gangliosides. Very recently the technique for the production of mouse monoclonal antibodies (MAbs) succeeded in the production of a large number of MAbs with high specificity for single gangliosides. Svennerholm (31) described the epitope specificity of three MAbs: In general 3—4 sugars can be compared with 1 to 2 sugars for the recognition group of a plant lectin. The animal lectin, the B-subunit of cholera toxin is an exception which requires at least 3 sugars in the recognition group, recognizing GM_1 in cell membranes. In the future, the B-subunit of cholera toxin and MAbs directed to specific gangliosides will certainly be widely used for the topological mapping of gangliosides in different structures within the brain or for the elucidation of ganglioside function both *in vivo* and *in vitro.*

Additionally, MAbs to several tumor-associated gangliosides have been of considerable medical importance for the early diagnosis of carcinomas and for monitoring cancer patients who had undergone surgery. (Concerning the possible use of monoclonal antiganglioside antibodies as therapeutic agents for tumor therapy comp. chapter V.).

IV. Gangliosides and Neuronal plasticity

The primary functional tasks that gangliosides have to fulfill within the nervous system are assumed to be very similar to those within peripheral organs, which are participation in cellular recognition, differentiation and specific cell function. The preferential role of gangliosides in cell surface functions, in general, seems to be primarly connected with reactions of a cell to changes in its environment, which in nerve cells are alterations in the electrical membrane surface potential. Concerning this, especially the synaptic terminals, in which transmission of electrically coded information from one neuron to the other takes place, are of particular interest, because they are known to be the most sensitive structures within the entire organism with regard to all kind of physical (temperature, pressure, charge) and/or chemical (ions, hormons, transmitters etc..) changes at their outer membrane surface. Since the exchange of information between the pre- and postsynaptic terminals of two adjacent neurons generally occurs far away, the molecular surface coat of the synaptic membranes is being primarily involved in the excitation process. The coding machinery (transcription, translation) within the nerve cell bodies is being involved in a second process following retrograde transport of substances (feed-back) from the nerve fiber terminals to the perikaryon.

On this background it is assumed that changes in the physico-chemical properties especially of the synaptic membranes are responsible for the most important functions of the neurons, which are short-term transmission of information and long-term adaptational changes including storage of information (= memory; 8—10). Therefore, special emphasis was laid on the investigations of those molecular events that might occur in connection with *calcium* in synaptic membranes following sensory stimulation. This is due to the fact that *calcium plays an essential role* as a primary messenger at the outer membrane surface as well as a second messenger within the synaptoplasm for the initiation of enzymatic cascades.

Having this in mind, already since about 1960 an increasing number of papers has been published, dealing with an *involvement of neuronal gangliosides in short-term excitation* of neurons by means of sensory, electrical or chemical stimulation *and during long-term adaptive processes.*

As reviewed by Rahmann (8) the following major trends of results were obtained:

- The brain ganglioside biosynthesis can be influenced in vivo by various *chemical agents*, like tetrazol, chlorpromazine, dinitrophenol, oestradiol, testosterone, anesthetics, puromycin etc.
- After *long-term neuronal adaptation* to behavioural conditioning, the brain ganglioside metabolism is being affected too.
- Following *thermal adaptation* of vertebrates (acclimatization to seasonal temperature changes or acclimation to experimentally induced temperature alterations) a general rule had been elaborated, indicating *"the lower the environmental temperature is, the higher is the degree of sialylation (= polarity) of brain gangliosides"*. This in some cases had been found to be due to alterations in the sensitivity of brain gangliosides to membrane-bound sialidases.

— *The bio-electrical activity* of neurons within brain slices was found to be significantly altered following either exogenous application of sialidases or different ganglioside species.

Summarizing these results are taken as evidence for the assumption that gangliosides in vertebrates fulfill the task as modulatory substances for functional membrane-bound proteins (comp. 10) for the regulation of synaptic transmission of information and long-term neuronal adaptation. These data complement one another due to the particular physico-chemical properties of gangliosides especially in connection with Ca^{2+} (32). In addition, they are being supported by several experimental data as reviewed by Ledeen et al. (18) and Skaper (17) indicating that not only naturally occurring gangliosides but also synthetic and ganglioside-related compounds when being applied exogenously are effective agents for neurite recruitment, especially of *neuritogenic activity* (promoting the outgrowth of fibers) and *neuronotrophic properties*, capable of influencing neuronal differentiation in vitro. These results are discussed to have possible functional significance for neuronal repair mechanisms *in vivo*.

Recently it has been shown, however, that the presence of calcium in the incubation medium plays an important role and that both effects (neuritogenic and neuronotrophic) can be induced also by several other compounds with amphiphilic properties, raising by this new intriguing questions concerning the mode of action of exogenous gangliosides.

With regard to a *possible mechanism of exogenous gangliosides* (as well as those of anti-glycolipid antibodies) *on cell growth* Hakomori (33) reviewed the results from current studies investigating the roles of gangliosides as possible modulators for functional membrane proteins as for instance for the function of the receptors for fibroblast growth factor (FGF), platelet-derived growth factor (PDGF), and epidermal growth factor and/or pericellular matrix proteins, like fibronectin, laminin or N-CAM (Fig. 8). From these data it is assumed that the target molecules at the cell surface could be growth factor receptors (or other functional membrane bound molecules) which are being influenced in a cell

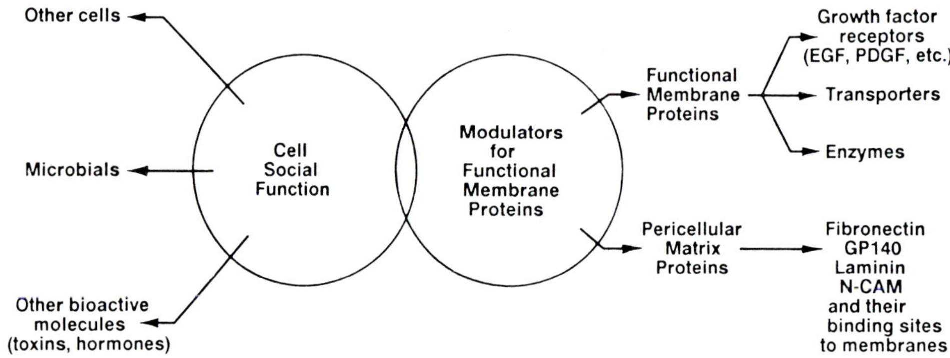

FIGURE 8

Bimodal function of gangliosides. Gangliosides in the cell membrane as sites for interaction with cells, microbials, and other molecules (left). On the other hand, they modulate various functional membrane proteins and the pericellular matrix (right). Many gangliosides can act in both ways, but some gangliosides, such as GM_3, mainly direct modulation of functional membrane proteins (According to 33)

specific way by different ganglioside species (e.g. GM_3 for FGF-receptor; GM_3 or GM_1 for PDGF-receptors and EGF-receptor).

Since the function of these receptors has been shown clearly to be regulated by phosphorylation of the receptor, the effect of gangliosides has been investigated with focus on the phosphorylation of these receptors. It was clearly shown that tyrosine phosphorylation of PDGF- and EGF-receptors was strongly inhibited by gangliosides. The inhibitory effect of GM_3 on tyrosine phosphorylation of EGF-receptor was demonstrated as well *in situ* in cells as *in vitro.*

In addition to the *receptor-associated functions of gangliosides*, these compounds were detected *to affect the function of adhesive proteins* or their receptors, since exogenously applied gangliosides inhibit cell adhesion on plates.

In extention of these data, Yu (34) reported on *gangliosides as differential modulators of protein phosphorylation:* In rat brain membrane preparations exogenous gangliosides stimulate the phosphorylation of several phosphoproteins in a pattern qualitatively similar to calmodulin. The stimulation was completely dependent on the presence of calcium. Di- and tri-sialogangliosides were more effective. Gangliosides also inhibited the phosphorylation on a 78 K protein in brain membrane and the myeline basic protein. The inhibition appeared to be an effect on protein C-Kinase. Thus gangliosides are assumed to be physiological agents which can differentiate between C-Kinase and calmodulin-dependant kinase.

Since 1987 further studies have suggested that gangliosides beside the physico-chemical and physiological peculiarities indeed possess specific trigger or modulatory properties for functional membrane proteins in synapses. As summarized by Nagai and Tsuji (35) gangliosides act as modulators for different types of protein kinases as well from the endo-type (e.g. Ca^{2+}/CAM-Kinase, Ca^{2+}/PS-kinase, J-kinase) as ecto-type (Gg-kinase) which activate membraneous channel proteins. Furthermore, gangliosides modulate the activity of different neuronal ATPases, like Na^+/K^+-ATPase, Ca^{2+}-Mg^{2+}, ATPase

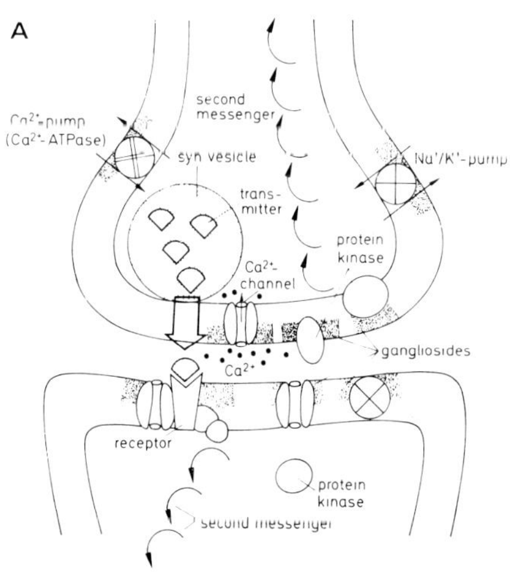

FIGURE 9A
Scheme of the synaptic terminal demonstrating the possible involvement of Ca^{2+}-ganglioside complexes in the process of synaptic transmission

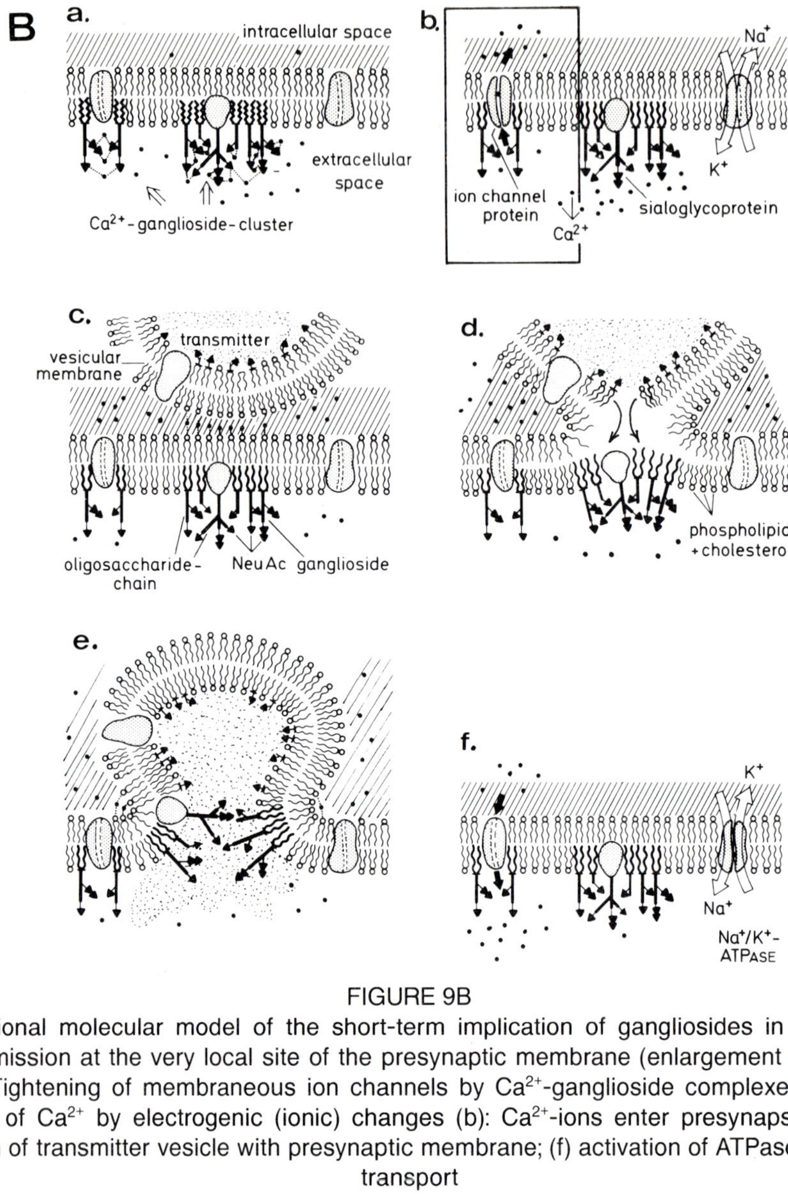

FIGURE 9B
Functional molecular model of the short-term implication of gangliosides in synaptic transmission at the very local site of the presynaptic membrane (enlargement from Fig. 9A). Tightening of membraneous ion channels by Ca^{2+}-ganglioside complexes (a); release of Ca^{2+} by electrogenic (ionic) changes (b): Ca^{2+}-ions enter presynapse; (c—e) fusion of transmitter vesicle with presynaptic membrane; (f) activation of ATPases for ion transport

and CA^{2+}-ATPase (for ref. 10), which function as ionic pumps in membranes for the maintenance of the membrane potential. In the mammalian brain, for instance, the Ca^{2+}-ATPase of synaptosomes becomes most significantly activated by the addition of gangliosides.

Summing up all data concerning the involvement of gangliosides in the phenomenon of neuronal plasticity it can be concluded that gangliosides can be regarded as to play a crucial role as trigger or modulatory substances for neuronal, especially synaptic membrane-bound processes like transmission of information and long-lasting adaptive events, like storage of information. This is due to the fact that gangliosides were found to modulate functional membraneous proteins like channel proteins (protein-kinases), ionic pumps (ATPases) or receptor molecules. For overview a summarizing schematic diagram of these possible ganglioside interactions on synaptic level and molecular level is given in Fig. 9A and 9B.

V. Biomedical Potential of Exogenous Ganglioside Application on Neuronal Function

In addition to the several reports concerning the chemistry, biochemistry, and biological phenomenology of gangliosides, especially with regard to their possible involvement in neuronal plasticity, the *pharmacological potentialities* of these glycosphingolipids came more and more into the center of discussion. Here the hot point appeared to be the multiple effects of exogenous gangliosides either in cell culture or *in vivo.* From this it was suggested that gangliosides might be useful as therapeutic agents, especially for peripheral neuropathies of various etiology. Therefore, it appears urgent to establish whether exogenously administered gangliosides are really able to modify nerve cell structures and organization, and to produce a recovery of anatomical integrity or at least of the functional activity when provided after lesions of the central or peripheral nervous system (36).

Four strategies were developed to investigate the efficacy of ganglioside action:

1) Analysis of the influence of gangliosides on cultured celles (neurons, glial cells, glioma, neuroblastoma);
2) investigation of the effect of gangliosides on the development of neuromuscular junctions, peripheral nerve sprouting and regeneration;
3) study of the recovery after central nervous system lesion following ganglioside application; and
4) the therapeutical trials applied in rather complex situations.

From these developments impetus had been expected for human peripheral neuropathies of various etiology.

a. *Animal Studies with Exogenous Ganglioside Applications*

As reviewed by Suzuki (37), Mandel (36) and Gorio (38), several studies concerning biomedical potentials of exogenously applied gangliosides (1 to 10 mg/Kg intraperitoncally) have been carried out with experimental animals following lesioning of peripheral and central nerves, chemical induction of diabetes or with genetic diabetes mutants. The ganglioside treament partly resulted in improvement in several meaningful parameters, like axon diameter, gain of conduction velocity or regain of behaviour. However, also contradictory results have been reported according to which no significant differences in the rate of regeneration among the animals were observed. In case of brain lesions it was hypothesized that the facilitated recovery of behavioral parameters might be due to a reduced tissue damage following ganglioside injection rather than to a regeneration phenomenon, for instance through protection against brain edema which follows the brain damage.

Summarizing these data, much more basic investigations are urgently needed, even the more as experimental evidence was obtained according to which an uptake of systemically administered gangliosides by the central nervous tissue is most unlikely due to the existence of a highly effective *blood-brain-barrier* (BBB) against gangliosides (19): By means of intra-ocular injections of radiolabeled gangliosides into one eye bulb of chicken it was shown by histo-autoradiographic documentation as well as radiochemical analysis that exogenous gangliosides when being offered directly to nerve cells (here retina ganglion cells) on the one side become incorporated into the cells where they undergo axonal transport to the nerve fiber terminals in the optic tectum within the brain. On the other side, however, those labeled gangliosides which after the intraocular injection enter the blood stream and become distributed in all peripheral organs fail to pass the BBB and by this become incorporated into the nerve cells neither of the CNS nor PNS (Fig. 10).

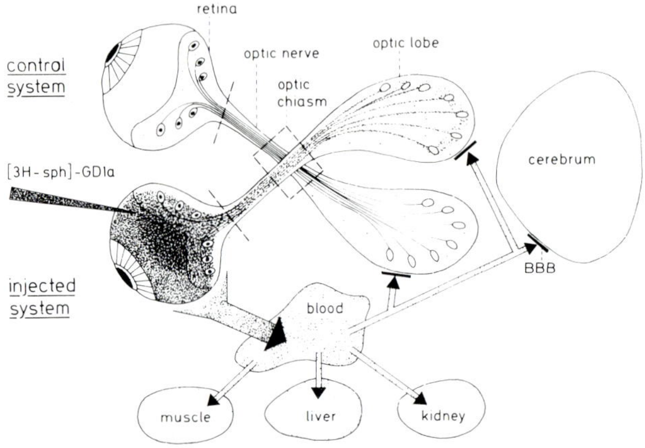

FIGURE 10

Scheme of the chicken optic system, demonstrating the mode of intra-ocular application of (^{3}H-sph)-GD_{1a} into one eye bulb, axonal transport in the entire visual pathway, penetration into the blood, uptake by non-neuronal organs, and the lack of accumulation from blood by central nervous tissue due to blood-brain-barrier (BBB) (According to 19).

b. Clinical Aspects of Ganglioside Therapy

were reviewed by Suzuki (37), Weisner (39) and Haferkamp (40). Most of the clinical trials were for diabetic, toxic (alcoholic, uremic) polyneuropathy and amyotrophic lateral sclerosis. In addition, patients with other neuropathies of divers etiology (idiopathic sensorimotor polyneurophathy, dorsal root ganglion degeneration, spinal muscular atrophy etc.) have been investigated.

In most cases the therapeutic trials were carried out with the double-blind controlled design with 20—40 patients per study, generally a half in the treated group and the remainder in the placebo group. Almost all studies used a bovine cerebral ganglioside mixture in an intramuscular dose between 20 or 40 mg/day and a duration of treatment ranging between 40 days to about 6 months. Criteria used for evaluation of results were for the most part clinical examinations and electrophysiological parameters which included nerve conduction velocity, distal latency, sensory and motor action potentials, evoked potentials, and EMG. Very often subjective "improvements" reported by patients were described.

The general consensus of the clinical trials to its present state (1987) was that the results were extremely variable and yet did not permit definite conclusions: On the one hand, the results were rather widely inconsistent from one study to another. On the other hand, almost all "improvements" recorded were relatively minor and required statistical analyses to ascertain significant differences from the controls.

There are several *possibilities* that might account *for the inconsistent results*: genetic diversity of patients, dose and duration of treatment, route of administration, type of neurophathy, evaluation criteria, and use of statistical treatment of data. Therefore, before coming to a definite conclusion concerning the possibility of beneficial effects of exogenous gangliosides, the following *still open questions* had been raised to stimulate future research activities:

— Are there any firm evidences for biological functions of *endogenous* gangliosides during the process of neuronal regeneration?
— Are there any indications that *exogenous* ganglioside administration in vivo might be harmful in any way?
— What is the fate of exogenously administered gangliosides within the body concerning penetration of blood-brain-barrier, distribution among different organs, blood level and possible withdrawal phenomena?
— How could selection and evaluation criteria especially concerning clinical trials be standardized?

c. Future Perspectives

Since the conference took place in 1986, it has become more apparent from relevant studies that gangliosides may indeed exert benificial effects in preventing nerve injuries. These include in particular diabetic, uremic and alcoholic neuropathies. An attractive explanation for some of the neuronotrophic properties thus shown by gangliosides, is offered by the observation of their modulatory influence on glutamate receptors (41). By an inhibition of excessive stimulation of these receptors gangliosides may counteract an otherwise excitotoxic CA^{2+}-influx via glutamate receptor-dependend channel.

During the past few years it has become increasingly evident that *anti-ganglioside antibodies* also may be of certain therapeutic value (comp. Svennerholm 31). With regard to this Dippold et al. (42) reported on their experiences with the treatment of patients with neuroectodermal tumors by monoclonal ganglioside antibodies. GD_3, GD_2 and GM_2 were recognized as tumor-associated antigens for neuroectodermal tumors. Expecially GD_3 represents one of these markers and monoclonal antibody R-24 recognizing GD_3 on malignant melanoma meets the criteria for tumor restriction well. In Ab R-24 infusions into patients with malignant melanoma or apudoma, all with metastatic disease, were found to induce inflammatory cutaneous responses around tumor nodules. In some patients, tumor regression was observed. These results indicate, that ganglioside antibodies are active in vivo and may be a basis for specific immunological tumor therapy.

Nevertheless, it has to be taken into consideration that according to Feizi (43) gangliosides can act as autoantigens and differentiation antigens. Using monoclonal autoantibodies (anti-T and anti-i) a large number of antigenic gangliosides with poly-N-acetyllactosamine backbones were discovered. These structures are both differentiation antigens and receptors for Mycoplasma pneumoniae. On the basis that this infection results in the production of autoantibodies with anti-I-specificities *a new concept was introduced according to which the interaction of infective agents with specific sac-*

charides of host cell membranes may be an important mechanism for eleciting autoantibodies.

Recently, some human monoclonal antibodies spontaneously formed against glycolipids are directed to neutral glycolipids, some other also to gangliosides. The findings of MAbs against ganglioside IV^3 NeuAc-Lc_4 Cer in *Guillain-Barrés syndrome* is an example of the deleterious effect of such an anti-ganglioside antibody. — This finding raises the question whether an exogenous ganglioside application in case of neurological diseases (neuropathies) might induce the formation of antiganglioside antibodies.

REFERENCES

1) Yamakawa, T. (1987): History of ganglioside research. In: Gangliosides and Modulation of Neuronal Functions. Rahmann, H. (ed.), NATO ASI Series H7, Springer-Verlag Berlin, Heidelberg, pp. 3—15.

2) Wiegandt, H. (1987): Fundamentals of ganglioside structures. In: Gangliosides and Modulation of Neuronal Functions. Rahmann, H. (ed.), NATO ASI Series H7, Springer-Verlag Berlin, Heidelberg, pp. 37—47.

3) Ando, S., Waki, H., Kon, K. and Kishimoto, Y. (1987): Up-to-date chromatography of gangliosides. In: Gangliosides and Modulation of Neuronal Functions. Rahmann, H. (ed.), NATO ASI Series H7, Springer-Verlag Berlin, Heidelberg, pp. 167—177.

4) Yu, R. K. (1987): Fundamentals of structural analysis of glycosphingolipids by proton nuclear magnetic resonance spectroscopy. In: Gangliosides and Modulation of Neuronal Functions. Rahmann, H. (ed.), NATO ASI Series H7, Springer-Verlag Berlin, Heidelberg, pp. 49—70.

5) Egge, H. and Peter-Katalinic, J. (1987): Fundamentals of the application of MS and NMR in the study of ganglioside structure. In: Gangliosides and Modulation of Neuronal Functions. Rahmann, H. (ed.), NATO ASI Series H7, Springer-Verlag Berlin, Heidelberg, pp. 71—85

6) Corti, M., Cantu, L. and Sonnino, S. (1987): Fundamentals of physico-chemical properties of gangliosides in solution. In: Gangliosides and Modulation of Neuronal Functions. Rahmann, H. (ed.), NATO ASI Series H7, Springer-Verlag Berlin, Heidelberg, pp. 101—118.

7) Grant, C. W. M. (1987): Fundamentals of physico-chemistry of glycolipids in membranes. In: Gangliosides and Modulation of Neuronal Functions. Rahmann, H. (ed.), NATO ASI Series H7, Springer-Verlag Berlin, Heidelberg, pp. 119—138.

8) Rahmann, H. (1987): Brain gangliosides, bio-electrical activity and post-stimulation effects. In: Gangliosides and Modulation of Neuronal Functions. Rahmann, H. (ed.), NATO ASI Series H7, Springer-Verlag Berlin, Heidelberg, pp. 502—521.

9) Rahmann, H. and Rahmann, M. (1988): Das Gedächtnis. Neurobiologische Grundlagen. Bergmann-Verlag München, Springer-Verlag New York, Berlin, Heidelberg, 320 pp.

10) Rahmann, H. (1990): Ca^{2+} — ganglioside interactions as modulators for synaptic transmission and long-term neuronal adaptation (memory). In: Fundamentals of Memory Formation: Neuronal Plasticity and Brain Function. Rahmann, H. (ed.), Progress in Zoology 37, pp. 349—368.

11) Körtje, K. H., Freihöfer, D. and Rahmann, H. (1990): Cytochemical localization of calcium in the central nervous system of vertebrates. Ultramicroscopy (in press).

12) Körtje, K. H., Freihöfer, D. and Rahmann, H. (1990): Cytochemical localization of high-affinity Ca^{2+}-ATPase activity in synaptic terminals. Histochem. Cytochem. (in press).

13) Seybold, V., Rösner. H., Greis, Ch., Beck, E. and Rahmann, H. (1989): Possible involvement of polysialogangliosides in nerve sprouting and cell contact formation: An ultracytochemical in vitro study. J. Neurochem. 52, pp. 1958—1961.

14) Schwarzmann, G., Marsh, D., Herzog, V. and Sandhoff, K. (1987): In vitro incorporation and metabolsm of gangliosides. In: Gangliosides and Modulation of Neuronal Functions. Rahmann, H. (ed.), NATO ASI Series H7, Springer-Verlag Berlin, Heidelberg, pp. 217—229.

15) Sandhoff, K., Schwarzmann, G., Sarmientos, F. and Conzelmann, E. (1987): Fundamentals of ganglioside catabolism In: Gangliosides and Modulation of Neuronal Functions. Rahmann, H. (ed.), NATO ASI Series H7, Springer-Verlag Berlin, Heidelberg, pp. 231—250.

16) Tettamanti, G., Ghindoni, R. and Trinchera, M. (1987): Fundamentals of brain ganglioside biosynthesis. In: Gangliosides and Modulation of Neuronal Functions. Rahmann, H. (ed.), NATO ASI Series H7, Springer-Verlag Berlin, Heidelberg, pp. 191—204.

17) Skaper, S. D. (1987): Modulation of neuronotrophic factor action by exogenous gangliosides. In: Gangliosides and Modulation of Neuronal Functions. Rahmann, H. (ed.), NATO ASI Series H7, Springer-Verlag Berlin, Heidelberg, pp. 481—490.

18) Ledeen, R. W. and Cannella, M. S. (1987): The neuritogenic effect of gangliosides in cell culture. In: Gangliosides and Modulation of Neuronal Functions. Rahmann, H. (ed.), NATO ASI Series H7, Springer-Verlag Berlin, Heidelberg, pp. 491—499.

19) Rösner, H., Willibald, C. J., Schwarzmann, G. and Rahmann, H. (1987): Uptake of exogenous gangliosides by the CNS ?. In: Gangliosides and Modulation of Neuronal Functions. Rahmann, H. (ed.), NATO ASI Series H7, Springer-Verlag Berlin, Heidelberg, pp. 582—592.

20) Nagai, Y., Sanai, Y. and Nakaishi, H. (1987): Fundamentals of genetic control of gangliosides: The enigma of corbohydrate chain diversity in glycosphingolipids. In: Gangliosides and Modulation of Neuronal Functions. Rahmann, H. (ed.), NATO ASI Series H7, Springer-Verlag Berlin, Heidelberg, pp. 275—292.

21) Suzuki, A., Hashimoto, Y., Yamakawa, T., Sakaizumi, M., Kiuchi, Y. and Moriwaki, K. (1987): Genetic regulation of ganglioside expression. In: Gangliosides and Modulation of Neuronal Functions. Rahmann, H. (ed.), NATO ASI Series H7, Springer-Verlag Berlin, Heidelberg, pp. 293—304.

22) Ledeen, R. W., Aquino, D. A., Sbaschnig-Agler, M., Gammon, C. M. and Vaswani, K. K. (1987): Fundamentals of neuronal transport of gangliosides. Functional implications. In: Gangliosides and Modulation of Neuronal Functions. Rahmann, H. (ed.), NATO ASI Series H7, Springer-Verlag Berlin, Heidelberg, pp. 259—274.

23) Rösner, H., Wiegandt, H. and Rahmann, H. (1973): Sialic acid incorporation into gangliosides and glycoproteins of the fish brain. J. Neurochem. 21, pp. 655—665.

24) Caputto, R., Caputto, B. L., Domowicz, M. S. and Kavatinitz, S. C. (1987): Gangliosides: biosynthesis, transport, and location in the plasma membrane. In: Gangliosides and Modulation of Neuronal Functions. Rahmann, H. (ed.), NATO ASI Series H7, Springer-Verlag Berlin, Heidelberg, pp. 251—258.

25) Hilbig, R. and Rahmann, H. (1987): Phylogeny of vertebrate brain gangliosides. In: Gangliosides and Modulation of Neuronal Functions. Rahmann, H. (ed.), NATO ASI Series H7, Springer-Verlag Berlin, Heidelberg, pp. 333—350.

26) Dennis, R. D., Geyer, R., Egge, H., Peter-Katalinic, J., Keller, M., Menges, H. and Wiegandt, H. (1987): Insects: animals without gangliosides — preliminary data. In: Gangliosides and Modulation of Neuronal Functions. Rahmann, H. (ed.), NATO ASI Series H7, Springer-Verlag Berlin, Heidelberg, pp. 350—358.

27) Felding-Habermann, B. and Wiegandt, H. (1987): Pre-organ glycolipids and pre-brain gangliosides. In: Gangliosides and Modulation of Neuronal Functions. Rahmann, H. (ed.), NATO ASI Series H7, Springer-Verlag Berlin, Heidelberg, pp. 359—372.

28) Rösner, H. and Rahmann, H. (1987): Ontogeny of vertebrate brain gangliosides. In: Gangliosides and Modulation of Neuronal Functions. Rahmann, H. (ed.), NATO ASI Series H7, Springer-Verlag Berlin, Heidelberg, pp. 373—390.

29) Baumann, N. A., Harpin, M. L., Baron-van Evercooren, A., Iwamori, M. and Maurin, Y. (1987): Brain gangliosides and neurological mutants. In: Gangliosides and Modulation of Neuronal Functions. Rahmann, H. (ed.), NATO ASI Series H7, Springer-Verlag Berlin, Heidelberg, pp. 391—407.

30) Fredman, P. (1987): Glioma-associated gangliosides. In: Gangliosides and Modulation of Neuronal Functions. Rahmann, H. (ed.), NATO ASI Series H7, Springer-Verlag Berlin, Heidelberg, pp. 435—444.

31) Svennerholm, L. (1987): Monoclonal antibodies against gangliosides. In: Gangliosides and Modulation of Neuronal Functions. Rahmann, H. (ed.), NATO ASI Series H7, Springer-Verlag Berlin, Heidelberg, pp. 423—433.

32) Probst, W. and Rahmann, H. (1987): Peculiarities of ganglioside-Ca^{2+}-interactions. In: Gangliosides and Modulation of Neuronal Functions. Rahmann, H. (ed.), NATO ASI Series H7, Springer-Verlag Berlin, Heidelberg, pp. 139—154.

33) Hakomori, S. (1987): Ganglioside-mediated modulation of growth factor receptor function and cell adhesion. In: Gangliosides and Modulation of Neuronal Functions. Rahmann, H. (ed.), NATO ASI Series H7, Springer-Verlag Berlin, Heidelberg, pp. 465—479.

34) Yu, R. K. (1987): Gangliosides as differential modulators of protein phosphorylation. In: Gangliosides and Modulation of Neuronal Functions. Rahmann, H. (ed.), NATO ASI Series H7, Springer-Verlag Berlin, Heidelberg, p. 529.

35) Nagai, Y. and Tsuji, S. (1988): Cell biological significance of gangliosides in neuronal differentiation and development: critique and proposals. In: New Trends in Ganglioside Research: Neurochemical and Neuroregenerative Aspects. Ledeen, R. W., Hogan, E. L., Tettamanti, G., Yates, A. J. and Yu, R. K. (eds.). Fidia Res. Series 14, pp. 329—350.

36) Mandel, P. (1987): Critical assessment on the transferability of in vitro-results to the in vivo-situation. An attempt to delimite the effects and the action mechanisms of exogenous gangliosides. In: Gangliosides and Modulation of Neuronal Functions. Rahmann, H. (ed.), NATO ASI Series H7, Springer-Verlag Berlin, Heidelberg, pp. 601—611.

37) Suzuki, K. (1987): Gangliosides and Neuropathy. In: Gangliosides and Modulation of Neuronal Functions. Rahmann, H. (ed.), NATO ASI Series H7, Springer-Verlag Berlin, Heidelberg, pp. 531—546.

38) Gorio, A. (1987): Gangliosides and functional recovery of injured nervous system. In: Gangliosides and Modulation of Neuronal Functions. Rahmann, H. (ed.), NATO ASI Series H7, Springer-Verlag Berlin, Heidelberg, p. 561.

39) Weisner, B. (1987): Clinical aspects on bio-medical potential of exogenous ganglioside application. In: Gangliosides and Modulation of Neuronal Functions. Rahmann, H. (ed.), NATO ASI Series H7, Springer-Verlag Berlin, Heidelberg, pp. 563—571.

40) Haferkamp, G. (1987): Present state of clinical experience of ganglioside application in man. In: Gangliosides and Modulation of Neuronal Functions. Rahmann, H. (ed.), NATO ASI Series H7, Springer-Verlag Berlin, Heidelberg, pp. 573—580.

41) Vanarino, F., Guidotti, A. and Costa, E. (1987): Ganglioside inhibition of glutamate-mediated protein Kinase C translocation in primary cultures cerebellar neurons. Proc. Nat. Acad. Sci. USA 84, pp. 8707—8711.

42) Dippold, W. G., Bernhard, H., Dienes, H. P. an Meyer zum Büschenfelde, K. H. (1987): Treatment of patients with neuroectodermal tumors by monoclonal ganglioside antibodies. In: Gangliosides and Modulation of Neuronal Functions. Rahmann, H. (ed.), NATO ASI Series H7, Springer-Verlag Berlin, Heidelberg, pp. 445—452.

43) Feizi, T. (1987): Gangliosides and autoantigens and differentiation antigens. In: Gangliosides and Modulation of Neuronal Functions. Rahmann, H. (ed.), NATO ASI Series H7, Springer-Verlag Berlin, Heidelberg, pp. 409—421.

REGULATORY MECHANISMS OF NEURON TO VESSEL COMMUNICATION IN THE BRAIN

Magnoni, M.S., Trabucchi, M.*, Battaini, F.* and Govoni, S.**

Inst. of Pharmacological Sciences, Univ. of Milan;
*Dept. of Exp. Med. and Biochemical Sciences, Univ. of Rome;
**Dept. Pharmacobiol., Univ. of Bari, Italy

The highest level of integrative brain function requires the existence of homeostatic mechanisms aimed to the maintenance of the internal cerebral milieu. In fact, marked fluctuations in the extracellular fluid concentration of neurotransmitters, amino acids, hormones, nutrients, ions, can interfere with the neuronal processes, which are dependent on the chemical composition of the cerebral microenvironment. For example, significant changes in K^+ concentration can adversely affect neuronal activity, and trigger a cascade of events which may ultimately be deleterious for the cell. The evolution in parallel of the high level of cerebral integrative processes and the efficiency of dynamic mechanisms for the regulation of brain milieu is supportive of the hypothesis that homeostasis and isolation are the principles that ensure a correct brain function.

ANATOMICAL VERSUS BIOCHEMICAL BLOOD BRAIN BARRIER

The anatomical basis for isolation and homeostasis is constituted by the cerebral capillary bed, which forms an interface between the blood and the brain parenchyma: the blood-brain barrier (BBB) (1).

Tight intercellular junctions between contiguous endothelial cells and absence of pores and fenestrae render the capillary wall relatively impermeable to hydrophilic molecules. Another characteristic of the BBB is the low frequency of pinocytotic vesicles, generally associated with transcapillary exchange in peripheral organs. In this respect, recent electron-microscopy studies have shown, by three-dimensional reconstructions based on consecutive ultrathin sections, that the vesicular structures observed in the brain endothelium are probably invaginations of the cell membrane, and not vesicles, and may represent diffusion pathways for hydrophilic substances. On the other hand, there is evidence indicating that in particular conditions, for example in the presence of histamine or after activation of adenylate cyclase, the number of vesicles is enhanced in parallel with the increase in capillary permeability. Further studies are therefore needed to clarify this controversial point.

Beside the concept of anatomic barrier, an enzymatic barrier can be defined on the basis of the identification of specific carrier systems for certain solutes, which facilitate

NATO ASI Series, Vol. H 51
Cell to Cell Signals in Plants and Animals
Edited by V. Neuhoff and J. Friend

the permeation of selected polar compounds, such as glucose and amino acids, across the capillary wall (2, 3).

In most cases, however, kinetic models of transport have considered only a single step in the movement of substances from the blood to the brain. The single-membrane model is an oversimplification of the processes that occur at the level of the capillary endothelium. In fact, both *in vivo* and *in vitro* data have provided evidence that the antiluminal and luminal membranes of the microvasculature have different permeability properties. For example, small neutral amino acids, potassium and iodine move more readily from brain to blood. These findings are confirmed by cytochemical techniques and biochemical studies on the purified and separated surfaces of the capillary endothelium revealing an asymmetric distribution of the content of various membrane markers Na, K ATPase activity is mostly localized to the antiluminal membrane, in line with the low permeability of the barrier for potassium in the blood to brain direction. The transport system for large neutral manino acids (L-system, leucine preferring) and the transport system for basic amino acids are ubiquitous whereas the transport system for small neutral amino acids (A-system, alanine prefering) can be observed only on the antiluminal side. There is evidence that in physiological conditions a cooperative mechanism involing both the A- und the L-systemes may participate to the regulation of the exchange of amino acids between the blood and the brain.

Complexively, these findings indicate that a comprehensive representation of the BBB must take into account the existence of two distinct membranes with different characteristics and roles as well as of the cytoplasma of the endothelial cells which may contain enzymes capable of reacting with the solutes while they are passing across the cell. Capillary polarity has important implications for BBB function. As in epithelial cells, it permits the active transport of solutes a concentration gradient. This may be of importance for the maintenance of a constant ion concentration in the extraneuronal compartment as well as for the inactivation of certain neurotransmitters and the removal of metabolic wastes. Furthermore, the transport of ions could result in fluid secretion by brain microvessels and provide an additional source of cerebrospinal fluid. An interesting aspect requiring further exploration is the possibility of a polarity for capillary receptors which would permit differential responses of the endothelium to neurotransmitters and hormones present in the brain as compared to those in the blood.

METABOLIC REGULATION OF THE BBB

Cerebral microvessels are dynamically regulated to respond to stimuli of different nature such as: hormones, neurotransmitters and local metabolic factors. Early observations by Roy and Sherrington in 1890 (4) reported that cerebral blood flow is coupled locally to neuronal activity via products of brain metabolism. By means of techniques, as autoradiography and, more recently, positron emission tomography (PET), it is possible to investigate the hemodynamic and metabolic activities of the brain in various physiological and pathological conditions and to acquire deeper knowledge on the relationship between blood flow, metabolism and higher mental functions.

These techniques, based on the measurement of local concentration of radioactive isotopes involved in specific biochemical processes, require the formulation of a kinetic model of the process under study and the derivation of an operational equation which utilizes the values of local tissue concentration of the isotope as well as other variable and parameters determined from measurement of the isotope in the blood (5). These principles have been applied to the measurement of the rate of glucose utilization in the structures of the central nervous system in conscious animals or, in the case of PET, in man. These methods have demonstrated a coupling between glucose consumption and local functional activity of the neurons (6). This requires that the supply of the substrate and its transport across the BBB are coordinated with neuronal glycolysis and oxidation of lactate and pyruvate. However, the regional "flow/transport/metabolism coupling" is not sustained by a simple correlation; several factors influence this process, as the kinetics of transport across the capillary wall and the neuronal membrane as well as the anatomical distribution of glucose carriers and pyruvate dehydrogenase. Along this line, autoradiographic studies using Cytochalasin B labeling, a high affinity ligand for the D-glucose carrier, have demonstrated a high concentration of sites in cerebral microvessels and in particulate subcellular fractions from neuronal homogenates. At neuronal level, a marked intra- and inter-regional difference in the presence of glucose transport proteins has been described. Within the hippocampus, for example, the highest density is found in the synaptic zones, whereas layers corresponding to the pyramidal cell bodies show lower degree of binding. This finding correlates with glucose phosphorylation rate, measured using the autoradiographic deoxyglucose technique. In contrast, the inter-regional comparison revealed a mismatch between the two parameters, suggesting that binding density is presumably related to the total available pool of carriers, only a fraction of which is functionally active. It is then possible that the various cerebral areas differ in the reserve capacity and in the translocation of the glucose-carrier from storage to functional pool. The observation that chronic hyperglycemia represses glucose transfer and that this process is increased in fasted rats suggests that plasma glucose concentration plays a primary role in the translocation process. A further event potentially affecting the capacity to transport glucose and other substrates across the BBB is the change in the number of perfused capillaries, i.e. the endothelial surface area and hence the number of available glucose carriers. The importance of this factor is suggested by the fact that in physiological conditions the amount of cerebral microvessels actively perfusing the tissue is estimated to be about 30 % of the total, indicating that capillary recruitment in the brain can be a major mechanism governing blood-brain transport phenomena.

In addition, a factor to be taken into account concerns the possibility of changes in the intrinsic resistance parameters relating to the movement of loaded and unloaded forms of the carrier within the capillary cells, which can determine modifications in the apparent maximal transport capacity and/or affinity of the substrate for the carrier.

The metabolism of glucose, under normal resting conditions is mostly oxidative, and involves the pyruvate dehydrogenase pathway. Immunogold technique has demonstrated that this enzyme is predominantly located in the neuronal cell bodies and cell processes, whereas the neurophils generally show much less immunoreactivity. This distribution contrasts with that of glucose transporter protein, confirming the clear distinction between the two underlying biochemical processes, i.e. the delivery of glucose and the oxidative metabolism. There is evidence that the local temporal uncoupling of the two processes is associated with physiological activation of brain structures. For example, it has been described that the transient increase in neuronal activity occurring during visual stimulation causes a glucose uptake rate in excess as compared to oxidative metabolism.

The measure of cerebral glucose metabolism or oxygen consumption as an index of neuronal function has found an extensive application in physiological and pathological aging (7).

In particular, it has been observed that in healthy elderly subjects cerebral blood flow and metabolism in resting conditions (reduced visual and auditory input) do not differ from that in young people, suggesting that the basal homeostatic mechanisms of the coupling between blood supply and neuronal energy requirement are not impaired in advanced age.

Mental functions, however, are mediated by networks of interconnected brain regions, and include visual, auditory, somatosensory perceptions for simple processes, and attention, memory, speech, logical thinking for higher cognitive activities. With the use of PET and the elaboration of a correlation matrix method (8, 9) it has been possible to examine the patterns of intra- and interhemispheric connections of regional blood flow and metabolism in various conditions: normal aging, pathological aging (Alzheimer's disease), psychiatric diseases.

The results show a decrease in the integrated functions between pairs of metabolic rates in ipsilateral parietal and frontal lobes, which may be related to the reduced "fluid intelligence" in senescence.

In pathological aging, not only regional glucose metabolism was decreased, but the number of correlations in frontal and parietal association areas appeared further impaired, whereas sensory and motor regions were spared. The disease was also characterized by abnormal metabolic right-left asymmetries in association areas, which parallel the deficits in cognitive functions.

It is important to note that the simple measure of regional glucose metabolism may not reflect underlying brain chronic diseases when the compensative mechanisms are operative. For example, it has been demonstrated that the unilateral auditory deprivation in immature mice initially reduces the rate of glucose utilization in the contralateral inferior colliculus. However, this parameter returns to normal values after some weeks, whereas the area remains histologically and functionally altered. Further evidence comes from clinical data showing that cortical metabolism after thalamic stroke can recover despite the maintained cortical deafferentation. This indicates that the correlation matrix method may provide more complete informations on the integrity of neuronal circuits. In this respect, it will be of extreme importance to evaluate data obtained not only in resting conditions, with reduced sensory inputs, but during performances involving specific cerebral networks.

Interesting data emerged from the study of the regional correlations in rats after callosectomy. It appears that the reduced values of interhemispheric connection due to the lesion are accompanied by higher levels of intrahemispheric correlations, suggesting that the corpus callosum exerts a functional inhibitory tone on the circuits linking the areas of the same side of the brain.

NEURONAL METABOLISM AND CEREBRAL BLOOD FLOW

The molecular mechanisms at the basis of the coupling between cerebral blood flow and neuronal metabolism are not completely known: the production of local metabolic metabolic factors or vasoactive substances, as oxygen, CO_2, H^+, adenosine, glycolitic intermediates, has been proposed in the mediation of the short-term coupling activity, whereas changes in capillary density are probably involved in long term events. On the other hand, a large amount of experimental evidence points to the importance of an alternative control of brain circulation exerted by neurotransmitter systems. The existence of neuronal circuits which influence vascular function represents an advantage in terms of speed and specificity of information and can guaranty a tighter maintenance of cerebral homeostasis.

Furthermore, the direct intervention of the neurons may account for the finding that brain metabolism and blood flow may be not always coupled, in contrast with the classical hypothesis by Roy and Sherrington, but, under some circumstances, flow may increase more than metabolism (10).

The adrenergic innervation is so far the best investigated, on the basis of anatomical, physiological and biochemical studies (11).

The existence of a close association between nerve terminals and endothelial cells has been documented by light and electron microscopy. The origin of the fibers is the pontine nucleus Locus Coeruleus. The stimulation (chemical or electrical) or destruction of this area results in a rapid alteration in CBF and capillary permeability to water, reversed by the intraventricular administration of the alfa-adrenergic antagonist phentolamine. This finding supports the concept that brain microvasculature do not constitute a static interface but a variable and alterable barrier, sensitive to neurogenic or pharmacological means.

The locus coeruleus receives a major set of cholinergic afferents from a pontomesencephalic region which also projects to the midbrain dopaminergic and serotoninergic neurons. The strategic location of this cholinergic nucleus, permitting the contact with both ascending and descending pathways, suggests that it may function as a switchboard which integrates, modulates and coordinates the input and output of the neuronal informations. In this view, the neuronal circuits in their complexity may be regarded as a central autonomic system, in parallel and analogy with the sympathetic nervous system, which "orchestrates" the needs of the brain and of the organism as a whole, not only in relation to the regulation of capillary tone and permability, but also influencing hormone release, temperature control, sleep-wake cycle etc.

The existence of an intrinsic *cholinergic innervation* of the brain circulation is supported by the anatomical identification of cholinergic nerve terminals associated with small intraparenchymal vessels and by the detection of high affinity muscarinic binding sites in preparations of cerebral capillaries. Furthermore, recent observations have revealed that the electrical stimulation of the dorsal medullary reticular formation causes an increase in regional blood flow partly reversed by local application of atropine and not coupled to EEG activation.

A large amount of studies have shown the importance of other neurotransmitters (serotonin, dopamine, GABA, neuropeptides) on the regulation of large cerebral vessels and microvasculature. In particular, many of the recent works deal with the emerging role of the *peptidergic innervation* (12). The vasoactive intestinal polypeptide (VIP) was the first observed to be present in perivascular nerve fibers probably originating from the neocortex and the sphenopalatine ganglion. The functional significance of this transmitter is not completely understood: in vitro it has been shown to dilate cerebral vessels, but *in vivo* its effects on brain perfusion are complicated by the interaction with cerebral metabolism; VIP-sensitive adenylate cyclase has been demonstrated also in capillary enriched preparations, suggesting a role of this peptide in BBB reactivity.

Less known is the function of CCK fibers originating from the cortex and innervating brain microvasculature.

Substance P, neurokinin A and calcitonin-gene related peptide (CGRP) are often co-localized in fibers deriving from the trigeminal ganglion and supplying the larger cerebral vessels, where they exert a dilator effect; in particular, the potency of GCRP and its sustained action *in vivo* point to the involvement of this transmitter in the restoration of vascular calibre after vasoconstriction.

Neuropeptide Y (NPY) is co-localized with norepinephrine in sympathetic nerves coming from the superior cervical ganglion and, in part, from the stellate ganglion. The fibers innervate the adventitia and reach the small arterioles, probably the precapillaries, and the veins.

The communication between BBB and neurotransmitter pathways is mediated by specific receptors located on cerebral endothelial cells (13, 14). In particular, biochemical studies based mainly on radioligand binding techniques have identified in capillary fractions the presence of several receptor types (alfa-1, alfa-2 and beta-2 adrenergic, muscarinic, histamine 2, adenosine 2, VIP, parathyroid hormone, prostacyclin, atrial natriuretic peptide) coupled to transducing mechanisms as adenylate and guanylate cyclase. The second messengers in turn activate cAMP- and cGMP-dependent protein kinases; also Ca^{++}-calmodulin-, Ca^{++}-/phospholipid- and diacylglycerol-dependent protein phosphorylation is reported in brain microvessels. The major substrates appeared to be synapsin I and a protein (50—55K) specific for the capillary fraction.

The major question is to know the functional role of capillary receptors. As reported above, evidence based on the experimental manipulation of the locus coeruleus indicates the involvement of adrenoreceptors in the regulation of BBB permeability. This is further confirmed by recent data suggesting that the adrenergic innervation is important in regulating the brain endothelial capacity to actively transport Na^{+} and K^{+} ions, contributing to the maintenance of water and electrolyte homeostasis in the central nervous system (15). In addition, it has been observed that the stimulation of capillary adenylate cyclase (linked to H2, adenosine and VIP receptors) and guanylate cyclase is associated to an increase in the macromolecular transport across the BBB and in the number of pinocytotic vesicles. Moreover, norepinephrine and histamine have been found to activate prostaglandin synthesis in brain microvessels. It seems probable, therefore, that in some pathological conditions (radiation injury, acute hypoxia, kainic acid induced epilepsy) a massive release of these transmitters may lead to an abnormal transendothelial passage of molecules and water, causing vasogenic brain edema; experimental evidence indicates that histamine H2-receptor blockers are effective in the prevention of this process (14).

On the other hand, recent observations have revealed that in cultures of endothelial cells derived from brain vessels norepinephrine activates glycolitic processes, suggesting a role of capillary adrenergic receptors also in the control of the cell metabolism (16).

Studies on pathological states associated with morphological and functional alterations of brain microvasculture (hypertension, diabetes,) have shown a reduction in capillary beta-adrenergic receptor number and adenylate cyclase activity, supporting the concept that adrenergic innervation plays an important role in BBB reactivity (13, 17).

Recent data have opened new perspectives on the organization of neuronal circuits controlling the brain microcirculation (17).

In particular, it has been demonstrated that ischemia, a pathologic condition associated with alterations in both neuronal function and capillary permeability, causes a decrease in microvascular beta-receptor density which is more pronounced and persistent in the left hemisphere, independently of the side of vascular occlusion. The machanisms at the basis of the peculiar sensitivity of the left side of the brain are unknown. Both the destruction of the central adrenergic system by i.ventricular injection of 6-OHDA and the lesion of the corpus callusum prevent the asymmetric response of cerebral microvasculature to the ischemic insult. The data provide evidence that capillary beta-receptors are, at least in part, directly regulated by neuronal pathways and suggest that the integrity of interhemispheric connections is essential for both the expression and the "polarity" of the vascualar response. The hypothesis of the lateralization in the neuronal circuits controlling the BBB, the functional significance of which is at present not known, is in line with anatomical, biochemical and behavioral studies which have extensively documented the asymmetrical specializations of the cerebral hemispheres.

In aged rats, the capillary beta-adrenergic response to ischemia is similar to that obtained with split-brain rats, suggesting that senescence impairs the transfer of neuronal informations between the two hemispheres.

In addition, the basal level of brain endothelial beta-receptor density is reduced during aging; although both animal and human studies report that there is no evidence for a relevant alteration in cerebral blood flow and metabolism in healthy senescence, indicative of a maintained equilibrium between neuronal and vascular function, the decreased receptor number may indicate that the regulatory mechanisms of microcirculation may be vulnerable to particular demands of energy and metabolic supply.

NEURON-VESSEL-GLIA COMMUNICATION: REGULATION OF NEURONAL MICROENVIRONMENT

In the last years the role of glia in the regulation of neuronal homeostasis has received increasing attention (18—20). The glial membranes constitute a huge surface area allowing a large capacity of passage/exchange for ions and small molecules. Electrogenic Na^+/K^+ ATPase, Na^+/glutamate symport, Na^+/HCO_3-cotransporter, Na^+/H^+ exchanger, passive K^+ Cl-entrance are some of the systems that operate in the maintenance of extracellular space homeostasis, acting as "buffer" mechanisms of neuronal metabolism. For example, the excess of K^+ released by the neurons after prolonged stimulation can be uptaken into the glial cells, thus preventing the deleterious accumulation of the ion in the extracellular fluid. Clearly, the redistribution of potassium influences neuronal membrane potential and, as a consequence, the overall information processing.

Among the constituent of the interstitial fluid, neurotransmitters and neuromodulators, besides K^+ and Ca^{++} ions, play an important role in neuronal excitability. The presence in the glial cells of enzymes for the transport and metabolism of glutamate, GABA, adenosine, catecholamines supports the importance of neuron-glia interactions for brain function.

Recently, voltage gated currents have been recorded at glial level, indicating the existence of a new pathway involved in the modulation of the neuronal microenvironment. In particular, Na^+-channel, sensitive TTX, Ca_2^+-activated K^+-channels and L-type Ca-channels, sensitive to dihydropyridines, have been identified. This finding suggests the possibility that during periods of increased neuronal activity a large amount of Ca_2^+ may enter the glial cells, mainly triggered by the increased K^+-concentration.

An important point still to be clarified is the voltage control of ion movements across glial membranes, since the cells are not normally depolarized by the 40—50 mV necessary to open the channel gate.

The control of ion and neurotransmitter homeostasis represents only one facet of glial functions. Emerging evidence indicates that the involvement of glia also in the regulation of the BBB. The astrocyte endfeet, which are strategically located around the synapses and the capillaries, seem to control the exchanges between CNS and vasculature. For instance, it has been observed that the macromolecular transport across the endothelium is suppressed by glial cells; in certain pathological states, as ischemia, this process may counteract the formation of edema.

By investigating the factors that influence the development and the expression of the unique properties of cerebral microvasculature, interesting observations have emerged on the determinant role of glial cells. In particular, experimental data with co-cultures of endothelial and glial cells indicate that astrocytes are the source of inductive signals for the appearance of the structural and biochemical characteristics of the BBB; further evidence comes from the observation that astrocytes purified from immature brain and implanted into the avascular cornea stimulate the growth of new microvessels which display BBB properties. Experiments performed on an artificial BBB, constituded by endothelial cells on one side of a filter membrane and C6 glioma cells on the other side, and exposed to a flux of a small neutral amino acid, show an asymmetric transport of the tracer. In the absence of glial cells, or in the presence of other types of cells, the tracer moved at equal rates in both directions, suggesting the role of astroglia in the induction of capillary polarity.

SOME ASPECTS OF BRAIN CAPILLARY ENDOTHELIUM IN PATHOLOGY

Cerebral ischemia due to vascular occlusion by thrombi or emboli is characterized by deep alterations in both neuronal and BBB function. Much of current knowledge on the pathophysiology of cerebrovascular disorders is based upon informations obtained from animal models of stroke or *in vitro* studies on cultured cells (21). It should be noted, however, that in both cases there may be limitations in extrapolating the experimental results to the clinical situations. Furthermore, animal models may present the disadvantage of technical difficulties and large variability and low reproducibility of the neuronal damage.

The biochemical events involved in the genesis and progression of the ischemic cell injury are extremely complex and include impairment of calcium homeostasis, acidosis and free radical formation (22). The identification of these mechanisms represents a basis for possible points of attack in the therapy of cerebrovascular disorders. In particular, the target of pharmacology could be, beside the vasculature, the neuron itself which participates to the metabolic exchanges between the blood and the cerebral parenchyma, directly controlling the brain microenvironment.

Recent experimantal data suggest a new, emerging role for brain capillary endothelium in controlling immunological response in the brain (23).

The concept of the central nervous system as a immunological privileged site protected by a highly selective BBB has been challenged by the observation that the cerebral endothelium may be involved in the development of inflammatory processes in the brain. In particular, brain capillaries of patients affected by autoimmune demyelinating diseases, as multiple sclerosis, show the presence of class II histocompatibility antigens (as e.g. myelin basic protein) to specifically-sensitized T cells. The activation of T limfocytes leads to the release of vasoactive transmitters which, in turn, by increasing BBB permeability facilitate the migration of blood elements into the brain parenchyma and favour the formation and growth of demyelinating lesions.

In conclusions, the concept of BBB in the course of few years has evolved from an almost static lining of cells interposed between the blood and the brain to a dynamic interface which regulates, and in turn is regulated by, neuronal and glial activity. The interplay neuron-glia-vessel reflects multiple interactions aimed to the maintenance of cerebral homeostasis. The flexibility, within certain limits, of this complex system provides the physiological basis for an integrated response to external stimuli. At the same time, alterations in these interacting processes may have a role in metabolic and neurodegenerative disorders of the central nervous system.

REFERENCES

1) E.M. Burns, T.W. Kruckeberg (1989): The ultrastructural basis of capillary permeability. In: Regulatory Mechanisms of Neuron to Vessel Communication in the Brain. (F. Battaini, S. Govoni, M.S. Magnoni, M. Trabucchi, Eds.) NATO ASI Series, Springer-Verlag, Berlin, Heidelberg, Vol. H33, pp. 1—20.

2) A.L. Betz (1989): Structural and functional polarity of the blood-brain barrier. In: Regulatory Mechanisms of Neuron to Vessel Communication in the Brain. (F. Battaini, S. Govoni, M.S. Magnoni, M. Trabucchi, Eds.) NATO ASI Series, Springer-Verlag, Berlin, Heidelberg, Vol. H33, pp. 21—36.

3) C. Cangiano, P. Cardelli-Cangiano (1989): Active and facilitated amino acid transport mechanisms across the brain capillary endotehlial cells in physiological and pathological conditon. In: Regulatory Mechanisms of Neuron to Vessel Communication in the Brain. (F. Battaini, S. Govoni, M.S. Magnoni, M. Trabucchi. Eds.) NATO ASI Series, Springer-Verlag, Berlin, Heidelberg, Vol H33, pp. 37—64.

4) C.S. Roy, C.S. Sherrington (1980): On the regulation of the blood supply to the brain. J. Physiol., (Lond), 11: 85—108.

5) L. Sokoloff (1989): Measurement of regional hemodynamic and metabolic changes in the central nervous system with imaging techniques. In: Regulatory Mechanisms of Neuron to Vessel Communication in the Brain. (F. Battaini, S. Govoni, M.S. Mangoni, M. Trabucchi, Eds.) NATO ASI Series, Springer-Verlag, Berlin, Heidelberg, Vol. H33, pp. 345—392.

6) V.J. Cunningham, J.E. Cremer and R.J. Hargreaves (1989): Relationships between neuronal activity, energy metabolism and cerebral circulation. In: Regulatory Mechanisms of Neuron to Vessel Communication in the Brain. (F. Battaini, S. Govoni, M.S. Magnoni, M. Trabucchi, Eds.) NATO ASI Series, Springer-Verlag, Berlin, Heidelberg, Vol. H33, pp. 325—344.

7) S.I. Rapoport and B. Horwitz (1989): Use of positron emission tomography to study patterns of brain metabolism in relation to age and disease: a correlation matrix approach. In: Regulatory Mechanisms of Neuron to Vessel Communication in the Brain. (F. Battaini, S. Govoni, M.S. Magnoni, M. Trabucchi, Eds.) NATO ASI Series, Springer-Verlag, Berlin, Heidelberg, Vol. H33, pp. 393—410.

8) B. Horwitz, R. Duara, S.I. Rapoport (1984): Intercorrelations of glucose metabolic rates between brain regions: application to healthy males in a state of reduced sensory input. J. Cerebr. Blood Flow Metab., 4: 484—499.

9) B. Horwitz, R. Duara, S.I. Rapoport (1986): Age differences in intercorrelations between regional cerebral metabolic rates for glucose. Ann. Neurol., 19: 60—67.

10) H.C. Lou, L. Edvinsson (1989): Innervation of the cerebral vasculature and its putative role in the coupling of flow to metabolism. In: Regulatory Mechanism of Neuron to Vessel Communication in the Brain. (F. Battaini, S. Govoni, M.S. Magnoni, M. Trabucchi, Eds.) NATO ASI Series, Springer-Verlag, Berlin, Heidelberg, Vol. H33, pp. 75—96.

11) B.K. Hartman (1989): Central noradrenergic innervation of cerebral microvessels: implications for an autonomic nervous system within the brain. In: Regulatory Mechanism of Neuron to Vessel Communication in the Brain. (F. Battaini, S. Govoni, M.S. Magnoni, M. Trabucchi, Eds.) NATO ASI Series, Springer-Verlag, Berlin, Heidelberg, Vol. H33, pp. 65—74.

12) J. McCulloch, L. Edvinsson (1989): Peptidergic innervation of the cerebral vasculature and its functional significance. In: Regulatory Mechanisms of Neuron to Vessel Communication in the Brain. (F. Battaini, S. Govoni, M.S. Magnoni, M. Trabucchi, Eds.) NATO ASI Series, Springer-Verlag, Berlin, Heidelberg, Vol. H33, pp. 97—112.

13) G.C. Palmer, S.A. McCreedy and L.R. Freedman (1989): Neurotransmitter coupled responses in the microvasculature of the brain under normal and pathological conditions. In: Regulatory Mechanisms of Neuron to Vessel Communication in the Brain. (F. Battaini, S. Govoni, M.S. Magnoni, M. Trabucchi, Eds.) NATO ASI Series, Springer-Verlag, Berlin, Heidelberg, Vol. H33, pp. 113—144.

14) F. Joo (1989): Physiological and pathological events mediated by capillary neurotransmitter receptors. In: Regulatory Mechanisms of Neuron to Vessel Communication in the Brain. (F. Battaini, S. Govoni, M.S. Magnoni, M. Trabucchi, Eds.) NATO ASI Series, Springer-Verlag, Berlin, Heidelberg, Vol. H33, pp. 145—168.

15) S.I. Harik (1986): Blood brain barrier sodium/potassium pump: modulation by central noradrenergic innervation. Proc. Natl. Acad. Sci. USA, 69: 4067—4070.

16) M. Spatz, B.B. Marsulja, B. Wroblewska, N. Merkel, J. Bembry (1986): Modulation of glycogen metabolism in cerebromicrovascular smooth muscle and endothelial cultures. Biochem. Biophys. Res. Comm., 134: 484—491.

17) M.S. Magnoni, S. Govoni, F. Battaini, M. Trabucchi (1989): Biochemical changes of brain microvessels during aging. In: Regulatory Mechanisms of Neuron to Vessel Communication in the Brain. (F. Battaini, S. Govoni, M.S. Magnoni, M. Trabucchi, Eds.) NATO ASI Series, Springer-Verlag, Berlin, Heidelberg, Vol. H33, pp. 169—186.

18) R.K. Orkand (1989): Role of glial cells in the control of the neuronal microenvironment. In: Regulatory Mechanisms of Neuron to Vessel Communication in the Brain. (F. Battaini, S. Govoni, M.S. Magnoni, M. Trabucchi, Eds.) NATO ASI Series, Springer-Verlag, Berlin, Heidelberg, Vol. H33, pp. 253—270.

19) L. Hertz (1989): Functional interactions between neurons and glial cells. In: Regulatory Mechanisms of Neuron to Vessel Communication in the Brain. (F. Battaini, S. Govoni, M.S. Magnoni, M. Trabucchi, Eds.) NATO ASI Series, Springer-Verlag, Berlin, Heidelberg, Vol. H33, pp. 271—306.

20) R. Pumain, J. Louvel and I Kurcewicz (1989): Dynamic variations of the extracellular space in the central nervous system. In: Regulatory Mechanisms of Neuron to Vessel Communication in the Brain. (F. Battaini, S. Govoni, M.S. Magnoni, M. Trabucchi, Eds.) NATO ASI Series, Springer-Verlag, Berlin, Heidelberg, Vol. H33, pp. 229—252.

21) S. Sampaolo, J. Cervos-Navarro (1989): Experimental models of ischemia. In: Regulatory Mechanisms of Neuron to Vessel Communication in the Brain. (F. Battaini, S. Govoni, M.S. Magnoni, M. Trabucchi, Eds.) NATO ASI Series, Springer-Verlag, Berlin, Heidelberg, Vol. H33, pp. 187—208.

22) B.K. Siesjo and M.L. Smith (1989): Mechanisms of ischemic damage to neurons, glial cells and vascular tissue. In: Regulatory Mechanisms of Neuron to Vessel Communication in the Brain. (F. Battaini, S. Govoni, M.S. Magnoni, M. Trabucchi, Eds.) NATO ASI Series, Springer-Verlag, Berlin, Heidelberg, Vol. H33, pp. 209—228.

23) U. Traugott (1989): Alteration of the blood-brain-barrier by components of the immune system. In: Regulatory Mechanisms of Neuron to Vessel Communication in the Brain. (F. Battaini, S. Govoni, M.S. Magnoni, M. Trabucchi, Eds.) NATO ASI Series, Springer-Verlag, Berlin, Heidelberg, Vol. H33, pp. 307—324.

THE AGING BRAIN

M.S. Magnoni, S.Govoni*, F.Battaini, M.Trabucchi****

Institute of Pharmacological Sciences, Universityof Milan;
* Dept.of Pharmacobiology, University of Bari;
**Chair of Toxicology, Second University of Rome, Italy

One of the major problems in the study of the mechanisms underlying the processes at the basis of cerebral aging is to define an unitary view of the vast array of scattered data and a rational approach to the interpretation of the specific neurobiological changes.

MORPHOLOGY OF THE AGING BRAIN

Morphological studies indicate that aging is accompanied by neuronal loss (1). This process is not uniform throughout the brain: in certain nuclei such as the locus coeruleus, the caudate-putamen and the nucleus basalis of Meynert the degeneration is particularly pronounced. At cortical level, the loss in the frontal areas exceeds that in parietal and occipital lobe; in the hippocampus, a linear decrease of 20% in neuronal density between the age of 45 and 90 years has been described. However, Geddes, Monaghan and Cotman (2) have reported a continued capacity of dendritic arborization in the surviving hippocampal cells, suggesting that neuronal plasticity is maintained in advanced age.

Several studies report alterations in the cell body during senescence, like reduction in nuclear size and slight decrease in RNA concentration. According to some authors, these changes reflect a reduction in overall production of proteins, perhaps in response to a decreased neuronal activity, rather than being indicative of a site of damage.

Also the dendritic processes and the synaptic contacts are reduced in the aging brain. This event may be ascribed to the decreased number of neurons or to insufficient axonal transport. Furthermore, the fiber conduction velocity in subcortical projections is reported to be lower.

Complexively, the data indicate a reduced rate of interneuronal communication which can account for the age-related modifications in cognitive functions.

BIOCHEMISTRY OF THE AGING BRAIN

So far as the biochemistry of the brain, senescence is characterized by deficits in specific neurotransmitter systems, due to the vulnerability of selective populations of neurons, which have been invoked to explain changes in cognitive and psychological behavior in the elderly.

NATO ASI Series, Vol. H 51
Cell to Cell Signals in Plants and Animals
Edited by V. Neuhoff and J. Friend

One of the limits of the studies utilizing animal models is the underestimation of the individual variance in the senescence process. For example, it has been shown that an apparently homogeneous group of rats can be divided into subgroups on the basis of their ability to perform in a water maze. This underlines the importance of evaluating the impact of neurochemical parameter modifications on the performance of a function and suggests caution in interpreting the biochemical data.

The attempt of the investigations has been to detect the most significant changes in the biochemistry of the aging brain, at different levels of organization and complexity, with particular attention to the mechanisms controlling some basic cellular processes (i. e. calcium homeostasis) and the identification of selectively susceptible neuronal systems.

CALCIUM HOMEOSTASIS

The decrease in neurotransmitter release, in particular acetylcholine and dopamine, observed in aged rats might be related to a modified ability of the cells to control calcium movements (3). The brain's dependence upon numerous Ca-regulated processes make it particularly vulnerable to alterations in the homeostasis of the ion.

The emerging data point to a deficit in the (k^+-induced) uptake of calcium in nerve terminals, both *in vitro* and *in vivo* (although in the latter case methodological problems make the interpretation of the data difficult). The mechanisms underlying the decline in total calcium uptake may be various: age-related increase in intracellular lactate, modifications in membrane composition and decreased fluidity, increase in calcium superficial binding, altered regulation of voltage-dependent calcium channels (VDCC). This last aspect has been investigated by Trabucchi et al. (4) by means of radioligand experiments.

Both the binding of tritiated nitrendipine (^{3}H-NTP) and tritiated verapamil (^{3}H-V) display modifications in senescent animals: the affinity of ^{3}H-NTP and the calcium sensitivity of the binding are decreased, whereas the maximum number of binding sites of verapamil and its ability to displace nitrendipine are increased. Complexively, the results indirectly suggest an age-related modification of the L-type neuronal calcium channel.

The concentration of calcium required to saturate ^{3}H-NTP binding is increased with aging. In addition, the ability of the ion to stimulate ^{3}H-glutamate binding to the hippocampus is reduced in senescent rats. These findings complexively suggest an impairment of cell membranes to transduce the signals supplied by the coupling of extracellular ligands to receptor sites.

Recently, a decrease in binding sites labelled by iodinated omegaconotoxin (which is thought to mark N-type channel) has been observed in the cortex and the striatum of aged rats. Electrophysiological studies assign a predominant role to N-type channels in neurotransmitter release (5), in normal conditions. In this light, the decrease in ^{125}I-omega CgTx binding sites appears to be functionally linked to the reduced calcium-dependent release of some transmitters, as acetylcholine, during aging. Notably, the treatment with 3,4-diaminopyridine, a promotor of calcium uptake, ameliorates this deficit. In addition, the concentration of ionophore A23187 required to promote neurotransmitter release in aged animals is higher than in young animals.

Not only is the regulation of calcium conductance through VDCC affected by senescence, but also the mechanisms that control the extrusion or compartmentation of the ion in the intracellular stores (calcium uptake in endoplasmatic reticulum and mitochondria) are depressed in old age.

The interaction of calcium with phospholipids, and in particular with the complex phosphatidylinositol cascade, has never been investigated during senescence and may represent an important subject of research.

Postsynaptic events, such as stimulation-induced after hyperpolarization and K^+ conductance are prolonged during aging. Since the exposure of cells to high calcium media alters these parameters in a manner resembling the senescence condition, it is possible to hypothesize that intracellular calcium is elevated in the postsynapse of the aged brain or that the age-related increase in Ca^{2+} superficial binding may affect postsynaptic events.

The changes in calcium metabolism may be at the basis of the biochemical and behavioral deficits in advanced age. In particular, beside the influence on neurotransmitter release, the activation of Ca-dependent enzymes may be altered, leading to the accumulation of abnormal cell proteins (lipofuscin, neurofilamentous proteins, etc.) which could be in part responsible for some neurodegenerative events. Furthermore, there is evidence that memory disfunctions in aged rats can be partially improved by 3,4-diaminopyridine treatment.

It is important to underline that the detection of age-related modifications in calcium homeostasis, although functionally important, does not necessarily identify the primary defect, because of the complex interactions which regulate the movements of the ion (cell entry, extrusion, sequestration in the intracellular stores).

Nevertheless, the fact that pharmacological treatment with calcium promotor agents partially reverses the alterations in calcium homeostasis and some related behavioral deficits encourages the concept that the aging organism is still capable of responding to appropriate stimuli.

Another consideration which can be drawn from the data is that caution is necessary in the use of calcium antagonists in the elderly, in view of a potential risk of neuronal depression (parkinson-like side effects by flunarizine in old patients are well known). On the other hand, these drugs can be beneficial in case of calcium overload and excessive cell excitation, like in ischemia, which are harmful to the neurons.

NEUROTRANSMITTER CHANGES IN THE AGING BRAIN

Studies on neurotransmitter systems in normal aging may reflect something of the state of interneuronal connections within the CNS.

The reports, coming from studies on both animal models and post mortem human brain tissues, are focussed on the measurement of "markers", as radioligand binding sites and content of neurotransmitters and related enzymes in various cerebral areas.

The main concepts arising from the widespread results are the differential vulnerability of neuronal pathways to the aging process and the "threshold phenomenon".

Even if the data available do not allow the identification of a single neuronal population controlling the aging of the brain (this would be an oversimplification of the problem), significant progresses have been made in the attempt to detect and interpret the most relevant changes in cerebral biochemistry.

The catecholaminergic system, especially the dopaminergic, appears to be highly sensitive to senescence (4, 6—9). In particular, the deficit in brain DA circuits involves both the ability of the terminals to synthesize and reuptake the transmitter and a complex pattern of changes at the level of postsynaptic receptors (depending also on the species examined) and associated transducing mechanisms (adenylate cyclase and cAMP-dependent phosphorylation of neuronal specific substrates).

The age-related modifications in receptor number and function may be caused by several factors, including alterations in the rate of synthesis, as shown by Roth (10), or degradation, changes in membrane lipid composition, cell loss; alternatively, they may represent a part of a normal adaptive process to changes in neuronal function.

Although the dopaminergic system is the most extensively investigated, a large amount of studies concerns the involvement of other transmitters (acetylcholine, serotonin, GABA, peptides etc.) in the aging process. For example, a substantial neuronal dropout in the cholinergic nucleus basalis occurs with age. The presynaptic marker choline acetyltransferase has been reported to be reduced particularly in the hippocampus and the cortex, whereas the loss of muscarinic receptors seems, according to some observations, to be mostly confined to the hippocampus. These data have relevance in view of the role of the cholinergic system, particularly at hippocampal level, in memory and learning processes, which are reduced in advanced age. In this line, an interesting observation has been recently published (11) on the increased level of dynorphin and prodynorphin mRNA in the hippocampus of aged rats displaying spatial learning deficits, as compared to animals of the same age without behavioral impairments. Since brief high frequency stimulation induces prolonged depression of the level and genetic expression of the peptide in the hippocampus, the elevation in dynorphin content during aging probably reflects a diminished excitatory input of the perforant path to the dentate gyrus.

An important subject of investigation is the role of excitatory amino acids (aspartate, glutamate) in the neurochemical events associated with senescence, in view of their involvement in physiological processes, like memory and learning, which are impaired with old age, and in neurotoxic reactions occurring in particular conditions like ischemia and, perhaps, neurodegenerative disorders.

Although the data on the biochemistry in the aging brain are scattered and sometimes controversial, it emerges that the functional balance between different transmitter systems is deeply modified with age. This may explain why the pharmacokinetic and pharmacodynamic properties of drugs acting at the level of the CNS may be changed. For example, it has been shown that in old male ICR mice the dose of morphine to elicit running and analgesia are increased and decreased respectively (12).

It is important to underline that the "physiological" decline in neurotransmitter content, especially DA and acetylcholine, does not reach the threshold level which determines the clinical signs of cerebral pathology. In the case of DA, for instance, the concentration of the transmitter in the substantia nigra has to fall below the critical value of ~80% before parkinsonism becomes manifest. The theory of the "threshold level" could be extended to other degenerative disorders as senile dementia and Huntington chorea.

NEUROENDOCRINE ASPECTS OF AGING

An important aspect of senenscence concerns the role of the neuroendocrine system, which exerts a commanding influence over body functions. Most of the hormones in the hypothalamus, pituitary and target glands are secreted in a rhythmic or pulsatile manner, and are regulated by circadian and ultradian mechanisms (the biological clocks) that may reside in the hypothalamus or other cerebral areas. The secretion of hormones can be deeply altered by external and internal environmental agents, as nutrition, temperature, stress, disease etc.

Important changes in the "setpoints" of neuroendocrine function occur not only in fetal life, pregnancy or puberty but also during senescence, with the initiation of processes leading to modifications of a variety of body functions, as loss of reproductive cycle in females, decline in testosterone secretion in males, decrease in GH production and protein synthesis. Some of these events can be corrected or hastened by manipulations at the level of the hippocampus. For example, the electrical stimulation of the medial preoptic area, which is involved in GnRH release, induces ovulation in old rats. It seems that the mechanism at the basis of the decline in GnRH release is the decrease in the hypothalamic norepinephrine activity. Along this line, Meites has shown that the administration of L-DOPA and other central catecholaminergic drugs to old rats reinitiates estrous cycles in females and increases testosterone secretion in males (13).

On the other hand, there is evidence that the action of gonadal hormones during a relative long portion of the life span can induce permanent changes at hypothalamic level associated to cycle lengthening (6). Chronic treatment with estradiol of young mice induces neuroendocrine dysfunctions that are similar to the usual age-related changes, including long lasting impairment of the LH surge, glial hyperreactivity, which is associated to neuronal degeneration, and deposition of lipofuscin in the dopaminergic arcuate nucleus. Conversely, some of the age-related neuroendocrine changes in female rodents can be delayed by ovariectomy.

The reciprocal interactions between ovary and hypothalamus and their impact on the various steps of the reproductive cycle during life span may represent a model for the identification of the molecular mechanisms related to the aging process.

An important aspects which needs to be investigated is the relationship between immune and endocrine systems, both of which are depressed in advanced age. It is well established that hormones (GH, TSH, ACTH, gonadotropins, glucocorticoids) modulate structurally and functionally the immune system; conversely, recent evidence indicates that the immune response may influence neuroendocrine activity, for example by eliciting a significant increase in electrical activity in certain hypothalamic neurons, by affecting norepinephrine turnover in the same area and promoting the secretion of some pituitary hormones. However, little is known about this interplay during senescence. It has been reported that the injection of mature lymphocytes markedly prolong the life of short-lived Snell-Bagg dwarf mice and correct some of the aging characteristics, as loss of hair, cataracts, reduced cellular turnover. The same authors showed also that grafts of neonatal thymus placed into old mice are able to correct abnormal serum levels of T3 and insulin (14, 15).

These observations suggest that more extensive research on this line can contribute to the knowledge of the mechanisms responsible for the decline of homeostatic processes during aging.

BRAIN METABOLISM AND PROTEIN SYNTHESIS IN AGING

The informations deriving from the morphology and the biochemistry of the aging brain can be integrated with functional studies on cerebral metabolism by means of *in vivo* 11Ctechniques (PET).

The parameters assessed are the consumption of oxygen or glucose, indicative of the energy supply to the cells in relation to their functional activation, and the neuronal protein synthesis, which reflects the long term maintenance of synaptic contacts and the remodeling capacity of the brain.

Using as a tracer the ^{11}C-(Me)-L-methionine, the computer processing of PET human images has revealed a 30% reduction in protein synthesis during aging (16). Quantitative autoradiography has confirmed the same result in rats. This finding may account for the reduction of cell density in normal aging, which is not completely compensated by neuronal adaptive mechanisms such as dendritic growth and branching, and is therefore indicative of a decreased capacity of the brain to maintain its cellular integrity.

PET studies using ^{18}F-2-fluoro-2-deoxy-D-glucose (^{18}F-DG) have demonstrated that cerebral metabolism in resting conditions is practically not modified in healthy elderly subjects, suggesting that the basal equilibrium between blood supply and neuronal energy requirement is still operative in advanced age (17, 18).

However, by examining the correlation coefficients between paired cerebral areas, Rapoport et al.(17) showed a certain degree of loss of integrated activity in the parietal and frontal associative neocortex. The functional region to region disconnection can be based on a decreased dendritic activity or lesser synaptic connectivity and may be related to the reduced "fluid intelligence" in the elderly.

Along this line, studies on the neuronal control of cerebral microcirculation have shown that in senescent rats the effects of experimental ischemia on brain microvessel response is reduced in the hemisphere contralateral to the side of injury, in a manner that resembles the condition of corpus callosum section in young animals (4). This suggests that aging may affect the interhemispheric communication and therefore influence the reciprocal interactions between the two sides of the brain.

NEURONAL PLASTICITY IN THE AGING BRAIN

In parallel with the morphological, biochemical and functional changes reported in the aging brain, a series of adaptive cellular responses takes place, suggesting that the neurons are endowed with the potential capability of compensating, to a certan extent, the age-related deficits. For example, old cells retain the ability of dendritic and axonal growth; furthermore, there is evidence of neuronal sprouting in the hippocampus and dentate gyrus of patients with Alzheimer's disease, indicating that structural plasticity is maintained not only in physiological aging but also in pathologically compromised aged brain (2).

At biochemical level, the decline in DA neurons in the substantia nigra of parkinsonian patients is compensated by the increase in both transmitter turnover and sensitivity of postsynaptic receptors, suggesting that the system is still capable of compensatory responses (7).

On the other hand, the production of factors capable of supporting survival and axonal growth of specific populations of neurons (nerve growth factors) seems to be reduced in aging. In particular, experimental evidence shows that the neuronotrophic activity of the dopaminergic system in response to partial unilateral lesion of the nigro-striatal pathway (reflected by the measure of tyrosine hydroxylase and by the capability of striatal extracts of enhancing survival and DA uptake in mesencephalic cells) is reduced in aged rats (19).

On this basis, it is possible to hypothesize that an insufficient trophic activity in the brain could have a role in human degenerative disorders associated with loss of selective groups of neurons (i.e. Alzheimer's and Parkinson's disease). In support of this view, recent evidence shows that sera obtained from parkinsonian patients contain immunoglobulins which specifically impair the expression of dopaminergic characteristics in primary mesencephalic-striatal cell co-cultures. However, it is not clear whether these antibodies play a role in the pathogenesis of Parkinson's disease or represent a secondary response to neuronal destruction.

GENERAL CONSIDERATIONS: FROM PHYSIOLOGICAL TO PATHOLOGICAL AGING

The mechanisms underlying the specific morphological and biochemical changes in old age are unknown. The cell- and transmitter-specificity of the alterations possibly implies a selective gene regulation at the roots of aging. Whether changes of old age result from a normal expression of a genetic program that begins at conception and ends in death or are due to error accumulation in functioning genes is matter of discussion. Presumably, the pacemakers that determine the timing of the age-related processes are the integrated result of intrinsic and extrinsic influences.

Along this line, senescence can be viewed as a bio-psychosocial event, characterized by high variance among the individuals. The intellectual and physical changes that accompany "normal" aging, although reflecting an impaired homeostatic reservoir, should be regarded as "physiological". Furthermore, it is clear that, due to the heterogeneity of the process, the assessment of the status of health in later age implies not only the evaluation of objective parameters (hypertension, hypercholesterolemia, etc.) but also the attention to subjective signs (20). The ability to perform activities necessary to ensure the personal well being and maintain a high quality of life, even in the presence of pathology, can be considered as a status of "good health" and depends on subjective criteria (the subjective improvement during clinical trials is what matters most of the patients and, according to some authors, should be included in the assessment of drug effects). In other words, health does not merely mean absence of disease.

However, objective criteria are fundamental for the definition of pathological aging, in particular for neurodegenerative disorders.

PATHOLOGY OF THE AGING BRAIN

Alzheimer's disease

Senile dementia of the Alzheimer type (AD) has emerged in the past decade as a major problem for public health (21—25). The disease is characterized by neuronal degenerative processes, severely affecting mental functions. The cholinergic system, whose cell bodies lie in the basal forebrain, seems to be the most affected, although also 5-HT neurons in the raphe nucleus and NE neurons in the locus coeruleus decline in number in a massive way. Neuronal degeneration is accompanied by structural and functional changes: extracellular clusters of amyloid, in the form of neuritic plaques and cerebromeningeal vascular deposits, and intracellular aggregates of "twisted" cytoskeletal components, called neurofibrillary tangles, appear routinely in the hippocampus and neocortical association areas, and rarely in sensorimotor, auditory and visual cortices. Thus, AD lesions are not randomly distributed but follow a selective pattern of diffusion.

Functional studies by means of *in vivo* techniques have shown a high reduction in protein synthesis (16), reflecting hypodensity of brain structures (revealed in a later stage by CT scanning). This is reported to be one of the most sensitive measurement for early detection of dementia. Brain metabolism is impaired: the interactions between cortical areas are reduced at a degree exceeding "physiological" aging, suggesting disruption of intracortical communications; the left-right metabolic asymmetries in neocortical areas, which correlate with neocortically mediated cognitive abnormalities in moderate dementia, are increased, indicating that the metabolic dysfunctions precedes measurable mental deficits (17). The functional impairment may be caused by cytoskeletal pathology, resulting in disturbed axonal transport of neurofilaments and axonal dystrophy.

We should take in account, however, that plaques and tangles are observed also in the brain of elderly subjects not affected by dementia, although they are confined generally in the hippocampus and only occasionally present in the cortex; furthermore, the lesion of subcortical nuclei does not explain sufficiently the neurotransmitter deficits and the cortical neuronal outfall in AD: in some cases of late onset of the disease the loss of these nuclei has not been observed. This poses the question on what are the minimal pathological and neurochemical lesions (threshold phenomenon) associated with the disorder that satisfy the diagnostic criteria of AD.

Some light is shed on this problem by the evidence, emerged in recent years, on the necessity of discriminating between late and early onset of AD, on the basis of a number of neurobiological features: neuronal loss, decreased activity of cholinergic markers, reduced protein synthesis and cerebral metabolism, density of plaques and tangles are more severe in early AD onset.

It has been proposed by Roth (23) that the frequency of plaques and tangles appears to correlate significantly with the extent of mental impairment, neuronal loss and transmitter (particularly acetylcholine) deficiency. In this view, the structural and neurochemical alterations in AD are likely to be the expression of a unitary etiological process underlying the disease, the nature of which is still unknown.

It is therefore necessary to consider the primary factor antecedent of structural changes. It has been hypothesized that the hallmarks of the AD brain contain protein filaments which can deeply alter the structural integrity of the neurons. Pair helical filaments (PHF) are the main constituent of the tangle; MAP 2, tau and amyloid proteins are also present, more or less tightly associated with the core structure. Immunohistochemical studies (24) have shown that in the core of the tangles there are peculiar proteins normally not present in neurons in a sufficient aboundance to be identified histologically or present in a form which is not recognized by the antisera. It has been proposed that the self-assembly of these abnormal constituents causes PHF formation. This suggests that some precise processes at molecular level are responsible for the expression of the structural modifications in AD. The intimate nature of the causes of these events is still an open question. A role of genetic factors operating in the predisposition to AD can be involved, at least for the familial subtype of this disorder. In fact, 10 to 15% of the cases have been found to be due to autosomal dominant inheritance.

Another line of research, developed with the application of molecular biology techniques, has lead to the identification of the genes encoding for two constituents of the neuritic plaque, the a-protein and the protease inhibitor α1-antichymotrypsin. Subsequently, the precursor of B-protein was discovered: its structural to other membrane receptors suggests a role in cell-cell and cell-substrate recognition or signaling.

The localization in the chromosome 21 of the gene for B-protein has raised considerable interest because of the association between AD and Down's syndrome (trysomy 21); furthermore, the gene has been found close to the locus for familial AD. The two genes do not coincide; however, there is evidence for linkage of DNA markers on chromosome 21 to both amyloid B-protein gene and autosomal dominant AD gene in some families. On the other hand, some studies report the possibility that the AD gene lies on a different chromosome in some cases, particularly those with late onset.

So far as α-antichymotrypsin (ACT), it has been found tightly associated with the B-protein in the amyloid plaque. However, increase in ACT mRNA and protein content seem to represent a general response of the brain to pathological states, as stroke, parkinson disease, Huntington Chorea, suggesting that the overexpression of the protein is not *per se a* characteristic of AD.

Interesting data come from the identification of protein A68 in the brain of AD patients and people with Down syndrome as well as in the cerebral tissue of new born and fetuses, although with some differences in the distribution pattern (26). This finding suggests the hypothesis that the early expression of this protein can be viewed as a marker for cells preprogrammed to grow and die in both the developing brain and neurodegenerative disorders of advanced age. The factors that lead to unmasking of the expression of A68 protein in the elderly neurons are unknown, as still obscure is the relation between protein expression and neurodegenerative processes.

One unsolved question is whether the amyloid plaques themselves exert a neurotoxic effect or rather molecular changes associated with their formation are responsible for neuronal injury.

Whether nongenetic environmental factors (toxins, infectious agents) are involved in the etiogenesis of non familial AD is unknown. The existence of different subtypes of the disease is important in view of the possibilities of therapeutical intervention and responsivity of the subgroups of patients.

Parkinson's disease

Parkinson's disease (PD) is characterized pathologically by degenerative processes which involve most markedly the pars compacta of the substantia nigra, resulting in akinesia, resting tremor, rigidity. The degree of dopaminergic neuron loss is correlated with the severity of the symptoms.

DA neuronal degeneration occurs normally during physiological aging, but it has been estimated that dopamine must fall to about 20% of normal before the clinical signs become manifest. At this threshold level, in fact, the ability of the surviving cells to compensate for degenerative processes (by increasing DA turnover and receptor affinity) is impaired.

The paucity of evidence supporting hereditary or infectious etiology for PD has stimulated the search for environmental factors at the basis of the disease.

Both endogenous and exogenous mechanisms have been hypothesized as etiological determinants. Barbeau (27) has proposed a vicious cycle of damage in DA cells, due to auto-oxidation processes and production of free radicals which are increased in parallel with the compensatory increase in the transmitter turnover.

The studies for the identification of specific exogenous toxins causing PD are recent. In 1983 Langston (28) reported the revolutionary observation that four drug addicts developed a parkinsonian syndrome after injection of synthetic narcotic. The toxic component was identified as MPTP, which seems to act through the metabolite MPP^+ (due to oxidation via MAO B). The reasons of the selective action of this agent on nigro-striatal cells are unknown.

Depending on the level of exposure, the symptoms may develop acutely; however, PET studies with asymptomatic addicts have shown less striatal ^{12}F-dopamine binding compared to controls. This provides evidence that *in vivo* image techniques may detect a subclinical level of the disease, which presumably becomes manifest progressively with increasing age (although a prospective evaluation is required to verify this). In support of this view, experiments on primate demonstrate that old animals are more vulnerable to MPTP toxic effects. Further evidence comes from the study of the parkinson-dementia-ALS Guam complex (28), indicating that a long latency may occur between exposure to specific neurotoxins (excitatory amino acids contained in the seeds of Cycas circinalis with glutamate agonist properties, probably acting on Q and K type receptors) and clinical signs of disease. Most commonly, in fact, exposure is not sufficient to precipitate symptoms acutely and must be combined with age-related compromission of the neurons in order to give rise to clinical expression.

Compounds with structures related to MPTP have been developed as herbicides and may be present in significant concentrations in the environment. The disease in fact paralleled industrial development, with a prevalence in industrial areas. It seems that the cases of early onset of PD are increasing, especially in populations living in rural villages, supporting the environmental hypothesis and assuming that the risk factors are becoming more common.

The growth of the elderly population and the high incidence of aging-related diseases has given a relevant and expanding role to gerontology in medicine. Management of elderly patients require specific knowledge, due to superimposition of several factors: coexistence of multiple pathologies, impairment of homeostatic processes, drug pharmacodynamic and pharmacokinetic modifications.

Geriatric intensive care units are still rare, although there is some evidence that intensiveness should be included in most instances of geriatric care (30). In fact, conditions which do not represent an emergency in young patients may become life-threatening in the elderly, because of the reduced resistance to injuries. An example is represented by surgery, which is associated in aged subjects to increased perioperative mortality rate. If patients receive intensive care, the prognosis is better and the hospital stay reduced.

So far as pharmacology, the development of drugs pointed to the "treatment" of aging (31), i.e. to slow down the biological events associated with senescence and improve the quality of life, is limited by the lack of knowledge of the mechanisms underlying the aging process. However, the identification of neuronal systems which are a target for risk factors of accelerated aging may provide a rationale for the design of drugs useful in the treatment of pathological dysfunctions. The best example probably comes from the history of the therapy of PD, aimed to the correction of a specific neurotransmitter deficits. However, the evidence that this approach (i.e. use of inhibitors of the enzyme acetylcholine esterase, like tetrahydroaminoacridine) is of some efficacy in Alzheimer's disease is still controversial. On the other hand, therapies based on the substitution of a defective transmitter are limited by the fact that they do not correct the primary defect and that transmitter alterations are frequently multiple. This suggests the importance of a basic knowledge of the pathogenesis of age-related dysfunctions and the role of prevention against environmental and iatrogenic factors of accelerated aging.

REFERENCES

1. A.N. Davison (1987): Functional morphology of neurons during normal and pathological aging. In: Modification of cell to cell signals during normal and pathological aging. (S. Govoni and F. Battaini, eds.). NATO ASI Series, Springer-Verlag, Berlin, Heidelberg, Vol. H9, pp. 1—16.

2. J.W. Geddes, D.T. Monaghan and C.W. Cotman. Plasticity of hippocampal circuitry in Alzheimer's disease. Science 230: 1179—1181, 1985.

3. G.E. Gibson and C. Peterson (1987): Changes in calcium homeostasis during aging. In: Modification of cell to cell signals during normal and pathological aging. (S. Govoni and F. Battaini, eds.). NATO ASI Series, Springer-Verlag, Berlin, Heidelberg, Vol. H9, pp. 123—140.

4. S. Govoni, F.Battaini, M.S. Magnoni, R.A. Rius, L.Lucchi and M.Trabucchi (1987): Example of age related changes of transducing mechanisms and of cortical circuits. In: Modification of cell to cell signals during normal and pathological aging. (S. Govoni and F. Battaini, eds.). NATO ASI Series, Springer-Verlag, Berlin, Heidelberg, Vol. H9, pp. 111—122.

5. L.D. Hirning, A.P. Fox, E.W. McCleskey, B.M. Olivera, S.A. Thayer, R.J. Miller, R.W. Tsien. (1988): Dominant role of N-type calcium channels in evoked release of norepinephrine from sympathetic neurons. Science 239: 57—61.

6. C.E. Finch (1987): Neurotransmitters, genetics and aging. In: Modification of cell to cell signals during normal and pathological aging. (S. Govoni and F. Battaini, eds.). NATO ASI Series, Springer-Verlag, Berlin, Heidelberg, Vol. H9, pp. 63—82.

7. O. Hornykiewicz (1987): Neurotransmitter changes in human brain during aging . In: Modification of cell to cell signals during normal and pathological aging. (S. Govoni and F. Battaini, eds.). NATO ASI Series, Springer-Verlag, Berlin, Heidelberg, Vol. H9, pp. 169—182.

8. C. Missale, M. Memo, P.F. Spano (1987): Dopaminergic system in the aged brain: evidence for a selective loss of D1 but not D2 receptors. In: Modification of cell to cell signals during normal and pathological aging. (S. Govoni and F. Battaini, eds.). NATO ASI Series, Springer-Verlag, Berlin, Heidelberg, Vol. H9, pp. 153—167.

9. C. O'Neil, J. Marcusson, A. Nordberg, B. Winblad (1987): The influence of age on nerotransmitters in the human brain. In: Modification of cell to cell signals during normal and pathological aging. (S. Govoni and F. Battaini, eds.). NATO ASI Series, Springer-Verlag, Berlin, Heidelberg, Vol. H9, pp. 183—198.

10. G.S. Roth and J.A. Joseph (1987): Dietary and neuroendocrine modulation of age related changes in dopaminergic action during aging. In: Modification of cell to cell signals during normal and pathological aging. (S. Govoni and F. Battaini, eds.). NATO ASI Series, Springer-Verlag, Berlin, Heidelberg, Vol. H9, pp. 141—151.

11. H.K. Jiang, V. Owyang, J.S. Hong and M. Gallagher (1989): Elevated dynorphin in the hippocampal formation of aged rats: relation to cognitive impairment on a spatial learning task. Proc. Natl. Acad. Sci. 86: 2948—2951.

12. B. Hoskins, C.K. Burton, I.K. Ho (1986): Differences in morphine-induced antinociception and locomotor activity in mature adult and aged mice. Pharmacol. Biochem. and Behav. 25: 599—605.

13. J. Meites (1987): Neuroendocrine aspects of aging. In: Modification of cell to cell signals during normal and pathological aging. (S. Govoni and F. Battaini, eds.). NATO ASI Series, Springer-Verlag, Berlin, Heidelberg, Vol. H9, pp. 83—99.

14. N. Fabris, W. Pierpaoli, E. Sorkin (1972): Lymphocytes, hormones and aging. Nature: 240: 557—559.

15. N. Fabris and L. Piantanelli (1982): Thymus-neuroendocrine interactions during development and aging. In: Endocrine and neuroendocrine mechanisms of aging. (RC Adelman, GS Roth, eds.). CRC Press, Boca Raton, FL, pp. 167—184.

16. P. Bustany and M. Moulin (1987): PET study of human brain metabolism in aging and dementia. In: Modification of cell to cell signals during normal and pathological aging. (S. Govoni and F. Battaini, eds.). NATO ASI Series, Springer-Verlag, Berlin, Heidelberg, Vol. H9, pp. 27—38.

17. S.I. Rapoport, B. Horwitz, C.L. Grady, J.V. Haxby (1987): Alzheimer's disease causes metabolic uncoupling of associative brain regions beyond that seen in the healthy elderly. In: Modification of cell to cell signals during normal and pathological aging. (S. Govoni and F. Battaini, eds.). NATO ASI Series, Springer-Verlag, Berlin, Heidelberg, Vol. H9, pp. 47—61.

18. C. Fieschi, F. Orzi, G. Diana, F. Pontieri (1987): Anexperimental study nof the functional derangment which follows aging of the brain and dementia. In: Modification of cell to cell signals during normal and pathological aging. (S. Govoni and F. Battaini, eds.). NATO ASI Series, Springer-Verlag, Berlin, Heidelberg, Vol. H9, pp. 39—45.

19. M.G. Nunzi and G. Toffano (1987): The neuronotrophic hypothesis of the aging brain. In: Modification of cell to cell signals during normal and pathological aging. (S. Govoni and F. Battaini, eds.). NATO ASI Series, Springer-Verlag, Berlin, Heidelberg, Vol. H9, pp. 101—109.

20. L. Israel, D.J. Kozarevic, J.M. Orgogozo (1987): Subjective and objective health in aging: problems of assessment. In: Modification of cell to cell signals during normal and pathological aging. (S. Govoni and F. Battaini, eds.). NATO ASI Series, Springer-Verlag, Berlin, Heidelberg, Vol. H9, pp. 261—285.

21. P.C. Emson, H. Arai (1987): Functional morphology during normal and pathological ageing with emphasis on peptidergic transmission. In: Modification of cell to cell signals during normal and pathological aging. (S. Govoni and F. Battaini, eds.). NATO ASI Series, Springer-Verlag, Berlin, Heidelberg, Vol. H9, pp. 17—25.

22. L. Amaducci, A. Lippi, S. Sorbi (1987): Senile dementias: an overview. In: Modification of cell to cell signals during normal and pathological aging. (S. Govoni and F. Battaini, eds.). NATO ASI Series, Springer-Verlag, Berlin, Heidelberg, Vol. H9, pp. 199—211.

23. M. Roth (1987): Growing points of the neurobiology of Alzheimer's disease. In: Modification of cell to cell signals during normal and pathological aging. (S. Govoni and F. Battaini, eds.). NATO ASI Series, Springer-Verlag, Berlin, Heidelberg, Vol. H9, pp. 213—226.

24. C.M. Wischik and R.A. Crowther (1987): The Alzheimer tangle and aging. In: Modification of cell to cell signals during normal and pathological aging. (S. Govoni and F. Battaini, eds.). NATO ASI Series, Springer-Verlag, Berlin, Heidelberg, Vol. H9, pp. 227—234.

25. R.E. Tanzi, P.H. St George-Hyslop, J. Gusella (1989): Molecular genetic approach to Alzheimer's disease. TINS 12: 152—158.

26. Wolozin B., Scicutella A. and Davies P. (1988): Reexpression of a developmentally regulated antigen in Down syndrome and Alzheimer disease. Proc. Natl. Acad. Sci. USA 85: 6202—6206

27. A. Barbeau (1984): Etiology of Parkinson disease: a research strategy. Can. J. Neurol. Sci. 11: 24—28.

28. Langston J.W. and P.A.Ballard (1983): Parkinson's disease in a chemist working with l-methyl-4-phenyl-1, 2, 5, 6-tetrahydropyridine. New Engl .J. Med. 309: 310—313.

29. R.F. Peppard and D.B. Calne (1987): New perspectives in Parkinson's disease. In: Modification of cell to cell signals during normal and pathological aging. (S. Govoni and F. Battaini, eds.). NATO ASI Series, Springer-Verlag, Berlin, Heidelberg, Vol. H9, pp. 235—247.

30. F. Antonini, N. Marchionni, L. Ferrucci, M. Di Bari (1987): Guidelines of the care of the aged patients. In: Modification of cell to cell signals during normal and pathological aging. (S. Govoni and F. Battaini, eds.). NATO ASI Series, Springer-Verlag, Berlin, Heidelberg, Vol. H9, pp. 249—259.

31. M. Ermini (1987): Drug development in geriatrics. In: Modification of cell to cell signals during normal and pathological aging. (S. Govoni and F. Battaini, eds.). NATO ASI Series, Springer-Verlag, Berlin, Heidelberg, Vol. H9, pp. 287—292.

MOLECULAR ACTION OF STEROIDS AND GROWTH FACTORS IN HUMAN BREAST TUMOR CELLS

Bart van der Burg and Siegfried W. de Laat

Hubrecht Laboratory, Netherlands Institute for Developmental Biology,
Uppsalalaan 8, 3584 CT, The Netherlands

INTRODUCTION

The Advanced Research Workshop on "Cell to Cell Signals in Mammalian Development" was held in Les Arcs, France, February 21—26, 1988 (NATO ASI Series H: Cell Biology, Vol. H26, Eds. S.W. de Laat, J.G. Bluemink, C.L. Mummery: 1988). At this stimulating meeting discussions focused on the regulatory mechanisms by which cell growth, differentiation, and tissue patterning are controlled. Particular emphasis was given on the role of polypeptide growth factors and hormones as long range signalling molecules, extracellular matrix and cell adhesion molecules as local signalling molecules, and their potentially interconnected action with tissue patterning genes and gene products, such as the homeogenes.

A substantial part of our present knowledge on these cell to cell signalling mechanisms has been derived from studies on *in vitro* and *in vivo* models of carcinogenesis, in particular since oncogenic action appears to be closely related to signal transduction by growth factor receptors. Human breast tumor cells provide such a model and, for obvious reasons, have been studied in great detail. They are of particular interest in that their growth and state of differentation are regulated by the combined action of steroid hormones and polypeptide growth factors. For that reason they provide a suitable system to unravel the intricate molecular pathways of endocrine, paracrine and autocrine signalling. Inspired by our own recent work in this area we review in the following several aspects of current knowledge of the molecular action of steroids and growth factors in human breast tumor cells. The concepts arising from these studies have implications for our views on cell to cell signalling mechanisms in general.

CLINICAL BACKGROUND

Breast cancer is the most frequent malignancy of females in many western countriers (1). Estrogens play an important role not only in the development of breast tumors, but also in their proliferation at later stages (2, 3). Approximately one third of these tumors are estrogen (E2)-dependent, showing a regression following a hormonal therapy (e.g. ovariectomy or therapy with antiestrogens (4). This hormonal control of proliferation may be a remnant of the normal regulation of mammary epithelial growth (5, 6).

NATO ASI Series, Vol. H 51
Cell to Cell Signals in Plants and Animals
Edited by V. Neuhoff and J. Friend

As E2 action is mediated by binding to specific cellular receptors (reviewed in 7, 8), the detection of E2 receptors (ER) in tumors can be used to predict their response to hormonal therapy (reviewed in 9). Furthermore, the additional presence of progesterone receptors (PgR) is predictive for successful hormonal therapy (10, 11). The PgR is a protein induced by E2 and its presence is therefore considered to be a marker for the functionality of the ER (12). When tumors possess both receptors a regression following hormonal therapy occurs in 70—80% of the cases (11, 13).

Unfortunately, this regression is often only temporary and tumor regrowth occurs that is frequently ER^- (13, 14). There are several other indications that the hormone-dependent phenotype is lost upon progression of the disease. First of all, the absence of both steroid receptors is more often observed in poorly differentiated tumors (9). In addition, these receptor-negative tumors in general proliferate more rapidly (15), while abnormal DNA content is also more commonly observed (16). Moreover, metastases are more often ER^- (10, 13, 14) and PgR^- (13, 17) compared to primary tumors. This progression from a hormone-dependent to a hormone-independent state strongly hampers the use of hormonal therapy in breast cancer treatment.

THE ROLE OF E2 IN THE PROLIFERATION OF HUMAN BREAST CANCER CELLS

Various human breast cancer cell lines strictly depend on E2 for *in vivo* tumorigenesis in nude mice (18—21). Results of *in vitro* studies, however, have been contradictory. Although E2 was reported to stimulate the proliferation of most of these cell lines (22—26), the extent of stimulation strongly depends on the culture conditions used (24, 25, 27—29), and may even be absent (18, 30, 31). The absence of E2-induced proliferation observed in several cases may be caused by the the presence of E2 in the cultures due to retention by the cells (32) or by ineffective removal from the serum used in the culture media (33). Furthermore, it has been claimed that serum contains an inhibitor for proliferation whose activity can be abolished by E2, favoring the use of high serum concentrations (29, 34—37), sometimes of a special type (29, 34, 35) when measuring *in vitro* effects of E2. To date, however, no data on the nature of this inhibitor have been presented. On the other hand also evidence for the existence of serum factors permissive for E2-induced proliferation has been presented (27, 28, 38, 39), which may explain the failure to demonstrate effects of E2 in serum-free media (40—42). The most important observation in this respect, however, seems to be that the omission of phenol red, that has estrogenic activity, from the culture medium leads to a substantial reduction of background proliferation (34). But even then, hormone dependent cell lines like MCF7 continued to grow in the total absence of E2, contradicting their *in vivo* strict E2-dependence.

In order to demonstrate E2 effects on breast tumor cells generally culture media are used supplemented with serum treated with dextran-coated charcoal to remove endogenous steroids (DCC-FCS). These sera, however, still contain serum-derived polypeptide growth factors, causing background mitogenic activity. We recently showed that the polypeptide growth factors in DCC-FCS can be inactivated by a treatment that leads to cleavage of disulfide bridges (43, 44). As a result of this mitogen depletion the MCF7 cells become quiescent in the G_1/G_0 phase of the cell cycle, while they can be stimulated to enter the cell cycle by using defined mitogens (Fig. 1). In other words, under these

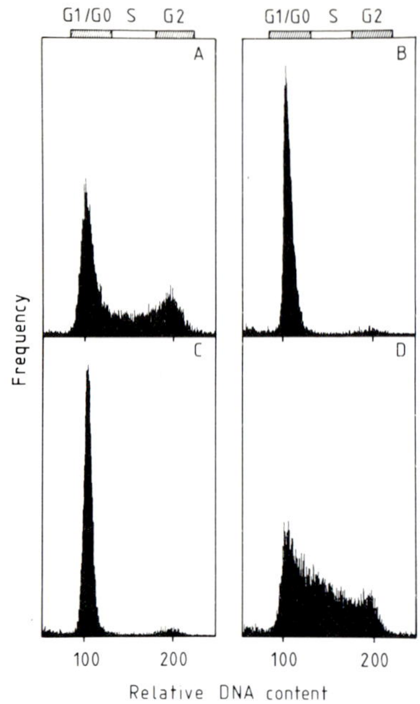

FIGURE 1
DNA histograms of MCF7 human breast cancer cells after growth stimulation in in medium supplemented with 10% serum treated to inactivate growth factors and steroids (DCC-SH-FCS) of quiescent MCF7 cells. Cells were made quiescent by a 24-hr incubation in serum-free medium. After mitogenic stimulation in 10% DCC-SH-FCS for the indicated periods of time, cells were harvested and analyzed by DNA flow cytometry. A: Control-exponentially growing cells. B: 0 hr. C: 30 hr, no additives. D: 30 hr, 10 μg/ml insulin, 100 ng/ml EGF, and 10^{-9} M E_2 (From ref. 44)

conditions MCF7 cells displayed their strict hormone dependent phenotype that has been established *in vivo*, under physiological conditions (18—21).

The ER functions as a transcription factor in the nucleus of target cells (reviewed in 8). It has been shown that E2-stimulation of breast cancer cells leads to the induction of a number of specific mRNAs (45—48) and proteins, among which the PgR (12), proteolytic enzymes (49, 50) and enzymes involved in DNA synthesis (46, 51, 52). It is unknown to what extent the induction of these proteins contribute to the mitogenic response of breast cancer cells to E2. In most of these studies, it has not been assessed whether these inductions are the direct result of a stimulation of transcription by the ER.

It has been hypothesized that estrogen-induced proliferation of breast cancer cells is mediated by stimulation of production and secretion of polypeptide growth factors that induce the proliferation of producing cell in an autocrine way (reviewed in 53). Externally added growth factors such as insulin may bypass E2 requirement for proliferation (41, 54), which may be an alternative explanation for the lack of estrogen response in serum-free media supplemented with polypeptide growth factors, and in serum containing media due to the serum-derived growth factors.

THE RÔLE OF POLYPEPTIDE GROWTH FACTORS IN THE PROLIFERATION OF BREAST CANCER CELLS

Polypeptide growth factors are important regulators of cellular proliferation (reviewed in 55, 56). Growth stimulation by these factors is mediated by specific receptors in the plasma membrane of target cells. While normal cells depend on external growth factors for proliferation, transformed cells generally have a strongly reduced growth factor requirement. An important mechanism leading to this autonomous growth behaviour may be the secretion of growth factors by tumor cells that stimulate the producing cell in an autocrine fashion (57). Since some polypeptide growth factors were found to be potent growth inhibitors for certain cells, the negative autocrine hypothesis was postulated which states that uncontrolled growth in addition may be caused by a diminished sensitivity of cells to these growth inhibitory factors (58). The evidence for a role of growth factors in the proliferation of human breast cancer cells will be discussed.

Estromedins

In 1978 Sirbasku (59) observed that E2-treatment of animals resulted in the induction of specific polypeptide growth factors in uterus, kidney and pituitary, which were growth stimulatory *in vitro* for E2-sensitive cells like rat mammary cancer cells. The existence of these potentially endocrine acting growth factors offered an explanation for the lack of direct growth stimulatory effects *in vitro* by E2 on these cells. Further purification of these factors resulted in the isolation of a number of apparently related factors that were different from all known growth factors in being resistant to reducing agents, much smaller and heat stable (reviewed in 60, 61). These factors were strongly mitogenic for rat and human mammary cancer cells in serum- and protein-free medium, much more so than known polypeptide growth factors. The estromedins were also mitogenic for the producer cells, and were found in extracts of rat mammary tumor cells. Since growth stimulating activity was in addition secreted by E2-dependent human breast cancer cell lines when assayed under serum-free conditions, a role for these estromedins in autocrine stimulation of mammary cancer was suggested (61). These results, however, have to be re-evaluated since the estromedin extracted from the pituitary cells was identified as being transferrin, an essential medium component for serum-free culturing of most cells, and also for the rat mammary cells used in the bioassay to detect the estromedins (62). Using more physiological culture conditions in the bioassay, some polypeptide growth factors such as insulin-like growth factor-I (IGF-I), IGF-II (63) and transforming growth factor α (TGFα; 64) were also found in pituitary extracts, possibly having a function in endocrine stimulation of breast cancer cells. The clear direct mitogenic action of E2 on hormone-dependent breast cancer cell lines that is observed when phenol-red is omitted from the culture medium (34), together with proof for a direct stimulatory action of E2 on these cells *in vivo* (65), have diminished the interest in E2-inducible endocrine acting factors.

52K and pS2 protein

Stimulation with E2 of hormone-dependent breast cancer cell lines leads to the induction of a number of proteins. The most prominent and therefore most extensively studied proteins are a 52K (66, 67) and a 6.5K (68, 69), also called pS2 protein. The expression of both proteins is transcriptionally regulated by E2 in the MCF7 breast cell line (70, 71). The pS2 protein is expressed only in a subset of E2-dependent breast tumors (72) and curiously enough also in stomach mucosa cells (73). It is closely related to a porcine pancreatic enzyme that inhibits gastrointestinal motility and gastric acid secretion (74, 75). The 52K protein has a broader tissue distribution (76, 77). Both proteins may be used as additional markers to identify subsets of E2-dependent breast cancer.

The 52K protein has been identified as the major growth promoting activity for MCF7 cells in conditioned medium of these cells, suggesting that it is an autocrine growth factor (78). This protein, however, was recently identified as the proteolytic enzyme cathepsin D (50, 79), and its apparent mitogenic activity may have been caused by its proteolytic activity. Since the mitogenic activity was assessed in serum-containing medium it may have been caused by an activation of a serum-derived growth factor or an inactivation of a growth inhibitor. Although probably not mitogenic by themselves, E2-induced production of proteolytic enzymes may be important to regulate the activity of growth factors from both autocrine and external sources.

Some structural similarities that the pS2 protein shares with IGFs, which are potent growth factors for MCF7 cells (see below), has led to the suggestion that it might be an E2-regulated autocrine growth factor (80). Recently, however, it was shown that purified pS2 protein is not mitogenic for MCF7 cells (81), while monoclonal antibodies against pS2 (72) do not inhibit E2-induced proliferation of MCF7 cells (B. van der Burg *et al.*, unpublished observation).

Insulin-like growth factors

Insulin-like growth factor I (IGF-I) and IGF-II are closely related in terms of protein structure, sequence, and biological activity (reviewed in 82, 83). They resemble the monomeric precursor form of insulin. Most cells express high affinity receptors for IGFs and insulin on their cell surface. These receptors preferentially bind one of the ligands, but they also bind the others with lower affinity, except for IGF-II and insulin that do not react with each others receptors. IGFs in blood are bound to various high affinity binding proteins (reviewed in 84). The total blood level of IGFs in adults is approximately 800 ng/ml, which is about 1000 times that of insulin. Therefore it has been proposed that the function of these binding proteins is the prevention of insulin-like activities by this large excess of IGFs (84). In addition, these binding proteins may modulate IGF activity, either positively or negatively, and they are produced by various cells (84).

In vitro studies have shown mitogenic activity of IGF-I, IGF-II (85, 86) and insulin (87, 88) on human breast cancer cell lines, and the latter factor is essential for growth in serum-free media of these cells (23, 89). The relative potency of these factors, with IGF-I being the strongest mitogen, suggests that the mitogenicity by all three is mediated through the IGF-I receptor, although receptors for the other factors are also present on breast cancer cell lines (87, 88).

Production of an immunoreactive IGF-I-related protein by human breast cancer cell lines has been measured in conditioned medium using radio immuno assays (90). Furthermore, expression of aberrantly sized (compared to liver transcripts) IGF-I mRNAs was detected (91), while a few lines express IGF-II mRNA (92). The production of this IGF-I related factor, but not its putative mRNA (91), was stimulated by E2 in E2-dependent MCF7 cells, while autonomously growing lines constitutively produced this factor (90), suggesting that it might be an important autocrine factor (90). The exact nature of this IGF-I-related immunoreactive factor has not been established yet. The biological activity of this factor may be regulated by IGF-binding proteins that were produced by all breast cancer lines tested (93).

Evidence for the *in vivo* relevance of this autocrine hypothesis comes from experiments in which it was shown that either concentrated conditioned medium, IGF-I (or EGF) is able to replace E2, although only temporarily, in inducing MCF7 tumors in E2-depleted nude mice (94). In addition, both immunoreactive IGF-I (95) and its IGF-I receptor (95, 96) were detected in human breast tumors. Since whole tumor homogenates were used, these studies do not prove that these proteins are produced by the tumor cells. In fact, stromal cells that normally are present in large quantities in breast tumors express IGF-I receptors and produce IGF-I (97).

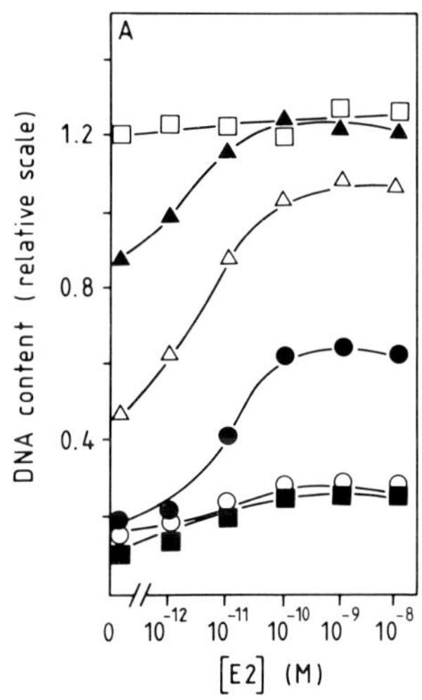

FIGURE 2

Synergism between E_2 and insulin in stimulating proliferation of MCF7 cells. Mitogens were added at the following concentrations: insulin, 10 μg/ml ( ); 1 μg/ml ( ); 100 ng/ml ( ); 10 ng/ml ( ); 1 ng/ml ( ); 0 ng/ml ( ); Each point represents the mean of three determinations after a 4-day incubation period. DNA content is expressed relative to DNA measured in cultures incubated with 10% FCS-supplemented medium. S.e.m. were less than 5%. (From ref. 44)

Recently, it has been shown that E2 alone is poorly mitogenic, but it synergistically stimulates proliferation together with low concentrations of insulin (Fig. 2) and IGFs (44). The latter factors are also present in DCC-treated fetal calf serum, explaining the strong mitogenicity of E2 (or phenol red) in the presence of DCC-FCS (44). We propose that this synergistic stimulation of MCF7 proliferation may reflect the *in vivo* situation in the nude mice model, in which an important role for both insulin and estrogen in MCF7 tumorigenesis has been established. MCF7 tumorigenesis and the capacity to metastasize can be prevented when nude mice are either ovariectomized or made diabetic (101, 102). In the diabetic mice daily injection of insulin completely restores the malignant behaviour of MCF7. More recently, synergism between E2 and IGFs has been established in hypophysectomized nude mice (100). The removal of the pituarity can be used as an effective form of hormonal therapy in patients. It will lower blood levels of E2, but also of various other hormones, among which IGF-I and prolactin. Although rodent mammary tumors are often prolactin-dependent, the role of prolactin in proliferation of both normal as well as neoplastic human mammary epithelium is uncertain (101, 102 and references therein). For MCF7 tumors in nude mice it was shown that the effect of hypophysectomy can be completely restored by a combination of both IGF-I and E2, synergistically inducing MCF7 tumorigenesis and proliferation (100). This implies that beside E2, endocrine acting IGFs may be of importance in stimulating hormone-dependent breast cancer proliferation.

Transforming growth factor α and epidermal growth factor

Although the homology in protein sequence of transforming growth factor α (TGFα) and epidermal growth factor (EGF) is only 30%, both factors show a nearly identical biological activity (reviewed in 106). Only one receptor for these factors is known, designated as the EGF receptor. The similarity of the three dimensional structure of these peptides, due to three conserved disulfide bridges, is likely the cause of their comparable binding capacity to the EGF-R.

A number of studies have shown a mitogenic effect of EGF or TGFα on some human breast cancer cell lines (104—108). E2-independent human breast cancer cell lines were not stimulated by these factors (104—106), while their effect on E2-dependent lines is usually relatively small compared to stimulation by serum or insulin, and can even be undetectable (89, 104—106). In contrast, EGF receptors seem to be present on almost all human breast cancer cell lines (106, 107, 109), and 30—70% of breast tumors in patients (110—112). Various studies have found an inverse relationship between the presence of high levels of EGF-R and the presence of ER in mammary tumors (110—112), and in cell lines (107). In a minority of the E2-independent cell lines expressing high numbers of EGF-R, amplification of the EGF-R gene has been detected (107, 113, 114).

Controversially, the E2-independent MDA468 cell line, expressing 1.9×10^6 EGF-R/cell, is growth inhibited by EGF *in vitro* (113), a situation reminiscent of the growth inhibition by EGF of another EGF-R overexpressing cell line, the vulvary epithelial cell line A431 (115). Only variants of both cell lines expressing lower receptor numbers can be stimulated *in vitro* by EGF (115, 116). The high levels of EGF-R, however, seem to be benificial *in vivo* since the growth rate of tumors in nude mice is higher in the variants expressing large amounts of receptors (116, 117). Interestingly, in mammary tumors in patients, expression of elevated levels of EGF-R also correlates with a more malignant character (112, 118) and a higher proliferation rate (119, 120). It remains to be proven that the faster proliferation *in vivo* is in fact due to stimulation by EGF or TGFα. Since EGF-R expression in breast cancer cells can be influenced by a large variety of agents, like progestins (121, 122), phorbol esters (123, 124), EGF self (123, 125), vitamin D3 (126), triiodothyronine (125), and E2 (127), the high level of EGF-R in tumors may reflect an altered response to one or more of these compounds.

One paper reports the production of EGF by MCF7 cells (128), but on subsequent purification this protein turned out to be the pS2 protein (129). ZR75-1 and T47D, two other E2-dependent breast cancer cell lines, however, both express EGF mRNA (130). In addition EGF is found in human milk (131—135), especially during lactation when epithelial cells proliferate rapidly (131, 132, 134). A role for this milk-derived EGF in human mammary epithelial proliferation has not been demonstrated, and it may in fact be more important to support intestinal growth and development shortly after birth of the offspring, as has been shown in rats (136).

On the other hand there is ample evidence for secretion of TGFα related factors (137, 138) and expression of TGFα mRNA (92, 139) by most breast cancer cell lines. Partial purification of these factors from MCF7 conditioned medium resulted in the isolation of multiple 30K forms closely related to TGFα as assessed by immunoreactivity, biological activity and EGF-R competition. Since TGFα mRNA is expressed by MCF7 cells these factors seem to be high molecular weight forms of TGFα (140). The production of this factor is 2- to 8-fold increased by E2 in estrogen-dependent lines (137, 138, 140). In MCF7 cells it was demonstrated that E2 increases TGFα mRNA levels (139). TGFα mRNA is found in around 50—70% of the homogenates of human breast tumors (139, 141). In one study its presence correlated with hormone independence (141), while others did not find such a correlation (139).

Recently, a new member of the EGF family has been isolated from TPA-treated MCF7 cells (142). This factor, called amphiregulin, has a spectrum of biological activities reminiscent of that of EGF or TGFα; it stimulates anchorage-independent proliferation of NRK cells in the presence of TGFβ, it stimulates murine keratinocytes and inhibits A431 cell proliferation (143). The relative potency of this factor differs from those of TGFα and EGF in most systems, which is in line with the observation that it possesses its own high affinity receptor, and binds more weakly to the EGF-R (143, 144). Both the gene coding for amphiregulin and its receptor have been cloned (144). Amphiregulin does not influence MCF7 proliferation (143).

Transforming growth factor β

Transforming growth factor β (TGFβ) was first identified as a factor that is able to induce, in combination with TGFα, a transformed phenotype in fibroblasts. Besides this activity, TGFβ is able to inhibit the proliferation of a large variety of cells, in particular normal and various transformed epithelial cell lines (reviewed in 145, 146). An important feature of TGFβ activity *in vivo* may be its angiogenic properties (150), since neovascularization is thought to be an important event in the development of tumors. TGFβ is produced by most cultured cells, but generally in a latent form. Little is known about the way latent TGFβ is activated under physiological conditions, but the activity of proteases like cathepsin D and plasmin could be involved. Three types of membrane receptors for TGFβ are present on virtually all cell lines (148). One of those, the type I receptor, appears to be essential for TGFβ action (149). Various closely related forms of TGFβ exist but in general these appear to have similar biological effects on cells.

Like most cells breast cancer cell lines express TGFβ mRNA (150, 151) and secrete TGFβ (150). Knabbe *et al.* (150) reported that a small fraction of TGFβ in conditioned medium of phenol red-stimulated MCF7 cells is in an active form. Upon treatment with growth inhibiting antiestrogens both the total amount of TGFβ and the part in an active form was elevated, suggesting that it might be an E2-regulated negative autocrine growth factor. These effects occurred in the absence of changes in mRNA levels (150). Other investigators have questioned the role of TGFβ in E2-induced proliferation, since they found no inhibition of proliferation of hormone-dependent breast cancer cell lines by TGFβ, although antiestrogens were inhibitory (152). In contrast, E2-independent lines were highly TGFβ sensitive and possessed specific high affinity TGFβ receptors, while no such receptors were detectable on the E2-dependent cells (152).

TGFβ mRNA is detectable in all breast tumors (141, 151). In an early study it was shown that a growth inhibitor produced by BSC-1 cells, found to be identical to TGFβ (153), inhibited proliferation when injected into tumors formed by a breast cancer cell line in nude mice (154).

Platelet-derived growth factor

Platelet-derived growth factor (PDGF), purified from human platelets consists of two homologous polypeptide chains encoded by different genes; one gene codes for the A-chain of PDGF, while a gene called c-*sis* codes for the B-chain of PDGF (reviewed in 155, 156). Homodimeric combinations of PDGF are also naturally occurring. Two receptors for PDGF exist which are predominantly found on cells from mesodermal origin such as fibroblasts and glia cells.

Breast cancer cell lines have been found to secrete PDGF (157, 158) and to express mRNA coding for PDGF-A and PDGF-B (92, 158). In the MCF7 cell line all these activities were E2-stimulated, while E2-independent lines produce large amounts of PDGF (158). Transcripts for both chains of PDGF were in addition found in all human breast tumors (141, 159), which in the case of *sis* was found to be due to expression by the epithelial cells (159). Since, as expected, breast cancer cells do not express receptors for these growth factors, production of this factor cannot lead to autocrine stimulation of these cells. A role for this factor in paracrine stimulation of fibroblastic cells that often are present in large quantities in breast tumors has been suggested instead (157, 158).

Fibroblast growth factor

Basic (bFGF) and acidic fibroblast growth factor (aFGF) are both members of the heterogeneous family of heparin-binding growth factors (reviewed in 160, 161). FGFs are strong stimulators of endothelial cell proliferation and are angiogenic *in vivo*. These factors lack signal peptides that direct secretion, and it has been postulated that they are mainly released after tissue damage. A number of genes such as *int*-2 and *hst* have been identified that code for putative other members of this family, probably coding for FGF-like factors that are secreted.

Both basic and acidic FGF have been reported to be mitogenic for the E2-dependent breast cancer cell lines T47D and MCF7, with bFGF being the most potent of the two (108). No reports on secretion of FGFs by breast cancer cells have been published so far, but expression of the *int*-2 gene has been detected in 18 out of 100 primary breast tumors (162). This expression correlated with recurrence and distal metastases. Although the *int*-2 gene has found to be amplified in around 20% of human breast tumors, no protein expression has been found (165). Interestingly, together with amplification of the *int*-2 locus, the nearby *hst* locus has been found to be co-amplified (164). Possibly the latter amplification is more relevant to human mammary tumorigenesis. Although a role of these factors in autocrine stimulation cannot be excluded, a possible role in the important process of neovascularization of tumors seems more relevant.

THE ROLE OF ONCOGENES IN THE PROLIFERATION OF BREAST CANCER CELLS

Oncogenes are genes that can induce transformation when transferred to cultured cells (reviewed in 156, 165—167). Such genes were detected in DNA extracted from tumor cells but were also found to be present in the genomes of acutely transforming tumor viruses. For many of these oncogenes normal cellular counterparts have been found, called proto-oncogenes. These proto-oncogenes can become oncogenic by overexpression or by specific mutations that alter their activity. Transformation of cells is thought to be a multistep process, and in most instances the cooperation of at least two oncogenes has been found necessary to lead to full transformation of normal primary cells, although the immortal cell lines often used to identify these oncogenes may become fully transformed by introducing a single oncogene (reviewed in 167). Some of the oncogenes incorporated in the viral genomes are exceptionally potent, and when highly expressed appear to be able to transform cells in a single step. The fact that the cell did not get rid of these potentially cancer causing proto-oncogene indicates that they have an important function in the cell. Interestingly, the function of many proto-oncogene products seem to be linked to the mitogenic pathway elicited by polypeptide growth factors. One proto-oncogene encodes for a growth factor (*sis*) , some others for growth factor receptors (e.g. *erb*B, *fms*). The *ras* proto-oncogene codes for a so-called G-protein that is activated by growth factor receptors, probably leading to activation of an important enzyme (phospolipase C; 168) that is located in the plasma membrane and that cleaves phosphatidyl inositol 4, 5-biphosphate, generating two second messengers that are involved in signal transduction (reviewed in 169, 170). Nuclear proto-oncogenes like c-*fos*, c-*jun* and c-*myc* are transcription factors that are induced shortly after stimulation of cells with growth factors. They are thought to be involved in the induction of changes in expression of many genes, ultimately leading to DNA synthesis and cellular proliferation (reviewed in 171, 172). Therefore, not only secretion of autocrine growth factors may be a possible mechanism by which tumor cells escape growth control, the deranged expression of intracellular oncogene products may have the same effect.

The *neu* oncogene is one of the examples that demonstrate a close linkage between oncogene products and growth factor signalling pathways. The product of the *neu* gene shows a strong similarity with the EGF receptor (173), its mRNA and protein is overexpressed in approximately 25% of human breast tumors (ref. 174 and references therein). This overexpression is associated with gene amplification, and correlates with poor prognosis. No correlation with steroid receptor expression has been found (174), suggesting that it is involved in a step important in determining the malignant nature of both estrogen-dependent and independent tumors.

A number of oncogenes called *int* genes were found to be involved in the mouse mammary tumor virus (MMTV)-induced mammary tumors of mice (reviewed in 175). These cellular genes are transcriptionally inactive in normal adult tissue but were found activated by proviral insertion in a majority of MMTV induced tumors. *Int*-1 and *int*-2 are the *int* genes studied most so far, but there is no evidence for involvement in human mammary tumorigenesis, since expression of the proteins encoded by these genes has not been detected in human breast tumors.

The *ras* family of oncogenes contains three extremely homologous genes: N-*ras*, Ki-*ras* and Ha-*ras* (reviewed in 176). Only the latter, the Ha-*ras* gene has been implicated in breast cancer (reviewed in 177). Elevated levels of *ras* and the *ras* encoded p21 protein were detected in approximately two-thirds of the breast tumors examined, although no evidence for gene amplification or rearrangements have been found. The *ras* genes in these human tumors seemed normal in that none of the specific point mutations that are involved in activation of these genes in certain types of chemically induced rat tumors (178) were detected (177). However, indications have been found that rare alleles of the Ha-*ras* gene are more frequently found in patients compared to normal controls (179, 180). In patients heterozygous for this gene a loss of one allele correlates with aggressiveness and hormone independence (180, 181). The human breast cancer cell line MCF7 loses its E2-dependence upon transfection with a mutated form of Ha-*ras*, the viral v-Ha-*ras*. In these cells this activated oncogene is overexpressed (182). Other investigators have found that when normal c-Ha-*ras* or activated c-Ha-*ras* was constitutively expressed to normal levels in MCF7 only the cells receiving the activated *ras* lost their hormonal dependence (183). In contrast to the results with the overexpressed v-Ha-*ras* which became totally E2-independent (182), these cells only lost their E2-dependence *in vitro* but not *in vivo* (183). These results suggest that the qualitative differences, e.g. in the state of activation, of various forms of *ras* may be important in their potential of altering the hormonal dependence of breast cancer cells, together with their level of expression.

In recent years it has become apparent that the receptors for retinoic acid, vitamin D3, thyroid and steroid hormones are all members of a superfamily of nuclear receptors (reviewed in 8). These receptors can bind to DNA at specific target sequences, termed hormone responsive elements. When activated, they directly influence the transcription of target genes. They are composed of a number of distinct domains that are required for their activity. The most obvious are the ligand binding, the DNA binding and one or more transcriptional activation domains. It has been postulated that the genes coding for steroid receptors may be proto-oncogenes that can be mutated in a way that constitutive active oncogenic hormone receptors are generated (184). Such activation in breast cancer cells may lead to constitutive production of E2-inducible growth factors. For certain steroid receptors it has been found that deletion of the hormone binding domain leads to such a hormone-independent activating receptor. Interestingly, the v-*erb* A oncogene was found to code for such a truncated form of a nuclear receptor, the thyroid hormone receptor. So far, there is little evidence that such a mechanism is important in the loss of E2-dependence of breast cancer cells. Deletion of the hormone binding domain of the estrogen receptor, or parts of it, resulted in a constitutively, but very poorly active receptor (185). Recently, the existence of variant sized mRNAs that may code for such truncated receptors has been reported in a small subset of breast tumors that also expressed wildtype receptors (186), although other investigators did not find such variant mRNAs (72). Since mostly a good correlation is found between the presence of immunoreactive receptors and hormone binding (187—189), the presence of non-hormone binding, constitutively active E2 receptors seems not to be a common phenomenon in breast tumors, and does not give an explanation for the loss of E2-dependence in ER^- tumors. However, if these variant mRNAs code for constitutively active receptors, this may contribute to the lack of response to a hormonal therapy which is observed in a subset of ER^+ breast tumors.

Nuclear proto-oncogenes like c-*fos* and c-*myc* are a primary target of the signal transduction that leads from growth factor receptors to the nucleus (reviewed in 171, 172). The products of these genes are among the first to be induced after mitogenic stimulation and are involved in transforming the incoming signal into a change in gene expression. The expression of these genes is, however, rapidly downregulated. Constitutive expression of such nuclear oncogenes may lead to cellular transformation; deletions in the *fos* oncogene that prevent downregulation, as found in v-*fos*, are important for its oncogenic potential (190).

Most studies report elevated expression of c-*myc* mRNA (191—195) and protein (196) in the majority of human breast tumors. The expression was mainly elevated in the epithelial cells (196), and the level of expression has been found to correlate with malignancy (192—195). In about 10% of the tumors with high expression levels gene amplification was found (195). Hormone-independent breast cancer cell lines constitutively express c-*myc* mRNA, but this is not caused by gene amplification, nor rearrangement (196). These results, although interesting, should not be overinterpretated; in an asynchronously proliferating population of cells a back ground level of *myc* will always be detected, and this level generally will rise as the cells proliferate more rapidly.

The products of both the c-*myc* (197) and c-*fos* (198—204) do not directly bind to DNA but they interact with other nuclear transcription factors. Recently it has been shown that the *fos* product (Fos) can form heterodimers with Jun, a major component of the transcription factor AP-1 (198—203). This interaction results in a greatly enhanced affinity for the AP-1 target sequence, the TPA-responsive element (TRE), and a stimulation of TRE-dependent transcriptional activity (198, 200, 203). The gene coding for Jun (*jun*) is also growth factor inducible (200, 205, 206). All these nuclear oncogenes are members of gene families with closely related genes (207—212). These genes may exert similar functions, but it is also conceivable that they may differ e.g. in their ability to form complexes with other nuclear factors, directing the regulation of specific subsets of genes. It has been shown recently in PC12 cells that in situations where they are either induced to proliferate or to differentiate, different combinations of these nuclear proto-oncogenes are induced (213). This led to the suggestion that the diversity and specificity of the cellular response to external stimuli is directed by the induction of distinct combinations of a relatively small number of early response genes (213).

The multistep process of carcinogenesis seems also to involve the loss of certain gene functions (reviewed in 167, 214). The putative genes involved are called anti-oncogenes or tumor suppressor genes. Just like oncogenes were not specially designed to cause transformation and have an essential function in the cell, these anti-oncogenes are also expected to have a role in the normal functioning of cells. Recently, p105-Rb, the product of the *Rb* gene, one of the few putative anti-oncogenes that have been identified so far, was shown to be able to bind to E1A (215). E1A is a protein of adenovirus that is involved in regulating transcription (216) after viral infection, and which appears to be a functional homologue of c-*myc*, directing expression of genes using the same target sequence (197). Mutations that inactivate the transforming activity of E1A also prevented p105-Rb binding, suggesting that the *Rb* gene product inactivates E1A (215). It has not been shown that p105-Rb binds to c-*myc*. Nevertheless, the above examples illustrute that anti-oncogenes may function as transcription factors that are either able to inhibit the function of oncogene products or alternatively are able to stimulate expression of genes that will lead to growth inhibition, e.g. by inducing the cell to differentiate. Evidence for the absence of *Rb* function has recently been found in some human breast carcinomas and cell lines (217, 218).

MECHANISM OF GROWTH STIMULATION OF HORMONE-DEPENDENT BREAST CANCER CELLS BY E2

The role of polypeptide growth factors

Early studies have demonstrated that highly concentrated conditioned media of the human breast cancer cell line MCF7 can stimulate proliferation of the producer cell (137, 219). This activity was elevated by E2, and subsequent studies have shown the E2-induced secretion of a variety of polypeptide growth factors-like activities in these and other breast cancer cell lines (see above). Among those factors, the most important candidates to serve an autocrine function seem to be a TGFα-like and an IGF-I immunoreactive protein that were found to be produced by breast cancer cell lines, since IGF-I and TGFα have been reported to stimulate these E2-dependent cell lines in the absence of E2. The hypothesis that E2-induced proliferation is mediated by autocrine acting polypeptide growth factors (53) therefore seems very attractive. In addition, the constitutive production of these growth factors by E2-independent breast cancer cell lines gives an explanation for their loss of E2-dependence (53).

An interesting set of experiments provides additional indications for this autocrine hypothesis to be correct. In these experiments it was shown that transfection of a viral Ha-*ras* oncogene into MCF7 cells leads to an increase in secreted growth factor activity (220) and abrogates E2-dependence for tumorigenicity (182). These transfected cells, injected at a distant site in nude mice, are able to support tumorigenicity of wildtype MCF7 cells in the absence of E2 (221). In addition, coculture experiments showed that the presence of MCF7-*ras* stimulated anchorage-independent growth of the wildtype cells (221). If breast cancer cells indeed depend on autocrine mechanisms to proliferate, this could be used in a clinical setting by inhibiting proliferation of both E2-dependent and independent breast tumors through interference with these autocrine mechanisms.

Caution, however, should be exerted in interpreting these results. The factor that was responsible for the mitogenic activity in the studies that were the first to demonstrate that E2-dependent MCF7 cells secrete autostimulatory activity (218, 219) has been identified to be the proteolytic enzyme cathepsin D (50, 79). Does this mean that the polypeptide growth factors that are produced by MCF7 cells are irrelevant? This cannot be concluded on the basis of these experiments. The assays to detect growth factor production by the cells do not address the question whether these factors are secreted in a biologically active form, that enables stimulation of the producer cell. Especially the activity of IGFs may be greatly diminished by binding to specific binding proteins (84). The studies to assess the secretion of biological active growth factors by MCF7 cells were carried out in the presence of externally added (serum-derived) growth factors. The activation of growth stimulating factors or inactivation of inhibitors, either serum-derived or produced in an autocrine fashion, may therefore explain the mitogenic activity of the proteolytic 52K protein. However, the serum-derived growth factors will also obscure the detection of externally added growth factors. Therefore, to determine the activity of growth factors (either purified, in conditioned media or in coculture experiments) defined culture conditions should be used without the presence of active external growth factors.

Another point that has to be established to prove the existence of an autocrine loop is that interference of this cycle e.g. by neutralizing the autocrine factor with specific antibodies, inhibits the proliferation of the cells. In other cells, stimulation by an induced autocrine mechanism has been shown to cause a delayed mitogenesis compared to direct stimulation with the autocrine mitogen (222, 223). When E2 stimulates breast cancer cells with such delayed kinetics, this would be indicative for an indirect mechanism of growth stimulation.

A role of EGF and TGFα in autocrine stimulation is doubtful, since constitutive production of TGFα by MCF7 cells after transfection of the gene for this factor into the cells, leaves their E2-dependence *in vivo* and *in vitro* completely unaltered (224). Furthermore, expression of TGFα does not at all correlate with proliferation of hormone dependent breast cancer cells (225) and tumorigenesis of transformed mammary epithelial cells (226). Although a partial and temporary inhibition has been reported of E2-induced proliferation by anibodies against TGFα or its receptor (140), others were unable to demonstrate any effect of such antibodies (227). These results show that TGFα and EGF are poor mitogens for most breast carcinoma cells not only *in vitro* but also *in vivo*. Possibly, the stimulatory effect on MCF7 tumorigenesis in nude mice of pharmacological dosages of externally added EGF (94) is not the result of a direct stimulatory effect on MCF7 cells.

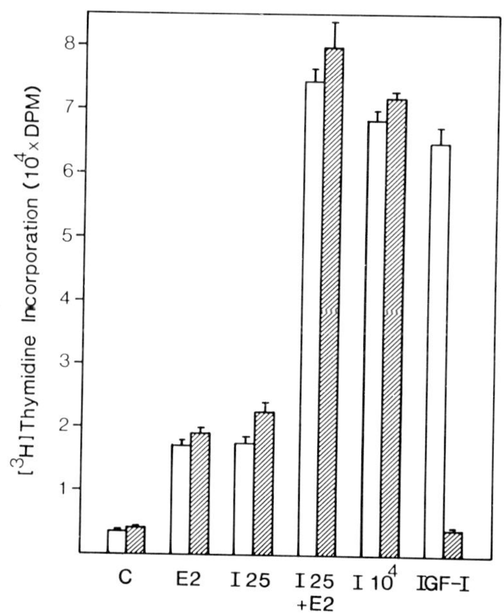

FIGURE 3

Effect of defined mitogens on the [QH]thymidine incorporation of MCF7 cells, both in the presence (hatched bars) and absence (open bars) of high concentrations of IGF-binding proteins. Mitogens were added at the following concentrations: E2, 10^{-9}M; insulin 25 ng/ml (I 25); insulin 10 μg/ml (I 10R); IGF-I 5 ng/ml (IGF-I); controls, vehicle only (C). All points represent the mean of three determinations (± s.e.m.). (From ref. 229)

IGFs, being highly mitogenic for MCF7 cells, appear to be more interesting candidates for an autocrine factor. However, the strong mitogenic effects of low concentrations of insulin or IGFs on E2 stimulated cells (Fig. 2) suggest that no significant production of biologically active IGFs is induced by E2. Moreover, E2 stimulates DNA synthesis without a delay-period compared to insulin, also suggesting that no autocrine loop is involved in E2-induced mitogenesis (228). In addition, as was shown recently (229), IGF-binding proteins that efficiently block IGF-induced mitogenesis do not inhibit E2-induced mitogenesis (Fig. 3). Moreover, we found no IGF-like biological activity in conditioned medium of E2-treated MCF7 cells (229). This is in line with results of Karey and Sirbasku (108), who measured IGF-I after column-fractioning of MCF7 conditioned medium and found levels that were far to low to cause growth stimulation of MCF7 cells, and were independent of E2-treatment. Only one study has demonstrated inhibition of both E2-induced and basal proliferation of MCF7 cultured in serum containing medium with use of antibodies against IGF-I (230). This inhibition was probably caused by neutralization of serum-derived IGF, since these results could not be repeated under serum-free conditions (231).

These data suggest that secreted autocrine growth factors are not the main mediators of E2-induced mitogenesis. The kinetic studies mentioned above make the possible existence of either an external or an internal autocrine loop unlikely. Moreover, induction of c-*jun*, characteristic of IGF-induced proliferation in these cells (232) does not occur after stimulation with E2 (Fig. 4). Therefore we conclude that autocrine mechanisms seem not to play an important role in E2-induced proliferation of MCF7 cells.

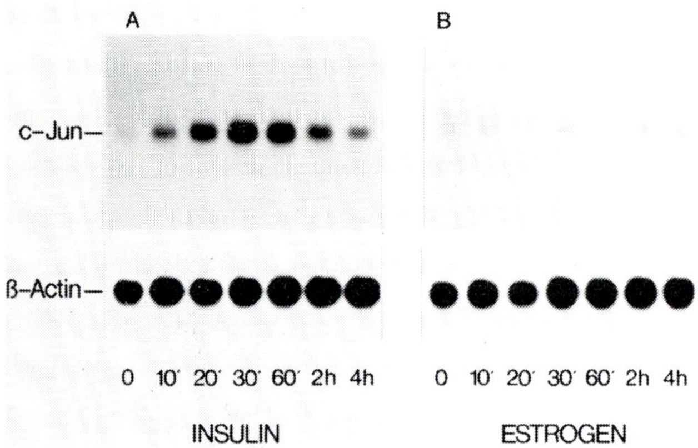

FIGURE 4
Time-course of c-*jun* expression after stimulation of quiescent MCF7 cells with insulin (10R ng/ml; A) or E2 (10^{-9}M; B). Northern blot analysis of total RNA (20 μg/lane). (From ref. 232)

MCF7 cells are dependent on external IGFs for their proliferation (see above). Possibly the large amount of stromal cells often present in breast tumors may be an important source of IGFs, since in human fibroblasts PDGF-stimulated IGF-I production has been observed (97). The production of growth factors for stromal cells like PDGF (92, 157, 158) and TGFα (90, 91, 138—140, 220) by the epithelial might lead to a mutual dependence of both cell types in the tumor. Since E2 stimulates production of both PDGF (158) and TGFα (91, 138, 140, 220) by breast cancer cells, it might be the major endocrine factor that is necessary to keep this process of mutual paracrine stimulation going. This process has been termed an "induced autocrine mechanism" since the induced production of IGF-I by the fibroblasts may be essential to keep the tumor cells in a rapidly proliferating state (229). In addition, endocrine acting IGFs may also be important in stimulating proliferation of MCF7 cells (see above). The reaction to external growth factors of the breast cancer cells may be elevated by a decreased production of inhibitory binding proteins and by elevated production of proteolytic enzymes that are able to activate latent growth factors. The relative importance of endocrine and paracrine processes in stimulating breast tumor cells may vary, and may in part be directed by the environment of the cells. For instance, the MCF7 cells which were isolated from a pleural effusion (233) may depend more heavily on endocrine mechanisms compared to a cell line that is isolated from a solid tumor.

The role of nuclear proto-oncogenes

We and others showed that E2 induces the expression of both the c-*fos* (228, 234) and c-*myc* (228, 235) proto-oncogenes. These nuclear transcription factors are also rapidly induced upon stimulation of cells by growth factors and are thought to be important in mediating the cellular response to these factors (reviewed in 171, 172). Since the induction by E2 is independent of protein synthesis it is not mediated by E2-induced autocrine growth factors, giving additional support to the idea that E2 directly stimulates proliferation without inducing autocrine loops (228).

Fos, the protein product of c-*fos* influences gene expression by binding to Jun, a major component of the transcription factor AP1 (198—203). This interaction results in a greatly enhanced affinity for the AP-1 target sequence, the TRE, and stimulation of TRE-dependent transcriptional activity (198, 200, 203). We recently showed that TRE activity is induced in the situation that MCF7 cells are optimally stimulated to proliferate, either by insulin at high concentrations or the synergistic combination of low concentrations of insulin together with E2 (Fig. 5). Induction of TRE activity is thought to be a critical step in the onset of cellular proliferation and transformation by polypeptide growth factors (reviewed in 172). In this process c-*fos* expression is essential since suppression of expression by antisense RNA abolishes the TRE-dependent induction of genes (204), and inhibits cellular proliferation and transformation (172). This is in line with our results that show that the induction of c-*fos* by E2 can enhance cooperatively both TRE activity and cellular proliferation induced by suboptimal concentrations of insulin, and stresses the importance of c-*fos* in mediating the response to E2. Sustained expression of Fos can lead to transformation of cells (190), therefore the relatively long induction of c-*fos* by E2, in the presence of insulin-induced c-*jun*, may be an explanation for the strong response of MCF7 cells to this combination of mitogens.

TRE activity is involved in regulating the expression of genes that may have a role in metastasis and tumor invasiveness, such as collagenase (204, 236) and stromelysin (236). The essential role for both insulin and E2 for the formation of metastasis by MCF7 cells in nude mice has been correlated to a stimulatory effect of these compounds on collagenase production (98), again suggesting that the synergistic stimulation of TRE activity by E2 and insulin is important in mediating their effect on MCF7 cells.

Stimulation of TRE activity in cells can also lead to elevated expression of growth factor genes. TPA, a potent inducer of TRE activity, induces the expression of many genes including the genes encoding for growth factors like PDGF B (237), and TGFα (238, 239). Therefore, the observed induction of growth factor genes after stimulation of MCF7 cells by E2 may be the result of induced TRE activity by E2 (232).

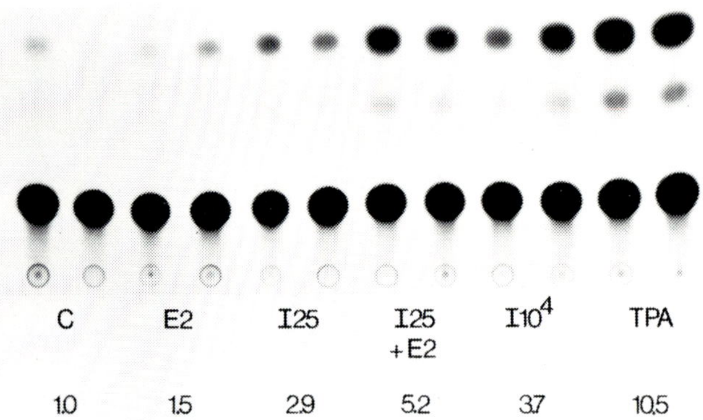

FIGURE 5

Hormonal induction of activity of TRE-CAT constructs transfected into MCF7 cells. MCF7 cells, transfected with TRE-CAT construct (10 µg/dish) were growth-arrested by a 24 h-incubation in 10% DCC-SH-FCS medium. Subsequently, the medium was replaced by fresh medium supplemented with the indicated compounds. After another 24 h-incubation the cells were harvested, and CAT activity was determined. TPA was added in a concentration of 100 ng/ml. The concentrations of the other compounds are given in Fig. 3. Transfection efficiencies varied less than 20% as measured by the activity of 1 µg of co-transfected pSV_2 PAP constructs. The mean value of the total percentage conversion of chloramphenicol is indicated. This is a representative experiment from three seperate experiments. (From ref. 232)

These results suggest that enhancement of TRE activity is a critical event in tumorigenesis and proliferation of MCF7 cells. Elevated TRE activity may be essential in the conversion towards hormone-independence. However, treatment of MCF7 cells with TPA leads to an enhanced TRE activity (Fig. 5), but not to growth stimulation. In contrast, all human (240—242) and canine (243) mammary cancer cell lines tested so far are strongly growth inhibited by TPA. Since MCF7 cells growth-arrested with TPA show characteristics of a differentiated mammary epithelial cell (240, 243), it would be interesting to see if induction of differentiation is a general response of breast cancer cells to TPA. Also in other cell lines TRE activity seems to be important not only in directing cellular proliferation and transformation, but also differentiation (reviewed in 172).

THE LOSS OF HORMONE-DEPENDENCE IN BREAST CANCER CELLS

The role of polypeptide growth factors

The fact that autocrine mechanisms seem not to be involved in the mediation of the E2-response of hormone dependent breast cancer cells like MCF7, does not exclude that autonomous growth factor production is responsible for the lack of E2-response of hormone-independent breast cancer lines. Since IGFs can bypass E2-requirement for proliferation of MCF7 cells, production of these factors by the cells may have the same effect. However, *in vitro* hormone-independent breast cancer cell lines do not produce significant levels of biologically active IGFs (229). In addition, it is has been shown recently that coinoculation of MCF7 with the E2-independent breast cancer cell line MDA-MB-231, reported to be a constitutive producer of high levels of both TGFα (140, 220) and IGF-I (90), does not support MCF7 tumorigenesis (245), also suggesting that if these hormone-independent cells produce high levels of growth factors they are biologically inactive. Thus the autonomous growth behaviour of these human E2-independent cell lines can not be explained by a constitutive production of IGFs. The possibility remains that these E2-independent cells depend on other growth factors for autocrine stimulation. So far, however, no study has demonstrated the existence of the putative autocrine loops in these human cell lines. We therefore speculate that direct autocrine mechanisms are of little importance in the growth regulation of human breast cancer cell lines. In addition, the proposed "induced autocrine" mechanisms that may be operative in inducing proliferation of the hormone-dependent MCF7 cell line (see above) seem not to be important for the proliferation of these progressed, E2-independent cell lines, since these cell lines proliferate independently of externally added growth factors.

The role of oncogenes

Since elevation of TRE activity seems important in E2-induced proliferation of hormone-dependent cells, we propose that elevated TRE activity may be responsible for the autonomous proliferation of hormone-independent breast tumor cells (232). Like in MCF7 cells (see above) this enhanced TRE activity may result in increased growth factor secretion by E2-independent cell lines (90, 140, 220), such as the v-Ha-*ras* transfected E2-independent MCF7 cells (221). Genes from the *ras* family, and especially oncogenic forms like v-Ha-*ras*, are highly efficient stimulators of TRE activity (reviewed in 246) which seems to be the prime reason for the reduced E2-dependence of v-Ha-*ras* transfected MCF7 cells. This notion is supported by the fact that constitutive expression levels of less oncogenic *ras* genes to more physiological levels only leads to loss of *in vitro* E2-dependence, while normal *ras* genes do not alter the hormonal response of MCF7 cells. These less oncogenic forms of *ras* probably are inefficient inducers of TRE activity, since it has been found that induction of TRE activity and oncogenicity of *ras* genes are strongly correlated (246). Moreover, the levels in which these *ras* genes were expressed were much lower than in the study with v-Ha-*ras*. Therefore, in this experimental model both quantitative and qualitative changes in *ras* expression may be important in the stimulation of TRE activity.

In patients *ras* genes may also have a role in elevating TRE activity, since in most studies elevated levels of the *ras* encoded proteins have been found in human breast tumors, although no evidence has been presented for frequently occurring *ras* mutations, (reviewed in 177). Many other ways, however, may lead to enhanced TRE activity, since it is at the receiving end of growth factor signalling pathways. Activation of other oncogenes that are involved in these pathways may therefore also lead to elevation of TRE activity (172).

It is a well known phenomenon that *ras* genes cooperate with other oncogenes to promote tumor development (reviewed in 167). They even cooperate with TPA, also a strong inducer of TRE activity. In the multistep process of tumorigenesis a gradual or step-wise elevation of TRE activity may be important, and in some cases even sufficient for full transformation when maximally induced. Other steps, involving for instance the loss of anti-oncogenes, the expression of *neu*, elevation of *myc*, and expression of genes involved in important processes such as angiogenesis and matrix degradation, may facilitate this tumorigenic process strongly. However, in the loss of E2-dependence, a part of the tumorigenic process in human breast cancer, the elevation of TRE activity may prove to be essential.

CONCLUSION

In conclusion the following model is proposed for the loss of hormone-dependence of breast cancer cells. Hormone-dependent cells depend on stimulation with the external, endocrine mitogens E2 and IGFs. This leads to enhanced TRE activity and subsequent growth factor secretion. The putative elevation of growth factor secretion may lead to an amplification of the TRE activity via "induced autocrine" or in some cases, in humans apparently rare, autocrine mechanisms. The loss of hormone dependence is initiated by intracellular lesions leading to constitutive TRE activation. The relative importance of both the signals caused by intracellular deregulation and the enhancement through external signals will vary between different tumors, and the importance of the intracellular deregulation will increase upon progression. This may lead to a further elevation of growth factor secretion although these factors are not essential for proliferation of the tumor cells.

ACKNOWLEDGEMENTS

We wish to thank Dr. E.J.J. van Zoelen for critical review of this manuscript.

This work was supported by the Netherlands Cancer Foundation (Koningin Wilhelmina fonds).

REFERENCES

1) Moolgavkar, S.H., Knudson, A.T. (1981): Mutation and cancer: a model for human carcigogenesis. J Natl Cancer Inst 66, 1037—1052.

2) Henderson, B.E., Ross, R.K., Pike, M.C., Casagrande, J.T. (1982): Endogenous hormones as a major factor in human cancer. Cancer Res. 42, 3232—3239.

3) Henderson, B.E., Ross, R.K., Bernstein, L. (1988): Estrogens as a cause of human cancer — the Richard and Hinda-Rosenthal Foundation Award Lecture. Cancer Res. 48, 246—253.

4) Lippman, M.E. (1985): Endocrine responsive cancers of man. In: "Textbook of Endocrinology" (R.H. Williams, ed.). W.B. Sanders, Philadelphia, pp. 355—411.

5) Topper, Y.J., Freeman, C.S. (1980): Multiple hormone interactions in the developmental biology of the mammary gland. Physiol. Rev. 60, 1049—1106.

6) Lippman, M.E., Dickson, R.B., Gelmann, E.P., Rosen, N., Knabbe, C., Bates, S., Bronzert, D., Huff, K., Kasid, A. (1987): Growth regulation of human breast carcinoma occurs through regulated growth factor secretion. J. Cell Biochem. 35, 1—16.

7) Jensen, E.V., DeSombre, E.R., Jongblut, P.W. (1967): Estrogen receptors in hormone responsive tissues and tumors. In: "Endogenous Factors Influencing Host-Tumor Balance" (R.W. Wissler, T.L. Dao, S. Wood, eds.). University of Chicago Press, 15—30.

8) Green, S., Chambon, P. (1988): Nuclear receptors enhance our understanding of transcriptional regulation. TIG 4, 309—314.

9) Sledge, G.W., McGuire (1983): Steroid hormone receptors in human breast cancer. Adv. Cancer Res. 38, 61—76.

10) McGuire, W.L., Horwitz, K.B., Pearson, O.H., Segaloff, A. (1977): Current status of estrogen and progesterone receptors in breast cancer. Cancer 39, 2934—2947.

11) McGuire, W.L., Osborne, C.K., Clark, G.W., Knight, W.A. (1982): Steroid hormone receptors and carcinoma of the breast. Am. J. Physiol. 243, E99—E102.

12) Horwitz, K.B., McGuire, W.L. (1978): Estrogen control of progesterone receptor in human breast cancer. J. Biol. Chem. 253, 2223—2228.

13) Maas, H., Jonat, W. (1983): Steroid receptors as a guide for therapy of primary and metastatic breast cancer. J. Steroid. Biochem. 19, 833—838.

14) Quazi, R., Chuang, J.-L., Drobyski, W. (1984): Estrogen receptors and the pattern of relapse in breast cancer. Arch. Intern. Med. 144, 2365—2367.

15) Silvestrini, R., Daidone, M.G., DiFronzo, G. (1979): Relationship between proliferative activity and estrogen receptors in breast cancer. Cancer 44, 665—670.

16) Olszewski, W., Darzynkiewics, Z., Rosen, P.P., Schwartz, M.K., Melamed, M.R. (1981): Flow cytometry of breast carcinoma: I. Relation of DNA ploidy level to histology and estrogen receptor. Cancer 48, 980—984.

17) Vihko, R., Jähne, O., Kotulo, K., Syrjälä, P. (1980): Female sex steroid receptor status in primary and metastatic breast carcinoma and its relationship to serum steroid and peptide hormone levels. Int. J. Cancer 26, 13—21.

18) Shafie, S.M. (1980): Estrogen and the growth of breast cancer: New evidence suggests indirect action. Science 209, 701—702.

19) Soule, M.D., McGrath, C.M. (1980): Estrogen responsive proliferation of clonal human breast carcinoma cells in athymic mice. Cancer Lett. 10, 177—189.

20) Leung, C.K.H, Shiu, R.P.C. (1981): Required presence of both estrogen and pituitary factors for the growth of human breast cancer cells in athymic nude mice. Cancer Res. 41, 546—551.

21) Osborne, C.K., Hobbs, K., Clarck, G.W. (1985): Effect of estrogens and antiestrogens on growth of human breast cancer cells in athymic nude mice. Cancer Res. 45, 584—590.

22) Lippman, M.E., Bolan, G., Huff, K. (1976): The effects of estrogens and antiestrogens on hormone responsive human breast cancer in long-term tissue culture. Cancer Res. 36, 4602—4609.

23) Allegra, J.L., Lippman, M.E. (1978): Growth of a human breast cancer cell line in serum-free hormone supplemented medium. Cancer Res. 38, 3823—3829.

24) Butler, W.B., Kelsey, W.H., Goran, N. (1981): Effect of serum and insulin on the sensitivity of the human breast cancer cell line, MCF7 to estrogen and antiestrogens. Cancer Res. 41, 82—88.

25) Chalbos, D., Vignon, F., Keydar, I., Rochefort, H. (1982): Estrogens stimulate cell proliferation and induce secretory proteins in a human breast cancer cell line. J. Clin. Endocrinol. Metab. 55, 276—283.

26) Natoli, C., Sica, G., Natoli, V., Serra, A., Iacobelli, S. (1983): Two new estrogen-supersensitive variants of the, MCF7 human breast cancer cell line. Breast Cancer Res. Treat 3, 23—32.

27) Dembinski, T.C., Green, C.D. (1983): Regulation of oestrogen responsiveness of, MCF-7 human breast cancer cell growth by serum concentrations in the culture medium. In: "Hormonally Defined Media" (G. Fischer, R.J. Wieser, eds.). Springer-Verlag, Berlin, pp. 439—442.

28) Page, M.J., Field, J.K., Everett, N.P., Green, C.D. (1983): Serum regulation of the estrogen responsiveness of the human breast cancer cell line, MCF7. Cancer Res. 43, 1244—1250.

29) Soto, A.M., Sonnenschein, C. (1984): Mechanism of estrogen action in cellular proliferation: Evidence for indirect and negative control on cloned breast tumor cells. Biochem. Biophys. Res. Commun. 122, 1097—1103.

30) Jozan, S., Moure, C., Gillois, M., Bayard, F. (1979): Effects of estrone on cell proliferation of a human breast cancer (MCF-7) in long term tissue culture. J. Steroid. Biochem. 10, 341—342.

31) Edwards DP, McGuire, W.L. (1982): Estrogen action in human breast cancer (review). Anticancer Res. 2, 297—308.

32) Strobl, J.W., Lippman, M.E. (1979): Prolonged retention of estradiol by human breast cancer cells in tissue culture. Cancer Res. 39, 3319—3327. 33) Vignon, F., Terqui, M., Westley, B., Derocq, D., Rocherort, H. (1980): Effects of plasma estrogen sulfates in mammary cancer cells. Endocrinol. 106, 1079—1086.

34) Berthois, Y., Katzenellenbogen, J.A., Katzenellenbogen, B.S. (1986): Phenol red in tissue culture media is a weak estrogen: implications concerning the study of estrogen-responsive cells in culture. Proc. Natl. Acad. Sci. USA 83, 2496—2500.

35) Foekens, J.A., Henkelman, M.S., Fukkink, J.F., Blankenstein, M.A., Klijn J.G.M. (1986): Combined effects of buserelin, estradiol and tamoxifen on the growth of, MCF-7 human breast cancer cells in vitro. Biochem. Biophys. Res. Commun. 140, 550—556.

36) Lykkesfeldt, A.E., Briand, P. (1986): Indirect mechanism of oestradiol stimulation of cell proliferation of human breast cancer cell lines. Br. J. Cancer 53, 29—35.

37) Soto, A.M., Murai, J.T., Siiteri, P.K., Sonnenschein, C. (1986): Control of cell proliferation: evidence for negative control on estrogen-sensitive T47D human breast cancer cells. Cancer Res. 46, 2271—2275. 38) Dembinski, T.C., Leung, C.K.H, Shiu, R.P.C. (1985): Evidence for a novel pituitary factor that potentiates the mitogenic effect of estrogen in human breast cancer cells. Cancer Res. 45, 3083—3089.

39) Jozan, S., Faye, J.C., Tournier, J.F., Tauber, J.P., David, J.F., Bayard, F. (1985): Interaction of estradiol and high density lipoproteins on proliferation of the human breast cancer cell line, MCF7 adapted to grow in serum-free conditions. Biochem. Biophys. Res. Commun. 133, 105—112.

40) Barnes (1984): Serum-free cell culture of, MCF-7 human mammary carcinoma. In: "Methods for Serum-Free Culture of Cells of the Endocrine System" (D.W. Barnes, D.A. Sirbasku, G.H. Sato, eds.). Alan R. Liss Inc, New York, pp. 201—216.

41) Iacobelli, S., Scambia, G., Natoli, V., Natoli, C., Sica G. (1984): Estrogen stimulates cell proliferation and the increase of a 52,000 dalton glycoprotein in human breast cancer cells. J. Steroid. Biochem. 20, 747—752.

42) Briand, P., Lykkesfeldt, A.E. (1986): Long-term cultivation of a human breast cancer cell line, MCF-7, in a chemically defined medium. Effect of estradiol. Anticancer Res. 6, 85—90.

43) Van Zoelen, E.J.J., van Oostwaard, T.M.J., van der Saag, P.T., de Laat, S.W. (1985): Phenotypic transformation of normal rat kidney cells in a growth-factor-defined medium: induction by a neuroblastoma-derived transforming growth factor independently of the EGF receptor. J. Cell Physiol. 123, 151—160.

44) Van der Burg, B., Rutteman, G.R., Blankenstein, M.A., de Laat, S.W., van Zoelen, E.J.J. (1988): Mitogenic stimulation of human breast cancer cells in a growth factor-defined medium: synergistic action of insulin and estrogen. J. Cell Physiol. 134, 101—108.

45) Adams, J.D., Edwards, D.P., McCuire, W.L. (1980): Estrogen regulation of specific messenger RNA's in human breast cancer cells. Biochem. Biophys. Res. Commun. 97, 1354—131.

46) Aitken, S.C., Lippman, M.E., Kasid, A., Schoenberg, D.R. (1985): Relationship between the expression of estrogen-regulated genes and estrogen-stimulated proliferation of, MCF-7 mammary tumor cells. Cancer Res. 45, 2608—2615.

47) Prud'homme, J.-F., Fridlansky, F., Le Cunff, M., Atger, M., Mercier-Bodart, C., Pichon, M.-F., Milgrom, E. (1985): Cloning of a gene expressed in human breast cancer and regulated by estrogen in, MCF-7 cells. DNA 4, 11—21.

48) May FEB, Westley BR (1988): Identification and characterization of estrogen-regulated RNAs in human breast cancer cells. J. Biol. Chem. 263, 12901—12908.

49) Butler, W.B., Kirkland, W. L., Jorgensen, T.L. (1979): Induction of plasminogen activator by estrogen in a human breast cancer cell line (MCF-7). Biochem. Biophys. Res. Commun. 90, 1328—1334.

50) Capony, F., Morriset, M., Barett AJ, Capony, J.P., Broquet, P., Vignon, F., Chambon, M., Louisot, P., Rochefort, H. (1987): Phosphorylation glycosilation, and proteolytic activity of the 52-kD estrogen-induced protein secreted by, M.C.F7 cells. J. Cell Biol. 104, 253—262. 51) Aitken, S.C., Lippman, M.E. (1985): Effect of estrogens and antiestrogens on growth-regulatory enzymes in human breast cancer cells in tissue culture. Cancer Res. 45, 1611—1620.

52) Edwards, D.P., Murthy, S.R., McGuire, W.L. (1980): Effect of estrogen and antiestrogen on DNA polymerase in human breast cancer. Cancer Res. 44, 1722—1726.

53) Dickson, R.B., Lippman, M.E. (1987): Estrogenic regulation of growth and polypeptide growth factor secretion in human breast carcinoma. Endocrine Rev. 8, 29—43.

54) Lippman, M.E. (1984): Definition of hormones and growth factors required for optimal proliferation and expression of phenotypic responses in human breast cancer cells. In: "Methods for Serum-Free Culture of Cells of the Endocrine System" (D.W. Barnes, D.A. Sirbasku, G.H. Sato, eds.). Alan R. Liss Inc., New York, pp. 183—200.

55) Heldin, C.-H., Westermark B. (1984): Growth factors: mechanism of action and relation to oncogenes. Cell 37, 9—20.

56) James, R., Bradshaw, R.A. (1984): Polypeptide growth factors. Ann. Rev. Biochem. 53, 259—292.

57) Sporn, M.B., Todaro, G.J. (1980): Autocrine secretion and malignant transformation. New Engl. J. Med. 303, 878—880.

58) Sporn, M.B., Roberts, A.B. (1985): Autocrine growth factors and cancer. Nature 313, 745—747.

59) Sirbasku, D.A. (1978): Estrogen induction of growth factors specific for hormone-responsive mammary, pituitary, and kidney tumor cells. Proc. Natl. Acad. Sci. USA 75, 3786—3790.

60) Ikeda, T., Liu, Q.-F., Danielpour, D., Officer, J.B., Iio, M., Leland, F.E., Sirbasku, D.A. (1982): Identification of estrogen-inducible growth factors (estromedins) for rat and human mammary tumour cells in culture. In Vitro 18, 961—979.

61) Sirbasku, D.A., Ikeda, T., Danielpour, D. (1985): Endocrine and autocrine estromedins for mammary and pituitary tumor cells. In: "Mediators in Cell Growth and Differentiation" (R.J. Ford, Maizel A.L., eds.). Raven Press, New York, 213—232.

62) Riss, T.L., Sirbasku, D.A. (1987): Purification and identification of transferrin as a major pituitary derived mitogen for MTW9/PL2 rat mammary tumor cells. In Vitro 23, 841—849.

63) Ogasawara, M., Karey, K.P., Sirbasku, D.A. (1988): Insulin-like growth factor I: Relationship between plasma and tissue sources, biological potency and N^a truncation. Proc. Am. Assoc. Cancer Res. 29, 52 (Abstract).

64) Riss, T.L., Sirbasku, D.A. (1989): Identification of a 15,000-molecular weight form of immunoreactive transforming growth factor α in extracts of porcine pituitary. J. Cell Physiol. 138, 393—404.

65) Huseby, R.A., Maloney, T.M., McGrath, C.M. (1984): Evidence for a direct growth-stimulating effect of estradiol on human, MCF-7 cells *in vivo.* Cancer Res. 44, 2654—2659.

66) Westley, B., Rochefort, H. (1980): A secreted glycoprotein induced by estrogen in human breast cancer cell lines. Cell 20, 353—362.

67) Capony, F., Garcia, M., Capdevielle, J., Rougeot, C., Ferrara, P., Rochefort, H. (1986): Purification and first characterization of the secreted and cellular 32-kDa protein regulated by estrogens in human-breast cancer cells. Eur. J. Biochem. 161, 505—512.

68) Masiakowski, P., Breathnach, R., Bloch, J., Gannon, F., Krust, A., Chambon, P. (1982): Cloning of cDNA sequences of hormone-regulated genes from the, MCF-7 human breast cancer cell line. Nucl. Acids Res. 10, 7895—7903.

69) Nunez, A.-M., Jakowlew, S., Briand, J.-P., Gaire, M., Krust, A., Rio, M.-C., Chambon, P. (1987): Characterization of the estrogen-induced pS2 protein secreted by the human breast cancer cell line, MCF-7. Endocrinol. 121, 1759—1765.

70) Brown, A.M.C., Jeltsch, J.-M., Roberts, M., Chambon, P. (1984): Activation of pS2 gene transcription is a primary response to estrogen in the human breast cancer cell line, MCF-7. Proc. Natl. Acad. Sci. USA 81, 6344—6348.

71) Cavailles, V., Augereau, P., Garcia, M., Rochefort, H. (1988): Estrogens and growth factors induce the mRNA of the 52K-pro-cathepsin-D secreted by breast cancer cells. Nucl. Acids Res. 16, 1903—1919.

72) Rio, M.-C., Bellocq, J.P., Gairard, B., Rasmussen, U.B., Krust, A., Koehl, C., Calderoli, H., Schiff, V., Renaud, R., Chambon, P. (1987): Specific expression of the pS2 gene in subclasses of breast cancers in comparison with expression of the estrogen and progesterone receptors and the oncogene, *ERBB2.* Proc. Natl. Acad. Sci. USA 84, 9243—9247.

73) Rio, M.-C., Bellocq, J.P., Daniel, J.Y., Tomasetto, C., Lathe, R., Chenard, M.P., Batzenschlager, A., Chambon, P. (1988): Breast cancer-associated pS2 protein: synthesis and secretion by normal stomach mucosa. Science 241, 705—708.

74) Baker, M.E. (1988): Oestrogen-induced pS2 protein is similar to pancreatic spasmolytic polypeptide and the kringle domain. Biochem. J. 253, 307—311.

75) Thim, L. (1988): A new family of growth factor-like peptides: "Trefoil" disulfide loop structures as a common feature in breast cancer associated peptide (pS2), pancreatic spasmolytic polypeptide (PSP), and frog skin peptides, plasmolysins. FEBS Lett. 250, 85—90.

76) Garcia, M., Salazar-Retana, G., Pages, A., Richer, G., Domergue, J., Pages, A.M., Cavalie, G., Martin, J.M., Lamarque, J.-L., Pau, B., Pujol, H., Rochfort, H. (1986): Distribution of the Mr 52,000 Estrogen-regulated protein in benign breast diseases and other tissues by immunohistochemistry. Cancer Res. 46, 3734—3738.

77) Cavailles, V., Garcia, M., Salazar, G., Domergue, J., Simony, J., Pujol, H., Rochefort, H. (1987): Immunodetection of estrogen receptor and 52,000 Dalton protein in fine needle aspirates of breast cancer tumors. J. Natl. Cancer Res. 79, 245—251.

78) Vignon, F., Capony, F., Chambon, M., Freiss, G., Garcia, M., Rochefort, H. (1986): Autocrine growth stimulation of the, M.C.F7 breast cancer cells by the estrogen-regulated 52K protein. Endocrinol. 118, 1537—1545.

79) Augereau, P., Garcia, M., Mattes, M.G., Cavailles, V., Depadova, F., Derocq, D., Capony, F., Ferrara, P., Rochefort, H. (1988): Cloning and sequencing of the 52K cathepsin, D. complementary deoxyribonucleic acid of, MCF7 breast cancer cells and mapping on chromosome 11. Mol. Endocrinol. 1, 186—192.

80) Jakowlew, S.B., Breathnach, R., Jeltsch, J.-M., Masiakowski, P., Chambon, P. (1984): Sequence of the pS2 mRNA induced be estrogen in the human breast cancer cell line, MCF-7. Nucl. Acids Res. 12, 2861—2878.

81) Kida, N., Yoshimura, T., Mori, K., Hayashi, K. (1989): Hormonal regulation of synthesis and secretion of pS2 protein relevant to growth of human breast cancer cells (MCF-7). Cancer Res. 49, 3494.

82) Clemmons, D.R. (1989): Structural and functional analysis of insulin-like growth factors. Br. Med. Bull. 45, 465—480.

83) Daughaday, W.H., Rotwein, P. (1989): Insulin-like growth factors I and II. Peptide, messenger ribonucleic acid and gene structures, serum, and tissue concentrations. Endocrine Rev 10, 68—91.

84) Baxter, R.C., Martin, J.C. (1989): Binding proteins for the insulin-like growth factors: structure, regulatio and function. Prog. Growth Factor Res. 1, 49—68.

85) Furlanetto, R.W., DiCarlo, J.N. (1984): Somatomedin-C receptors and growth effects in human breast cells maintained in long-term tissue culture. Cancer Res. 44, 2122—2128.

86) Myal, Y., Shiu, R.P.C., Bhaumick, B., Bala, M. (1984): Receptor binding and growth-promoting activity of insulin-like growth factors in human breast cancer cells (T47D): in culture. Cancer Res. 44, 5486—5490.

87) Osborne, C.K., Bolan, G., Monaco, M.E., Lippman, M.E. (1976): Hormone responsive human breast cancer in long-term tissue culture: effect of insulin. Proc. Natl. Acad. Sci. USA 73, 4536—4540.

88) Osborne, C.K., Monaco, M.E., Lippman, M.E., Kahn, C.R. (1978): Correlation among insulin binding, degradation, and biological activity in human breast cancer cells in long-term tissue culture. Cancer Res. 38, 94—102.

89) Barnes, D., Sato, G. (1979): Growth of a human mammary tomour cell line in a serum-free medium. Nature 281, 388—389.

90) Huff, K.K., Kaufman, D., Gabbay, K.H., Spencer, E.M., Lippman, M.E., Dickson, R.B. (1986): Secretion of an insulin-like growth factor-I-related protein by human breast cancer cells. Cancer Res. 46, 4613—4619.

91) Huff, K.K., Knabbe, C., Lindsey, R., Kaufman, D., Bronzert, D., Lippman, M.E., Dickson, R.B. (1988): Multihormonal regulation of insulin-like growth factor-I-related protein in, MCF-7 human breast cancer cells. Mol. Endocrinol. 2, 200—208.

92) Péres, R., Betsholtz, C., Westermark, B., Heldin, C.-H. (1987): Frequent expression of growth factors for mesenchymal cells in human mammary carcinoma cell lines. Cancer Res. 47, 3425—3429.

93) De Leon, D.D., Wilson, D.M., Bakker, B., Lamsom, G., Hintz RL, Rosenfeld, R.G. (1989): Characterization of insulin-like growth factor binding proteins from human breast cancer cells. Mol. Endocrinol. 3, 567—574.

94) Dickson, R.B., McManaway, M.E., Lippman, M.E. (1986): Estrogen-induced factors of breast cancer cells partially replace estrogen to promote tumor growth. Science 232, 1540—1543.

95) Foekens, J.A., Portengen, H., Janssen, M., Klijn, J.G.M. (1989): Insulin-like growth factor-1 receptors and insulin-like growth factor-1-like activity in human primary breast cancer. Cancer 63, 2139—2147.

96) Pollak, M.N., Perdue, J.F., Margolese, R.G., Baer, K., Richerd, M. (1987): Presence of somatomedin receptors on primary huaman breast and colon carcinomas. Cancer Lett. 38, 223—230.

97) Clemmons, D.R., Underwood LE, van Wyk, J.J. (1981): Hormonal control of immunoreactive somatomedin production by cultured human fibroblasts. J. Clin. Invest. 67, 10—19.

98) Shafie, S.M., Liotta, L.A. (1980): Formation of metastasis by human breast carcinoma cells (MCF-7) in nude mice. Cancer Lett. 11, 81—87.

99) Shafie, S.M., Grantham, F.H. (1981): Role of hormones in the growth and regression of human breast cancer cells (MCF-7) transplanted into athymic nude mice. J. Natl. Cancer Inst. 67, 51—56.

100) Huff, K.K., McManaway, M., Paik, S., Brünner, N. (1988): In vivo effects of insulin-like growth factor-I (IGF-I) on tumor formation by, MCF-7 human breast cancer cells in athymic mice. Program of the 70th Annual Meeting of The Endocrine Society, New Orleans, LA, p. 238 (Abstract).

101) Leung, C.K.H, Shiu, R.P.C. (1981): Required presence of both estrogen and pituitary factors for the growth of human breast cancer cells in athymic nude mice. Cancer Res. 41, 546—551.

102) Kleinberg, D.L., Niemann, W., Flamm, E., Cooper, P., Babitsky, G., Valensi, Q. (1985): Primate mammary development. J. Clin. Invest. 75, 1943—1950.

103) Derynck, R. (1986): Transforming growth factor-α: structure and biological activities. J. Cell Biochem. 32, 293—304.

104) Osborne, C.K., Hamilton, B., Titus, G., Livingston, R.B. (1980): Epidermal growth factor stimulation of human breast cancer cells in culture. Cancer Res. 40, 2361—2366.

105) Imai, Y., Leung, C.K.H, Friesen, H.G., Shiu, R.P.C. (1982): Epidermal growth factor receptors and effect of epidermal growth factor on growth of human breast cancer cells in long-term tissue culture. Cancer Res. 42, 4394—4398.

106) Fitzpatrick, S.L., LaChance, M.P., Schultz, G.S. (1984): Characterization of epidermal growth factor receptor and action on human breast cancer cells in culture. Cancer Res. 44, 3442—3447.

107) Davidson, N.E., Gelmann, E.P., Lippman, M.E., Dickson, R.B. (1987): Epidermal growth factor receptor gene expression in estrogen receptor-positive and negative human breast cancer cell lines. Mol. Endocrinol. 1, 216—223.

108) Karey, K.P., Sirbasku, D.A. (1988): Differential responsiveness of human breast cancer cell lines, MCF-7 and T47D to growth factors and 17a-estradiol. Cancer Res. 48, 4083—4092.

109) Osborne, C.K., Hamilton, B., Nover, M. (1982): Receptor binding and processing of epidermal growth factor by human breast cancer cells. J. Clin. Endocrinol. Metab. 55, 86—93.

110) Fitzpatrick, S.L., Brightwell, J., Wittliff, J.L., Barrows, G.H., Schultz, G.S. (1984): Epidermal growth factor binding by breast tumor biopsies and relationship to estrogen receptor and progestion receptor levels. Cancer Res. 44, 3448—3453.

111) Pérez, R., Pascual, M., Macias, A., Lage, A. (1984): Epidermal growth factor receptors in human breast cancer. Breast Cancer Res. Treatm. 4, 189—193.

112) Sainsbury, J.R.C., Farndon, J.R., Sherbet, G.V., Harris, A.L. (1985): Epidermal-growth-factor receptors and oestrogen receptors in human breast cancer. Lancet 1, 364—366.

113) Filmus, J., Pollak, M.N., Cailleau, R., Buick, R.N. (1985), MDA-468, a human breast cancer cell line with a high number of epidermal growth factor (EGF) receptors, has an amplified EGF receptor gene and is growth inhibited by EGF. Biochem. Biophys. Res. Commun. 128, 898—905.

114) Kraus, M.H., Popescu, N.C., Amsbaugh, S.C., King, C.R. (1987): Overexpression of the EGF receptor-related proto-oncogene *erb*B-2 in human mammary tumor cell lines by different molecular mechanisms. EMBO J. 6, 605—610.

115) Kawamoto, T., Mendelsohn, J., Le, A., Sato, G.H., Lazar, C.S., Gill, G.N. (1984): Relation of epidermal growth factor receptor concentration to growth of human epidermoid carcinoma A431 cells. J. Biol. Chem. 259, 7761—7766.

116) Filmus, J., Trent, J.M., Pollak, M.N., Buick, R.N. (1987): Epidermal growth factor receptor gene-amplified, MDA-468 breast cancer cell line and its nonamplified variants. Mol. Cell Biol 7, 251—257.

117) Santon, J.B., Cronin, M.T., MacLeod, C.L., Mendelsohn, J., Masui, H., Gill, G.N. (1986): Effect of epidermal growth factor receptor concentration on tumorgenicity of A431 cells in nude mice. Cancer Res. 46, 4701—4705.

118) Macias, A., Azavedo, E., Pérez, R., Rutqvist, L.E., Skoog, R.L. (1986): Receptors for epidermal growth factor in human mammary carcinomas and their metastases. Anticancer Res. 6, 849—852.

119) Walker, R.A., Camplejohn, R.S. (1986): DNA flow cytometry of human breast carcinomas and its relationship to transferrin and epidermal growth factor receptors. J. Pathol. 150, 37—42.

120) Spitzer, E., Grosse, R., Kunde, D., Schmidt, H.E. (1987): Growth of mammary epithelial cells in breast cancer biopsis correlates with EGF binding. Int. J. Cancer 39, 279—282.

121) Murphy, L.J., Sutherland, R.L., Stead, B., Murphy, L.C., Lazarus, R.L. (1986): Progestin regulation of epidermal growth factor receptor in human mammary carcinoma cells. Cancer Res. 46, 728—734.

122) Murphy, L.C., Murphy, L.J., Shiu, R.P.C. (1988): Progestin regulation of EGF-receptor mRNA accumulation in T-47D human breast cancer cells. Biochem. Biophys. Res. Commun. 150, 192—196.

123) Bjorge, J.D., Kudlow, J.E. (1987): Epidermal growth factor receptor synthesis is stimulated by phorbol ester and epidermal growth factor. J. Biol. Chem. 262, 6615—6622.

124) Lee S.C.L., Koga, M., Sutherland, R.L. (1989): Modulation of estrogen receptor and epidermal growth factor receptor mRNA by phorbol ester in MCF 7 breast cancer cells. Biochem. Biophys. Res. Commun. 162, 415—421.

125) Fernandez-Pol, J.A., Hamilton, P.D., Klos, D.J. (1989): Transcriptional regulation of proto-oncogene expression by epidermal growth factor, transforming growth factor $\beta 1$, and triiodothyronine in MDA-468 cells. J. Biol. Chem. 264, 4151—4156.

126) Koga, M., Eisman, J.A., Sutherland, R.L. (1988): Regulation of epidermal growth factor receptor levels by 1,25-Dihydroxyvitamin D_3 in human breast cancer cells. Cancer Res. 48, 2734—2739.

127) Berthois, Y., Dong, X.F., Martin, P.M. (1989): Regulation of epidermal growth factor-receptor by estrogen and antiestrogen in the human breast cancer cell line, MCF-7. Biochem. Biophys. Res. Commun. 159, 126—131.

128) Mori, K., Kurobe, M., Furukawa, S., Kubo, K., Hayashi, K. (1986): Human breast cancer cells synthesize and secrete an EGF-like immunoreactive factor in culture. Biochem. Biophys. Res. Commun. 136, 300—305.

129) Mori, K., Fujii, R., Kida, N., Ohta, M., Hayashi, K. (1988): Identification of a polypeptide secreted by human breast cancer cells (MCF-7) as the human estrogen-responsive gene (pS2) product. Biochem. Biophys. Res. Commun. 155, 366—372.

130) Murphy, L.C., Murphy, L.J., Dubik, D., Bell,, G.I., Shiu, R.P.C. (1988): Epidermal growth factor gene expression in human breast cancer cells: regulation of expression by progestins. Cancer Res. 48, 4555—4560.

131) Carpenter G. (1980): Epidermal growth factor is a major growth-promoting agent in human milk. Science 210, 198—199.

132) Breadmore, J.M., Lewis-Jones, D.I., Richards, R.C. (1983): Urogastrone and lactose concentrations in precolostrum, colostrum and milk. Pediat. Res. 17, 825—828.

133) Shing, Y.W., Klagsbrun, M. (1984): Human and bovine milk contain different sets of growth factors. Endocrinol. 115, 273—282.

134) Jaspar, J.M., Franchimont, P. (1985): Radioimmunoassay of human epidermal growth factor in human breast cyst fluid. Eur. J. Cancer Clin. Oncol. 21, 1343—1348.

135) Petrides, P.E., Hosang, M., Shooter, E., Esch, F.S., Böhlen, P. (1985): Isolation and characterization of epidermal growth factor from human milk. FEBS Lett. 187, 89—95.

136) Berseth, C.L. (1987): Enhancement of intestinal growth in neonatal rats by epidermal growth factor in milk. Am. J. Physiol. 253, G662—G665.

137) Dickson, R.B., Huff, K.K., Spencer, E.M., Lippman, M.E. (1986): Induction of epidermal growth factor-related polypeptides by 17 β-estradiol in, MCF-7 human breast cancer cells. Endocrinol. 118, 138—142.

138) Perroteau, I., Salomon, D., DeBortoli, M., Kidwell, W., Hazarika, D., Pardue, R., Dedman, J., Tam, J. (1986): Immunological detection and quantitation of alpha transforming growth factors in human breast carcinoma cells. Breast Cancer Res. Treatm. 7, 201—210.

139) Bates, S.E., Davidson, N.E., Valverius, E.M., Lippman, M.E. (1988): Expression of transforming growth factor α and its messenger ribonucleic acid in human breast cancer: its regulation by estrogen and its possible functional significance. Mol. Endocrinol. 2, 543—555.

140) Bates, S., McManaway, M.E., Lippman, M.E., Dickson, R.B. (1986): Characterization of estrogen responsive transforming activity in human breast cancer cell lines. Cancer Res. 46, 1707—1713.

141) Travers, M.T., Barrett-Lee, P.J., Berger, U., Luqmani, Y.A., Gazet, J.-C., Powles, T.J., Coombes, R.C. (1988): Growth factor expression in normal, benign, and malignant breast tissue. Br. Med. J. 296, 1621—1624.

142) Shoyab, M., Plowman, G.D., McDonald, V.L., Bradley, J.G., Todaro, G.J. (1989): Structure and function of human amphiregulin: a member of the epidermal growth factor family. Science 243, 1074—1076.

143) Shoyab, M., McDonald, V.L., Bradley, J.G., Todaro, G.J. (1988): Amphiregulin: a bifunctional growth-modulating glycoprotein produced by the phorbol 12-myristate 13-acetate-treated human breast adenocarcinomal cell line, MCF-7. Proc. Natl. Acad. Sci. USA 85, 6528—6532.

144) Plowman, G., Shoyab, M. (1989): Amphiregulin and its receptor: cloning of novel homologous to EGF and the EGF receptor. Proceedings NATO/EEC Course: "The Molecular Basis of Cell Growth Regulation", Mallorca, Spain, A25.

145) Moses, H.L., Coffey, R.J., Leof, E.B., Lyons, R.M., Keski-Oja, J. (1987): Transforming growth factor β regulation of cell proliferation. J. Cell Physiol. 5, 1—7.

146) Keski-Oja, J., Postlethwaite, A.E., Moses, H.L. (1988): Transforming growth factors in the regulation of malignant cell growth and invasion. Cancer Invest. 6, 705—724.

147) Roberts, A.B., Sporn, M.B., Assoian, R.K., Smith, J.M., Roche, N.S., Wakefield, L.M., Heine, U.I., Liotta, L.A., Falanga, V., Kehrl, J.H., Fauci, A.S. (1986): Transforming growth factor type β: rapid induction of fibrosis and angiogenesis in vitro and stimulation of collagen formation *in vitro.* Proc. Natl. Acad. Sci. USA 83, 4167—4171.

148) Massagué, J., Cheifetz, S., Ignotz, R.A., Boyd, F.T. (1987): Multiple Type-β transforming growth factors and their receptors. J. Cell Physiol. Suppl. 5, 43—47.

149) Boyd, F.T., Massagué, J. (1989): Transforming growth factor-β inhibition of epithelial cell proliferation linked to the expression of a 53-kDa membrane receptor. J. Biol. Chem. 264, 2272—2279.

150) Knabbe, C., Lippman, M.E., Wakefield, L.M., Flanders, K.C., Kasid, A., Derynck, R., Dickson, R.B. (1987): Evidence that transforming growth factor-β is a hormonally regulated negative growth factor in human breast cancer cells. Cell 48, 417—428.

151) Zajchowski, D., Band, V., Pauzie, N., Tager, A., Stampfer, M., Sager, R. (1988): Expression of growth factors and oncogenes in normal and tumor-derived human mammary epithelial cells. Cancer Res. 48, 7041—7047.

152) Arteaga, C.L., Tandon, A.K., Von Hoff, D.D., Osborne, C.K. (1988): Transforming growth factor β: potential autocrine growth inhibitor of estrogen receptor-negative human breast cancer cells. Cancer Res. 48, 3898—3904.

153) Tucker, R.F., Shipley, G.D., Moses, H.L., Holley, R.W. (1984): Growth inhibitor from MBC-1 cells closely related to platelet type β transforming growth factor. Science 226, 705—707.

154) Holley, R.W., Armour, R., Baldwin, J.H., Greenfield, S. (1983): Activity of a kidney epithelial cell growth inhibitor on lung and mammary cells. Cell Biol. Int. Rep. 7, 141—147.

155) Heldin, C.-H., Wasteson, A., Westermark, B. (1985): Platelet-derived growth factor. Mol. Cell Endocrinol. 39, 169—187.

156) Heldin, C.-H., Betsholtz, C., Claesson-Welsh, R.L., Westermark, B. (1987): Subversion of growth regulatory pathways in malignant transformation. Biochim. Biophys. Acta 907, 219—244.

157) Rozengurt, E., Sinnett-Smith, J. (1985): Production of PDGF-like growth factor by breast cancer cell lines. Int. J. Cancer 36, 247—253.

158) Bronzert, D.A., Pantazis, P., Antoniades, H.N., Kasid, A., Davidson, N., Dickson, R.B., Lippman, M.E. (1987): Synthesis and secretion of platelet-derived growth factor by human breast cancer cell lines. Proc. Natl. Acad. Sci. 84, 5763—5767.

159) Ro, J., Bresser, J., Ro, J.Y., Brasfield, F., Hortobagyi, G., Blick, M. (1989): SIS/PDGF-B expression in benign and malignant human breast lesions. Oncogene 4, 351—354.

160) Folkman, J., Klagsbrun, M. (1987): Angiogenic factors. Science 235, 442—447.

161) Folkman, J., Klagsbrun, M. (1987): A family of angiogenic peptides. Nature 329, 671—672.

162) Lidereau, R., Callahan, R., Dickson, C., Peters, G., Escot, C., Ali, I.U. (1988): Amplification of the *int*-2 gene in primary human breast tumors. Oncogene Res. 2, 285—291.

163) Schuuring, E., Verhoeven, E., Peterse, H., Thunnissen, E., Michalides, R. (1989): The activation of oncogenes during progression of breast cancer. Proceedings NATO/EEC Course: "The Molecular Basis of Cell Growth Regulation", Mallorca, Spain, D19.

164) Ali, I.U., Merlo, G., Callahan, R., Lidereau, R. (1989): The amplification unit on chromosome 11q13 in aggressive primary human breast tumors entails the *bcl*-1, *int*-2 and *hst* loci. Oncogene 4, 89—92.

165) Weinberg, R.A. (1985): The action of oncogenes in the cytoplasm and nucleus Science 230, 770—776.

166) Bishop, M. (1987): The molecular genetics of cancer. Science 235, 305—311.

167) Weinberg, R.A. (1989): Oncogenes, antioncogenes, and the molecular bases of multistep carcinogenesis. Cancer Res. 49, 3713—3721.

168) Konkel, D.A. (1988): What do ras oncogenes do? Mol. Endocrinol. 2, 883—885.

169) Rozengurt, E. (1986): Early signals in the mitogenic response Science 234, 161—166.

170) Berridge, M.J. (1987): Inositol lipides and cell proliferation Biochim. Biophys. Acta 907, 33—45.

171) Rollins, B.J., Stiles, C.D. (1988): Regulation of c-*myc* and c-*fos* proto-oncogene expression by animal cell growth factors. In Vitro 24, 81—84.

172) Herrlich, P., Ponta, H. (1989): 'Nuclear' oncogenes convert extracellular stimuli into changes in the genetic program. TIG 5, 112—116.

173) Schechter, A.L., Stern, D.F., Vaidyanathan, R.L., Decker, S.J., Drebin, J.A., Greene, M.I., Weinberg, R.A. (1984): The *neu* oncogene: an *erb*-B-Related gene encoding a 185,000-M_r tumour antigen. Nature 312, 513—516.

174) Slamon, D.J., Godolphin, W., Jones, L.A., Holt, J.A., Wong, S.G., Keith, D.E., Levin, W.J., Stuart, S.G., Udove, J., Ullrich, A., Press, M.F. (1989): Studies of the HER-2/*neu* proto-oncogene in human breast and ovarian cancer. Science 244, 707—244.

175) Nusse, R. (1988): The *int* genes in mammary tumorigenesis and in normal development TIG 4, 291—295.

176) Barbacid, M. (1987): *Ras* Genes. Ann. Rev. Biochem. 56, 779—827.

177) Thor, A., Ohuchi, N., Horan-Hand, P., Callahan, R., Weeks, M.O., Theillet, C., Lidereau, R., Escot, C., Page, D.L., Vilasi, V., Schlom, J. (1986): *Ras* Gene alterations and enhanced levels of ras p21 expression in a spectrum of benign and malignant human mammary tissues. Lab. Invest. 55, 603—615.

178) Sakumar, S., Notario, V., Martin-Zanca, D., Barbacid, M. (1983): Induction of mammary carcinomas in rats by nitrosamethyl urea involves the malignant activation of the H-*ras*-1 locus by single point mutation. Nature 306, 658—661.

179) Lidereau, R., Escot, C., Theillet, C., Champeme, M.-H., Brunet, M., Gest, J., Callahan, R. (1986): High frequency of rare alleles of the human c-Ha-*ras*-1 Proto-oncogene in breast cancer patients. J. Natl. Cancer Inst. 77, 697—701.

180) Theillet, C., Lidereau, R., Escot, C., Hutzell, P., Brunet, M., Gest, J., Schlom, J., Callahan, R. (1986): Loss of a c-H-*ras*-1 Allele and aggressive human primary breast carcinomas. Cancer Res. 46, 4776—4781.

181) Ali, I.U., Lidereau, R., Theillet, C., Callahan, R. (1987): Reduction to homozygosity of genes on chromosome 11 in human breast neoplasia. Science 238, 185—188.

182) Kasid, A., Lippman, M.E., Papageorge, A.G., Lowy, D.R., Gelmann, E.P. (1985): Transfection of v-Ha-*ras*^{H} DNA into, MCF-7 human breast cancer cells bypasses dependence on estrogen for tumorigenicity. Science 228, 725—728.

183) Sukumar, S., Carney, W.P., Barbacid, M. (1988): Independent molecular pathways in initiation and loss of hormone responsiveness of breast carcinomas. Science 240, 524—526.

184) Sluyser, M., Mester, J. (1985): Oncogenes homologous to steroid receptors? Nature 315, 546.

185) Kumar, V., Green, S., Stack, G., Berry, M., Jin, J.-R., Chambon, P. (1987): Functional domains of the human estrogen-receptor. Cell 51, 941—951.

186) Murphy, L.C., Dotzlaw, H. (1989): Variant estrogen receptor mRNA species detected in human breast cancer biopsy samples. Mol. Endocrinol. 3, 687—693.

187) McCarthy, K.S., Miller, L.S., Cox, E.B., Konradt, J., McCarthy, K.S. (1985): Estrogen receptor analyses: correlation of biochemical and immunohistochemical methods using monclonal anti-receptor antibodies. Arch. Pathol. Lab. Med. 109, 716—721.

188) Jonat, W., Maas, H., Stegner, H.E. (1986): Immunohistochemical measurement of estrogen receptors in breast cancer samples. Cancer Res. 46, 4296S—4298S 189) Giri, D.D., Dangerfield, V.J.M., Lonsdale, R., Rogers, K., Underwood, J.C.E. (1987): Immunohistology of oestrogen receptor and D5 antigen in breast cancer: correlation with oestrogen receptor content of adjacent cryostat sections assayed by radioligand binding and enzyme immunoassay. J. Clin. Pathol. 40, 734—740.

190) Raymond, V., Atwater, J.A., Verma, I.M. (1989): Removal of an mRNA destabilizing element correlates with the increased oncogenicity of proto-oncogene *fos*. Oncogene Res. 5, 1—12.

191) Slamon, D.J., de Kernion, J.B., Verma, I.M., Cline, M.J. (1984): Expression of cellular oncogenes in human malignancies. Science 224, 256—262.

192) Crepin, M., Thierry, D., Zafrani, B., Magdelenat, H. (1985): Expression of c-erb, B., c-myc and c-myb oncogenes in human primary breast tumors. Bull. Cancer (Paris) 72, 145—152.

193) Whittaker, J.L., Walker, R.A., Varley, J.M. (1986): Differential expression of cellular oncogenes in benign and malignant human breast tissue. Int. J. Cancer 38, 651—655.

194) Spandidos, D.A., Pintzas, A., Kakkanas, A., Yiagnisis, M., Mahera, H., Patra, E., Agnantis, N.J. (1987): Elevated expression of the *myc* gene in human benign and malignant breast lesions compared to normal tissue. Anticancer Res. 7, 1299—1304.

195) Guérin, M., Barrois, M., Terrier, M.-J., Spielmann, M., Rion, G. (1988): Overexpression of either c-*myc* or c-*erb*B-2/neu proto-oncogenes in human breast carcinomas: correlation with poor prognosis. Oncogene Res. 3, 21—31.

196) Dubik, D., Dembinski, T.C., Shiu, R.P.C. (1987): Stimulation of c-*myc* oncogene expression associated with estrogen-induced proliferation of human breast cancer cells. Cancer Res. 47, 6517—6521.

197) Onclercq, R., Gilardi, P., Lavenu, A., Cremisi, C. (1988): c-myc products *trans*-activate the adenovirus E4 promoter in EC stem cells by using the same target sequence as E1A products. J. Virol. 62, 4533—4537.

198) Chiu, R., Boyle, W.J., Meek, J., Smeal, T., Hunter, T., Karin, M. (1988): The c-Fos protein interacts with c-Jun/AP-1 to stimulate transcription of AP-1 responsive genes. Cell 54, 541—552.

199) Halazonetis, T.D., Georgopoulos, K., Greenberg, M.E., Leder, P. (1988): c-Jun dimerizes with itself and with c-Fos, forming complexes of different DNA binding affinities. Cell 55, 917—924.

200) Lamph, W.W., Wamsley, P., Sassone-Corsi, P., Verma, I.M. (1988): Induction of proto-oncogene JUN/AP-1 by serum and TPA. Nature 334, 629—631.

201) Nakabeppu, Y., Ryder, K., Nathans, D. (1988): DNA binding activities of three murine Jun proteins: stimulation by Fos. Cell 55, 907—915.

202) Rauscher, F.J., Sambucetti, L.C., Curran, T., Distel, R.J., Spiegelman, B.M. (1988): Common DNA binding site for Fos protein complexes and transcription factor AP-1. Cell 52, 471—480.

203) Sassone-Corsi, P., Lamph, W.W., Kamps, M., Verma, I.M. (1988): Fos-associated cellular p39 is related to nuclear transcription factor AP-1. Cell 54, 553—560.

204) Schöntal, A., Herrlich, P., Rahmsdorf, H.J., Ponta, H. (1988): Requirement for fos expression in the transcriptional activation of collagenase by other oncogenes and phorbol esters. Cell 54, 325—334.

205) Quantin, B., Breathnach, R. (1988): Epidermal growth factor stimulates transcription of the c-jun proto-oncogene in rat fibroblasts. Nature 334, 538—539.

206) Ryder, K., Nathans, D. (1988): Induction of protooncogene c-jun by serum growth factors. Proc. Natl. Acad. Sci. USA 85, 8464—8467.

207) Cohen, D.R., Curran, T. (1988): Fra-1: A serum-inducible, cellular immediate-early gene that encodes a Fos-related antigen. Mol. Cell Biol. 8, 2063—2069.

208) Ryder, K., Lau, L.F., Nathans, D. (1988): A gene activated by growth factors is related to the oncogene v-jun. Proc. Natl. Acad. Sci. USA 85, 1487—1491.

209) Ryseck, R.-P., Hirai, S.I., Yaniv, M., Bravo, R. (1988): Transcriptional activation of c-jun during the G1/G0 transit in mouse fibroblasts. Nature 334, 535—537.

210) Hirai, S.-I., Ryseck, R.-P., Mechta, F., Bravo, R., Yaniv, M. (1989): Characterization of junD: a new member of the jun proto-oncogene family. EMBO, J. 8, 1433—1439.

211) Ryder, K., Lanahan, A., Perez-Albuerne, E., Nathans, D. (1989): Jun-D: A third member of the Jun gene family. Proc. Natl. Acad. Sci. USA 86, 1500—1503.

212) Zerial, M., Toschi, R.L., Ryseck, R.-P., Schuermann, M., Müller, R., Bravo, R. (1989): The product of a novel growth factor activated gene, fos, B., interacts with JUN proteins enhancing their DNA binding activity. EMBO, J. 8, 805—813.

213) Bartel, D.P., Sheng, M., Lau, L.F., Greenberg, M.E. (1989): Growth factors and membrane depolarization activate distinct programs of early response gene expression: dissociation of fos and jun induction. Genes. Dev. 3, 304—313.

214) Klein, G. (1987): The approaching era of the tumor suppressor genes. Science 238, 1539—1544..

215) Whyte, P., Buchkovich, K.J., Horowitz, J.M., Friend, S.H., Raybuck, J., Weinberg, R.A., Harlow, E. (1988): Association between an oncogene and an anti-oncogene: the adenovirus E1A proteins bind to the retinoblastoma gene product. Nature 334, 124—129.

216) Lillie, J.W., Green, M.R. (1989): Transcription activation by the adenovirus E1a protein. Nature 338, 39—44.

217) Lee, EY.-HP., To, H., Shew, J.-Y., Bookstein, R., Scully, P., Lee, W.-H. (1988): Inactivation of the retinoblastoma susceptibility gene in human breast cancers. Science 241, 218—221.

218) T'Ang, A., Varley, J.M., Chakraborty, S., Murphree, A.L., Fung Y.-KT. (1988): Structural rearrangement of the retinoblastoma gene in human breast carcinoma. Science 242, 263—266.

219) Vignon, F., Derocq, D., Chambon, M., Rochefort, H. (1983): Estrogen-induced proteins secreted by, MCF-7 human breast cancer cells stimulate their proliferation., C., R. Acad. Sci. Paris 296, 151—156. 220) Dickson, R.B., Kasid, A., Huff, K.K., Bates, S.E., Knabbe, C., Bronzert, D., Gelmann, E.P., Lippman, M.E. (1987): Activation of growth factor secretion in tumorigenic states of breast cancer induced by 17β-estradiol or v-Ha-*ras* oncogene. Proc. Natl. Acad. Sci. USA 84, 837—841.

221) Kasid, A., Knabbe, C., Lippman, M.E. (1987): Effect of v-ras^{H} oncogene transfection on estrogen-independent tumorigenicity of estrogen-dependent human breast cancer cells. Cancer Res. 47, 5733—5737.

222) Shipley, G.D., Tucker, R.F., Moses, H.L. (1985): Type β transforming growth factor/growth inhibitor stimulates entry of monolayer cultures of AKR-2B cells into, S phase after a prolonged prereplicative interval. Proc. Natl. Acad. Sci. USA 82, 4147—4151.

223) Leof, E.B., Proper, J.A., Goustin, A.S., Shipley, G.D., Dicorletto, P.E., Moses, H.L. (1986): Induction of c-sis mRNA and activity similar to platelet-derived growth factor by transforming growth factor β: A proposed model for indirect mitogenesis involving autocrine activity. Proc. Natl. Acad. Sci. USA 83, 2453—2457.

224) Clarke, R., Brünner, N., Katz, D., Glanz, P., Dickson, R.B., Lippman, M.E., Kern, F.G. (1989): The effects of constitutive expression of the transforming growth factor-α on the growth of, MCF-7 human breast cancer cells *in vitro* and *in vivo.* Mol. Endocrinol. 3, 372—380.

225) Murphy, L.C., Dotzlaw, H. (1989): Regulation of transforming growth factor α and transforming growth factor β messenger ribonucleic acid abundance in T-47D, human breast cancer ceils. Mol. Endocrinol. 3, 611—617.

226) Valverius, E.M., Bates, S.E., Stampfer, M.R., Clark, R., McCormick, F., Salomon, D.S., Lippman, M.E., Dickson, R.B. (1989): Transforming growth factor α production and epidermal growth factor receptor expression in normal and oncogene transformed human mammary epithelial cells. Mol. Endocrinol. 3, 203—214.

227) Arteaga, C.L., Coronado, E., Osborne, C.K. (1988): Blockade of the epidermal growth factor receptor inhibits transforming growth factor α-induced but not estrogen-induced growth of hormone-dependent human breast cancer. Mol. Endocrinol. 2, 1064.

228) Van der Burg, B., Van Selm-Miltenburg, A.J.P., De Laat, S.W., Van Zoelen, E.J.J. (1989): Direct effects of estrogen on c-*fos* and c-*myc* protooncogene expression and cellular proliferation in human breast cancer cells. Mol. Cell Endocrinol. 64, 223—228.

229) Van der Burg, B., Isbrücker, R.L., Van Solm-Miltenburg, A.J.P., De Laat, S.W., Van Zoelen, E.J.J. (1990): The role of estrogen-induced insulin-like growth factors in the proliferation of human breast cancer cells (submitted).

230) Rohlik, Q.T., Adams, D., Kull, F.C., Jacobs, S. (1987): An antibody to the receptor for insulin-like growth factor I inhibits the growth of, MCF-7 cells in tissue culture. Biochem. Biophys. Res. Commun. 149, 276—281.

231) McGuire, W., R.L.., Dickson, R.B., Osborne, C.K., Salomon, D. (1988): The role of growth factors in breast cancer. Breast Cancer Res. Treatm 12, 159—166.

232) Van der Burg, B., de Groot, R.P., Isbrücker, R.L., Kruijer, W., de Laat, S.W. Stimulation of TRE activity by a cooperative action of insulin and estrogen in human breast cancer cells, submitted.

233) Soule, H.D., Vazquez, J., Long, A., Albert, S., Brennan, M. (1973): A human cell line from a pleural effusion derived from a breast carcinoma. J. Natl. Cancer Inst. 51, 1409—1416.

234) Weisz, A., Bresciani, F. (1988): Estrogen induces expression of c-*fos* and c-*myc* protooncogenes in rat uterus. Mol. Endocrinol. 2, 816—824.

235) Dubik, D., Shiu, R.P.C. (1988): Transcriptional Regulation of c-*myc* oncogene expression by estrogen in hormone-responsive human breast cancer cells. J. Biol. Chem. 263, 25, 12705—12708.

236) Angel, P., Imagawa, M., Chiu, R., Stein, B., Imbra, R.J., Rahmsdorf HJ, Jonat, C., Herrlich, P., Karin, M. (1987): Phorbol ester inducible genes contain a common cis element recognized by a TPA-modulated trans-acting factor. Cell 49, 729—739. 237) Colamonici, O.R., Trepel, J.B., Vidal, C.A., Neckers, L.M. (1986): Phorbol ester induces c-sis gene transcription in stem cell line K-562. Mol. Cell Biol. 6, 1847—1850.

238) Bjorge, J.D., Paterson, A.J., Kudlow, J.E. (1989): Phorbol ester or epidermal growth factor (EGF) stimulates the concurrent accumulation of mRNA for the EGF receptor and its ligand transforming growth factor-α in a breast cancer cell line. J. Biol. Chem. 264, 4021—4027.

239) Pittelkow, M.R., Lindquist, P.B., Abraham, R.T., Graves-Deal, R., Derynck, R., Coffey, R.J. (1989): Induction of transforming growth factor-α expression in human keratinocytes by phorbol esters. J. Biol. Chem. 264, 5164—5171.

240) Abe, M., Kufe, D. (1986): Effects of maturational agents on expression and secretion of two partially characterized high molecular weight milk-related glycoproteins in, MCF-7 breast carcinoma cells. J. Cell Physiol. 126, 126—132.

241) Issandou, M., Bayard, F., Darbon, J.-M. (1986): Activation by phorbol esters of protein kinase, C in, MCF-7 human breast cancer cells. FEBS Lett. 200, 337—342.

242) Roos, W., Fabbro, D., Küng, W., Costa, S.D., Eppenberger, U. (1986): Correlation between hormone dependency and the regulation of epidermal growth factor receptor by tumour promoters in human mammary carcinoma cell lines. Proc. Natl. Acad. Sci. USA 83, 991—995.

243) Van der Burg, B., Isbrücker, R.L., Blankenstein, M.A., Misdorp, W., de Laat, S.W. Growth factor secretion by autonomously growing canine mammary cancer cell lines, submitted.

244) Valette, A., Gas, N., Jozan, S., Dupont, M.A., Bayard, F. (1987): Influence of 12—O tetradecanoyl-phorbol-13-acetate on proliferation and maturation of human breast carcinoma cells (MCF-7): Relationship to cell cycle events. Cancer Res. 47, 1615—1620.

245) Osborne, C.K., Ross, C.R., Coronado, E.B., Fuqua, S.A.W., Kitten, L.J. (1988): Secreted growth factors from estrogen receptor-negative human breast cancer do not support growth of estrogen receptor-positive breast cancer in the nude mice model. Breast Cancer Res. Treat. 11, 211—219.

246) Imler, J.-L., Wasylyk B. (1989): AP1, A composite transcription factor implicated in abnormal growth control. Prog. Growth Factor Res. 1, 69—77.

PLASTICITY, RECOGNITION AND DIFFERENTIATION IN CELLULAR SYSTEMS

THEORETICAL MODELS FOR CELL TO CELL SIGNALLING

Albert Goldbeter

Faculté des Sciences
Université Libre de Bruxelles
Campus Plaine, C.P. 231
B-1050 Brussels, Belgium

Cell to cell signals govern the development of multicellular organisms and most of their functions. Given the complexity of dynamic phenomena involved in intercellular communication, it is often advantageous to complement their experimental study by a theoretical approach. It was thus a welcome suggestion of the panel of the special NATO programme "Cell to Cell Signals in Plants and Animals" that an Advanced Research Workshop be held on the contribution of theoretical models to the understanding of cell to cell signalling processes. Such a Workshop, organized in collaboration with John Rinzel (Mathematical Research Branch, NIH, Bethesda, USA) and Lee A. Segel (Department of Applied Mathematics, Weizmann Institute of Science, Rehovot, Israel), took place in September 1988 in Knokke-Zoute (Belgium) in the presence of some 75 participants. This unusually large audience was due to the scope of the project.

The resulting volume, entitled "Cell to Cell Signalling: From Experiments to Theoretical Models" was published in October 1989. The aim of the Workshop and of the subsequent book was to present an up-to-date account of the contribution and usefullness of theoretical models in the main fields of cell to cell signalling. A peculiarity of the volume, reflected in its title, is to link experimental and theoretical work. This link is emphasized in all 44 contributions written by leading experts in the fields. Models considered are primarily discussed with respect to their biological rather than mathematical interest. The goal of the meeting was indeed to help bridging the gap between experimental and theoretical approaches in biology.

Rather than writing a general essay on the role of modelling in this domain of biology, it seemed more appropriate to give here a succinct account of the content of the volume which reflects the themes discussed in the Workshop. This will allow to present specific examples of the interplay between theory and experiment, while illustrating at the same time the variety of topics that are presently approached by means of theoretical models.

Although far from being exhaustive, the volume is divided into seven Parts which cover most domains where cell to cell signals play a prominent role. These parts will be dealt with sequentially; the titles of the different contributions to the book are given in an Appendix. Each section begins with an experimental paper that sets the stage for the

NATO ASI Series, Vol. H 51
Cell to Cell Signals in Plants and Animals
Edited by V. Neuhoff and J. Friend

following contributions which are of a theoretical and/or experimental nature. Both aspects are thus closely related so that a dialogue is established over the various themes considered in the volume.

1. FROM NERVE CELLS TO NEUTRAL NETWORKS

It is only fitting that the first Part be devoted to nerve cells and neural networks. After all, the fertility of theoretical models in neurobiology has long been recognized since the work of Hodgkin and Huxley on the nerve impulse: the series of papers that they published in 1952 dealt with experimental aspects as well as with the mathematical modelling of the action potential. This seminal work still represents a prototypic combination between the experimental and theoretical approaches in biology.

As a follow-up of these classical studies devoted to excitability of the squid axon, papers in Part 1 start at the cellular level by addressing more complex modes of behavior such as bursting or multiple modes of neuronal oscillations. The electophysiological properties of central neurones are discussed by Llinas, who relates these properties to the global behavior of the brain. Particular attention is given to the intrinsic oscillatory properties of nerve cells, and to the relation of these properties to the global, cyclic activity of the central nervous system. Among the cells discussed by Llinas are thalamic neurones whose bursting behavior is analyzed in a mathematical model by Hindmarsh and Rose. These authors first discuss a three-variable, qualitative model accounting for bursting in the form of trains of high-frequency spikes separated by silent phases. This is followed by the analysis of a three-variable, quantitative model based on ionic currents observed in thalamic cells. The latter model accounts for many properties of these neurones observed in the experiments.

One of the most studied cells in neurobiology is the *R15* burster of *Aplysia* whose rhythmic behavior is considered from an experimental and theoretical point of view by Adams and Benson. Also discussed are endogenous burster cells in crustacea and in the vertebrate CNS. The authors provide a computer program for numerical simulations of bursting.

Another topic is that of neurotransmitter release. H. and I. Parnas use the kinetics of release as a tool to distinguish between alternative mechanisms for neurotransmitter secretion. A first mechanism is based on the hypothesis that Ca^{2+} accumulation is the only factor controlling release; a second mechanism stipulates that another factor, together with Ca^{2+}, accumulates during stimulation. The two possibilities are discussed in the light of experiments and examined by means of computer simulations.

At the multicellular level, several issues are considered. The first is how the coupling between secretory cells modifies their dynamic behavior. Thus, Sherman and Rinzel examine the behavior of insulin-secreting β-cells by extending a model proposed by Chay and Keizer on the basis of equations of the Hodgkin-Huxley type. This theory, which is appropriate for describing the behavior of a cell in a perfectly synchronized pancreatic islet, is modified to account for the disorganized spiking of a single isolated cell. The analysis of the global behavior of such stochastic, single-cell models shows how regular bursting may arise in clusters of β-cells, as a result of channel sharing through electrical coupling.

A major issue in multicellular interactions in the nervous system is how neural networks behave as central pattern generators that control, for example, locomotion in invertebrates or vertebrates. This topic is addressed from an experimental and theoretical point of view by Grillner et al. who discuss the segmental burst-generating network of the lamprey. The neural mechanisms generating swimming, and the effect of various neurotransmitters are discussed and investigated by means of computer simulations. The mathematical modelling of central pattern generators if further addressed by Ermentrout and Kopell who consider a class of models based on a chain of oscillators, each of which represents a local neural subnetwork capable of producing a stable, periodic output. These models are used to discuss experimentally observed properties such as "fictive swimming" produced by the isolated cord of lampreys, the induction of different local frequencies by exposing different parts of the cord to baths with different ionic concentrations, as well as the effect of periodic forcing.

Finally, the process of learning is discussed experimentally and in terms of neural networks. Dudai reviews the molecular and cellular aspects of learning, insisting on its multilevel organization. Molecular and cellular mechanisms for the encoding of representations are presented, with reference to *Aplysia* and *Drosophila*. Particular attention is give to the neurogenetic approach which indicates the role of cyclic AMP and adenylate cyclase as a convergence locus for the cellular codes of associative stimuli, the cyclic-AMP dependent protein kinase being a candidate retention locus for cellular memory. At the other extreme of studies on learning, the contribution of Demongeot et al. illustrates how this problem is addressed in terms of the collective behavior of networks of excitable elements. This approach has received increased attention in recent years, partly because of its implications for the development of new types of computers. Applications of neurocomputing methods to neuromodelling are presented, with particular reference to a sensory process, namely, auditory nerve activity.

2. MORPHOGENESIS AND DEVELOPMENT

Part 2 is devoted to morphogenesis and development. Here also, a seminal paper published in 1952 by A.M. Turing opened the way for understanding the theoretical bases of pattern formation. This study was later extended in the studies of Prigogine and coworkers on the thermodynamics of irreversible processes; the latter studies led to the concept of *dissipative structures*, by explaining how spatial or spatiotemporal patterns may originate beyond a critical point of instability, far from equilibrium, in open systems comprising chemical reactions and diffusion.

The section opens with a contribution by Bonner on the evolution of morphogens. This starts with the role of volatile morphogens, exemplified by NH_3 in the cellular slime mold *Dictyostelium discoideum*. Larger morphogens are discussed thereafter, such as cyclic AMP in the same species of amoeba. Morphogens exchanged through gap junctions are considered in turn. Throughout this paper, evolutionary aspects are emphasized. The role of positional information is addressed by Wolpert, who raises the issue of how this information is interpreted in the developing organism. The ideas are applied to the growth of limbs in the chick embryo. Development of the early amphibian embryo is discussed by Cooke who reviews the experimental aspects of the process and considers a minimal model for the interactions that polarize and proportion the body plan in amphibian mesoderms.

Current theories of morphogenesis divide between two main types, namely, those based on reaction-diffusion mechanisms, and those that involve mechanochemical instabilities due to cell traction. Murray and Maini provide a comparative discussion of the two types of structure-generating mechanisms, and apply them to the appearance of patterns in limb development and in the formation of feathers or scales. A mechano-chemical model for the budding of membranes is presented by Oster and Moore. The model is applied to the budding of vesicles from the Golgi, and to budding of viruses from a host cell's membrane.

Another major developmental system is that of the *Drosophila* embryo. The emergence of the band pattern characteristic of that system is discussed in a series of theoretical papers which are both closely related to experimental data on mutants. First, Meinhardt discusses how positional information may arise in reaction-diffusion systems involving autocatalysis and lateral inhibition. He then addresses the problem of how particular genes are activated in multi-gradient systems, and deals with borders between domains as organizing regions during development. These ideas are then applied to the control of segmentation in *Drosophila*. The same problem is considered by Goodwin and Kauffman on the basis of reaction-diffusion mechanisms giving rise to patterns of different wavelength in the developing embryo. The sequence of harmonics generated is related to the different segmentation stages observed in the experiments. Hunding presents the results of simulations in three curvilinear coordinates, performed on supercomputers. Stable patterns of stripes reminiscent of the segmentation patterns seen in the insect are obtained.

The closing paper of this section deals with the modelling of plant growth. De Reffye et al. present the method that they have developed to simulate the development of various species of trees. The modelling is based on measurements of probabilities of meristem growth, branching out, and mortality; these data are incorporated into a computer algorithm, referred to as "growth engine", which resorts to Monte Carlo methods for simulating the time evolution of the system. This procedure generates pictures that closely resemble actual plants.

3. CELL TO CELL SIGNALS IN IMMUNOLOGY

Immunology is one of the fastest growing fields for the application of theoretical ideas in biology. In Part 3, some of the most salient examples of immunological modelling are presented. In the introductory Chapter, Urbain et al. review the major problems facing a general theory of immunoregulation. These are the mechanism of self — non self discrimination, the selection of antibody repertoires, and the regulation of the temporal response that follows introduction of an antigen. Particular attention is given to regulation by idiotypic cascades. A model based on two populations of developing B lymphocytes, one of which only plays a regulatory role, is presented, which provides a clue to the problem of self-non self discrimination.

Theoretical aspects of immune networks are dealt with in the following contributions. Perelson presents a topological view of idiotypic networks of increasing complexity. The dynamics of such networks viewed as rooted trees are studied via Monte Carlo computer simulations. Of particular interest are findings of behavioral modes reminiscent of phase transitions. In the next paper, Segel and Perelson outline their approach of immune networks in shape space, in which cells and molecules of the immune system are characterized by a "generalized shape" variable. An equation is proposed for the time evolution of the population of lymphocytes corresponding to a particular shape. Instabilities in shape space may occur as a result of cell to cell interactions. The model is further developed to include antibody secretion and binding to the cell receptors.

De Boer examines immunological networks with regard to dynamic properties such as memory, connectivity thresholds, tolerance, and self-non self discrimination. Models based on clonal selection as well as on idiotypic regulation are considered, while interactions between B and T cells are taken into account. Kaufman and Urbain follow by examining the dynamics of the immune response in terms of a model based on idiotypic regulation. Primary and secondary responses are compared with experimental observations. The authors show how the various modes of humoral response can be interpreted in terms of transitions between multiple steady states.

How the immunological system controls tumor growth is discussed by Lefever et al. who present a model for the interaction of cytotoxic cells with freely evolving tumors. By analysis of a mathematical model for these interactions, the authors account for multiphasic patterns of tumor growth as a function of time (e.g., an initial reduction followed by an irreversible increase in the size of the tumor). The analysis points to the existence of an "immune response dilemma": in certain instances, any further increase in the number or in the affinity of cytotoxic cells for the target tumor cells should only result in a stabilization of the tumoral state.

The last paper in that section investigates the complex immune response that follows infection with HIV. Anderson and May investigate the dynamic behavior of a simple model in which the four variables are the population of free HIV virus at time t, and a population of T lymphocytes subdivided into populations of immature non-activated cells, activated but uninfected cells, and infected cells. Incorporated in the model are cell death after exposure to the virus, and killing of the latter by the immune system. Complex patterns of dynamic behavior are observed when the effect of a second infection is taken into account. In particular chaotic oscillations in HIV concentration occur, which might account for the erratic antigenic patterns observed in some patients infected with HIV.

4. HORMONAL SIGNALLING: THE REPRODUCTIVE SYSTEM

Hormonal signalling is one of the primary modes of intercellular communication. Rather than covering all types of hormonal control, Part 4 focuses on the signals which govern the reproductive function in mammals. Knobil reviews the circhoral hypothalamic clock that governs the menstrual cycle. This clock consists of a pulse generator that triggers the release, every hour or so, of a signal of gonadotropin-releasing hormone (GnRH) which induces the secretion of the gonadotropic hormones LH and FSH by the pituitary.

Another aspect of the ovarian cycle is dealt with by Lacker et al. who examine the mechanism of follicle selection in the mammalian ovary, by studying a mathematical model based on cell to cell signalling through circulatory feedback. The analysis shows how this feedback through circulating oestradiol and gonadotropin hormones regulates follicle growth and the mean ovulation number which is a characteristic of each mammalian species (the latter phenomenon is known as Lipschütz's law of follicular constancy). The model allows to simulate this follicle selection process as well as the spontaneous transition from cyclic to steady anovulatory behavior. On the other hand, Thalabard et al. present numerical simulations of a related model subjected to noise, and show how some follicles which are selected for ovulation at an early stage may become atretic and be replaced by others.

Subsequent papers in this section go back to the role of the pulsatile secretion of GnRH which represents a prototype for the other hormones or growth factors which exert their physiological effect only when delivered in a pulsatory manner. Two papers are devoted to the clinical importance of this phenomenon in humans. Filicori addresses the function of the pulsatile GnRH signal in women, and discusses various reproductive disorders due to deranged pulsatile LH secretion. The role of episodic hormone secretion in reproductive disorders in man is discussed by Wagner et al. The "slow pulsing oligospermia syndrome" identified by these authors might be the first syndrome described where a pulse frequency disorder is of pivotal importance.

The molecular bases for the efficiency of periodic signalling in intercellular communication are discussed by Goldbeter and Li, in the light of results obtained in a theoretical model of a receptor subjected to desensitization and stimulated in a periodic manner. This analysis indicates the existence of an optimal pattern of pulsatile stimulation, and shows how signals can be encoded in terms of their frequency, on the basis of target cell desensitization. A link is made between pulsatile patterns of hormone secretion, as exemplified by GnRH, and intercellular communication by pulses of cyclic AMP in *Dictyostelium* cells.

5. SIGNAL TRANSDUCTION BASED ON INTRACELLULAR CALCIUM OSCILLATIONS

An increasing number of hormones and neurotransmitters are found to act on target cells by triggering intracellular calcium oscillations. The experimental evidence and the theoretical bases for this phenomenon are discussed in a series of three papers. Cuthbertson reviews the experimental results obtained for calcium oscillations in hepatocytes stimulated by vasopressin and in mouse oocytes after fertilization. In the former system, the frequency of calcium transients increases with the extent of hormonal stimulation. Possible mechanisms of calcium oscillations and the biological advantages of signalling based on repetitive calcium transients are discussed.

Berridge provides another view of signalling through cytoplasmic calcium oscillations. A table showing the types of cells where the phenomenon has been observed in recent years demonstrates its wide occurrence. Models of cytosolic oscillations of calcium fall into two classes, depending on whether or not they rely on the concomitant, periodic variation of inositol-(1,4,5)trisphosphate ($Ins(1,4,5)P_3$), a calcium-mobilizing messenger whose synthesis is triggered by external stimulation. Thus, models are either based on $Ins(1,4,5)P_3$ pulsing or on an endogenous oscillation of calcium transport between the cytosol and an intracellular store. The functional significance of calcium spiking is discussed, and a comparison is made between cellular responses based on amplitude or frequency control of the signal.

In the last paper of this section, Dupont and Goldbeter analyze the conditions in which calcium oscillations may develop in the absence of $Ins(1,4,5)P_3$ pulsing, solely on the basis of transport between the cytosol and an intracellular store. Theoretical arguments indicate that the process of calcium-induced calcium release observed in excitable cells and oocytes may suffice to produce calcium oscillations, provided that this regulation takes the form of an activation by cytosolic calcium of calcium transport from the $Ins(1,4,5)P_3$-insentive intracellular store into the cytosol. Encoding of the external signal in terms of the frequency of calcium transients could be achieved through protein phosphorylation by a calcium-dependent protein kinase.

6. INTERCELLULAR COMMUNICATION IN DICTYOSTELIUM

Another system in which intercellular communication occurs through pulsatile signals is the slime mold *Dictyostelium discoideum.* These amoebae aggregate after starvation by a chemotactic response to pulses of cyclic AMP (cAMP) emitted with a periodicity of several minutes by cells behaving as aggregation centers. Gundersen et al. review their experimental results on the molecular mechanism of cAMP oscillations, with particular emphasis on the role of desensization through reversible phosphorylation of G-protein-coupled cAMP receptors. The possible existence of a second, cAMP-independent oscillator in *Dictyostelium discoideum* is discussed by Nanjundiah and Wurster, on the basis of light-scattering measurements which demonstrate the occurrence of two distinct types of oscillations in the course of development.

After these papers which deal with the temporal aspects of signalling, two contributions deal with the mechanism of wavelike aggregation of the amoebae after starvation. First, Müller and Hess relate this phenomenon to experimental examples of spatio-temporal order in chemical systems. Thus, these authors present results obtained on spiral waves of chemical activity in the Belousov-Zhabotinsky reaction which is the prototype of chemical oscillatory system. The quantitative analysis of these chemical waves throws light on the origin of spiral patterns, on the nature of the spiral core and on the interaction between colliding wavefronts. The effect of hydrodynamic flow is shown to produce deformations in the chemical patterns generated through the nonequilibrium coupling between reaction and diffusion. Preliminary results on the extension of these experimental techniques to the analysis of spiral patterns in aggregating *Dictyostelium* amoebae are presented.

The physico-chemical bases of wavelike aggregation of *Dictyostelium* amoebae are dealt with by Tyson in a paper which provides a review of general theories of wavelike propagation in excitable biological media. The specific aspects of cAMP waves in *Dictyostelium* are considered by means of a theoretical analysis based on a two-variable model for cAMP signalling. The relation between wave speed and temporal period obtained in that analysis is compared with experimental data. The link is made with wavelike propagation in other excitable media such as the Belousov-Zhabotinsky reaction and the heart.

7. SIGNAL PROPAGATION IN THE HEART

The last Part of the volume deals with signal propagation in the heart. In few tissues are cell to cell signals so essential for the maintenance of life! In this section also, experimental and theoretical aspects are closely intertwined. Goldberger and Rigney first present an overview of cardiac dynamics, insisting on the nonlinear motions of the heart. Fractal aspects of this muscle are encountered both in its spatial organization and in its temporal dynamics. Heart rate time series are discussed with respect to the occurrence of chaos, and the role of neurohumoral control of this phenomenon is underlined. Also discussed in this light is the dynamics of sudden cardiac death.

Guevara et al. address the dynamics of isolated ventricular myocytes subjected to periodic stimulation. This experimental and theoretical paper provides insights into the origin of the phenomenon of alternans, which consists in the beat to beat alternation in the intensity of the heart beat. The experimental results are simulated by means of a model developed by Beeler and Reuter for the dynamic behavior of ventricular fibres. The results demonstrate that the phenomenon of alternans may be generated already at the unicellular level: intrinsic alternation of electrical activity can indeed be produced in a single myocyte. The results are further explored theoretically by analysis of a one-dimensional map proposed for the dynamics of cardiac cells. The authors conclude by relating their findings on alternans to ventricular fibrillation.

Allessie et al. devote their experimental analysis to the role of anisotropic impulse propagation in ventricular tachycardia. Of particular importance is the process of re-entry, which refers to the failure of the propagation of the depolarization wave to extinguish after having excited all cardiac fibres. This process is responsible for the majority of cardiac arrhythmias. The authors review the different mechanisms of re-entry, with particular emphasis on the recently discovered mechanism based on the anisotropic propagation of the impulse. The experiments designed to study this phenomenon in a two-dimensional preparation of ventricular myocardium are described. Van Capelle and Allessie then show how these results can be simulated by means of a model consisting of an arbitrary number of electrically excitable elements which can be connected in various ways in a two-dimensional matrix. Anisotropic conduction in a rectangular sheet of tissue is simulated by connecting elements with different resistances in horizontal and vertical directions; this difference results in a marked propagation of the impulse in the two directions. Conditions for the occurrence of circulating impulses are obtained and the role of an excitable gap, observed in the experiments of Allessie et al., as well as the position of the conduction block, are discussed.

The cellular bases of ventricular tachycardia in the myocardium are further addressed by Keener who presents a mathematical model for propagation of the action potential and re-entry in a discrete excitable medium. A simplified description for the generation of the action potential is used in the form of the cubic, Nagumo equation. This allows to analyze the conditions for the propagation of wavelike solutions to these equations in the case of continuous coupling between the excitable elements. With anisotropic coupling coefficients, propagation along a single axis is shown to occur in certain conditions. The simulations indicate that in a discrete anisotropic excitable medium there is a period of vulnerability during which a premature stimulus will elicit a re-entrant pattern of excitation. The author shows how such theoretical results account for experimental observations.

In the last paper of this section, which also concludes the volume, Winfree further discusses in a more intuitive manner the process of re-entry which plays a central role in fibrillation and sudden cardiac death. First, the author discusses the myocardium as a generic exitable medium, and stresses common properties also encountered in other excitable media such as *Dictyostelium* or the Belousov-Zhabotinsky reaction. Second, the author shows how similar results on the dynamics of impulse propagation can be obtained by means of different theoretical paradigms. Non-uniform dispersion of refractoriness in cellular anisotropic tissue is again at the core of re-entry and fibrillation.

CONCLUSIONS AND PERSPECTIVES

The variety of topics addressed in this Workshop and in the resulting volume of proceedings illustrates the wide range of application of theoretical models to cell to cell signalling processes. Is there a common theme that emerges from the presentation of experimental and theoretical aspects of intercellular communication in such different fields of cell biology? Looking for general, unifying principles can be misleading and somewhat ludicrous, since the usefulness of models is directly related to their predictive power; the latter, in turn, requires their being highly specific rather than vague. However, despite the diversity of the phenomena considered, some common themes do appear. Thus, excitable and oscillatory behavior occur for nerve as well as cardiac cells. Pulsatile signalling is a neuronal property which is also found for an increasing number of hormones and for *Dictyostelium* amoebae. Encoding of an external signal in terms of its frequency is observed in all these instances of intercellular communication and might also underlie signal transduction based on intracellular calcium oscillations. These examples of frequency encoding are brought together for the first time in a single volume. On the other hand, wavelike phenomena are observed in *Dictyostelium* as in the myocardium. Wave propagation in excitable media has become a subject of choice for modelling studies in biology.

Being closely related to experimental data, the models discussed in the Workshop retain the specificities of their respective fields. Most studies nevertheless deal with dynamic events which often lead to instabilities and nonequilibrium self-organization. Therefore, bifurcations associated with oscillations or with transitions between multiple steady states are recurrent themes, from neurobiology to immunology and hormonal regulation, while spatial pattern formation arises in morphogenesis as in cardiac physiology or slime mold aggregation.

In each case, models have proved highly sucessful in performing one or more of the following functions: (i) accounting for puzzling observations which escape sheer intuition, (ii) providing a unified explanation of experimental observations in terms of physico-chemical processes, (iii) suggesting new experiments which will in turn corroborate the model or lead to its necessary modification. Knowledge progresses through such ever-continuing process of mutual feedback between theory and experiment.

One of the most important preliminary questions in modelling is to determine the optimal level of biological details to be incorporated at the stage of model building. Since models are maps that help finding a way in the maze of the complex dynamics of living systems, they should be drawn at an appropriate scale, neither too minute (they would be as intricate at the systems under study, and would not be amenable to thorough analysis), nor too vast (they would be too vague and would lose any predictive power if too distant from reality).

In a lithograph entitled “Ophtalmologie”, reproduced in the book, the Belgian painter Pierre Alechinsky represented a human face probing the world with two pairs of eyes. This figure could serve as metaphor to illustrate the purport of this Workshop and of the resulting volume, which is to demonstrate the added insight that the combined approaches of experiment and theory provide into the dynamics of living systems and, in particular, into the various modes of intercellular communication.

ACKNOWLEDGMENTS

I would like to express my gratitude to the members of the panel of the special NATO programme "Cell to Cell Signals in Plants and Animals" for their invitation to organize this Workshop, and to Dr. Alain Jubier, in charge of this programme at the NATO Division for Scientific Affairs, for his constant help. Thanks are also due to the European Communities, the Belgian National Fund for Scientific Research (FNRS) and the Belgian Ministry for Education, for further financial support. Finally, I am deeply grateful to the two co-organizers of this Workshop, John Rinzel, from the Mathematical Research Branch of the NIH (Bethesda, USA), and Lee A. Segel, from the Weizmann Institute of Science (Rehovot, Israel) for their most precious collaboration in selecting themes and contributors.

PLANT HORMONE RECEPTORS

A.M. Mennes[1], A. Maan[1], and M.A. Hall[2]

[1]Department of Plant Molecular Biology,
Rijksuniversiteit Leiden
The Netherlands

[2]Department of Biological Sciences,
University College of Wales, Aberystwyth,
Dyfed SY23 3DA, Wales, U.K.

INTRODUCTION

The NATO ARW on Plant Hormone receptors held in August 1986 was probably the first conference to deal exclusively with this subject, yet barely three years later in 1989 there were two international conferences devoted entirely to such studies and four others where it was given a high profile. This reflects not only an acceleration of work in the field but also greatly increased interest amongst the scientific community as a whole, although it is probably still true to say that there are fewer laboratories working on receptors for all the plant hormones put together than those involved in studies on any one animal hormone receptor. Given this proliferation of information it is impossible in a short article such as this to do complete justice to the field and readers are referred to the recent reviews and symposium articles noted in the references for further information.

Auxins

A. Membrane-bound auxin receptors

The most far-reaching results can be reported for studies on membrane-bound auxin-binding proteins (ABPs), in particular those from maize. In order to raise antibodies against the maize membrane-bound ABP and to sequence it, the protein was solubilised and extensively purified. Napier, et al. (1988) described protocols based on commercially available chromatographic materials yielding receptor preparations of over 50% purity. Such preparations were used to produce monoclonal antibodies. Native PAGE of the semi-purified preparations yielded a homogeneous ABP. Its native molecular weight is ca. 44kD which consists of a glycosylated homodimer of two 22 kD subunits. The latter was used to produce a polyclonal rabbit antiserum. Two monoclonal antibodies specifically recognize the major 22kD binding protein polypeptide and could be mapped to epitopes within a C-terminal 1 kD region (Venis and Napier, 1990). This region appeared to be involved in a conformational change which is induced when auxin binds. In addition, three other monoclonals recognize a minor 21 kD species, and one of these mono-

NATO ASI Series, Vol. H 51
Cell to Cell Signals in Plants and Animals
Edited by V. Neuhoff and J. Friend

clonals has been mapped to an epitope within 7 kD of the N-terminus (Venis and Napier, 1990).

Löbler and Klämbt (1985) reported a single 20 kD polypeptide from SDS-PAGE of their ABP preparation while Shimomura et al., (1986) found a major 21 kD subunit but also a 1 kD smaller subunit still showing full auxin-binding activity. Treatment with endoglycosidase H showed that the ABP contains a single glycan chain of 2 kD (Löbler et al., 1987; Napier et al., 1988; Inohara et al., 1989).

All the monoclonal antibodies isolated by Napier et al. (1988) recognized the polypeptides rather than the glycan side chain and two monoclonals even showed an apparently higher affinity for deglycosylated ABP. Immunoprecipitation with the two monoclonal antibodies which specifically recognize the 22 kD ABP removed auxin-binding activity from the partially purified receptor preparations. The polyclonal antiserum immunoprecipitated the 22 kD polypeptide, inhibited auxin-binding and cross-reacted with polypeptides in wide range of other species, both mono- and dicotyledons (Venis and Napier, 1990). The polyclonal serum also exhibited approximately equal affinities for glycosylated and deglycosylated forms of the ABP.

Functional evidence for an auxin receptor localised at the plasmamembrane of tobacco mesophyll protoplasts was reported by Barbier-Brygoo et al., (1989). They used auxin-induced hyperpolarisation of membrane potential (E_m) in tobacco protoplasts as a functional test to show (see Table 1) the effects of polyclonal antibodies directed against the auxin receptor ($IgG_{anti\text{-}ABP}$) in maize (Löbler and Klämbt, 1985). Removal of the vacuole from the protoplast did not affect the membrane hyperpolarisation induced by auxin, thus indicating that the observed response resides at the plasmamembrane alone. The involvement of the plasmamembrane ATPase in the E_m of isolated tobacco protoplasts was revealed by using $IgG_{anti\text{-}ATPase}$ (Table 1).

TABLE 1

Effects of $IgG_{anti\text{-}ATPase}$ and $IgG_{anti\text{-}ATP}$ on the variation in membrane potential difference in tobacco protoplasts induced by auxin and fusicoccin

Addition to medium	Em, mV	NAA- or FC-induced Em, mV
None	− 5.5 ± 0.2	—
NAA (5 μM)	−11.2 ± 0.2	−5.7
$IgG_{anti\text{-}ABP}$ (4 nM)	− 5.4 ± 0.2	—
$IgG_{anti\text{-}ABP}$ (4 nM) + NAA (5 μM)	− 5.9 ± 0.2	−0.5
None	− 6.1 ± 0.3	—
$IgG_{anti\text{-}ATPase}$	+ 0.8 ± 0.6	—
NAA (5 μM)	−12.5 ± 0.4	−6.4
$IgG_{anti\text{-}ATPase}$ + NAA (5 μM)	− 0.2 ± 0.5	−1.0
FC (1 μM)	−10.8 ± 0.4	−4.7
$IgG_{anti\text{-}ATPase}$ + FC (1 μM)	+ 0.9 ± 0.6	+0.1

Combined data from Barbier-Brygoo et al. (1989). Functional evidence for an auxin receptor at the plasmalemma of tobacco mesophyll protoplasts. Proc. Natl. Acad. Sci. USA 86, 891—895.

These results show that an auxin-receptor is present at the tobacco plasmamembrane and that it shows strong cross-reactivity with maize-derived antibodies. Also, direct evidence was obtained that ATPase is involved in the auxin-induced membrane response. Further support for the localisation of the auxin receptor on the plasmamembrane was obtained by using impermeant auxin analogues having the same effect as NAA, but only when the cell walls were removed (Venis and Napier, 1990).

Guern et al. (1990) using protoplasts from an auxin-resistant tobacco mutant selected by M. Caboche have demonstrated by immunotitration with $IgG_{anti\text{-}ABP}$ that the mutant had fewer receptors. Similarly, protoplasts isolated from root tips of plants transformed by *Agrobacterium rhizogenes* show greatly enhanced sensitivity to auxin and a corresponding increase in the concentration of receptors. In both cases, membrane hyperpolarisation was used as the functional test. This work indicates that the number of functional receptors at the plasmalemma is a major determinant of sensitivity to auxin.

Further evidence that the maize ABP functions as a receptor comes from the report of Löbler and Klämbt (1985) showing that $IgG_{anti\text{-}ABP}$ blocks the auxin-induced growth response of maize coleoptiles. This ABP was found by indirect immunofluorescence to reside primarily at the plasma membrane of epidermal cells. However, Shimomura et al. (1986) failed to inhibit cell elongation using an $IgG_{anti\text{-}ABP}$ raised against a purified ABP from maize shoot membranes. This binding protein was found to be present in the ER and to be concentrated in the outer epidermis (Shimomura et al., 1988).

The group of Shimomura was the first to succeed in the molecular cloning of the ABP and in publishing the complete primary structure (Inohara et al., 1989). A cDNA clone derived form mRNA encoding the auxin-binding protein was isolated and sequenced. The nucleotide sequence of the 822-bp cDNA includes a 603-bp open reading frame. The predicted precursor for the binding protein contains 201 amino acid residues and has a molecular weight of 21976 D. The NH_2-terminal sequence of 38 residues is hydrophobic and may be a signal peptide for translocation of the protein across the membrane of the ER. The mature ABP contains a potential N-glycosylation site and the COOH-terminal tetrapeptide (KDEL) may be a signal for retention of the protein in the lumen of the ER. These sequence data are in accord with results showing that the main auxin-binding activity is associated with ER vesicles (Shimomura et al., 1988).

At almost the same time two other reports showing similar results appeared. Hesse et al. (1989) purified several ABPs to homogeneity and the major binding protein from maize coleoptiles was investigated by extensive amino acid sequence analysis. These sequences were then used to design oligonucleotides to be used as probes to isolate cDNA encoding this protein. Tillmann et al. (1989) selected an ABP cDNA clone from a λgt 11 cDNA library from maize coleoptiles with highly purified $IgG_{anti\text{-}ABP}$. The full length *in vitro* transcript directs the *in vitro* synthesis of a 22 kD protein which is precipitated by $IgG_{anti\text{-}ABP}$. Antibodies raised against a fusion protein detect the ABP as double band on Western blots. Only one of the two bands is recognised by $IgG_{anti\text{-}KDEL}$. The ABP species which is not recognised by $IgG_{anti\text{-}KDEL}$, and thus probably lacks the ER-sorting signal, may be secreted to exercise a receptor function at the outer face of the plasma membrane, in accordance with other data (Löbler and Klämbt, 1985; Barbier-Brygoo et al., 1989).

Besides the use of poly- and monoclonal antibodies to show the presence of ABPs, a new approach is the use of a photoaffinity labelling technique since in this way non-dissociating ligand-receptor complexes can be isolated. With the photoaffinity analogue 7-[^{3}H]-5N_3-IAA it was found that a polypeptide double of 40 and 42 kD was specifically labelled in membrane preparations of tomato (Hicks et al., 1989b) and in zucchini (Hicks et al., 1989a). With the same labelling agent in maize two peptides with subunit molecular weights of 24 and 22 kD were identified (Jones and Venis, 1989), and it was found that the bulk of the 22 kD ABP is localised in the ER (Jones et al., 1989).

A possible physiological function for the membrane-bound ABP (MABP) in tobacco has been suggested by Maan et al. (1985) who showed that the apparent absence of this binding protein was correlated with the inability of cells to differentiate roots. Based on these results Nakamura et al. (1988) studied the levels of MABP and, as a marker, a root-specific peroxidase (RSP) in several tobacco cell lines including an auxin-resistant variant lacking the ability to differentiate roots. It was shown that this variant has neither detectable levels of MABP nor RSP whereas in other cell lines there exists a correlation between the presence of MABP and of RSP.

In another paper Nakamura and Ono (1988) described success in solubilising MABP from tobacco cell suspension cultures. Comparison of the properties of the tobacco ABPs showed no significant differences between the solubilised and the bound MABP but the MABP and the cytoplasmic ABP (see section B) are two distinct auxin-binding proteins.

B. Soluble auxin receptors

While the membrane-bound ABPs as described above are probably involved in regulation of cell elongation or cell differentiation, a nuclear/cytoplasmic ABP is thought to be involved in regulation of cell division. The mechanism of signal transduction from MABP to physiological effect is either a direct coupling to a membrane-bound effector or via a 'second messenger' while the nuclear/cytoplasmic ABP may directly regulate transcription of auxin responsive genes. The unravelling of the latter process and the steps immediately following the interaction of ABP and DNA has been approached from two sides:

1) Isolation of soluble ABPs and their effect(s) on transcription.
2) Isolation of RNA and proteins formed specifically after induction by auxin.

An overall stimulation of RNA synthesis has already been reported by van der Linde et al. (1984). In isolated nuclei from tobacco they found an increase in *in vitro* mRNA synthesis only after a purified ABP preparation and 10^{-6}M IAA were supplied but not with either ABP alone or IAA alone. A similar experiment was performed using mung bean nuclei (Sakai et al., 1986). They used two purified ABPs (ABP-I and ABP-II) to determine overall RNA synthesis. Although they found 25% and 40% increases using ABP-I and ABP-II, respectively, IAA did not have any effect, whether added alone or in combination with any of the two ABPs.

Later, Kikuchi et al. (1989) looked at a qualitative rather than a quantitative stimulation of transcription. They found that upon *in vitro* transcription and translation in nuclei treated with ABP-I plus IAA one peptide (MW 79 kD) and with ABP-II plus IAA another peptide (MW 18 kD) were specifically formed. These products were not formed when IAA was absent or when either of the two ABPs was absent. The transcription products were shown to be newly synthesised by using [γ-S]GTP which labels 5' ends of transcripts exclusively and can be isolated by HG-sepharose affinity chromatography.

Apart from mung bean and tobacco, soluble ABPs have also been isolated from pea by Jacobsen and his co-workers (Jacobsen, 1984) and more recently from soybean cell suspensions (Herber et al., 1988). In both cases two ABPs can be detected upon IEF.

Both soybean-ABPs are modulated during the growth cycle and have almost completely disappeared from the cytosol after 4 days (Herber et al., 1988).

A different approach is that of van der Zaal et al. (1987) who isolated 7 cDNA clones of mRNA that were specifically transcribed after addition of auxin to a suspension of tobacco cells which had been starved of auxin for several days. The induction of one of these mRNAs in particular shows a good correlation with cell division with regard to specificity of auxins, analogues and other plant hormones on the one hand and dose response the other.

Using some of these cDNA clones, genomic clones have been isolated (van der Zaal, 1989). The promoters of two clones primary auxin-responsive genes were fused to the 5' end of the coding sequence of the β-D-glucuronidase gene (GUS). In transgenic tobacco plants and transgenic hairy roots the constructs give rise to specific GUS expression in the meristemic region of the root tip. In protoplasts derived from transgenic cell cultures GUS expression can be induced by exogenous auxin, thus providing a system for probing the function of cell-surface receptors in auxin-induced gene expression (van der Zaal, 1989).

In order to investigate the possible involvement of a nuclear/cytoplasmic ABP in the induction of the 7 auxin-responsive mRNAs, a method similar to that of Kikuchi et al. (1989) was used by Mennes et al. (1990) to demonstrate that reinitiation of transcription of 4 of the 7 mRNAs takes place. They isolated nuclei from tobacco cell suspensions that had been exposed to auxin for 60 min prior to isolation. These nuclei were able to synthesize [γ-S]RNA that hybridized to blots containing the cDNA clones mentioned above. The next step is obviously, to use untreated cells for the isolation of nuclei and incubate these with auxin and purified ABP.

NPA receptors

A different class of receptors are those of N-1-naphtylphthalmic acid (NPA), an auxin transport inhibitor. These receptors are membrane-bound and are predominantly present in the basal parts of the cell, as shown by Jacobs and Gilbert (1983) using monoclonal antibodies. Even though these monoclonal antibodies have been available for some time now, only a few steps toward purification have been reported. One promising step is the synthesis of a photolabile NPA analogue, 5'-azido-N-1-naphthylphtalmic acid (Voet et al., 1987) that is able to bind competitively, albeit with a 10-fold lower affinity, to the NPA receptor.

More emphasis has been put on structure-activity correlations and a model has been described that takes chemical and steriochemical characteristics into consideration (Katekar et al., 1987a,b). These authors also determined that the NPA receptor is present throughout maize plants (Katekar and Geissler, 1989) and has similar affinities for the phytotropins and analogues tested. Recently, a second type of NPA receptor has been described that binds with a lower affinity, but its function and characteristics are still unknown (Michalke et al., 1990).

A number of workers have attempted to isolate the endogenous ligands corresponding to the artificial phytotropins. These studies have not as yet produced definitive answers although Rubery (1990) has suggested that such compounds may include certain types of flavonoid aglycone such as quercetin.

Fusicoccin

Although fusicoccin, a phytotoxin from the fungus *Fusicoccum amygdali* is only likely ever to come into contact with a few plant species under natural conditions, the fact that its physiological effects are widespread (Marrè, 1979) and that binding sites for the ligand have been detected in all plants so far examined (Ballio & Aducci, 1986) has led to the proposition that it is an analogue of a natural endogenous ligand — a situation very much like that proposed for NPA and its binding sites.

Purification of the fusicoccin binding protein (FCBP) has proved difficult (see e.g. Aducci & Ballio, 1990). Like the ethylene binding protein the FCBP is hydrophobic and exhibits low rate constants of association and dissociation; it is also relatively unstable.

In a recent paper Boer et al. (1989) have demonstrated purification to homogeneity of a FCBP from oat root plasmamembrane. The key step in this procedure was the use of affinity chromatography employing fusicoccin-linked adipic acid dihydrazide agarose. On SDS-PAGE the active fractions contained two prominent bands with an Mr of 29,700 & 30,800, the stoichiometry of which (1:2) indicate that the native protein occurs as a heterotrimer with an apparent molecular weight of 92 kD. This figure is in the same range as other reported values (e.g. Aducci et al., 1984).

Perhaps the greatest step forward in this area has been the demonstration of functionality for the FCBPs. Aducci et al. (1988) incorporated partially purified FCBP and plasmamembrane H^+ ATPase from maize into liposomes by the freeze-thaw method. They then measured intravesicular acidification in response to various treatments using fluorescence quenching of 9-amino-6-chloro-2-methoxyacridine as a marker. Treatment of such preparations with fusicoccin significantly enhanced H^+ translocating activity in the presence of ATP (Fig. 1). Because both the FCBP and H^+ ATPase preparations were

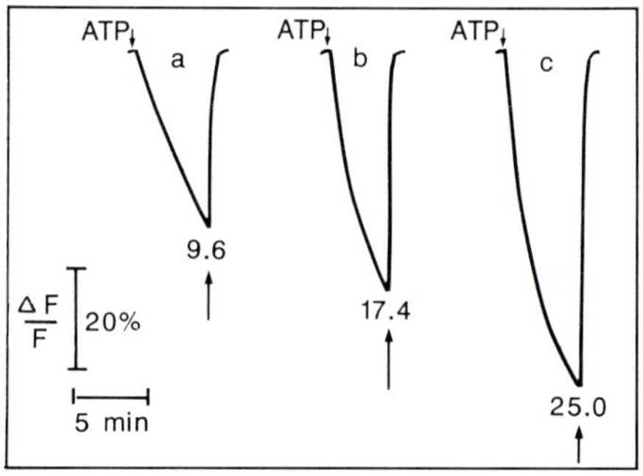

FIGURE 1
Effect of fusicoccin on ATP-driven H^+-transport in reconstituted liposomes. Acidification of proteoliposomes is measured by fluorescence (F) quenching of the dye 9-amino-6-chloro-2-methoxyacridine. The initial rate of quenching, expressed as the percentage of total fluorescence per minute, is indicated on the curves. Acidification was initiated by adding ATP (arrowheads), and the consequent gradient was discharged by addition of 0.5 μM valinomycin and 5 μM carbonyl cyanide m-chlorophenylhydrazone (arrows). Curves: a, proteoliposomes containing ATPase and FCBP (control); b, same proteoliposomes but briefly incubated with 1 μM FC before ATP addition; c, same proteoliposomes but preloaded with 1 μM FC.
From Aducci et al. (1988) Proc. Natl. Acad. Sci. USA 85, 7849—7851.

impure it is unclear whether the interaction(s) between these two components is direct or whether other elements of a transduction system exist. Clearly, however the system not only demonstrates the functionality of the FCBP but also provides a tool both for investigating other components of the transduction chain (if they exist) and relationships with the ABP which certainly shares the ultimate transduction step (Barbier-Brygoo et al. 1989) (see Table 1).

This work has been followed up by a comparison of the properties of FCBPs from maize and from spinach incorporated into liposomes (Aducci et al. 1989). Although in general the properties are very similar, differences emerged in terms of rates of association and dissociation, pH dependence and sensitivity to β-mannosidase indicating that the proteins from the two tissues are somewhat different.

The search for the endogenous ligand(s) corresponding to fusicoccin has continued apace. Fractionation of tissue using anti-fusicoccin antibodies bound to Sepharose has resulted in the isolation of substances able to compete effectively with fusicoccin for binding to the FCBP. However, the low concentration of this substance(s) and its increasing lability on purification have thus far precluded structural characterisation (Aducci & Ballio, 1990).

Ethylene

Despite the technical problems associated with studies on ethylene receptors two groups have succeeded in detecting ethylene binding (Jerie, Shaari & Hall, 1979; Sisler, 1979; Bengochea et al. 1980a,b) and the properties of the sites derived from *Phaseolus vulgaris* and from mung bean were described at the Bonn workshop. Both sites show affinities for ethylene and its physiologically active structural analogues in accord with those expected of functional receptors. On the other hand there are good reasons for questioning whether they are indeed receptors. In *Phaseolus vulgaris* the site was isolated from a tissue where there is no known response to ethylene, namely developing cotyledons, and both of the sites have low rate constants of association and dissociation. This latter constitutes something of a problem since many ethylene effects have short response times, which implies that the appropriate receptors will have high rate constants of association and dissociation. On the other hand, both tissues are relatively rich in the binding proteins and the low rate constants of dissociation permit the monitoring of fractionation by bound ^{14}C-ligand over an extended period. Moreover, given the simplicity of the ethylene molecule and the high and appropriate affinities of the sites for the gas and its analogues, then, by analogy with animal systems (e.g. Lindstrom et al., 1979) it seemed probable that any ethylene binding site/receptor would show homology with the binding domains. Purification of binding proteins such as those from *Phaseolus* and mung bean would therefore permit other approaches, for example, via immunology, leading to the isolation of ethylene binding proteins — including those with high rate constants of association and dissociation — from other tissues.

Both binding sites are highly hydrophobic integral membrane proteins which has greatly hampered efforts to purify them; this has now however been achieved with the protein from *Phaseolus vulgaris* (Hall et al., 1990). Charge shielding by detergent and hydrophobic interactions preclude molecular weight estimation of native protein via gel electrophoresis, but on SDS-PAGE gels two subunits of 10 and 12 kD are observed (see below).

Centrifugation on isokinetic sucrose gradients and gel permeation chromatography indicate that the size of the protein detergent complex is 55—60 kD (Thomas et al., 1985) suggesting that the native protein is a heterotrimer or heterotetramer.

An alternative approach to that outlined above has been the development of techniques for the measurement of ethylene binding sites having high rate constants of association and dissociation (e.g. Sisler, 1979; Sisler & Wood, 1987; Sanders et al., 1989a). While these techniques have serious technical drawbacks they do offer a means of investigating ethylene binding in many tissues and a wide range of plants have been so studied (for a review see Hall et al., 1990).

The system investigated most fully to date is the pea epicotyl, although studies on other tissues including *Arabidopsis*, rice and petioles and abscission zones of *Phaseolus vulgaris* have yielded similar results. What emerges is that in peas there are two classes of binding site showing a curious and, in higher plants, unprecedented, difference in properties. One class of sites shows very high rate constants of association and dissociation; the other class having very low rate constants similar to those in *Phaseolus vulgaris* and mung bean. In other respects such as affinity for ethylene and its analogues the two classes are essentially indistinguishable (Table 2). The concentration of the sites in peas is much lower than in *Phaseolus vulgaris* cotyledons and the concentration of the slow associating sites is higher than that of the fast associating sites.

TABLE 2

Properties of ethylene binding sites from various species as determined by in vivo assay

	Pea epicotyl			*Phaseolus*		*Arabidopsis*	*Rice*	
	Fast	Slow	Growth inhibition	Petiole/abscission zones (both sites)		(both sites)	(both sites)	
K_D (K_m) ethylene (M)	1.12×10^{-10}	1.97×10^{-10}	1.8×10^{-10}	1.92×10^{-10}		3.61×10^{-10}	5.01×10^{-10}	
K_i (K_m) propylene (M)	4.78×10^{-8}	4.25×10^{-8}	1.92×10^{-8}	2.95×10^{-8}				
K_i (K_m) carbon monoxide (M)	3.2×10^{-8}		3.6×10^{-8}					
K_i (K_m) vinyl chloride (M)	7.6×10^{-7}		1.45×10^{-7}					
K_i (K_m) acetylene (M)	3.7×10^{-7}		3.3×10^{-7}					
K_i (K_m) propyne (M)	1.37×10^{-6}		1.14×10^{-6}					
K_i (K_m) 1-butene (M)	4.66×10^{-5}		5.04×10^{-5}					
K_i norbornadiene (M)	1.18×10^{-6}		1.76×10^{-6}					
K_i Ag(I) (M)	3.1×10^{-7}		6×10^{-7}				4.4×10^{-6}	
				Fast	Slow		Fast	Slow
$t_{\frac{1}{2}}$ assic.	8 min	13 h	20 min[2]	17 min	12 h		15 min	12 h
$t_{\frac{1}{2}}$ dissoc.	32 min	22 h	15 min[3]	20 min	6 h			
Concentration of binding sites (pmoles [g F.W.]$^{-1}$)	2.5	8.0		1.5—3.0		0.6	0.6	

All molarities quoted are those in the liquid phase

1. Calculated at infinite dilution
2. Lag time for observable response after exposure
3. Lag time for observable response after removal

On a fresh weight basis there is a higher concentration of both classes of site in the tip (the most ethylene-sensitive region) than further down the epicotyl but on a cell basis the amounts of both classes are strikingly similar; although of course this means that the density of sites is much greater per unit cell volume at the tip. By analogy with the results of Guern et al. (1990) on the density of auxin receptors described above, these differences may have significance for sensitivity to ethylene.

Experiments with abscission zone and petiolar tissue from *Phaseolus vulgaris* have yielded similar results, namely the presence of two classes of site differing only in their rate constants of association and dissociation. Notably, however, the concentration of both classes of site is higher in abscission zones than in petioles.

Other interesting results haven arisen from work on mutants of *Arabidopsis*, wholly or partially insensitive to ethylene (Hall et al., 1990a; Hall et al., 1990b). Bleecker et al. (1988) have characterised similar mutants. Although the experiments are technically extremely difficult, direct measurements of ethylene binding indicate that one or more of these mutants are receptor deficient. Supporting evidence to this end has also been adduced from experiments using antibodies to the *Phaseolus* binding protein (see below).

At the Bonn meeting the question of ethylene metabolism and its possible role in the mediation of ethylene effects (see Smith & Hall, 1984) was discussed. It is now clear that such suggestions are unfounded. Recent work (Sanders et al., 1989a,b) indicates that not only is it unlikely that the enzymes responsible for ethylene metabolism in most plants are specific for the growth regulator, but in addition complete inhibition of metabolism by CS_2 has no effect on the response of tissue to ethylene.

Success in purifying the ethylene binding protein from *Phaseolus vulgaris* to homogeneity has permitted the production of polyclonal antibodies. The antibodies so raised are highly specific and in *Phaseolus* appear to recognise only the ethylene binding protein (Connern et al., 1990; Hall et al., 1990a). Using silver-enhanced immunogold dot-blots and an ELISA it has been demonstrated that a higher concentration of the protein is present in abscission zones than in petioles — a finding in good agreement with the results of direct measurements noted above. Other work has indicated that treatment of plants with the ethylene releasing chemical ethephon (2-chlorethane phosphoric acid) increases the concentration of binding sites in abscission zones.

The antibodies recognize homologous proteins in peas, tomatoes and *Arabidopsis*. In peas the subunit structure appears to be the same as in *Phaseolus* and recent work indicates that one at least of the ethylene-insensitive mutants of *Arabidopsis* is deficient in the antigen recognised by the antibodies.

A c-DNA expression library from abscission zones is currently being prepared which will be used in efforts to clone the gene for the binding protein.

In terms of their properties it seems likely that the ethylene binding proteins do in fact represent receptors. In the case of those sites with high rate constants of association and dissociation it is difficult to assign any other function, especially since the possibility that they represent metabolising systems can be excluded. The sites having low rate constants may be receptors or perhaps receptor precursors. Proof for such roles must however depend on the results of functional tests such as those applied successfully to auxin and fusicoccin binding proteins.

Gibberellins, cytokinins and abscisic acid

In contrast to the substantial progress made with receptors for auxins, ethylene and fusicoccin, very little which is new has emerged for other growth regulators.

In the case of gibberellins, cytosolic binding sites from peas and cucumbers were described at Bonn (Srivastava, 1986). Such sites showed affinities for a range of gibberellins in accord with those expected for a receptor. However, little has appeared since that time and the lack of a functional assay in elongating systems appears to preclude rapid progress. Recent exciting developments do however show great promise (Hooley et al., 1990). Using aleurone layer protoplasts from *Avena fatua* and impermeant gibberellin constructs (GA_4-17-Sepharose) these workers have provided strong evidence for the existence of gibberellin receptors at the plasmalemma which, perhaps surprisingly, appear to be involved with the induction of α-amylase activity.

In parallel work, anti-idiotypic sera raised against monoclonals recognising specifically the A/B ring epitope of GA_4 and GA_1 (Knox et al., 1987) compete with $[^3H]$-GA_4 for binding to the latter. Moreover these anti-idiotypic sera act as antagonists of gibberellin action in aleurone protoplasts (Hooley et al., 1989) again suggesting that receptors for the growth regulator reside at the plasmalemma. Such work is, of course, analogous to that described above in relation to auxin and fusicoccin receptors but differs in that in aleurone tissue the effect is measured upon transcription and translation. It also cast serious doubt — at least in this case — on the use of the animal steroid system as a model for the action of gibberellins.

In the case of abscisic acid it appeared that Hornberg & Weiler (1984) had succeeded in photoaffinity labelling in ABA receptor in guard cells of *Vicia faba* and this work raised expectations of an early isolation of ABA receptors. Since that time however no further publications have appeared.

The cytokinin binding proteins from cereal grains described by Fox and his coworkers have been described extensively, specific antibodies raised to them and the genes cloned. These proteins have doubtful receptor status however, both because they have a low affinity for the most active cytokinin — zeatin — and because of their excessively high concentration in the tissue ($\sim$ 20 nmole $[\text{g f.w}]^{-1}$).

Transduction systems

Although transduction of hormonal effects beyond the receptor level was not dealt with intensively at the Bonn meeting — and indeed at that time little was known — the increased emphasis on receptor work and the encouraging results being obtained have led to developments in this area also. The details of the work lie outwith the scope of this article but, in addition to the role of H^+-ATPases referred to above in connection with auxin and fusicoccin receptors, there is growing evidence that transduction of receptor-mediated responses in plants may well reflect many of the systems described in animals including G-proteins, protein kinases, calcium-calmodulin processes and inositol triphosphate systems. Equally, work on oligosaccharides is showing greater promise. Again, the reader is referred to a number of recent reviews (Hanke, 1990; Scherer, 1990; Fry, 1990; Saunders, 1990).

REFERENCES

1) ADUCCI, P., BALLIO, A., FIORUCCI, L. AND SIMONETTI, E. (1984). Inactivation of solubilised fusicoccin binding sites by endogenous plant hydrolases. Planta 160, 422—427.

2) ADUCCI, P., BALLIO, A., BLEIN, J.-P., FULLONE, M.R., ROSSIGNOL, M. AND SCALL,A R. (1988). Functional reconstitution of proton-translocating system responsive to fusicoccin. Proc. Natl. Acad. Sci. USA 85, 7849—7851.

3) ADUCCI, P., FULLONE, M.R. AND BALLIO, A. (1989). Properties of proteoliposomes containing fusicoccin receptors from maize. Pl. Physiol. 91,

4) ADUCCI, P. & BALLIO, A. (1990). Fusicoccin: Receptors & endogenous ligands. SEB Symposium no.Hormone Perception & Signal Transduction in Animals and Plants (in press).

5) BALLIO, A. & ADUCCI, P. (1986). Search for endogenous ligands to fusicoccin binding sites. In: Plant Hormone Receptors. D. Klämbt, ed. NATO ASI Series Vol. H10. Springer-Verlag, pp. 125—130.

6) BARBIER-BRYGOO, H., EPHRITIKHINE, G., KLÄMBT, D., GHISLAIN, M. AND GUERN, J. (1989). Functional evidence for an auxin receptor at the plasmalemma of tobacco mesophyll protoplasts. Proc. Natl. Acad. Sci. USA 86, 891—895.

7) BENGOCHEA, T., DODDS, J.H., EVANS, D.E., JERIE, P.H., NIEPEL, B., SHAARI, A.R. AND HALL, M.A. (1980a). Studies on ethylene binding by cell-free preparations from cotyledons of *Phaseolus vulgaris* L. I. Separation and Characterisation. Planta 148, 397—406.

8) BENGOCHIA, T., ACASTER, M.A., DODDS, J.H., EVANS, D.E., JERIE, P.H. AND HALL, M.A. (1980b). Studies on ethylene binding by cell-free preparations from cotyledons of *Phaseolus vulgaris* L. II. Effects of structural analogues of ethylene and of inhibitors. Planta 148, 407—411.

9) BLEECKER, A.B., ESTELLE, M.A., SOMERVILLE, C. AND KENDE,H. (1988). Insensitivity to ethylene conferred by a dominant mutation in *Arabidopsis thaliana*. Science 241, 1086—1088.

10) CONNERN, C.P.K., SMITH, A.R., TURNER, R. AND HALL, M.A. (1990). Putative ethylene binding proteins from abscission zones of *Phaseolus vulgaris*. NATO/ASI Series H35, Cell Separation in Plants (eds. D.J. Osborne & M.B. Jackson), pp. 351—356. Springer-Verlag.

11) DE BOER, A.H., WATSON, B.A. & CLELAND, R.E. (1989). Purification and identification of the fusicoccin binding protein from oat root plasma membrane. Pl. Physiol. 89, 250—259.

12) FRY, S.C., MCDOUGALL, G.J., LORENCES, E.P., BIGGS, K.J. AND SMITH, R.C. (1990). Oligosaccharins from xyloglucan and cellulose: modulators of the action of auxin and H^+ on plant growth. SEB Symposium, no. Hormone Perception and Signal Tranduction in Animals and Plants (in press).

13) GUERN, J., BARBIER-BRYGOO, H., EPHRITIKHINE, G., MAUREL, C. AND DELBARRE, A. (1990). Use of mutants and transformed plants fo study the action of plant hormones. SEB Symposium no. Hormone Perception and Signal Transduction in Animals and Plants (in press).

14) HALL, M.A., CONNERN, C.P.K., HARPHAM, N.V.J., ISHIZAWA, K., ROVEDA-HOYOS, G., RASKIN, I., SANDERS, I.O., SMITH, A.R., TURNER, R. AND WOOD, C.K. (1990). Ethylene: Receptors and Action. SEB Symposium no. Hormone Perception & Signal Transduction in Animals and Plants (in press).

15) HALL, M.A., BELL, M.H., CONNERN, C.P.K., HARPHAM, N.V.J., RASKIN, I., SANDERS, I.O., SMITH, A.R. AND WOOD, C.K. (1990). Ethylene, first or second messenger? In: Signal Perception and Transduction in Higher Plants, NATO ASI Series (ed. A.M. Boudet & R. Ranjeva), (in press)

16) HALL, M.A., BELL, M.H., CONNERN, C.P.K., RASKIN, I., ROBERTSON, D., SANDERS, I.O., SMITH, A.R. AND WOOD, C.K. (1990). Ethylene receptors. In: Proceedings of the 14th International Congress of Biochemistry, Prague (in press).

17) HANKE, D.E., BIFFEN, M. AND THIEL, G. (1990). Phosphoinositides and plant growth substance action. SEB Symposium no. Hormone Perception and Signal Transduction in Animals and Plants (in press).

18) HERBER, B., ULBRICH, B. AND JACOBSEN, H.-J. (1988). Modulation of soluble auxin-binding proteins in soybean cell suspensions. Plant Cell Rep. 7, 178—181.

19) HESSE, T., FELDWISCH, J., BALSHUSEMANN, D., BAUW, G., PUYPE, M., VANDEKERCKHOVE, J., LÖBLER, M., KLÄMBT, D., SCHELL, J. AND PALME, K. (1989). Molecular cloning and structural analysis of a gene from *Zea mays* L. coding for a putative receptor for the plant hormone auxin. EMBO J. 8, 2453—2461.

20) HICKS, G.R., RAYLE, D.L., JONES, A.M. AND LOMAX, T.L. (1989a). Specific photoaffinity labelling of two plasma membrane polypeptides with an azido auxin. Proc. Natl. Acad. Sci. USA 86, 4948—4952.

21) HICKS, G.R., RAYLE, D.L. AND LOMAX, T.L. (1989b). The *diageotropica* mutant of tomato lacks high specific activity auxin binding sites. Science 245, 52—54.

22) HOOLEY, R., BEALE, M.H. AND SMITH, S.J. (1990). Novel probes for the study of gibberellin receptors. SEB Symposium no. Hormone Perception and Signal Transduction in Animals and Plants (in press).

23) HORNBERG, C. AND WEILER, E.W. (1984). High-affinity binding sites for abscisic acid on the plasmalemma of *Vicia faba* guard cells. Nature (London) 310, 321—324.

24) INOHARA, N., SHIMOMURA, S., FUKUI, T. AND FUTAI, M. (1989). Auxin-binding protein located in the endoplasmic reticulum of maize shoots: molecular cloning and complete primary structure. Proc. Natl. Acad. Sci. USA 86, 3564—3568.

25) JACOBS, M. AND GILBERT, S.F. (1983). Basal localisation of presumptive auxin transport carrier in pea stem cells. Science 220, 1297—1300.

26) JACOBSEN, J.-J. (1984). Two different soluble cytoplasmic auxin-binding sites in etiolated pea epicotyls. Plant Cell Physiol. 25, 867—873.

27) JERIE, P.H., SHAARI, A.R. AND HALL, M.A. (1979). The compartmentation of ethylene in developing cotyledons of *Phaseolus vulgaris* L. Planta 144, 503—507.

28) JONES, A.M., LAMERSON,, P. AND VENIS, M.A. (1989). Comparison of site I auxin binding and a 22 kilodalton auxin-binding protein in maize. Planta 179, 409—413.

29) JONES, A.M. AND VENIS, M.A. (1988). Photoaffinity labelling of indole-3-acetic acid-binding proteins in maize. Proc. natl. Acad. Sci. USA 86, 6153—6156.

30) KATEKAR, G.F., GEISSLER, A.E., KENNARD, C.H.L. AND SMITH, G. (1987a). Recognition of phytotropins by the receptorfor 1-N-naphthylphthalamic acid. Phytochem. 26, 1257—1267.

31) KATEKAR, G.F., WINKLER, D.A. AND GEISSLER, A.E. (1987b). A conformational study of the topographical requirements of a phytotropin recognition site on the naphthylphthalamic acid receptor. Phytochem. 26, 2881—2889.

32) KATEKAR, G.F. AND GEISSLER, A.E. (1989). The distribution of the receptor for 1-N-naphthylphtalamic acid in different tissues of maize. Physiol. Plant. 76, 183—186.

33) KIKUCHI, M., IMASEKI, H. AND SAKAI, S. (1989). Modulation of gene expression in isolated nuclei by auxin-binding proteins. Plant Cell Physiol. 30, 765—773.

34) KNOX, J.P., BEALE, M., BUTCHER, G.W. AND MACMILLAN, J. (1987). Preparation and characterisation of monoclonal antibodies which recognise different gibberellin epitopes. Planta 170, 86—91.

35) LINDSTRSOM, J., WALTER, B. AND EINARSON, B. (1979). Immunochemical similarities between subunits of acetylcholine receptors from *Torpedo*, *Electrophorus* and mammalian muscle. Biochemistry 18, 4470—4480.

36) LÖBLER, M. AND KLÄMBT, D. (1985). Auxin-binding protein from coleoptile membranes of corn (*Zea mays* L.) II. Localization of a putative auxin receptor. J. Biol. Chem. 260, 9854—9859.

37) LÖBLER, M., SIMON, K., HESSE, T. AND KLÄMBT, D. (1987). Auxin receptors in target tissue. In: Molecular biology of plant growth control, pp. 279—288, Fox, J.E.& Jacobs, M. eds., UCLA Symposia vol. 44, Alan R. Liss, New York.

38) MAAN, A.C., VAN DER LINDE, P.C.G., HARKES, P.A.A. AND LIBBENGA, K.R. (1985). Correlation between the presence of membrane-bound auxin binding and root regeneration in cultured tobacco cells. Planta 164, 376—378.

39) MARRE, E. (1979). Fusicoccin: a tool in plant physiology. Ann. Rev. Plant Physiol. 30, 273—288.

40) MENNES, A.M., QUINT, A., BOOT, C.J.M., VAN DER ZAAL, E.J., MAAN, A.C. AND LIBBENGA, K.R. (1990). Specific transcription and reinitiation of 2,4-D-induced genes in tobacco nuclei (submitted for publication).

41) MICHALKE, W., KATEKAR, G.F. AND GEISSLER, A.E. (1990). Phytotropin binding sites and auxin transport: evidence for two recognition sites. Planta (in press).

42) NAKAMURA, C.AND ONO, H. (1988). Solubilization and characterization of a membrane-bound auxin-binding protein from cell supension cultures of *Nicotiana tabacum*. Plant Physiol. 88, 685—689.

43) NAKAMURA, C., VAN TELGEN, H.-J., MENNES, A.M., ONO, H. AND LIBBENGA, K.R. (1988). Correlation between auxin resistance and the lack of a membrane-bound auxin binding protein and a root-specific peroxidase in *Nicotiana tabacum*. Plant Physiol. 88, 845—849.

44) NAPIER, R.M., VENIS, M.A., BOLTON, M.A., RICHARDSON, L.I. AND BUTCHER, G.W. (1988). Preparation and characterisation of monoclonal and polyclonal antibodies to maize membrane auxin-binding protein. Planta 176, 519—526.

45) RUBERY, P.H. (1990). Phytotropins: receptors and endogenous ligands. SEB Symposium no. Hormone Perception and Signal Transduction in Animals and Plants (in press).

46) SAKAI, S., SEKI, J. AND IMASEKI, H. (1986). Stimulation of RNA synthesis in isolated nuclei by auxin-binding proteins-I and -II. Plant Cell Physiol. 27, 635—643.

47) SANDERS, I.O., SMITH, A.R. AND HALL, M.A. (1989a). The measurement of ethylene binding and metabolism in plant tissue. Planta 179, 97—103.

48) SANDERS, I.O., SMITH, A.R. AND HALL, M.A. (1989b). Ethylene metabolism in *Pisum sativum* L. Planta 179, 104—114.

49) SAUNDERS, M.J. (1990). Calcium and plant hormone action. SEB Symposium no. Hormone Perception and Signal Transduction in Animals and Plants (in press).

50) SCHERER, G. (1990). Activated protein kinases in plants; coupled to phospholipase A_2. SEB Symposium No. Hormone Perception and Signal Transduction in Animals and Plants (in press).

51) SHIMOMURA, S., SOTOBAYASHI, T., FUTAI, M. AND FUKUI, T. (1986). Purification and properties of an auxin-binding protein from maize shoot membranes. J. Biochem. 99, 1513—1524.

52) SHIMOMURA, S., INOHARA, N., FUKUI, T. AND FUTAI, M. (1988). Different properties of two types of auxin-binding sites in membranes from maize coleoptiles. Planta 175, 558—566.

53) SISLER, E.C. (1979). Measurement of ethylene binding in plant tissue. Pl. Physiol. 64, 538—542.

54) SISLER, E.C. AND WOOD, C. (1987). Ethylene binding and evidence that binding *in vivo* and *in vitro* is to the physiological receptor. In: Plant Hormone Receptors, NATO ASI Series H, Cell Biology, Vol. 10 (ed. D. Klämbt), pp. 239—248. Springer-Verlag, Berlin.

55) SMITH, A.R. AND HALL, M.A. (1984). Mechanisms of ethylene action. Plant Growth Regul. 2, 151—165.

56) SRIVASTAVA, L.M. (1986). The gibberellin receptor. NATO ASI Series H10, Plant Hormone Receptors (ed. D. Klämbt), pp. 11—228 Springer Verlag, Berlin.

57) THOMAS, C.J.R., SMITH, A.R. AND HALL, M.A. (1985). Partial purification of an ethylene-binding site from *Phaseolus vulgaris* L. cotyledons. Planta 164, 272—277.

58) TILLMANN, U., VIOLA, G., KAYSER, B., SIEMEISTER, G., HESSE, T., PALME, K., LÖBLER, M. AND KLÄMBT, D. (1989). cDNA clones of the auxin-binding protein from corn coleoptiles (*Zea mays* L.): isolation and characterisation by immunological methods. EMBO J. 8, 2463—2467.

59) VAN DER LINDE, P.C.G., BOUMAN, H., MENNES, A.M. AND LIBBENGA, K.R. (1984). A soluble auxin-binding protein from culture tobacco tissue stimulates RNA synthesis *in vitro*. Planta 160, 102—106.

60) VAN DER ZAAL, E.J., MEMELINK, J., MENNES, A.M., QUINT, A. AND LIBBENGA, K.R. (1987). Auxin-induced mRNA species in tobacco cell cultures. Plant Mol. Biol. 10, 145—157.

61) VAN DER ZAAL, E.J. (1989). Auxin-induced gene expression in tobacco. PhD thesis, Univerity of Leiden.

62) VENIS, M.A. AND NAPIER, R.M. (1990). Characterisation of auxin receptors. SEB Symposium no. Hormone Perception and Signal Transduction in Animals and Plants (in press).

63) VOET, J.G., HOWLEY, K.S., SHUMSKY, J.S. (1987). 5'-azido-N-1-naphthylphthalamic acid, a photolabile analog of N-1-naphthylphthalamic acid. Plant Physiol. 85, 22—25.

MOVEMENTS AND MORPHOGENETIC CHANGES IN RESPONSE TO SIGNAL RECEPTION IN HIGHER PLANTS

B. Millet

Botanical Laboratory,
University of Franche-Comte,
F-25030 Besançon Cedex

INTRODUCTION

Plants respond to internal or external stimuli through processes that lead to morphological (e.g., tropic or nastic curvatures, thigmomorphogenesis) or physiological (e.g., flowering, growth correlations, rhizogenesis) changes. Such processes can be understood by analysing the steps of the classical paradigm used to approach the underlying mechanisms: signal reception → transduction → transmission → response.

Based on the level of the organization, such responses raise the problem of cell-cell communications in plant organs. Therefore, the question of messengers is of importance. The role and significance of the cellular matrix and apoplast in signalling were examined both in animals and plants at the NATO ARW held in Besançon in March 1986. Three years later, new consistent results can be added in all fields. However, this overview will be focussed only on the study of mechanisms encountered in higher plants.

I. Recent advances in analysis of stimulated leaf movements

In many species, roots, shoots, leaves, petals or stamens are known to move either with a circadian or ultradian rhythm or upon stimulation by a signal. Often, both phenomena occur in the same organ.

Mechanisms of leaf movement have been studied extensively in leguminous plants, *Albizzia, Cassia, Mimosa, Phaseolus* and *Samanea* being examples. Each species shows particular potentialities. For example, *Samanea* exhibits no response to a mechanical stimulation whereas *Mimosa* does so. We will start this overview by analysing the successive steps of the sequence of events of the light-stimulated leaf movement of *Samanea saman*. Most of the recent data were collected by Satter's group and the new developments in the knowledge of the signal transduction are coming from studies on this species.

The question arises whether the transduction of an external (or internal) stimulus follows a similar pattern to the one used in animal systems (1).

NATO ASI Series, Vol. H 51
Cell to Cell Signals in Plants and Animals
Edited by V. Neuhoff and J. Friend

a. Signal reception

Despite the fact that the plant cell protoplasm is surrounded by the cell wall, the membranes remain the site of the signal reception. However, to date, the nature of the pigment whose absorption of light promotes activation of inositolphospholipid turnover is unknown. Phytochrome as well as blue-absorbing pigment might be the photoreceptor (2). In *Cassia,* it has been shown (3) that far-red-absorbing pigment is also involved in light-induced leaflet movements. In *Phaseolus,* blue light induces an upward movement of the lamina when the upper part of pulvinus is irradiated and a downward movement when the lower part is irradiated (4, 5).

Occurrence of G protein in plants is discussed (6, 7, 8).

b. Signal transduction

In *Samanea saman,* the pulvinus contains all of the components of the inositol-phospholipid cycle (9, 10, 11). Morse et al. (1) showed that the effects of light on leaflet movement are mediated by increased inositolphospholipid turnover. So, after a light pulse of 15 seconds or longer, labeled phosphatidylinositol-4-phosphate and phosphatidylinositol 4,5-bisphosphate decreased and their labeled metabolic products, inositol 1,4-bisphosphate and inositol 1,4,5-trisphosphate, increased. An increase in diacylglycerol level was detected after only 30 s of light treatment (12). Such data support the idea that inositolphospholipid turnover can be a primary effect of light treatment and the first step in the cascade of events coupling light absorption and the ensuing physiological processes (1). Implication of phosphatidylinositol turnover in plants was also reported in other systems (13, 14, 15, 16).

Some differences, however, have been seen in the phosphatidylinositol cycle in plants compared to that of animal cells (1, 2). For instance, in plants, inositol is involved in many reactions of intermediate metabolism and phosphatidylinositol 4,5-bisphosphate/ phosphatidyl inositol ratio is very low. According to Boss et al. (17), the phosphatidylinositol 4,5-bisphosphate in most plant cells is less than 1 % of the (^{3}H) inositol-labeled lipid and is about 10-fold less than the percentage of phosphatidylinositol-phosphate.

In the animal paradigm, inositol 1,4,5-trisphosphate operates as second messenger by binding to a specific membrane receptor inducing release of Ca^{2+} from stores (18). Recently, such a protein has been isolated and purified from mouse cerebella (19) or brain membranes (20).

In plant cells, vacuoles, ER, mitochondria and walls contain high Ca^{2+} levels (21, 22). Inositol 1,4,5-trisphosphate triggers calcium release from membrane vesicles of carrots (23) or oat (24, 21) but not from endoplasmic reticulum (25). The increase of cytoplasmic calcium can activate Ca^{2+}- and Ca^{2+}-calmodulin-dependent enzymes and protein kinases (22). According to Kauss (26), two principal types of active mechanisms exist in higher plants for maintenance of a low cytoplasmic free level. The first one is a Ca^{2+}/ Mg^{2+}-ATPase, the other one a Ca^{2+}/H^+ antiport.

Involvement of Ca^{2+} as second messenger in response to a signal has been demonstrated by use of Ca^{2+} inophore (A 23187), Ca^{2+} chelator (EDTA or EGTA) and Ca^{2+} channel blockers. Such a result has not yet been found in *Samanea* motor organs, but in the study of the light-stimulated leaflet movement of *Mimosa* and *Cassia,* Roblin's group reported a possible mobilization of Ca^{2+} from internal storage through calcium channels (27).

The role in plant cells of diacylglycerol resulting from cleavage of inositol 1,4,5-trisphosphate is not completely elucidated. A protein kinase C has been described by Schafer, et al. (28) and Harmon, et al. (29), but the protein phosphorylation by this enzyme has not yet been demonstrated. By contrast, it has been established that Ca^{2+}-calmodulin regulates protein phosphorylation (22). Boss (30) depicted several pathways which were possibly involved in signal transduction which took into account the data reported for different plant materials.

c. Signal transmission

In the light-stimulated leaf movement as well as in other types of leaf movement, curvature of the pulvinus appears to be a consequence of turgor changes: it either increased or decreased in the extensor cells according to which way the organ was bending. The coupling between illumination and changes in cellular volume has to be explained. Working on excised tissue, Iglesias and Satter (31) had shown that loss of water in flexor cells was correlated with the alkalinization of the medium containing flexor and with the efflux of K^+. Conversely, uptake of water in extensor cells was correlated with an acidification of the medium and a K^+ influx. Several explanations can be proposed to understand the H^+ and K^+ migrations. Evidence for a link between inositol-1,4,5-trisphosphate, Ca^{2+} release and K^+ channel activation has been brought out by Ranjeva, et al. (32) and then by Canut, et al. (33) and Colombo, et al. (34) from isolated vacuoles or microsomes fractions from *Acer* cell suspension cultures. In *Samanea,* further investigations are necessary to test whether inositol 1,4,5-trisphosphate promotes the release of Ca^{2+} from intracellular stores and, to date, no evidence has been presented to implicate cytoplasmic free Ca^{2+} in the gate opening of Ca^{2+} activated channels (35, 36). However, a patch-clamp study has been performed on motor cell protoplasts by Moran, et al. (37). The results provide evidence for the existence of K^+ channels in the plasma membranes and for their role in the leaf movement. The channels can provide a pathway for the passive efflux of K^+ ions. According to the authors, the mechanism for cell shrinking would be the following: membrane depolarization → K^+ channel opening → K^+ efflux → cell shrinkage. K^+ fluxes are accompanied by Cl^- fluxes. The question arises whether Cl^- channels occur in the plasma membrane. Cl^- ions fluxes have been monitored in *Samanea* pulvinus during leaflet movement. When the paired leaflets open, extensor cells take up K^+ and Cl^- while flexor cells lose both ions. The reverse processes occur at the leaflets closing. Evidence for Cl^- channels was recently found in some plant cells (38) but not yet in *Samanea* motor cells.

Previous observations by Iglesias and Satter (31) led them to consider that K^+ fluxes into and out of motor cells are coupled to oppositely directed H^+ fluxes. Due to the migration of H^+ ions through symplasm and apoplasm, some pH variations can be expected. A method has been developed for measuring apoplastic pH *in situ* (39, 40). The pH of the extensor cell apoplast was higher than that of the flexor cell apoplast upon illumination. Such a result is of interest because the pH difference (about 1.0 units) may influence the plasma membrane permeability to ions and the enzyme reactions in the cell wall. Moreover, apoplast could be a route for cell-cell communications between extensor and flexor parts of pulvinus with respect to ions and water.

An X-ray microanalysis was achieved so far as 1982 by Satter, et al. (41) showing K^+ and Cl^- movements between extensor and flexor cell walls in the pulvinus during leaflet movement. Ca^{2+} is barely detectable. In *Samanea,* the water flux is considered to follow ion flux (40). So, the osmotic regulation could be controlled by the water and ions through the apoplast.

d. Response

The volume changes of cells in extensor and flexor caused by the osmostic potential changes are responsible for the pulvinus curvature.

Despite the fact that some gaps have to be filled, the understanding of mechanisms involved in the light-stimulated leaf movement of *Samanea* is better today. Some steps of the cascade of events can be compared to those seen in other plant materials.

Mimosa pudica offers many advantages for studies in signal transduction in plants. Pulvinar movements occur even after growth has ceased and then are responsive to different kinds of signals such as light, gravity, electric shocks, mechanical perturbation (42). Moreover, these movements are regulated by a circadian clock. New data came from Roblin, et al. (3). They showed, by the use of calcium chelator (EGTA), calcium inophore (A 23187) and calcium channel blockers (verapamil, nifedipin) that Ca^{2+} was involved not only in the light-stimulated movement but also in seismonastic, skotonastic, gravitropic and IAA-induced movement.

The kinetics of ion redistribution into the pulvinus upon gravitropic stimulation has been followed for K^+, Cl^- and Ca^{2+}. Treatment with valinomycin altered K^+ transport in the seismonastic but not in the gravitropic curvature. Ca^{2+} accumulation could influence directly membrane transport characteristics. Roblin and Fleurat-Lessard (42) concluded that ions do not have the same functional importance in the regulation of the pulvinar movements induced by gravity or mechanical stimulation. Moreover it has to be pointed out that the results suggest a release of extracellular Ca^{2+} in seismonasty, skotonasty and gravitropism and a mobilization from internal storage in photonasty and IAA-induced movement (43). Further investigations should develop the arguments for more than one system of signal transduction in *Mimosa* pulvinar movements and for a possible implication of cAMP pathway (3).

Identification of K^+ and Ca^{2+} channels in plasmalemma of *Mimosa* protoplasts has been reported by Stoeckel and Takeda (44).

In *Cassia,* chlorpromazin and W7 increased the rates of leaflet opening induced by illumination during the skotophase (45). A reverse situation was expected. The authors explain such results by occurrence of an inhibition of phosphodiesterase and an increase of cAMP level.

The leaf movements of *Phaseolus* have been extensively studied especially in relation to circadian and ultradian rhythms. Light or other signals are often used to obtain a phase-shift of the rhythm. The curvature induced by a signal is apparently similar to that of the circadian up and down movement. So, the terminal steps of signal transduction can be analyzed according to the same protocol as circadian movement. Recent advances relative to the leaf movement in *Phaseolus* came from studies of Freudling, et al. (46) and Starrach & Mayer (47) at Tübingen University. The changes in the volume of the pulvinar cells are based on transport of osmotica between upper and lower part of the pulvinus and, as a consequence, some changes in apoplastic pH and K^+ concentration might be measured. In extensor and flexor cells, the apoplast operates as a K^+ reservoir which is depleted when protoplasts gain osmotica and the cells swell. The apoplast is filled up when protoplasts lose osmotica and the cells shrink. According to Freudling, et al. (46), 60 % of K^+ is transported from extensor to flexor and conversely through apoplastic and symplastic routes. Our own results give confirmation to those of Freudling (48).

According to the same authors, depletion of K^+ apoplastic reservoir during cell swelling is brought about by the electrogenic proton efflux pump which drives a K^+/H^+ exchange at the plasmalemma and a K^+/H^+ exchange in Donnan free space. The apoplastic pH measurement with a H^+-sensitive double-barrelled microelectrode was monitored simultaneously with the leaf movement. The pH increased during the leaf downward movement and decreased during the leaf elevation. A difference of 0.8 pH units could be detected between both positions (47).

It has to be pointed out that in *Samanea* pulvinus, the apoplastic K^+ level was high and the pH was low when cells were large (41, 40). A reverse situation occurred when cells were shrunken. Such data mean that we must be very cautious before envisioning a generalization of results obtained from one species.

In the twining shoot of *Phaseolus vulgaris* showing an ultradian rhythmicity, the cells located in a given point of the curvatore zone alternatively swell and shrink during one revolution (49). It is striking to note that the evolution of K^+ content and pH of the cells parallels those of pulvinus motor cells (50). However, it should be remembered that, by contrast with what occurs in the leaf movement, turgor is constantly changing (51) in cells of the beding zone. Therefore, it has been postulated (49) that an internal signal is driving this variation in turgor in an helicoidal pattern as the shoot grows.

II. New developments in other fields

We leave the pulvini of leguminous plants to consider other curvatures or morphogenetic responses of higher plants upon external stimulation. **Gravitropism** is very well documented. Many experiments have been performed to understand how gravity could induce root curvature (for review, see 52, 53). Some of them deal with the role of Ca^{2+} in tho gravitropic curvator. Poovaiah, et al. (54) showed that Ca^{2+}, Ca^{2+}-dependent protein phosphorylation and inositol phospholipid turnover may play a role in the light signal transduction required to develop gravisensitivity. Falke, et al. (55) postulated that the transduction step of both geotropism and thigmotropism is the opening of stretch-activated ion channels which permit the flow of Ca^{2+} across the plasmalemma. By pretreatment with 250 μM $GdCl_3$ (Fig. 1a), a positive orthogeotropism response was inhibited about 80 % and somewhat less with lower concentrations (56). Li^+ treatment induced a negative curvature in the first minutes following gravitropic stimulation (57).

In **thigmotropism,** which is a tropic curvature in response to a mechanical perturbation, analogous results have been obtained. When caps of maize seedling roots grown between wet paper towels under bright light were gently, unilaterally stroked with a smooth glass probe and then clinostated at 100 % relative humidity for development of curvature, they bent vigorously towards or away from the stimulus and were almost as uniform as that of geotropic controls if the mucigel was removed before stroking (56, 58). The curvature was inhibited after treatment in $GdCL_3$ (110–250 μM) for 30 minutes (Fig. 1 b).

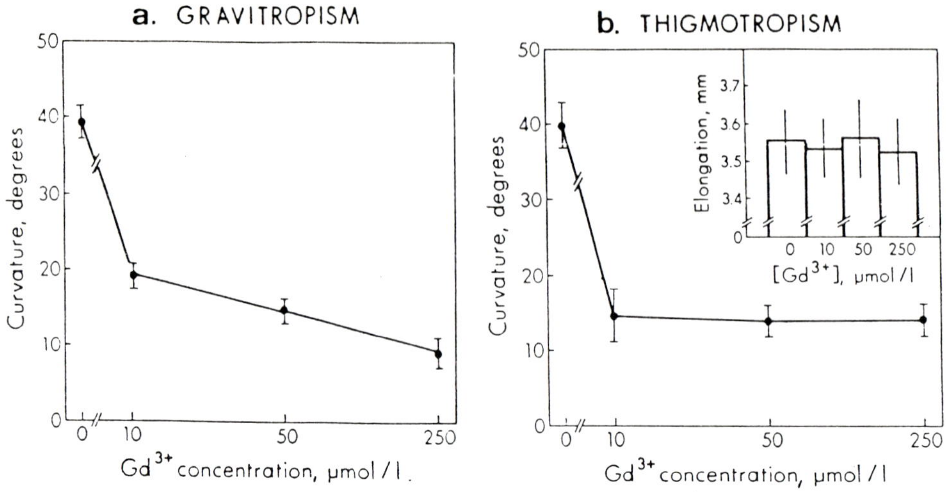

FIGURE 1
Effect of a Gd^{3+} pretreatment on the gravitropic (a) and thigmotropic (b) response of maize roots cv Merit.

Insert: elongation of roots as a function of Gd^{3+} concentration.

Gd^{3+} might possibly have a general affinity for stretch-activated ion channels or might have general utility for discriminating different types of Ca^{2+} permeable channels.

Thigmomorphogenesis is related to the irreversible changes in growth and development consequent upon a mechanical perturbation. The study of the underlying mechanisms is relevant to the previous approach. How can a mechanical stimulus induce a morphogenetic response? Such analysis has been undertaken on *Bryonia dioica,* a plant highly responsive to rubbing (59). De Jaegher and Boyer (60) depicted a model involving the role of inositol phospholipid turnover to explain the increased level of cytosolic Ca^{2+} and the subsequent steps. They postulate a flow of Ca^{2+} into the cytosol, the activation of the phospholipase C by Ca^{2+} and a release of vacuolar Ca^{2+} through inositol 1,4,5-trisphosphate. Several arguments corroborate the implication of phosphoinositides turnover and Ca^{2+} as a messenger. A treatment by Ca^{2+} promoted the thigmomorphogenesis; by contrast, EGTA and Li^{+} prevented it. The microviscosity of plasmalemma was increased by rubbing the internode (61). Ca^{2+} is similarly involved in coiling of pea tendrils induced by mechanical perturbation. Jaffe (62) using Ca^{2+} channel blockers and agonists and showed that the response was mediated by Ca^{2+}. Coming back to *Bryonia,* the cytosolic Ca^{2+} elevation would induce an activation of 1-aminocyclopropane-1-carboxylic acid (ACC) synthase, an accumulation of ACC, an increase of ACC conversion to ethylene and finally a growth inhibition of the rubbed internode (60).

The action potentials elicited by the mechanical perturbation could reflect the shuttle of ions Ca^{2+}, K^{+}, Cl^{-} between symplasm and apoplasm (63).

Long-distance communication could be achieved through the transmitted changes in membrane potential and/or apoplasm (63). Most cells in a plant are connected by plasmodesmata that show some similarities with gap junctions in their transport properties (64). The continuity of cytoplasms and plasma membranes allows us to understand that the diffusion of ions and molecules through apoplast needs a long time and, moreover, that ions and these molecules remain under control of the plasma membrane to enter the cell. The diffusive processes being temperature-dependent, the reaction time will be strongly influenced by temperature. However, temperature can by itself be a signal. Minorsky and Spanswick (65) reported evidence that plants do sense rapid drops in temperature and that a plasmalemma associated Ca^{2+} channel may be involved in this process.

Growth responses to a signal (other than thigmomorphogenetic responses) involving long-distance transport have been also examined. Before initiating adventitious roots, cuttings of grape vine show two peaks in ethylene production 2 and 10—12 h, respectively, after explanting (zero time) (66). In parallel, a diminution of specific peroxidase activity and an increase of 0-dihydroxyphenols level was observed with a minimum arising at 12 h. The first peak of ethylene might correspond to wound ethylene and initiate the sequence of events leading to root formation. The second might result from the beginning of the increase of endogenous IAA.

The *Bidens* model has been developped because the growth response of axillary buds to the cotyledonary pricking illustrates the storage and retrieval of morphogenetic messages (67). The recent advances in analysis of underlying mechanisms led to modelling of the successive steps. Remember that wounding elicits a wave of electric depolarization progressing towards the axillary bud and induces a growth inhibition of the bud. It has been shown (68) that lowering the cation exchange capacity of the cell wall with acidic pH decreased the potential difference variations induced by KC1. Thus, the apoplastic potential plays a major role in transmission of message. Where the message is stored is not still known. The adult cells located at the base of the cotyledonary buds could be the target cells (69).

In Jerusalem artichoke, **dormancy** of the buds is broken by an exposure to 4°C for 16 weeks. It is accompanied by some changes in the intracellular compartmentation of ATP, an increase of sucrose diffusion, a lowering of intracellular pH and an electrogenic-proton pump less active (70, 71). The rate of H^+ secretion was 5-fold greater in dormant tubers than in non-dormant material. In such a system (71) the signal could be the exposure to a low temperature and parenchyma cells as site of signal reception. The site of message storage remains unknown.

Flowering of photoperiodic plants is under control of the length of the day as shown by Ruiz Fernandez & Wagner in *Chenopodium rubrum* (72). It has been shown since long ago that the leaves are the site of signal reception while the apex is the site of the response. The expression of the response requires a message transmission. A lot of work has been done on changes in the shoot apex during floral evocation. The floral stimulus is unable to go through a water- or an oil-gap between an induced and a non-induced plant part (73). That means that the cel-to-cell connections are required for. According to Greppin et al. (74), light signal transduction would be linked to the modulation of membrane-bound electron transport and proton translocation at the cellular and tissue level. Changes in membrane thickness might be the structural evidence of its chemical modification (75, 76). The signal would be transmitted through plasma membrane and phloem cells. In *Spinacia,* the signal reception by phytochrome induces a cascade of events involving major components of signal transduction in *Samanea* (Ca^{2+}, K^+ movement, electric potential variations) or peroxidases as in *Bryonia*. Peroxidase activity would be linked to the consequences of interactions between phytochrome, membranes and alkaline peroxidases.

CONCLUSIONS

The barriers raised between animal and plant kingdom are falling one by one. The transduction of external signals is one among many other examples that show the similarities in the functioning of living cells.

The overview emphasizes the gaps in our knowledge of steps leading to the response of plants to light and to other signals. Taken together, the technological advances and the cooperation with scientists working on animal cells should led to a better understanding of mechanisms of signal transduction in plants. Elucidation of events occurring in the first steps of light-stimulated leaf movement should stimulate investigations dealing with the effects of external signals on plant morphogenesis and physiology. Of prime interest is the understanding of mechanisms such as flowering for agriculture. This analysis of events related to the signal transduction highlights the role of the apoplast as a route for ions and molecules involved as messengers in the movement of organs. We could also have pointed out the major interest of some molecules of the cell wall (e.g. oligosaccharins) in signalling and recognition.

Next, a better understanding of mechanisms driving the rhythmic processes in plants and animals should be expected from the studies achieved on signal transduction. In fact, the studies on leaf movements showed the overlap between signal transduction and rhythmicity. It is striking to note several convergences in the metabolic pathways related to both phenomena. Therefore, this is an additional argument for developing this field of research.

REFERENCES

1) MORSE, M.J., SATTER, R.L., CRAIN, R.C. & COTE, G.G. (1989). Signal transduction and phosphatidylinositol turnover in plants. Physiol. Plant., 76, 118—121.

2) MORSE, M.J., CRAIN, R.C. & SATTER, R.L. (1987b). Light-stimulated inositolphospholipid turnover in *Samanea saman* leaf pulvini. Proc. Nat. Acad. Sci., 84, 7075—7078.

3) ROBLIN, G., FLEURAT-LESSARD, P. & BONMORT, J. (1989). Effects of compounds affecting calcium channels on phytochrome — and blue pigment — mediated pulvinar movements of *Cassia fasciculata*. Plant Physiol. 90, 697—701.

4) NISHIZAKI, Y. (1986). Rhythmic and blue light-induced turgor movements and electrical potential in the laminar pulvinus of *Phaseolus vulgaris* L. Plant Cell Physiol., 27 (1), 155—162.

5) NISHIZAKI, Y. (1988). Blue light pulse-induced transient changes of electrical potential and turgor pressure in the motor cells of *phaseolus vulgaris* L. Plant Cell Physiol., 29 (6), 1041—1046.

6) BLUM, W., HINSCH, K.D., SCHULZ, G. & WEILER, E.W. (1988). Identification of GTP-binding proteins in the plasma membrane of higher plants. Biochem. Biophys. Res. Commun., 156, 954—959.

7) HORWITZ, B.A. (1989). The potential for second messengers in light signalling. In: "Second messengers in plant growth and development". W.F. Boss & D.J. Morré Eds, New York: Alan R. Liss, Inc., 289—313.

8) GAUL, G., ADUCCI, P. & MARME, D. (1989). G proteins in plants? NATO Internat. Symposium, Signal perception and transduction in higher plants, Toulouse, Abstracts, L 8.

9) MORSE, M.J., CRAIN, R.C. & SATTER, R.L. (1987a). Phosphatidylinositol cycle metabolites in *Samanea saman* pulvini. Plant Physiol., 83, 640—644.

10) COTE, G.G., DEPASS, A.L., QUARMBY, L.M., TATE, B.F., MORSE, M.J., SATTER, R.L., CRAIN, R.C. (1989). Separation and characterization of inositol phospholipids from the pulvini of *Samanea saman*. Plant Physiol., 90, 1422—1428.

11) IRVINE, R.F., LETCHER, A.J., LANDER, D.S., DROBAK, B.K., DAWSON, A.P. & MUSGRAVE, A.(1989). Phosphatidylinositol (4,5) biphosphate and phosphatidylinositol (4) phosphate in plant tissues. Plant Physiol, 89, 888—892.

12) MORSE, M.J., COTE, G.G., CRAIN, R.C. & SATTER, R.L. (1988). Light-modulated phosphatidylinositol turnover in *Samanea saman*. Plant Physiol., 86 (4) suppl. 93.

13) HELM, S., BAULEKE, A., WYLEGALLA, C., WAGNER, K.G. (1987). Evidence of phosphatidylinositol and diacylglycerol kinases in suspension cultured plant cell. Plant Science, 49, 159—165.

14) ETTLINGER, C. & LEHLE, L. (1988). Auxin induces rapid changes in phosphatidylinositol metabolites. Nature, 331, 176—178.

15) SCHERER, G.F.E. (1989). Phospholipid-stimulated protein kinase. NATO Internat. Symposium, Signal perception and transduction in higher plants, Toulouse, Abstracts, L 6.

16) CHO, M., MEMON, A.R. & BOSS, W.F. (1989). Identification and characterization of polyphosphoinositides in thetonoplast membrane. Plant Physiol., 89 (4), suppl., 149.

17) BOSS, W.F., MEMON, A. & CHEN, Q. (1989). Phospholipid-derived messengers. NATO Internat. Symposium, Signal perception and transduction in higher plants, Toulouse, Abstracts, L 11.

18) STREB, H., IRVINE, R.D., BERRIDGE, M.J. & SCHULZ, I. (1983). Release of Ca^{2+} form a non-mitochondrial intracellular store in pancreatic acinar cells by inositol-1,4,5-triphosphate. Nature, 306, 67—69.

19) FURUICHI, T., YOSHIKAWA, S., MIYAWAKI, A., WADA, K., MAEDA, N. & MIKOSHIBA, K. (1989). Primary structure and functional expression of the inositol 1,4,5-trisphosphate-binding protein P_{400}. Nature, 342, 32—38.

20) FERRIS, C.D., HUGANIER, R.L., SUPATTAPONE, S. & SNYDER, S.H. (1989). Purified 1,4,5-trisphosphate receptor mediates calcium flux in reconstituted lipid vesicles. Nature, 342, 87—89.

21) SCHUMACHER, K.S. & SZE, H. (1987). Inositol-1,4,5-triphosphate releases Ca^{2+} from vacuolar membrane vesicles of oat roots. J. Biol. Chem., 262, 3944—3946.

22) POOVAIAH, B.W. & READDY, A.S.N. (1987). Calcium messenger system in plants. (C.R.C. Crit. Rev. Plant Sci., 6, 47—103.

23) RINCON, M. & BOSS, W.F. (1987). Myo-inositol trisphosphate mobilizes calcium from fusogenic carrot (*Daucus carota* L.)protoplasts. Plant Physiol, 83, 395—398.

24) DROBAK, B.K. & FERGUSON, I.B. (1985). Release of Ca^{2+} from plant hypocotyl microsomes by inositol-1,4,5-triphosphate. Biochem. Biophys. Res. Commun., 130, 1241—1246.

25) LEW, R.R. (1989). Calcium activates an electrogenic proton pumps in *Neurospora* plasma membrane. Plant Physiol, 91, 213—216.

26) KAUSS, A. (1987). Some aspects of calcium-dependent regulation in plant metabolism. Ann. Rev. Plant Physiol., 38, 47—72.

27) MOYEN, C. & ROBLIN, G. (1989). Calcium involvement in the glycine uptake by leaf motor organs of *Mimosa pudica*. NATO Internat. Symposium, Signal perception and transduction in higher plants, Toulouse, Abstracts, A-13.

28) SCHAFER, A., BYGRAVE, F., MATHENAUER, S. & MARME, D. (1985). Identification of a calcium- and phospholipid-dependent protein kinase in plant tissue. FEBS Letters, 187, 25—28.

29) HARMON, A.C., PUTNAM-EVANS, C. & CORMIER, M.J. (1987). A calcium-dependent but calmodulin-independent protein kinase from soybean. Plant Physiol., 83, 830—837.

30) BOSS, W.F. (1989). Phosphoinositide metabolism: Its relation to signal transduction in plants. In: "Second messengers in plant growth and development". W.F. Boss & D.J. Morré, Eds. New York: Alan R. Liss, Inc., 29—56.

31) IGLEASIAS, A. & SATTER, R.L. (1983). H^+ fluxes in excised *Samanea* motor tissue. I. Formation by light. Plant Physiol., 72, 564—569.

32) RANJEVA, R., CARRASCO, A., BOUDET, A.M. (1988). Inositol trisphosphate stimulates the release of calcium from intact vacuoles isolated from *Acer* cells. FEBS Letters, 230 (12), 137—141.

33) CANUT, H., CARRASCO, A., GRAZIANA, A., BOUDET, A.M. & RANJEVA, R. (1989). Inositoltrisphosphate-stimulated calcium release from *Acer* microsomal fractions involves the uptake of potassium. FEBS Letters, 253 (1,2), 173—177.

34) COLOMBO, R., CERANA, R., LADO, P., PERES, A. (1989). Regulation by calcium of voltage-dependent tonoplast K^+ channels. Plant Physiol. Biochem., 27 (4), 557—562.

35) SATTER, R.L. & MORAN, N. (1988). Ionic channels in plant cell membranes. Physiol. Plant., 72, 816—820.

36) HEDRICH R. & SCHROEDER, J.I. (1989). The physiology of ion channels and electrogenic pumps in higher plants. Ann. Rev. Plant Physiol, 40, 539—569.

37) MORAN, N., EHRENSTEIN, G., IWASA, K., MISCHKE, CH., BARE, CH. & SATTER, R.L. (1988). Potassium channels in motor cells of *Samanea saman*. A patch-clamp study. Plant Physiol, 88 (33), 643—646.

38) TAZAWA, M., SHIMMEN, T. & MIMURA, T. (1987). Membrane control in the Characeae. Ann. Rev. Plant Physiol., 38, 95—117.

39) ZUCKER, C.A. & SATTER, R.L. (1988). Use of ion-sensitive microelectrodes to measure light induced changes in K^+ activity in the *Samanea* pulvinar apoplast. Plant Physiol, 86 (4), suppl., 93.

40) LEE, Y. & SATTER, R.L. (1989). Effects of white, blue, red light and darkness on pH of the apoplast in the Samanea pulvinus. Planta, 178, 31—40.

41) SATTER, R.L., GARBER, R.C., KHAIRALLAH, L. & CHENG, Y.S. (1982). Elemental analysis of freeze-dried thin section of *Samanea* motor organs: barriers to ion diffusion through the apoplast. J. Cell Biol, 95, 893—902.

42) ROBLIN, G. & FLEURAT-LESSARD, P. (1987). Redistribution of potassium, chloride and calcium during the gravitropically induced movement of *Mimosa pudica* pulvinus. Planta, 170, 242—248.

43) EVERAT-BOUBOULOUX, A., FLEURAT-LESSARD, P. & ROBLIN, G. (1989). Auxin-induced pulvinar movement of *Cassia fasciculata* leaflets. NATO Internat. Symposium, Signal perception and transduction in higher plants, Toulouse, Abstracts, A-12.

44) STOECKEL, H. & TAKEDA, K. (1990). Are ions channels involved in the perception of stimuli and in signal transduction? New perspectives offered by the patch-clamp technique Intra- and intercellular communications in plants (B. Millet & H. Greppin, eds) (in press).

45) ROBLIN, G., FLEURAT-LESSARD, P. & BONMORT, J. (1988). Effects of calmodulin antagonists on the dark- and light-induced leaflet movements in *Cassia fasciculata* Michx. C.R. Acad. Sci., 306, ser. II, 179—184.

46) FREUDLING, C., STARRACH, N., FLACH, D., GRADMANN, D. & MAYER, W.E. (1988). Cell walls as reservoirs of potassium ions for reversible volume changes of pulvinar motor cells during rhythmic leaf movements. Planta, 175, 193—203.

47) STARRACH, N. & MAYER, W.E. (1989). Changes of the apoplastic pH and K^+ concentration in the *Phaseolus* pulvinus *in situ* in relation to rhythmic leaf movements. J. exp. Bot., 40, n°217, 865—873.

48) DEGLI AGOSTI, R., BLAISE, M.-O., MILLET, B. (1989). Redistributions du potassium dans le pulvinus secondaire de *Phaseolus vulgaris* L. au cours du mouvement foliaire circadien. Bot. Helvet., 99 (2), 179—188.

49) MILLET, B., MELIN D. & P.-M. BADOT (1987). Circumnutation: model for signal transduction. In: The cell surface in signal transduction. E. Wagner, H. Greppin, B. Millet eds., Springer-Verlag, Berlin, 169—179.

50) BADOT, P.-M. (1987). Approche cellulaire du mecanisme du mouvement revolutif des tiges volubiles. Etude de quelques paramètres physico-chimiques. Ann. Sci. Univ. Besancon, Biologie, 4ème sér. (8), 53—110.

51) BOTTON, A.M., MILLET, B. & MERCIER, J. (1989). Structure du pulvinus secondaire de *Phaseolus vulgaris,* L. au cours du mouvement foliaire circadien. Ann. sci. Univ. Besancon, Biologie-Ecologie, 5 (1), (in press).

52) PICKARD, B.G. (1985). Roles of hormones, protons and calcium in geotropism. In: Encyclop. of Plant Physiol.: New Series (R.P. Pharis and D.M. Reid, eds), vol. 11, 193—281. Springer-Verlag, Berlin.

53) MOORE R. & EVANS, M.L. (1986). How roots perceive and respond to gravity. Amer. J. Bot., 73 (4), 574—587.

54) POOVAIAH, B.M., REDDY, A.S.N., FRIEDMANN, M., RAGHOTAMA, K.G., MC FADDEN, J.J., CHENGAPPA, S. & WANG, Z.Q. (1988). The role of calcium in the transduction of light signal. Plant Physiol., 86 (4), suppl. 67.

55) FALKE, L., EDWARDS, K.L., MISLER, S. & PICKARD, B.G. (1986). A mechanotransductive ion channel in patches from cultured tobacco cell plasmalemma. Plant Physiol., 80 (4), 40.

56) MILLET, B. & PICKARD, B.G. (1988b). Gadolinium ion is an inhibitor suitable for testing the putative role of stretch-activated ion channels in geotropism and thigmotropism. Biophys. J., 53, 155a.

57) MILLET, B. & PICKARD, B.G. (1988a). Early wrong-way response occur in orthogravitropism of maize roots treated with lithium. Physiol. Plant., 72, 555—559.

58) MILLET B. & PICKARD, B.G. (1988c). Reproducible thigmotropic response by maize roots. Plant Physiol, 87 (4) (suppl n° 29).

59) BAILLAUD, L. (1967). Variations d³uni périodicité endogène normalement circadienne affectant le dégagement des entre-noeuds de la Bryone, *Bryonia dioica,* en fonction de facteurs agissant sur la vitesse de croissance. Z. Pflanzenphysiol, 57, 203—205.

60) DE JAEGHER, G. & BOYER, N. (1990). On the role of membranes and calcium in signal perception and transduction in thigmomorphogenesis of *Bryonia dioica,* Intra- and intercellular communications in plants (B. Millet & H. Greppin, eds.) (in press).

61) CUBIZOLLES, L., DE JAEGHER, G., BOURGEADE, P., MOTTA, C. & BOYER, N. (1988). Use of fluorescence polarization for the study of membrane properties in mechanically stressed *Bryonia dioica* internodes. In: 6th Congress of F.E.S.P.P., Split, Abstracts, 501.

62) JAFFE, M. (1988). Calcium as a messenger in the thigmoperception of tendrile. Plant Physiol, 86 (4), suppl., 41.

63) DAVIES, E. (1987). Action potentials as multifunctional signals in plants: a umfying hypothesis to explain apparently disperate wound responses. Plant, Cell and Environment, 10, 623—631.

64) MEINERS, S., BARON-EPEL, D. & SCHINDLER, M. (1988). Intercellular communication. Filling in the gaps. Plant Physiol., 88, 791—793.

65) MINORSKY, P.V. & SPANSWICK, R.M. (1989). Electrophysiological evidence for a role for calcium in temperature sensing by roots of Cucumber seedlings. Plant, Cell and Environment, 12, 137—143.

66) GASPAR, TH., MONCOUSIN, CH. & GREPPIN, H. (1990). The place and role of exogenous and endogenous auxin in adventitious root formation. Intra- and intercellular communications in plants (B. Millet & H. Greppin, eds.) (in press).

67) DESBIEZ M.-O. & THELLIER, M. (1990). The *Bidens* plantlet: bilateral symmetry and possible breaking of the symmetry during bud growth. Intra- and intercellular communications in plants (B. Millet & H. Greppin, eds.) (in press).

68) FRACHISSE, J.-M. & DESBIEZ, M.-O. (1988). Extra- and intracellular measurement of the wave of electric depolarization induced by wounding in *Bidens pilosus* L. In: 6th Congress of F.E.S.P.P., Book of abstracts, 540.

69) DESBIEZ, M.-O., THELLIER, M. & CHAMPAGNAT, P. (1987). Storage and retrieval of morphogenetic messages in plantlets of *Bidens pilosus* L. In: The cell surface and signal transduction (E. Wagner, H. Greppin and B. Millet, eds.) NATO ASI Series, Vol. H 12, Springer-Verlag, Berlin, 189—203.

70) CANDELIER, P., DAUPHIN, G. & GENDRAUD, M. (1989). "In vivo" ^{31}P nuclear magnetic resonance spectroscopy of different *Helianthus tuberosus* organs during vegetative cycle. Plant Physiol. Biochem., 27 (2), 281—288.

71) GENDRAUD, M., & PETEL, G. (1990). Modifications in intercellular communications cellular characteristics and change in morphogenetic potentialities of Jerusalem artichoke tubers (*Helianthus tuberosus* L.) Intra- and intercellular communications in plants (B. Millet & H. Greppin, eds.) (in press).

72) RUIZ FERNANDEZ, S. & WAGNER, E. (1989). Flowering in *Chenopodium rubrum*: light control of stem elongation rate (S.E.R.) as systemic marker for flower induction. FLowering Newslett., 8, 15—19.

73) BERNIER, G. (1988). The control of floral evocation and morphogenesis. Ann. Rev. Plant Physiol. Plant Mol. Biol, 39, 175—219.

74) GREPPIN, H., BONSON, M., CRESPI, P., CRÈVECOEUR M., DEGLI-AGOSTI, R. & PENEL, C. (1990). Physiological macrofunctions and indicators of the flowering process. Intra- and intercellular communications in plants (B. Millet & H. Greppin, ed.s) (in press).

75) PENEL, C., AUDERSET, G., BERNARDINI, N., CASTILLO, F.J., GREPPIN, H. & MORRÉ, D.J. (1988). Compositional changes associated with plasma membrane thickening during floral induction of spinach. Physiol. Plant., 73, 134—146.

76) MALATIALY, L., GREPPIN, H. & PENEL, C. (1988). Calcium uptake by tonoplast and plasma membrane vesicles from spinach leaves. FEBS Letters, 233 (1), 196—200.

THE NEURONAL NICOTINIC RECEPTOR FAMILY

Bice Chini, Paola Tarroni and Francesco Clementi

CNR Center of Cytopharmacology
Dept. of Medical Pharmacology
University of Milano, Milano, Italy

Acetylcholine (ACh) was one of the earliest neurotransmitters to appear during the course of evolution and acetylcholine receptors (AChRs) are one of the most common receptors in the nervous system (1). Several molecules have evolved as Acetylcholine receptors and they have been classically divided into two large families: muscarinic and nicotinic receptors. Nicotinic receptors (nAChRs) mediate chemical communication between cells in many parts of the body: at the neuromuscular junction, in autonomic ganglia of the peripheral nervous system and in several areas of the brain.

nAChRs are activated by the binding of ACh released from the nerve terminals, which induce an influx of mainly Na^+ ions through the cell membrane, with a consequent depolarization of the target cell. They belong to the same receptor superfamily as $GABA_A$, Glycine and Glutamate receptors. All of them have a common architecture based on four or five subunits of similar molecular structure that define an ionic channel through the plasma membrane. In each subunit it is possible to identify four membrane spanning regions, an extracellular domain and a main cytoplasmic loop. The most conserved parts are the four membrane spanning regions, while the neurotransmitter binding site and the cytoplasmic sequences that modulate the activity of the receptors are the parts which have most evolved over time.

Muscle and neuronal nicotinic receptors

The nAChRs present in muscles and those present in neurons are different. Their pharmacological specificity has been recognized since the early days of pharmacology, while their molecular peculiarities have been unraveled only quite recently. Muscle nAChR was the first receptor to be completely understood from the pharmacological, functional and structural points of view (2,3).

The molecular structure of this receptor was discovered as a result of cloning experiments and, from these studies, various models for its function and structure have been proposed (2,3,4,5). The muscle nicotinic receptor is a pentameric complex composed of four different subunits α, β, γ, ∂, in a stoichiometric arrangement of $(\alpha)_2\ \beta\ \gamma\ \partial$. In plasma membrane, these subunits delimit a cationic channel opened up by nicotinic agonists and blocked by curare and α-Bungarotoxin (αBgtx) (6).

NATO ASI Series, Vol. H 51
Cell to Cell Signals in Plants and Animals
Edited by V. Neuhoff and J. Friend

Neuronal nAChRs have a heterogeneous structure and their functions and pharmacological properties are different from those of muscle receptors. They can be grouped into at least three subtypes:

1) *Nicotinic Acetylcholine Receptors* (nAChRs), which have a high affinity (nM range for nicotine, consist of a cationic channel which is gated by nicotinic agonists and not blocked by αBgtx although they may be blocked by neuronal Bungarotoxin (NBT) (7,8). NBT is a polypeptide isolated from the venom of the snake *Bungarus multicinctus* that is a potent, competitive antagonist of a variety of neuronal nicotinic receptors. In particular this toxin blocks nAChR in rat retinal ganglion, rat superior cervical ganglion, chick ciliary ganglia (8), chick optic lobe (9,10) and bovine chromaffin cells (11). Another toxin has been isolated from the same venom, namely the P15 toxin (12,13) that affects some ganglionic nicotinic receptors.

2) *α-Bungarotoxin sensitive nicotinic Acetylcholine Receptors* (B-nAChRs), which have a low affinity (μM range) for nicotine and a cationic channel which is opened up by nicotinic agonists and blocked by αBgtx. This receptor is pharmacologically similar to muscle nicotinic receptor (14).

3) *α-Bungarotoxin Binding Sites* (αBgtxR), which have a low affinity (μM range) for nicotine and bind αBgtx and other nicotinic agents. They are the most intriguing among the components of the neuronal nicotinic receptors because they do not seem to possess any ionic channel gated by cholinergic agonists.

Molecular structure of neuronal nicotinic receptors

nAChRs

a) Purification
nAChRs have been purified from the brain of various vertebrate species by using various mAb immunoaffinity procedures. mAb 35, which is specific for the main immunogenic region (MIR), a region exposed on the extracellular surface of the muscle nAChR a subunit, has been used for the isolation of a nAChR from chick brain (15), a receptor which is also recognized by NBT . Antisera raised in rats against this immunoaffinity purified neuronal nAChR specifically block the electrophysiological function of nAChRs in chick ciliary ganglionic neurons but do not inhibit the function of nAChRs in chick muscle cells (16). mAbs raised against the immunopurified chick nAChR have also been used to investigate nAChRs from other species. One of these, mAb270, has been particularly useful for the characterization of rat nAChR (17), while mAbs 290 and 295, raised against rat brain nAChR, have been important for studying human and cow neuronal nAChRs (18).

Immunopurified receptors are made up of two types of subunit: a major Mw subunit, labeled by ^{3}H-MBTA (4-(N-maleimido)benziltrymethylammonium iodide) and Bromo Acetylcholine, and thus called the ACh binding or α subunit, and a subunit which does not bind cholinergic ligands and is thus called the structural or β subunit. The Mw of these subunits varies according to the species from which they have been purified (see Table 1).

TABLE 1
Molecular weights of nAChR subunits isolated from different animal brains (18)

	chick	rat	human, bovine
α	59—75	79	75
β	49	51	50

From these data it has been proposed that nAChRs are different among the species and that in the same species more than one type of α and β subunits may be present.

b) Molecular biology
Several structural and ACh binding subunits have been cloned from the central nervous system and their combination has been proved to create different subtypes of functional nAChRs.

Five cDNAs — α_2, α_3, α_{4-1}, α_{4-2} (19) and α_5 (S. Heinemann, personal communication) — corresponding to ACh binding subunits, in that their deduced amino acidd sequences include the cysteines involved in ACh binding, have been cloned in rats. In situ hybridization has revealed that each of these cDNAs has a unique pattern of distribution in the brain. cDNAs corresponding to structural subunits — β_2, β_3, β_4 (20,21,22) — have also been cloned and shown to be differently distributed in the rat brain.

The Lindstrom group had determined the N-terminal amino acid sequences of the immunopurified nAChR subunits and has thus been able to correlate the cDNAs with the isolated subunits (23): the 79 kD subunit corresponds to rat α_4 and the 51 kD to rat β_2.

The stoichiometry of the nAChR subunits still remains an open question, given that the supramolecular complex may be either tetrameric or pentameric with a possible configuration of $(\alpha)_2 (\beta)_2$ or $(\alpha)_2(\beta)_3$.

The electrophysiological properties of nAChR subtypes have been studied in Xenopus oocytes after the microinjection of cDNAs or in vitro synthesized mRNA that code for the neuronal nicotinic subtypes. The combinations $\alpha_2\beta_2$, $\alpha_3\beta_2$, $\alpha_4\beta_2$, $\alpha_2\beta_4$, $\alpha_3\beta_4$, $\alpha_4\beta_4$ form functional nAChRs activated by nicotine and resistant to αBgtx. However, only the $\alpha_3\beta_2$ and the $\alpha_4\beta_2$ subtypes are blocked by NBT (21,23,24). The existence of a NBT insensitive nAChR subtypes is also suggested by autoradiographic studies showing that some brain regions with high affinity binding sites for nicotine are not labeled by NBT (24). Furthermore, no functional channels are detected by the combination of β_3 with any α subunit, thus raising the problem of the functional role of this subunit (20).

Four genes for ACh binding subunits — α_2, α_3, α_4, α_5, — and two genes for structural subunits — nα, nα_4 — have also been cloned from chick brain (25). These genes code for subunits which are highly homologous to the corresponding subunits in the rat, with the least conserved region residing in the cytoplasmic loop between the third and the fourth transmembrane spanning regions. This domain is poorly conserved among nAChR subunits in the same species as well, suggesting that it is probably responsible for pharmacological and functional differences among nAChR subtypes. The study of the function of newly cloned subunits has been enhanced by a new method for cDNA expression in oocytes (26). cDNAs were placed in plasmid vectors under the control of a heat shock promoter and these constructs were microinjected into oocyte nuclei. A high expression of nAChRs was then obtained by exposing the cells to a shock temperature.This system is a powerful tool for the study of the functional properties of cloned neuronal receptors.

Finally, two human α subunits, α_3 (27) and α_5 (B. Chini, unpublished results), have recently been cloned, and a very high degree of homology was found between the human and the corresponding rat subunits, underlining the fact that the conservation of these molecules throughout evolution reflects their crucial role in the nervous system.

B-nAChRs

αBgtx sensitive nAChRs have been demonstrated in rat cerebellar interneurons (14,28), in the insect CNS (see below), in the goldfish brain (29) and in the chick optic lobe.

The B-nAChR from the chick optic lobe has been purified by affinity chromatography, using αBgtx as ligand (30,31,32). It has been shown to have a sedimentation coefficient of 10.5 S and to be composed of three to four major polypeptides. The lowest molecular weight component, corresponding to the toxin binding polypeptide, has been submitted to N-terminal amino acid sequencing and a high degree of homology with known nAChR subunits has been found (30). A B-nAChR, with characteristics similar to those of the molecule isolated from chick optic lobe, has also been purified from total chick brain (30,31).

Insect nicotinic receptor

The αBgtx sensitive nAChR present in insects merits separate mention since it is very different from any vertebrate receptor. It has been extensively studied in the *Locusta,* in the CNS of the *Drosophila* and in the *Cockroach.*

From the *Locusta* CNS, a pentameric receptor with a Mw of 235—300 k and composed of five identical subunits of Mw 65 k has been purified and reconstituted in liposomes (33). It probably represents an ancient precursor of vertebrate neuronal nicotinic receptors. A cDNA capable of originating this receptor when microinjected into oocytes has also been cloned (34). Other nicotinic related cDNAs have been isolated from the *Drosophila* CNS, suggesting that, as in vertebrates, a number of neuronal nicotinic genes might be present in insects (31,35). The transcripts of one of these genes — the *ard* gene — were studied during *Drosophila* development and a specific neuronal expression was found which was significantly higher during periods of neuronal differentiation (36).

αBgtxRs

These molecules, which bind αBgtx and nicotinic drugs, but do not apparently contain any cationic channel, are the most elusive neuronal nicotinic receptors. They have been studied by several groups using molecular biology techniques but none of them yet has been successfully cloned. Thus the classical biochemical approach has been used for its isolation and characterization.

αBgtxRs have been studied in ganglionic and chromaffin cells as well as in related tumoral cell lines, which are particularly rich in these molecules.

αBgtxRs have been purified from PC12, a rat adrenal pheochromocytoma cell line, and was found to consist of four polypeptides of Mw 72 k, 64 k, 56k and 46k. The 56 k subunit is thought to be the agonist binding subunit since it is labeled by tritiated BromoAcetylcholine (37). However, a partial amino acid sequence of this band indicates that it could have arisen from a contaminant.

αBgtxRs have also been purified from IMR32, a human neuroblastoma cell line (38). It is made up of three major peptides, Mw 67 k, 60 k and 52 k, the last of which binds αBgtx and does not seem to be coded by a muscular α subunit gene. Immunological, pharmacological and biochemical experiments also show that this molecule is different from muscle nAChR, from neuronal nAChRs and B-nAChRs (38).

A similar protein has been purified from rat brain; it is composed of three subunits (Mw 55 k, 53 k and 49 k), one of which, 55 k, being the subunit which binds the ligand (39).

The number and molecular weight of the polypeptides forming this type of receptor is different in the various cell types but in all of them the subunit that binds αBgtx and cholinergic agents has a Mw of 52—55 k.

The function of this protein in brain and ganglia is still unknown.

Distribution of nicotinic receptors in the nervous system

The distribution of nicotinic receptors in rat brain has been clarified by means of autoradiographic studies using labeled compounds (40,41,42). High affinity binding sites for ^{3}H-Nicotine and for ^{3}H-Acetylcholine, which should correspond to nAChRs, show a markedly concordant distribution in the brain, whereas ^{125}I-αBgtx, which labels both B-nAChRs and αBgtxRs, has a different brain localization (40). Comparative studies in rats, mice and monkeys have revealed that the labeling patterns in different species are similar, although not identical.

Further information for the localization of nAChRs comes from the use of NBT. In avian ciliary and sympathetic ganglia, in rat sympathetic ganglia and in bovine chromaffin cells, it has been demonstrated that NBT binds to nAChRs and blocks ACh-induced depolarization. These effects are prevented by the competitive antagonists mecamylamine, d-tubocurarine and trimetaphan. However, NBT does not identify all nAChR subtypes, as has been shown by autoradiographic studies in the rat brain, where only some of the ^{3}H-nicotine high affinity binding sites were also labeled by NBT.

On the basis of these studies it is now becoming possible to draw up a first map of areas and connections rich in nAChRs for the rat brain. These include:

- The nigrostriatal and mesolimbic dopaminergic areas
- The thalamic nuclei
- Layers III and IV of the cortex and the thalamic nuclei projecting there
- The retinal ganglionic cells, the superior colliculus, the dorso lateral geniculated nucleus and the visual cortex
- The medial habenula and the interpeduncular nucleus, connected via the fasciculus retroflexus.

αBgtx binding was found to be high in anterior olfactory bulb, in medial nucleus of habenula, in hippocampus, caudate, putamen, colliculi and some nuclei of the midbrain.

Very low level of αBgtx binding was detected in cerebellum.

In goldfish optic tectum αBgtx binding is correlated with nicotine binding and this is in agreement with the physiological data that have shown that in this area αBgtx binding sites correspond to a nicotinic channel (42).

The specific distribution of nAChR subtypes in rat brain has been studied by means of in situ hybridization using in vitro synthesized mRNA as probes (43). Among the α subunitis, α_4 is the most widely distributed, α_3 is present mainly in the medial habenula, thalamus and cortex; and α_2 in the interpeduncular nucleus. The codistribution of α_3 and α_4 in some regions such as the thalamus, suggests that there is a microheterogeneity in some brain nuclei. Among the β subunits, β_2 is the most widely expressed in the brain; β_3 is expressed in the medial habenula, substantia nigra pars compacta, thalamus and ventral tegmental area; and β_4 is present only in the ventral medial habenula. This specific distribution of the different subunits of nAChRs suggests that in different areas of the brain different types of receptor are expressed. The distribution of human nAChR subunits α_3 and α_5 is also under study and this could have profound implications on understanding physiology and pathology of the nicotinic transmission in the human brain.

Subcellular localization

Both cholinergic and non cholinergic nerve pathways have been shown to possess presynaptic high affinity binding sites for ^{3}H-Nicotine, which are located on both neuronal cell bodies and dendrites. The subcellular distribution of nAChRs has been studied by using NBT. In cultured chick sympathetic neurons, in which both nAChRs and αBgtxRs are present, ^{125}I-NBT appears to label nAChRs concentrated in the synaptic membranes, with a density of 500 sites/μ^2. This labeling is completely prevented by the addition of cold NBT and is also partially prevented (50%) by αBgtx and d-tubocurarine, indicating the coexistence of both nAChRs and αBgtxRs in the synaptic regions of chick sympathetic neurons. On the other hand, in rat dissociated sympathetic neurons, NBT labels 5000 sites/μ^2 in the synaptic membranes even in the presence of cold αBgtx, revealing a clustering of nAChRs in the synaptic region. In these neurons the αBgtxRs have mainly an extrasynaptic localization. Thus nAChRs are concentrated in the synaptic region but with a density of 10 and 100 times lower (for rats and chicks respectively) than that of the nAChR present at the neuromuscular junction (44).

αBgtx does not appear to bind to synaptic regions in chick ciliary ganglia (45); however, in retina and sympathetic ganglia it is localized in synaptic structures (46,47,48).

These morphological studies are of difficult interpretation because it is difficult to discriminate between high and low affinity sites and between specific and non specific binding.

In vitro and in vivo regulation of neuronal nicotinic receptors

The regulation of nAChRs in chick ganglionic neurons has been investigated in detail. After both post-ganglionic axotomy and pre-ganglionic denervation, a reduction in the total number of nAChRs has been observed, with a parallel decrease in the expression of the mRNA for the α_3 subunit (49). This finding is completely different from the well known up-regulation of the muscle nAChR seen after denervation.

After chronic exposure to anticholinesterase compounds, a decrease in both nicotinic and muscarinic receptors has been demonstrated in several brain areas and this correlates well with a reduction in the antinociceptive effect of nicotine. Chronic nicotine treatment increases nAChR density in many brain regions (50). This increase takes place together with a shift of nAChRs into a desensitized, and thus inactive, state. In rats chronically exposed to nicotine, it has also been observed that the increase in high affinity binding sites in the cortex correlates with a reduction in the number of low affinity binding sites (51). However, a chronic nicotinic treatment reduces the number of presynaptic nicotinic receptors in a reversible manner (52).

It is interesting to note that some behavioral effects elicited by nicotine are enhanced after chronic nicotine treatment, and this is in accordance with the increased number of nAChRs. On the other hand, some studies have shown that tolerance to some of the behavioural effects of nicotine are strictly correlated to the increase in nicotinic receptor density. It is possible that the behavioral effects elicited by nicotine are obtained by the excitation of different types of nAChRs. Furthermore, these receptor subtypes are differently regulated during chronic drug exposure (52,53).

Direct visualization of human nicotinic receptors in vivo has been obtained by means of PET (positron emission tomography). In smokers' brains a major accumulation of nicotine is seen, which might reflect changes in the number of nAChRs (53).

Few data are available on the modulation of αBgtxRs in vivo. During development, αBgtxRs in the brain appear very early and precede other markers of cholinergic innervation (54). Chronic nicotinic treatment does increase the number of αBgtxRs in some areas of the brain, particularly in those areas in which there is an increase in nAChRs; however, the time course of the modifications of the two receptors is different (54). After dietary choline and in the newly born offspring of rats in which estradiol had been administered during pregnancy, an up-regulation of αBgtxRs has been reported, the significance of which is still unclear.

The in vitro regulation of αBgtxRs has been analyzed in the neuroblastoma cell line IMR32 and in adrenal medullary chromaffin cells. An increase in the number of αBgtxRs has been observed after cell exposure to dibutyryl cAMP or 5-Bromodeoxyuridine. These drugs allow the cells to differentiate in vitro, suggesting an involvement of αBgtxRs in this process (55,56). In chromaffin cells, d-tubocurarine induces a large up-regulation of αBgtx binding sites which can be modulated by second messengers and chronic depolarization (54,57).

It is interesting to note that the modulation of αBgtxR in cells in vitro does not always parallel that of nAChR.

Function of neuronal nicotinic receptors

The functional role of neuronal nicotinic receptors has been studied using different approaches.

Rubidium effluxes, taken as a measure of the activation of functional nAChRs, have been measured in PC12 after exposure to various nicotinic drugs. It is thus possible to analyze the effects of various agonists and antagonists on the ionic channel of nAChRs (58).

The presence of functional nAChRs in living animals has been identified by the method of local glucose utilization which is enhanced by nicotine. In this way it is possible to correlate the effects of nicotine with the distribution of nAChRs identified by ^{3}H-nicotine binding (59,60).

The data show that nicotine has a functional role in the visual pathways, in the substantia nigra, in the thalamus, in the median habenula and in the median eminence.

The best evidence for the role of the different members of the nicotinic receptor family has been provided by studies of inbred mouse strains with different numbers of both nAChRs and αBgtxRs, and whose sensitivity to various nicotine induced effects is different (61,62). A high correlation between the number of hippocampal αBgtxRs and nicotine induced seizure sensitivity was found. Moreover, genetic analysis revealed that both seizure sensitivity and the increased number of αBgtxRs segregate together as if they were regulated by a single gene. On the other hand, the number of nAChRs correlates with motor activity and the regulation of body temperature. It has been shown that nicotine induced effects on the respiratory rate and on startle response are regulated by both nAChRs and aBgtRs. Acute tolerance to nicotine-induced seizures rapidly develops in some mouse strains, while in others (where there is an increased number of hippocampal αBgtxRs and a high degree of sensitivity to seizures), acute tolerance does not develop. This indicates that αBgtxRs with different desensitization rates might be present in different mouse strains. Chronic exposure to nicotine increases the number of nAChRs (at low doses) and of αBgtxRs (at higher doses) and subsequently induces chronic tolerance. However, not only the number but also the desensitization rate of the different nicotinic subtypes appears to be crucial in the determination of nicotine sensitivity. From these studies it is clear that nAChRs and αBgtxRs mediate different central effects of nicotine.

The pharmacology of presynaptic nAChRs responsible for neurotransmitter release has been studied using synaptosomes prepared from diferent brain regions. The Dopamine release from the striatum and the GABA release from the hippocampus are under nicotinic regulation. In both cases nicotine, cytisine, 1,1-dimethyl-4-phenylpiperazinium (DMPP) and ACh were able to induce neurotransmitter release, which could be prevented by antagonists, such as mecamylamine and pempidine (63).

The presynaptic modulation by nicotinic receptors of the release of several neurotransmitters may explain the complex behavioural effects of nicotine.

Role of nicotinic receptors in aging and pathology

There is evidence that the cholinergic system is impaired by aging and, to a greater extent, by Alzheimer's disease (AD) (64). The Maynert nucleus, which is the major source of the ascending cholinergic projections from the basal forebrain to the neo- and archicortical areas, is particularly affected. The loss of neurons in this nucleus is probably responsible for the significant reduction in choline acetyltransferase activity in the hippocampus and cortex following the decreased synthesis of acetylcholine in these areas.

Brain nicotinic receptors change during aging and related pathological conditions.

A loss of hippocampal ^{3}H-nicotine binding sites has been demonstrated during aging, in conjunction with a reduction in choline acetyltranferase (ChAT) activities. To elucidate the possible presynaptic localization of such receptors and to correlate their loss with the degeneration of cortical projections, fimbria-fornix lesioned rats have been studied. In this condition, no decrease in the number of hippocampal nicotinic receptors was found, despite a substantial loss in ChAT activities, suggesting that, in the hippocampus, only a small fraction of nicotinic receptors is presynaptic and/or that a postsynaptic supersensitivity might occur (64,65).

A loss in presynaptic nicotinic cortical receptors, probably regulating ACh release, accompanied by a reduction in ChAT activities, has been demonstrated in Alzheimer's disease and senile dementia of Alzheimer type (AD/SDAT), Parkinson's disease and in middle aged subjects suffering from Down's syndrome. Furthermore antibodies to cholinergic neurons, which are able to destroy presynaptic nerve endings, have been found in the sera of patients with AD (66). Moreover, in both AD/SDAT and Parkinson patients, a decreased number of nicotinic receptors, probably localized on dopaminergic terminals, has been found in the caudate nucleus (65). In AD/SDAT a reduction in an endogenous factor which inhibits the nicotine binding has been described. It is a low molecular weight compound (1000 Daltons) of unknown identity and function (65).

A decrease in the number of high affinity binding sites during aging has been observed in the cortex, whereas in other brain regions, such as the thalamus, an increase has been reported to occur (53). In the thalamus a reduction in αBgtxRs has been demonstrated, suggesting that an interconversion between different types of nicotinic receptors might occur during aging. In AD/SDAT frontal cortex, a marked drop in the number of high affinity sites has been found, coexisting with a growth in the number of low affinity sites. Moreover, a selective presynaptic loss of nicotinic receptors in AD/SDAT has been suggested by release experiments on autoptic brain slices (53).

Similar studies have been performed in bioptic frontal cortices of normal subjects and AD/SDAT patients. It is claimed that bioptic studies offer advantages over autoptic studies mainly as a result of the better preservation of neuronal functional activities. It has been shown that with 1 cm^2 of bioptic tissue a battery of tests can be performed, including pathology, enzyme assays like ChAT activity, neurotransmitter release and receptor binding determination. These studies confirm a selective loss of nAChRs in the brains of Alzheimer patients, and demonstrate that there is no decrease in the release of ACh in patients treated with Physiostigmine by intraventricular infusion (67).

The decrease of nAChRs in Parkinson disease and the increase of these receptors after chronic nicotine treatment (at least in some brain areas) may explain the apparently protective effect of cigarette smoking on the onset of Parkinson's disease (68). Stimulation of nicotinic receptors, in fact, increases the firing of dopaminergic nigrostriatal neurons and causes an increase in striatal dopamine release (63). An increased stimulation of these receptors, after chronic nicotinic exposure, might thus enhance the residual dopaminergic activity and prevent the emergence of clinical symptoms in subjects whose dopaminergic function is impaired. However, there is evidence that in MPTP-induced Parkinson's disease in mice, a chronic nicotine treatment has a positive influence on locomotor activity without causing any change in brain amine levels (69).

Over the last ten years therapeutic approaches for the enhancement of cholinergic activity have been developed and tested in AD patients, as reviewed by Hollander (70). Some encouraging results have been achieved using anticholinesterase drugs, such as tetrahydroaminoacridine, given in combination with lecitin (71,72). This therapeutic treatment provides a higher availability of ACh at the synaptic sites.

Future directions include the development of specific M1 agonists and M2 antagonists, the former acting postsynaptically and the latter blocking presynaptic autoreceptors and thus preventing ACh from inhibiting its own release. Nicotinic receptors are the other possible site of action of cholinergic drugs. A preliminary trial with intravenous nicotine has been carried out in an AD patient (70), but more work is necessary to develop specific drugs possibly acting on specific subtypes of nicotinic receptors that can reverse the nicotinic deficit demonstrated in this disease.

However, it is difficult to foresee a large success of cholinergic therapy in SDAT/AD since in advanced cases the cholinergic deficit is too severe to be treated, and in any case too many other systems are involved to benefit from a treatment targeted only on the cholinergic system (73).

CONCLUSIONS

Neuronal nicotinic receptors are a family of related but different molecules. We know the complete structure of several nAChRs, and at least partially the protein composition of αBgtx sensitive nAChRs and of α-Bungarotoxin binding sites. However we are still unable to assign to each one of these molecules particular effects among the several elicited by nicotine in the CNS, and multidisciplinary studies are needed to unravel this complex problem. However the physiological and pharmacological evidence obtained so far has shown the fundamental role of these molecules in processes such as thermoregulation, seizure sensitivity, memory and motor control.

The regionally selective decrease in the number of nAChRs in some CNS degenerative pathologies further underlines the importance of these molecules. Data on the positive effects of some drugs active on the nicotinic pathways in degenerative diseases suggest that new therapies might be found for these distressing conditions.

REFERENCES

1) Wächter K. (1986) Phylogeny of the cholinergic synapse In: Whittaker V.P. (Ed.) The cholinergic synapse. Handbook Exp. Pharm., Springer Verlag, Berlin, 86: 57—80.

2) Conti-Tronconi B.M., Raftery M.A. (1982) The nicotinic cholinergic receptor: correlation of molecular structure with functional properties. Ann. Rev. Biochem. 51: 491—530.

3) Popot J.L.,Changeaux J.P. (1984) Nicotinic receptor of acetylcholine: structure of an oligomeric integral membrane protein. Phisyol. Rev. 64: 1162—1239

4) Hucho F., Hingenfeld R. (1989) The selectivity filter of a lingand-gated ion channel. FEBS Lett 257: 17—23.

5) McCarthy M.P., Earnest J.P., Young E.F., Choe S. & Stroud R.M. (1986) The molecular neurobiology of the acetylcholine receptor. Ann. Rev. Neurosci. 9: 383—413.

6) Low B.W. (1979) The three-dimensional structure of post-synaptic snake neurotoxins: consideration of structure and function. In: Lee C.Y. (Ed.) Snake venoms. Handbook Exp. Pharmacol. 52: 213—257, Springer Verlag, Berlin.

7) Chiappinelli V.A., Dreyer S.E., Sorenson E.M., Wolf K.M., Grant G.A., Chen S.J., Nooney J.M., Lambert J.J., Hider R.C. (1988) Functional studies of the neuronal nicotinic receptors utilizing k-neurotoxin. In: Clementi et al. (Eds.) Nicotinic acetylcholine receptors in the nervous system., Springer Verlag, Berlin, pp. 15—29.

8) Halvorsen S.W., Berg D.K. (1986) Identification of a nicotinic acetylcholine receptor on neurons using an a-neurotoxin that blocks receptor function. J. Neurosci. 6: 3405—3412.

9) Chiappinelli V.A. (1985) Actions of snake venom toxins on neuronal nicotinic receptors and other neuronal receptors. Pharmacol. Ther. 31: 1—32.

10) Wolf K.M., Ciarleglio A., Chiappinelli V. (1988) K-Bungarotoxin binding of a neuronal nicotinic receptor antagonist to chick optic lobe and skeletal muscle. Brain Res. 439: 249—258.

11) Higgins L.S. & Berg D.K. (1987) Immunological identification of a nicotinic acetylcholine receptor on bovine chromaffin cells. J. Neurosci. 7: 1792—1798.

12) Gotti C., Omini C., Berti F., Clementi F. (1985) Isolation of a polypeptide from the venom of Bungarus multicinctus that binds to ganglia and blocks the ganglionic transmission in mammals. Neuroscience 15: 563—575.

13) Saiani L., Kageyama H., Conti-Tronconi B.M., Guidotti A. (1984) Purification and characterization of a bungarotoxin polypeptide which blocks nicotinic receptor functions in primary culture of adrenal chromaffin cells. Mol. Pharmacol. 25: 327—334.

14) Garza R. de la, McGuire T.J., Freedman R., Hofler B.J. (1987) Selective antagonism of nicotine actions in the rat cerebellum with α-Bungarotoxin. Neuroscience 23: 887—891.

15) Whiting P.J., Lindstrom J.M. (1986) Purification and characterization of a nicotinic acetylcholine receptor from chick brain. Biochemistry 25: 2082—2093.

16) Stollberg J., Whiting P.J., Lindstrom J.M., Berg D. (1986) Functional blockade of neuronal acetylcholine receptors by antisera to a putative receptor from brain. Brain Res. 376: 179—182.

17) Whiting P.J. & Lindstrom J.M. (1987) Purification and characterization of a nicotine acetylcholine receptor from rat brain. Proc. Natl. Acad. Sci. USA 84: 595—599.

18) Lindstrom J.M., Schoepfer R., Whiting P., (1987) Molecular studies of the neuronal nicotinic acetylcholine receptors family. Mol. Neurobiol. 1: 218—337.

19) Heinemann S., Boulter J., Deneris E., Conolly J., Gardner P., Wada E., Wada K., Ballivet M., Swanson L., Patrick J. (1988) The nicotinic acetylcholine receptor gene family. In: F. Clementi et al. (Eds.) Nicotinic acetylcholine receptors in the nervous system., Springer Verlag, Berlin, pp. 173—191.

20) Deneris E.S., Connolly J., Boulter J. Wada E., Wada K. Swanson L.W., Patrick J., Heinemann S. (1988) Primary structure and expression of β2: a novel subunit of neuronal nicotinic acetylcholine receptors. Neuron 1: 45—54.

21) Deneris E.S., Boulter J., Swanson L.W., Patrick J., Heinemann S. (1989) β3: a new member of nicotinic acetylcholine receptor gene family is expressed in brain. J. Biol. Chem. 264: 6268—6272.

22) Duvoisin R.M., Deneris E.S., Patrick J., Heinemann S. (1989) The functional diversity of the neuronal nicotinic acetylcholine receptors is increased by a novel subunit: β4. Neuron 3: 487—496.

23) Lindstrom J., Whiting P. Schoepfer R., Luther M., Casey B. (1988) Structure of neuronal nicotinic receptors. In: F. Clementi et al. (Eds.) Nicotinic acetylcholine receptors in the nervous system., Springer Verlag, Berlin, pp. 159—172.

24) Deneris E.S., Boulter J., Connolly J., Wada E., Wada K., Saldman D., Swanson L.W., Patrick J., Heinemann S. (1989) Genes encoding neuronal nicotinic acetylcholine receptors. Clin. Chem. 35/5: 731—737.

25) Nef P., Oneyser C., Alliod C., Couturier S., Ballivet M. (1988) Neuronal nicotinic acetylcholine receptor genes in the avian genome. In: F. Clementi et al. (Eds.) Nicotinic acetylcholine receptors in the nervous system., Springer Verlag, Berlin, pp. 193—198.

26) Ballivet M., Nef P., Couturier S. Rungger D., Bader C.R., Bertrand D. & Cooper E. (1988) Electrophysiology of chick neuronal nicotinic acetylcholine receptors expressed in Xenopus oocytes after cDNA injection. Neuron 1: 847—852.

27) Fornasari D., Chini B., Tarroni P. & Clementi F. (1990) Molecular cloning of human neuronal nicotinic receptor α3-subunit. Neuroscience Lett., in press.

28) Garza R.de la, Hoffer B.J. & Freedman R., Heterogeneity of nicotine actions in the rat cerebellum. (1988) In: F. Clementi et al. (Eds.) Nicotinic acotylcholine receptors in the nervous system., Springer Verlag, Berlin, pp. 137—141.

29) Henle J.,Lindstrom M.J.,Oswald R.E.(1988) Interaction of monoclonal antibodies with α-Bungarotoxin and (-)-nicotine binding sites in goldfish brain. J. Biol. Chem. 263: 9686—9691.

30) Conti Tronconi B.M., Kokla A., Maelicke A., Milius R., Raftery M.A., Spencer S., Tang F., Tzartos S. (1988) Identification of surface domains on the nicotinic acetylcholine receptor molecule by the use of antibodies and toxins. In: F. Clementi et al. (Eds.) Nicotinic acetylcholine receptors in the nervous system., Springer Verlag, Berlin, pp. 119—136.

31) Hermans-Borgmeyer I., Sawruk E., Schloss P., Gundelfinger E., Betz H. (1988) Biochemical and molecular biology approaches to central nicotinic acetylcholine receptors. In: F. Clementi et al. (Eds.) Nicotinic acetylcholine receptors in the nervous system., Springer Verlag, Berlin, pp. 77—88.

32) Norman R.I., Mehraban F., Barnard E.A. & Dolly J.O. (1982) Nicotinic acetylcholine receptor from chick optic lobe. Proc. Natl. Acad. Sci. USA 79: 1321—1325.

33) Breer H., Hanke W. (1988) Identification of nicotinic acetylcholine receptors from the nervous system of insects. In: F. Clementi et al. (Eds.) Nicotinic acetylcholine receptors in the nervous system., Springer Verlag, Berlin, pp. 213—226.

34) Marshall J., David J.A., Darlison M.G., Barnard E.A., Sattelle D.B. (1988) Pharmacology, cloning and expression of insect nicotinic acetylcholine receptors. In: F. Clementi et al. (Eds.) Nicotinic acetylcholine receptors in the nervous system., Springer Verlag, Berlin, pp. 257—281.

35) Bossy B., Ballivet M., Spierer P. (1988) Conservation of neural nicotinic acetylcholine receptors from Drosophila to vertebrate central nervous systems. EMBO J. 7: 611—618.

36) Hermans-Borgmeyer I., Hoffmeister S., Sawruk E., Betz H., Schmitt B., Gundelfinger E.D. (1989) Neuronal acetylcholine receptors in Drosophila: mature and immature transcripts of the ard gene in the developing central nervous system. Neuron. 2: 1147—1156.

37) Kemp G. (1988) Characterization of the α-Bungarotoxin binding protein from PC12 cells. In: F. Clementi et al. (Eds.) Nicotinic acetylcholine receptors in the nervous system., Springer Verlag, Berlin, pp. 105—117.

38) Gotti C., Esparis Ogando A., Clementi F. (1989) The α-Bungarotoxin rceptor purified from a human neuroblastoma cell line: biochemical and immunological characterization. Neuroscience 32: 759—767.

39) Kemp G., Bentley L., McNamee M.G., Morley B.J. (1985) Purification and characterization of the α-Bungarotoxin binding protein from rat brain. Brain Res. 347: 274—282.

40) Clarke P.B.S. (1987) Recent progress in identifying nicotinic cholinoceptors in mammalian brain. Trends Pharmacol. Sci. 8: 32—35.

41) Pauly J.R., Stitzel J.A., Marks M.J., Collins A.C. (1989) An autoradiographic analysis of cholinergic receptors in mouse brain. Brain Res. Bull. 22: 453—459.

42) Schwartz M., Axelrod D., Feldman L.E., Agranoff B. (1980) Histological localization of binding sites of α-Bungarotoxin and of antibodies specific to acetylcholine receptor in goldfish optic tectum. Brain Res. 194: 171—180.

43) Wada E., Wada K., Boulter J., Deneris E., Heinemann S., Patrick J., Swanson L.W. (1989) Distribution of a2, a3, a4 and β2 neuronal nicotinic receptor subunit mRNAs in the central nervous system: a hybridization histochemical study in the rat. J. Comp. Neurol. 284: 314—335.

44) Zigmond R.E., Loring R.H. (1988) Characterization and localization of ganglionic nicotinic receptors using neuronal bungarotoxin. In: F. Clementi et al. (Eds.) Nicotinic acetylcholine receptors in the nervous system., Springer Verlag, Berlin, pp. 31—39.

45) Jacob M.H., Berg D.K. (1983) The ultrastructural localization of α-Bungarotoxin binding sites in relation to synapses on chick ciliary ganglion neurons. J. Neurosci. 3: 260—271.

46) Marshall L.M. (1981) Synaptic localization of α-Bungarotoxin binding which blocks nicotinic transmission at frog sympathetic neurons. Proc. Natl. Acad. Sci. USA 78: 1948—1952.

47) Smolen A.J. (1983) Specific binding of α-Bungarotoxin into synaptic membranes in rat sympathetic ganglia. Brain Res. 289: 177—188.

48) Daniels M.P., Vogel Z. (1980) Localization of α-Bungarotoxin binding sites in synapses of the developing chick retina. Brain Res. 281: 45—56.

49) Berg D.K., Boyd R.T., Halvorsen S.W., Higgins L.S., Jacob M.H., Margiotta J.F., McEachern A.E. (1988) Regulation of neuronal Acetylcholine receptors by cell-cell interactions. In: F. Clementi et al. (Eds.) Nicotinic acetylcholine receptors in the nervous system., Springer Verlag, Berlin, pp. 283—297.

50) Costa L.G., Murphy S.D. (1983) ^{3}H-nicotine binding in rat brain: alteration after chronic acetylcholinesterase inhibition. J. Pharm. Exp. Ther. 226: 392—397.

51) Nordberg A., Wahlström G., Arnelo V., Larsson C. (1985) Effect of longterm nicotine treatment on ^{3}H-nicotine binding sites in the rat brain. Drug and Alcohol Dependence 16: 9—17.

52) Lapchack P.A., Araujo D.M., Quirion R., Collier B. (1989) Effect of chronic nicotine treatment on nicotinic autoreceptors function and ^{3}H-methylcarbamylcholine binding sites in the rat brain. J. Neurochem. 52: 403—491.

53) Nordberg A., Adem A., Nilsson L., Romanelli L., Ziang X. (1988) Heterogeneous cholinergic nicotinic receptors in CNS. In: F. Clementi et al. (Eds.) Nicotinic acetylcholine receptors in the nervous system., Springer Verlag, Berlin, pp. 331—350.

54) Quitk M., Geertsen S. (1988) Neuronal nicotinic α-Bungarotoxin sites. Can. J. Physiol. Pharmacol. 66: 971—979.

55) Clementi F., Cabrini D., Gotti C., Sher E. (1986) Pharmacological characterization of cholinergic receptors in a human neuroblastoma cell line. J. Neurochem. 47: 291—297.

56) Gotti C., Sher E., Cabrini D., Bondiolotti G., Wanke E., Mancinelli E., Clementi F. (1987) Cholinergic receptors, ion channels, neurotransmitter synthesis and neurite outgrowth are independently regulated during the in vitro differentiation of a human neuroblastoma cell line. Differentiation 34: 144—155.

57) Quik M., Geersten S., Trifarò J.M. (1987) Marked upregulation of the α-Bungarotoxin site in adrenal chromaffin cells by specific nicotinic antagonists. Mol. Pharmacol. 31: 385—391.

58) Lukas R.J. (1988) Evidence for functional and structural diversity of nicotinic acetylcholine receptors. In: F. Clementi et al. (Eds.) Nicotinic acetylcholine receptors in the nervous system., Springer Verlag, Berlin, pp. 61—75.

59) London E.D., Conolly R.J., Szikszay M., Wamsley J.K., Dam M. (1988) Effects of nicotine and local cerebral glucose utilization in the rat. J. Neurosci. 8: 3920—3928.

60) London E.D., Dam M., Fanelli R.J. (1988) Nicotine enhances cerebral glucose utilization in the central components of the rat visual system. Brain Res. Bull. 20: 381—385.

61) Collins A.C., Marks M.J. (1988) Regulation of behavioural and physiological sensitivity to nicotine by brain nicotinic receptors. In: F. Clementi et al. (Eds.) Nicotinic acetylcholine receptors in the nervous system., Springer Verlag, Berlin, pp. 317—329.

62) Marks J.M., Stitzel J.A., Collins A.C., (1989) Genetic influences on nicotine responses. Pharmacol. Biochem. Behav. 33: 667—678.

63) Wonnacott S., Irons J., Lunt G.G., Rapier C.M., Albuquerque E.X. (1988) α-Bungarotoxin and presynaptic nicotinic receptors: functional studies. In: F. Clementi et al. (Eds.) Nicotinic acetylcholine receptors in the nervous system., Springer Verlag, Berlin, pp. 41—60.

64) Perry E.K., Perry R.H., Smith C.J., Dick D.J., Candy J.M., Edwardson J.A., Fairbairn A., Blessed G. (1987) Nicotinic receptor abnormalities in Alzheimer's and Parkinson's diseases. J. Neurol. Neurosurg. Psychiatr. 50: 806—809.

65) Perry E.K., Smith C.J., Xuareb J.H., Keith A.B., Perry R.H. (1988) Cholinergic neuropathology and nicotinic receptor binding in the human brain. In: F. Clementi et al. (Eds.) Nicotinic acetylcholine receptors in the nervous system., Springer Verlag, Berlin, pp. 351—361.

66) Foley P., Bradford H.F., Docherty M., Fillit H., Luine V.N., McEven B., Bucht G., Winblad B., Hardy J. (1988) Evidence for the presence of antibodies to cholinergic neurons in the serum of patients with Alzheimer's disease. J. Neurol. 235: 466—471.

67) Giacobini E., De Sarno P., McIlhany M., Clark B. (1988) The cholinergic receptor system in the frontal lobe of Alzheimer patients. In: F. Clementi et al. (Eds.) Nicotinic acetylcholine receptors in the nervous system., Springer Verlag, Berlin, pp. 367—378.

68) Baron J.A. (1986) Cigarette smoking and Parkinson's disease. Neurology 36: 1490—1496.

69) Sershen H., Hashim A., Lajtha A. (1987) Behavioural and biochemical effects of nicotine in an MPTP-induced mouse model of Parkinson's disease. Pharmacol. Biochem. Behav. 28: 299—303.

70) Hollander E., Mohs R.C., Davis K.L. (1986) Cholinergic approaches to the treatment of Alzheimer's disease. Br. Med. Bull. 42: 97—100.

71) Editorial: Cholinergic treatment in Alzheimer's disease: encouraging results. (1987) Lancet I: 139—141.

72) Summers W.K., Majovski L.V., Marsh G.M., Tachiki K., Kling A. (1986) Oral tetrahydroaminoacridine in long term treatment of senile dementia, Alzheimer Type. N. Engl. J. Med. 315: 1241—1245.

73) Ashford J.W., Sherman K.A., Kumar V. (1989) Advances in Alzheimer therapy: cholinesterase inhibitors. Neurobiol. Aging 10: 99—105.

GLIAL-NEURONAL COMMUNICATION IN DEVELOPMENT AND REGENERATION

Hans H. Althaus[1] and Wilfried Seifert[2]

[1]Max-Planck-Institut für experimentelle Medizin
Hermann-Rein-Straße 3, 3400 Göttingen, FRG

[2]Max-Planck-Institut für biophysikalische Chemie
Am Faßberg, 3400 Göttingen-Nikolausberg, FRG

INTRODUCTION

At the beginning of this century popular books on the function of the nervous system depicted the brain more mechanistically as a fabrique, which was organized in a number of functional units (1) and supported by pipelines transporting the nutrients.This simplistic view saw mainly the neurons as the predominant elements at work, whereas the glial cells, which rule out the neurons in number by 10:1, had to manage their life as underdogs. Despite the efforts to demonstrate partnerships the hierarchic view prevailed until nowadays (2). Kölliker, Lugaro, Spemann and others outlined already decades ago that the nervous system will only function on the basis of cooperativity: different cellular elements have to fulfill different tasks and have to interact. The guidance of neurons by radial glial cells, the intercellular K+ regulation, the detoxification and the axonal ensheathment by oligodendroglial cells might serve as examples. Research progressed only slowly to support this concept of glial-neuronal interrelationships. A major step forward was achieved by culturing glial cells as mixed or pure populations. This allowed a better characterisation of glial properties. Together with the refinement of the experimental tools at our disposal, we are till now faced with exciting new facets of the glial-neuronal communication in development and regeneration. In the following sections properties of the individual neural cells will be presented and their response to environmental influences described.

GLIAL CELLS: PRECURSORS, SUBTYPES AND INVERTEBRATE GLIA

Cells which belong to the astrocyte family are: radial glial cells, Müller cells, Bergmann glia, protoplasmic and fibrous astrocytes, ependymal cells, and pinealocytes (3). The astrocytes are basically epithelial cells and exhibit epithelial properties such as polarity, affinity for the basal lamina, and formation of cell junctions. The latter seems to occur during differentiation; gap junctions are formed not only with themselves but also with other cells such as the oligodendrocytes (3). Recently, evidence was given that the astro-

NATO ASI Series, Vol. H 51
Cell to Cell Signals in Plants and Animals
Edited by V. Neuhoff and J. Friend

cytes develop essentially in two lines: the astrocyte type 1 which is GFAP+ and A2B5– and the O-2A progenitor cell which is A2B5+ and GFAP– (4). Depending on the cultural environment or on substances secreted by the astrocyte type 1 such as PDGF or CNTF (5) oligodendrocytes or astrocytes type 2 develop (6). However, it has been pointed out that regional differences could occur as observed in studies where astroglial precursors from the cerebellum were cultured in a chemically defined medium (7).

Reactive astrocytes which might derive from astrocytes type 1 (8) appear in the CNS after injury or disease. It is larger than normal astrocytes and has a larger nuclear diameter; it contains GFAP and vimentin and exhibits an increase in glycogen (9).This astrocytic type is involved in scar tissue formation and probably to some extent in phagocytosis. Reactive astrocytes exhibit an increased GFAP immunoreactivity; their activation is mediated by substances such as lymphokines or prostanoids (9).

Phagocytosis is also attributed to another glial element, the microglial cells, which have initiated much debate about their origin. Most of the macrophages present in brain after e.g. injury seem to be derived from mesodermal origin and have entered the brain via the blood stream (10). However, a small fraction of microglial cells may also stem from the ectoderm (11). Due to explant cultures from neurectomized facial nuclei it was possible to distinguish 2 populations by their histochemical and immunological properties and by their functions: macrophages and microglial cells (12).

Oligodendrocytes, concerning their ontogenetic development, their properties and regulation of functions, will be discussed elsewhere in this volume (13). From primary cultures of fetal rat brain it was concluded that myelinogenesis appears to involve multiple, parallel, but distinct biochemical pathways. The pool of oligodendroglial progenitor cells seems to remain small in late fetal development and undergoes extensive proliferation and differentiation subsequently (14). This view has been supported and elaborated by findings from the optic nerve (15).

Glial cells are also present in the PNS where several distinct types can be distinguished. Among these are the enteric glia, which surround the neurons in the myenteric and submucosus plexus of the gastro-intestinal system; the satellite cells, which surround the neurons in the sensory, sympathetic and parasympathetic ganglia, and the Schwann cells, which manufacture myelin in the PNS. These types differ in their content of GFAP, RAN-2, A5E3, and N-CAM (16). For example, GFAP and A5E3 is expressed on all non-myelin forming cells but not on myelin producing Schwann cells.

With increasing complexity of the animal the number of glia relative to neurons increases and differentiation into glial cell types occurs. In the coelenterates most nerve nets are without supporting cells (17). In the invertebrates (e.g. annelida, mollusca) neuronal perikarya can be covered with overlapping glial processes (17). Loosely packed sheaths surrounding axons and dendrites are found in annelids, arthropods and molluscs. Glial lamellae can also group synaptic contacts, for example in arachnids and in Limulus. In leeches glial cells package nerve cells and invaginate the neurons with glycogen containing processes (18).

Receptors, Antigens, Recognition and Adhesion Molecules Expressed by Glial/ Neuronal Cells during Ontogenesis

A series of monoclonal antibodies was furnished which recognized various neuronal and glial specific antigens during the ontogenesis of the cerebellum. The presence or absence of a specific antigen could indicate a certain stage of development and could thus be insufficient for subtyping cells (19). These antigens were localized in the cell membrane but being only accessible on the cytoplasmic face of the membrane.

The presence of lectin binding proteins at the surface of cerebellar neurons suggests a specific role during cell recognition (20). Such a recognition seems to be partly necessary at the period of synaptogenesis where a specific recognition of partner cells has to occur. Cerebellar glycoproteins, termed C-1, which bind to lectins have been found to be developmentally regulated (20).

Outgrowth of neurites during development and regeneration involves neurite-promoting molecules. Adhesion of neurons to other cells or to the surrounding extracellular matrix (ECM) is also important in the outgrowth of neurites (21). To study the effect of various matrices and soluble substances N 18 neuroblastoma cells were used (22). Adhesiveness as such did not seem to play the major role for initiating neurite outgrowth rather than some specific cell surface structures. Neurite-promoting activity was particularly prominent on laminin surfaces; interestingly, its neurite-promoting effect was partly inhibited by an excess of gangliosides in the culture medium. The activity of laminin was abolished by oxidation with N-bromosuccinimide which indicated that aromatic amino acids might be important for the neurite-enhancing activity of laminin. Lectins however, which interacted with the cell surface, did not enhance neurite outgrowth to a similar extent. Since laminin binds heparin, serum was investigated for a heparin-binding protein fraction. One dimensional SDS-PAGE revealed a 65—70 kDa protein which had a similar effect as laminin or fibronectin on N 18 neuroblastoma cells (22).

Constituents of the ECM have been grouped into 4 categories: collagens, noncollagenous glycoproteins, glycosaminoglycans and proteoglycans (23). From these components glycosaminoglycans (GAG's) did not support nerve fibre growth, instead they inhibited neuron-substratum adhesion. The relative concentration of collagens and noncollagenous glycoproteins to GAG's and proteoglycans is lower in the CNS than in the PNS of adult mammals.The question arose whether this difference could be responsible for the worse neuronal regeneration in the CNS. In addition, laminin, developmentally expressed by radial glial cells, is essentially absent in the adult CNS. After an injury it can be only transiently synthesized by astrocytes. In contrast to laminin fibronectin did not support nerve fibre growth of CNS neurons (24). However, different types of neurons respond differently to ECM molecules. It has been found that ECM components can guide nerve fibre growth and can exhibit hormone-like activities, which results in an alteration of the cellular phenotype (23).

Cell surface glycoproteins play a dominant role in neuron-neuron and neuron-glial adhesion and interaction. N-CAM and Ng-CAM have been identified as molecules which mediate these interactions (25). The biosynthesis of N-CAM was studied in primary cultures of rat cerebral glial cells and cerebellar granule neurons. Of the 4 polypeptides precipitable with anti-N-CAM antibodies neuronal cultures incorporated 35S-methionine in polypeptides of 200 kDa and 135 kDa whereas glial cultures incorporated it mainly in 135 kDa and 115 kDa components, however, minor amounts were also observed in 200 kDa and 97 kDa polypeptides. N-CAM polypeptides also incorporated 35S-sulphate while only the neuronal N-CAM was phosphorylated within the time studied (26).

A protein, which belongs also to the family of cell adhesion or cell recognition molecules, is the myelin associated glycoprotein (MAG). It is now clear that it is an intrinsic constituent of the myelin (27). Its presence has been demonstrated in periaxonal regions of the myelin sheath (PNS and CNS), in myelinating oligodendrocytes and in Schwann cells (28). By using MAG antibodies and electron microscopical immunocytochemistry it was shown that dorsal root ganglion cells express MAG at early embryonic stages in chickens. MAG remained present in small sensory neurons. These results suggest that MAG might play also a role during cell migration (29).

For fulfilling the various tasks neural cells require receptors at the plasma membrane by which the cells receive their signals. Neurons have been extensively characterized in their response to neurotransmitters or other informational substances. However, glial cells have been greatly neglected in this respect. Evidence was provided within the last 5—8 years that also astrocytes possess receptors for a variety of physiologically active agents such as dopamine (30), adenosine (31), GABA (32), glutamate (33), norepinephrine (34) and serotonin (35). Binding of 3H-noradrenalin and 3H-histamine (H1 & H2) was demonstrated for astrocytes by means of light microscopic autoradiography (36). Electrophysiologically it was shown that noradrenaline and the β-agonist phenylephrine caused a depolarisation of the astrocytic membrane whereas the a-agonist isoprenaline produced a hyperpolarization.

Müller cells which belong to the astrocytic family were investigated for their role in GABA uptake in the rabbit retina (37). The results found questioned the role of Müller cells in the uptake and inactivation of synaptically released GABA. Furthermore, GFAP immunoreactivity was only shown to be associated with Müller cells of goldfish and frog while rat and rabbit retinas gave a negative result. Some GFAP immunoreactivity was expressed after an insult (37).

In contrast to astrocytes oligodendrocytes are less intensely studied. Previous work indicated the presence of receptors for serotonin (38) and prostaglandins (39). Recent results showed an increase of the intracellular cAMP on exposure of cultured adult oligodendrocytes to PGE1 (40). Electrophysiological studies indicated that glutamate and GABA can depolarize young oligodendrocytes (41). During development O—2A progenitor cells (see chapter 'Cellular and Molecular Biology of Myelination') respond to PDGF and CNTF which appear to be secreted by astrocytes type 1 (42).

An investigation of potential clinical importance dealt with the action of desipramine, a tricyclic antidepressant, on glial cells (43). An already known side effect of desipramine is its ability to induce modification of cellular lipids. By using C6 glioma cells evidence was provided that an additional lipid occurs while treating the cells with desipramine. This lipid was identified as sulfogalactosylceramide, a lipid which might act as opiate receptor (44).

Influence on Glial/Neuronal Cells by Hormones and Gangliosides-Trophic Factors and Informational Substances

A particular advantage of tissue culture methods is to allow the investigation of various factors which influence the proliferation and differentiation of neural cells under controlled conditions. Concerning the survival and maturation of cerebellar granule neurons (45) it was found that survival of the neurons needed the supplementation of the culture medium with fetal calf serum and a depolarizing concentration of K+ ions. Cerebellar interneurons can also be grown in a serum-free medium with the N-2 supplement of Bottenstein & Sato (46). Under these conditions depolarizing K+ concentrations were not necessary for long term survival, however, the maturation of the cells was substantially retarded. This was not due to the absence of thyroid hormones, since granule cells appear not to be targets for thyroid hormones as other nerve cells e.g. Purkinje cells. The ganglioside content of granule cells grown under serum or serum-free conditions differed markedly, however the developmental changes of the ganglioside pattern were broadly similar (45). Medium supplemented with GM1 (1 or 10 μM) resulted in a doubling of neurite length in the first 20 h in culture; the elongation was promoted rather than the formation of side branches. Addition of GM1 could also elicit a significant increase in the cAMP response to isoproterenol. The effect of gangliosides seem to be of a dual nature: they may support survival and cell growth and/or they stimulate neurite outgrowth. According to their ganglioside response 4 categories of neurons can be classified: 1) those, which are unresponsive to added gangliosides, 2) those, where gangliosides amplify the cellular response, 3) cells, which require extrinsic signals to initiate outgrowth, addition of gangliosides facilitates this effect, 4) those where the extension of neurites is solely promoted by gangliosides (47).

Treating PC 12 cells with NGF resulted in a release of several glycoproteins with the most prominent band of 52 kDa; a major glycoprotein band of 78 kDa was released after adding gangliosides to the cells. The release of glycoproteins seemed to be dependent on the cellular differentiation induced by either NGF or gangliosides. Taking together the results from this and other laboratories the interactions of gangliosides with various cellular compartments can be diagrammatically depicted, (47). To elucidate the role of cAMP in NGF induced neurite outgrowth from PC 12 cells, modified cyclic nucleotides (adenosine 3',5'-phosphorothioate and its diastereomers) and forskolin were used (48). The data obtained indicated a specific cAMP effect however, cAMP did not act as second messenger for NGF induced neurite outgrowth. Neurite initiation by NGF was dependent on RNA synthesis, whereas neurite regeneration of 'primed' cells was a transcription independent process (49). PC 12 cells cannot be primed by Bt2cAMP. The simultaneous action of NGF and cAMP analogues, which potentiates neurite outgrowth synergistically, was obviously mediated by different pathways.

In studies which also used PC 12 cells it was found that NGF induces the phosphorylation of 2 proteins of 90 and 70 kDa (50). Phosphotyrosine antibodies revealed that these proteins were at least in part phosphorylated at tyrosine residues.

The physiological counterpart of the pheochromocytoma cells are the chromaffin cells. These were investigated for their response to neuronotrophic and neurite promoting factors (51). NGF is the best known member of a family of neuronotrophic factors (NTF's) to which also the brain-derived neuronotrophic factor (BDNF) and the ciliary-neuronotrophic factor (CNTF) belong. These factors are operatively defined as substances which support for example the neuronal survival or induce transmitter synthesizing enzymes (52). Other factors such as the neurite-promoting factors (NPF's) comprise molecules such as laminin, fibronectin or the proteoglycans. Chromaffin cells store a large number of different neuropeptides (prominently enkephalins) in addition to the catecholamines. Their expression might be regulated by NTF activities. The most obvious effect of NTF's on chromaffine cells cultured from early postnatal rats was a change of their morphological phenotype. Chromaffine cells seem to be the target but also the source of NTF's and NPF's.

New neurotrophic factors which were secreted by astroglial cells and which acted on CNS neurons. According to their biological activities 3 components were distinguished: they supported either neuronal survival or neurite initiation or neurite elongation (53). NPF's have been isolated from muscle conditioned medium (54). Spinal motoneurons depend on these factors for survival and axonal outgrowth at least at a certain stage of their development.

Studies on trophic and metabolic couplings between astrocytes and neurons have been carried out over years by using refined culture techniques (55). The data obtained provided evidence for protective, metabolic and trophic contributions of astroglial cells to their partner neurons.

Only a few factors have been found which control the development of glial cells (see also section 1). A glial maturation factor (GMF) (56) and an astroglial growth factor (AGF) have been characterized. AGF consists of 2 components, AGF1 with a PI of 5.5 and AGF2 with a PI of 9.5 (57). The AGF's were purified from bovine brain and screened for their effects on astroblasts. The main difference was observed in the process formation. Immunohistochemical studies demonstrated their presence in neuronal cells. Sequence analysis of the AGF's revealed their homology to the fibroblastic growth factors (FGF's).

Astrocytes themselves might produce polypeptides which stimulate their proliferation (58). A polypeptide of 15 kDa could fulfill such a function.

Another growth factor, the epidermal growth factor (EGF) was tested for its effect on glial cells present in aggregating brain cell cultures derived from fetal rat telencephalon (59). Only proliferating glial cells seem to be sensitive for EGF. EGF concentrations in the culture medium being comparable to physiological concentrations stimulated the differentiation and maturation of glial precursor cells, whereas high EGF concentrations caused an increase and prolongation of their mitotic activity. These results were somewhat in contrast to the findings of other laboratories where astroglial monolayer cultures exhibited proliferation even under low EGF concentration.

Glial-Neuronal Interactions

Some examples for an intimate partnership between neurons and astroglial cells were already given in the section above.

The cerebellum is well suited for these kind of studies, since it offers several advantages such as specific neuronal subsets which are generated at predetermined periods, in addition, several mutants are available. It was found that neurons as well as glial cells can release plasminogen activators (60). The neuronal-glial interdependence was also demonstrated by using mutants which are characterized by a varying loss of Purkinje cells: glycerol-3-phosphate dehydrogenase (GPDH; EC 1.1.1.8) expression in Bergmann glia depends upon their interaction with adjacent Purkinje cells (61). The metabolic coupling of neurons and glial cells has often been postulated, but it is still on the way to obtain detailed evidence for that. A useful system for these investigations are the segmental ganglia of the leech. Studies of Pentreath et al. (62) showed that glial cells respond to signals released from neurons by alterations in their metabolism. For example, neuronal activity leads to an increase of extracellular K+ in the millimolar range, which stimulated the energy metabolism in the glial cells. Neurotransmitter such as noradrenaline, dopamine, 5-hydroxytryptamine and histamine elevated cAMP and induced glycogenolysis in glial cells, which may result in a supply of glucose to the neurons. Neurotransmitters such as L-glutamate (excitatory) and GABA (inhibitory) are terminated in their activity by synaptic events and by uptake into astrocytic processes which enwrap the synapses. The regulation of the astrocytic high affinity uptake of these transmitters by neuronal signalling was investigated by Schousboe et al. (63). Cultured astrocytes were exposed to conditioned media of cerebellar granular cells which led to a selective stimulation of the GABA uptake. Treatment of astrocytes with conditioned media from cortical neurons led to a stimulation of the high affinity glutamate uptake. In combination with other results it seems likely that a combination of intrinsic and extrinsic stimuli is needed to regulate the astrocytic phenotype.

A direct exchange of substances between glia and neurons was observed in the squid giant axon (64). A panel of glial polypeptides was found to be transferred into the axoplasm. Functionally, these glial proteins may help the axon to survive unfavourable conditions; this assumption was based on the fact that two of the proteins transferred were in their MW and PI similar to heat shock proteins.

When talking about glial-neuronal cooperativity one has to keep in mind that also competition might exist. The question whether glial cells compete with afferent fibres for apposition to the neuronal surface was investigated in the avian ciliary ganglion (65). The results suggested that nerve fibres and glial cells may compete with each other.

Myelination-Remyelination

One of the best known examples of an neuronal-glial interrelationship is that during myelination. Oligodendrocytes wrap their processes around the axons to manufacture the myelin sheath (66). Astrocytes type-2 are probably active informing the Ranvier node (67). After ensheathment by glial cells the axon membrane structure is locally modulated at the node. In contrast, the P-face ultrastructure of the axolemma changed during glial ensheathment but was not dependent on that (68).

Remyelination can occur after noxious events have damaged the myelin sheath. However, it is more efficient in the PNS than in the CNS. The components which are important for remyelination are now on the way to be analyzed. On this line, it was shown that lower levels of a monoclonal antibody directed against galactosylcerebroside (GC) interfered with the elongation of the mesaxon (culture system: Schwann cells and DRG neurons). Higher levels of this antibody inhibited axonal ensheathment. The results suggested that GC is involved in the generation of the myelin spiral (69). Schwann cells proliferate by contact with axons. For Schwann cell division a proteoglycan on the neuronal surface is required which interacts with a specific receptor on the Schwann cell surface. However, the adhesion of the neuron and Schwann cell depended on a separate set of molecules (70). Whether proliferation of oligodendrocytes occurs has been controversially discussed (71). Experiments with mice placed on a Cuprizone diet or exposed to a cerebral trauma (after anesthesia) revealed that even mature oligodendrocytes can undergo mitosis (72). In the experiments on cerebral trauma, oligodendrocytes divided, but at no time myelin basic protein (MBP) was expressed in the cytoplasm of these mature cells.

Isolated mature oligodendrocytes can produce various substances necessary for myelination in culture (73). Until recently, myelin itself was not manufactured in these cultures. By adding carbon fibres to substitute axons it was possible to obtain a myelin fraction (74) similar to authentic myelin, which indicated that mature oligodendrocytes are capable to remyelinate. Diseases which exhibit demyelinated areas in the CNS (e.g. Multiple Sclerosis) could benefit from approaches which substitute a loss of the cell number. One possible way is to transplant oligodendrocytes into the required area, where they remyelinate (75). Some of the transplanted cells seemed to migrate quite a distance apart from the locus of the application.

Repair Mechanisms during Neuronal Regeneration

Damage of the nervous system interrupts the normal functional units. If it affects an axon of the PNS, a physiological relevant repair is possible. In the CNS the axonal regeneration is very much restricted. Even worse is the situation when neurons are lost: only fishes and birds are capable of a neurogenesis also in the adulthood but not primates. Transplantation techniques were already developed at the beginning of this century and the special function of Schwann cells was recognized (76). However, more recently experimental techniques were refined by Aguayo's and Björklund's groups. PNS tissue has been found to support growth of CNS neurons when implanted (77). Transplanted monoaminergic mesencephalic neurons can grow extensively (several mm) into a sciatic nerve 'bridge' connected with the host striatum. Thus, Schwann cells may provide substances which are important for regeneration. However, other factors such as the ECM or astrocytosis have to be considered. By using different transplants and loci of grafting conditions were elaborated which have major influence on neuronal survival and growth (78). Furthermore, the results provided clear evidence that central neurons can form new synaptic connections after their normal ontogenetic development (79). While it is now established that donor tissue taken from fetal rat CNS can be successfully transplanted to neonatal or adult host brain and where the grafted tissue survives for a long time, it seems necessary to develop new strategies for the identification of donor cells. In this respect, several techniques such as cell labeling with 3H-thymidine, in situ hybridisation or a plasma clot method to reaggregate dissociated cells were introduced (80).

The role of glia in CNS axonal growth was also investigated (81). The hippocampal terrain was used which is supplied by perivascular sympathetic fibres. Damage to the septo-hippocampal pathway resulted in sprouting of sympathetic axons along vessels which enter the hippocampal formation. Glial elements seemed to be involved in this sprouting. They could produce and release NGF-like substances or provide a favourable terrain.

Glial cells form a scar after the CNS has been lesioned. This glial scar has been considered as a mechanical impediment for the regrowth of axons across this barrier. Attempts have been made to reduce the extent of this glial scar. If one intends to use a pharmacologic agent one has to beware of not inhibiting neuronal regrowth at the same time. Treatment of lesioned rats with cell-cycle specific antimitotic agents such as cytosine arabinoside, cyclophosphamide or vincristine reduced the glial proliferation. Also immunosuppression seemed to be of help. On the other hand, recent results indicate that it would be most important to regulate the astroglial response in terms of laminin production which seems to be only transiently occurring in the adult brain. Furthermore, myelinating oligodendrocytes could also release or present substances which are not favourable for an axonal regrowth (82); coculturing of oligodendrocytes and neurons indicated that oligodendrocytes could also provide a siutable milieu for axonal regrowth (83).

Altogether, the experiments thus far performed have elucidated the role of several factors in the complex situation of neuronal regeneration in the CNS. However, a direct benefit from these investigations for the clinical routine work will not be a realistic view for the near future. On the other hand, first attempts have been made to use the recent transplantation techniques for treating patients suffering from severe Parkinson's disease.

REFERENCES

References marked as "GNC" are taken from "Glial-Neuronal Communication in Development and Regeneration" (Althaus, H.H., Seifert, W. eds.) NATO ASI Series H-2 (1987).

1) KAHN, F. (1929). Das Leben des Menschen, Band IV, Kosmos, Franckh'sche Verlagshandlung, Stuttgart

2) ALBERTS, B., BRAY, D., LEWIS, J., RAFF, M., ROBERTS, K., WATSON, I.D. (1989). 'The Molecular Biology of the Cell', 2nd ed. Garland Publ. Inc., New York and London, p. 1064.

3) FEDEROFF, S. From neuroepithelium to mature astrocytes, GNC, 3—25.

4) RAFF, M.C., MILLER, R.H., NOBLE, M.D. (1983). A glial progenitor cell that developes in vitro into an astrocyte or an oligodendrocyte depending on the culture medium. Nature 303, 390—396.

5) RICHARDSON, W.D., PRINGLE, N., MOSLEY, J.D., WESTERMARK, B. DUBOIS-DALQ, M. (1988). A role for platelet-derived growth factor in normal gliogenesis in the central nervous system cell 53, 303—319.

6) NOBLE, M. MURRAY, K., STROOBANT, P., WATERLIELD, M.D., RIDDLE, P. (1988). Platelet-derived growth factor promotes division and motility and inhibits prenature differentiation of the oligodendrocyte/type-2-astrocyte progenitor cell. Nature 333, 560—562.

7) FISCHER, G.: Differentiation of astrocyte precursors from cerebellum. GNC, 433—449.

8) MILLER, R.H., ABNEY, E.R., DAVID, S., FFRENCH-CONSTANT, C.LINDSAY, R., PATEL, R., STONE, J., RAFF, M.C. (1986). Is reactive gliosis a property of a distinct subpopulation of astrocytes? J. Neurosci. 6, 22—29.

9) ENG, L.F.: Experimental models for astrocyte activation and fibrous gliosis, GNC, 27—39.

10) OEHMICHEN, M. (1978). Mononuclear phagocytes in the central nervous system. Springer-Verlag, Berlin.

11) LING, E.A. (1981). The origin and nature of microglia. In: 'Advances in Cellular Neurobiology' (S. Federoff & L. Hertz, eds.). Acad. Press, New York, Vol. 2, 33—82.

12) RIESKE-SHOWS, E., TETZLAFF, W., CZONKOWSKA, A., GRÄBER, M. KREUTZBERG, G.W.: Microglia in culture, GNC, 41—51.

13) JESERICH,G., ALTHAUS, H.H., WAEHNELDT, T.V.: 'Cellular and Molecular Biology of Myelination' V.

14) BANSAL, R., BARBERESE, E., BHAT, S., CARSON, J., EDGAR, A., FRIEDRICH, V., MACKLIN, W., PFEIFFER, S.E., SINGH, H. WOODIEL, F.: Oligodendrocyte differentiation: quantitative studies in primary cultures of dissociated fetal rat brain, GNC, 737—754.

15) MILLER, R.H., DAVID, S., PATEL, R., ABNEY, E.R., RAFF, M.C. (1985). A quantitative immunohistochemical study of macroglial cell development in the rat optic nerve: in vivo evidence for two distinct astrocyte lineages, Develop. Biol. 111, 35—41.

16) MIRSKY, R., JESSEN, K.R.: Molecular properties of peripheral glia. GNC, 53—62.

17) ROOTS, B. (1978). A phylogenetic approach to the anatomy of glia. In: 'Dynamic Properties of Glial Cells' (Schoffeniels, E., Franck, G., Towers, D.B., Hertz, L. eds), Pergamon Press, Oxford, New York, 45—54.

18) PENTREATH, V.W., PENNINGTON, A.J., SEAL, L.H., SWIFT, K.: Modulation by neuronal signals of energy substrate in the glial cells of leech segmental ganglia. GNC, 211—229.

19) GHANDOUR, M.S., FONCOAUD, B., GOMBOS, G.: Developmentally regulated glial and neuronal antigens detected by monoclonal antibodies, GNC, 131—144.

20) ZANETTA, J.P., DONTENWILL, M. REEBER, A., VINCENDON, G.: Expression of recognition molecules in the cerebellum of young and adult rats. GNC, 91—104.

21) LETOURNEAU, P.C. (1975). Cell to substratum adhesion and guidance of axonal elongation. Dev. Biol. 44, 92—101.

22) RAUVALA, H. MÄHÖNEN, Y., JOUSIMAA, J., MERENMIES, J., LINDHOLM, D., VUENTO, M. Neurite outgrowth induced by adhesive proteins. GNC, 159—181.

23) CARBONETTO, S., TURNER, D.C., DE GEORGE, J. Neuronal adhesion to components of the extracellular matrix and control of nerve fiber growth. GNC, 183—207.

24) ROGERS, S.L., LETOURNEAU, P.C., PALM, S.L., MCCARTHY, J., FURCHT, L.T. (1983). Neurite extension by peripheral and central nervous systems neurons in response to substratum bound fibronectin and laminin. Develop. Biol. 98, 212—220.

25) GRUMET, M., HOFFMAN, S., EDELMAN, G.M. (1984). Two antigenically related neuronal cell adhesion molecules of different specificities mediate neuron-neuron and neuron-glia adhesion. Proc. Natl. Acad. Sci. USA 81, 267—271.

26) ALBRECHTSEN, M., MØLLER, S.P., C.J., BOCK, E.: Biosynthesis of the neural cell adhesion molecule (N-CAM) glial, muscle and neuronal forms of N-CAM studied in primary rat cell cultures. GNC, 147—157.

27) MORELL, P. (1984). Myelin, 2nd. Ed. Plenum Press, New York, London.

28) QUARLES, R.H., EVERLY, J.L., BRADY, R.O. (1973). Evidence for the close association of a glycoprotein with myelin in rat brain. J. Neurochem. 21, 1177—1191.

29) OMLIN, F.X., ROCHAT, KA., PHILIPPE, E., MATTHIEU, J.-M., DROZ, B.: Myelin-associated glycoprotein (MAG) in chicken neuronal crest derivatives, including neurons. GNC, 105—115.

30) HANSSON, E., RÖNNBACK, L., SELLSTRÖM, A. (1984). Is there a 'dopaminergic glial cell'? Neurochem. Res. 9, 679—689.

31) VAN CALKER, D., MÜLLER, M., HAMPRECHT, B. (1979). Receptors regulating the level of cyclic AMP in primary cultures of perinatal mouse brain. In: Natural Growth and Differentation. E. Meisami and M.A.B. Brazier eds. Raven Press, New York, 11—25.

32) HÖSLI, E., MÖHLER, H., RICHARDS, J.G., HÖSLI, L. (1980). Autoradiographic localization of binding sites for (^{3}H)γ-aminobutyrate, (^{3}H) muscimol, (+) (^{3}H) bicuculline methiodide and (^{3}H) flunitrazepam in cultures of rat cerebellum and spinal cord. Neuroscience 5, 1657—1665.

33) HERTZ, L., KVAMME, E., MCGEER, E., SCHOUSBOE, A. (eds.) 1983). Glutamine, Glutamate and GABA in the Central Nervous System; Alan Liss Inc., New York.

34) HÖSLI, L., HÖSLI, E., ZEHNTNER, LEHMANN, R., LUTZ, T.W. (1982). Evidence for the existence of α- and β-adrenoreceptors on cultured glial cells — an electrophysiological study. Neuroscience 7, 2867—2872.

35) TARDY, M., COSTA, M.F., FAGES, C., BARDAKDJIAU, J., GONNARD, P. (1982). Uptake and binding of serotonin by primary cultures of mouse astrocytes. Dev. Neurosci. 6, 19—26.

36) HÖSLI, L., HÖSLI, E.: Adrenergic- and histamine-receptors on cultured astrocytes: electrophysiological and autoradiographic binding studies, GNC, 65—79.

37) OSBORNE, N.N.: Neurone-glia relationship in the retina: localisation of GABA uptake and endogenous GFAP and GABA immunoreactivities. GNC, 117—129.

38) MURRAY, M.R. (1965). Nervous tissues in vitro. In: Cells and Tissues in Culture (Willmer EN ed.), Adacemic Press, London and New York.

39) MCCARTHY, K.D., DE VELLIS, J. (1978). Alpha-adrenergic receptor modulation of beta-adrenergic, adenosine and prostaglandin E, increased adenosine 3':5'-cyclic monophosphate levels in primary cultures of glia. J. Cycl. Nucl. Res. 4, 15—26.

40) SIEPL, C. (1986). Oligodendroglia, Syntheseort und Ziel von Prostaglandinen. PhD-Thesis, Göttingen-Hannover.

41) KETTENMANN, H., GILBERT, P., SCHACHNER, M. (1984). Depolarisation of cultured oligodendrocytes by glutamate and GABA. Neurosci. Let. 47, 271—276.

42) ALBERTS, B., BRAY, D., LEWIS, J. RAFF, M., ROBERTS, K., WATSON, J.D. (1989). The Molecular Biology of the Cell, 2nd ed., Garland Publ. Inc. New York & London, 909—911.

43) CABON, F., ALBONZ, S., TOCQUE, B., ZALC, B.: Desipramine treated C_6 glioma cells express opiate receptors; unmasking or induction? GNC, 81—89.

44) CRAVES, F.B., ZALC, B., LEYBIN, L., BAUMANN, N. LOH, H.H. (1980). Cerebroside sulfate antibodies inhibit the effects of morphine and β-endorphin. Science 207, 75—76.

45) BALAZS, R., GALLO, V., KINGSBURY, A., THANGNIPON, W., SMITH, R., ATTERWILL, C., WOODHAMS, P.: Factors affecting the survival and maturation of nerve cells in culture. GNC 285—302.

46) BOTTENSTEIN, J.E., SATO, G.H. (1979). Growth of rat neuroblastoma cell line in serumfree supplemented medium. Proc. Natl. Acad. Sci. USA 76, 514—517.

47) YAVIN, E., GIL, S., GUROFF, G., HAMA, T., RICHTER-LANDSBERG, C.: Glycoconjugate metabolism, accretion and release during neurite outgrowth in normal and transformed cells of neuronal origin. GNC, 303—321.

48) RICHTER-LANDSBERG, C., JASTORFF, B.: Effect of more growth factor and CAMP analogues on neuronal differentation in PC 12 cells: studies with an inhibitor of cAMP-dependent protein kinases GNC, 339—354.

49) BURSTEIN, D.E., GREENE, L.A. (1978). Evidence for both RNA-synthesis dependent and independent pathways in stimulation of neurite outgrowth by nerve growth factor. Proc. Natl. Acad. Scie. USA 75, 6059—6063.

50) MARCHISIO, P.C., TOUSCO, F., CORBASCIO, G.C.: Are phosphotyrosine-containing proteins induced by nerve growth factor stimulation of PC 12 cells? GNC, 355—364.

51) UNSICKER, K., LIETZKE, R.: Chromaffin cells: modified neurons that are both targets and storage sites of neuronotrophic and neurite promoting factors. GNC, 365—384.

52) THONEN, H., BARDE, Y.-A. (1980). Physiology of nerve growth factor. Physiol. Rev. 60, 1284—1335.

53) BECKH, S., MÜLLER, H.W., SEIFERT, W.: Neurotrophic and neurite promoting activities in astroglial conditioned medium. GNC, 385—406.

54) HENDERSON, C.E.: Neurite-promoting factors for spinal neurons. GNC, 407—417.

55) VARON, S., SKAPER, S.D., FACCI, L., RUDGE, J.S., MANTHORPE, M.: Trophic and metabolic couplings between astroglia and neurons. GNC, 491—504.

56) LIM, R., MILLER, J.F. (1984). An improved procedure for the isolation of glia maturation factor. J. Cell. Physiol. 119, 255—259.

57) PETTMANN, B., GENSBURGER, C., WEIBEL, M. PERRAUD, F., SENSENBRENNER, M., LABOURDETTE, G.: Isolation of two astroglial growth factors from bovine brain; comparison with other growth factors; cellular localisation. GNC, 451—478.

58) MICHLER-STUKE, A.: Putative autoregulative growth factor in conditioned medium of serum-free grown astrocytes. GNC, 419—432.

59) HONEGGER, P., GUENTERT-LAUBER, B.: Glial cell development and reactivity in reaggregating brain cell culture. GNC, 479—489.

60) MOONEN, G., SELAK J, GRAU-WAGEMANS, M.-P.: In vitro analysis of glial-neuronal communication during cerebellum ontogenesis. GNC, 323—338.

61) FISHER, M. (1984). Neuronal influence on glial enzyme expression: Evidence from mutant mouse cerebella. Proc. Natl. Acad. Sci. USA 81, 4414—4418.

62) PENTREATH, V.W., KAI-KAI, M.A. (1982). Significance of the potassium signal from neurons to glial cells. Nature 295, 59—61.

63) SCHOUSBOE, A., DREJER, J., LARSSON, O.M., MEIER, E.: Regulation of astrocytic high affinity uptake of transmitter amino acids by neuronal signaling. GNC, 231—245.

64) TYTELL, M.: Characterization of glial proteins transferred into the squid giant axon. GNC, 247—261.

65) MUGNAINI, E., FIORI, M.G.: Do glial cells compete with afferent fibers for apposition to the neuronal surface in development and aging of the nervous system? A study in the avian ciliary ganglion with references to other neurons. GNC, 263—282.

66) BUNGE, R.P. (1981). The development of myelin and myelin-related cells. Trends Neurosci. 4, 175—177.

67) FFRENCH-CONSTANT, C., MILLER, R.H., KRUSE, J., SCHACHNER, M. RAFF, M.C. (1986). Molecular specialization of astrocyte processes at modes of Ranvier in rat optic nerve. J. Cell Biol. 102, 844—852.

68) WAXMAN, S.G.: Molecular organisation of the cell membrane in normal and pathological axons: relation of glial contact. GNC, 709—736.

69) RANSCHT, B., WOOD, P.M., BATES, M., BUNGE, R.P.: Role of galactocerebroside in the formation of peripheral myelin. GNC, 665—681.

70) RATNER, N., WOOD, P., GLASER, L., BUNGE, R.: Further characterization of the neuronal cell surface protein mitogenic for Schwann cells. GNC, 683—698.

71) ALTHAUS, H.H., MONTZ, H., SCHWARTZ, P., NEUHOFF, V. (1984). Isolation and cultivation of mature oligodendroglial cells. Naturwissenschaften 71, 309—315.

72) LUDWIN, S.K.: Oligodendrocyte proliferation: its relationship to central nervous system remyelisation. GNC, 799—816.

73) SZUCHET, S.: Myelin palingenesis: the reformation of myelin by mature oligodendrocytes in the absence of neurons. GNC, 755—777.

74) ALTHAUS, H.H., BURGISSER, P., KLÖPPNER, S., ROHMANN, A., SCHRÖTER, J. SCHWARTZ, P. SIEPL. C., NEUHOFF, V.: Oligodendrocytes ensheath carbon fibres and produce myelin in vitro. GNC, 779—798.

75) GUMPEL, M., LACHAPELLE, F., BAULAC, M., BARON VAN EVERCOOREN, A., LUBETZKI, C., GANSMULLER, A., LOMBRAIL, P., JACQUE, C., BAUMANN, N.: Myelination in the mouse by transplanted oligodendrocytes. GNC, 817—830.

76) BALLANCE, C.A., STEWART, P. (1901). The Healing of Nerves. MacMillan & Co. London.

77) AGUAYO, A.J., BENFEY, M., DAVID, S. (1983). A potential for axonal regeneration in neurons of the adult mammalian nervous system. In: Nervous system regeneration (Haber B., Perez-Polo, J, Hashim, J., Stella G. eds) Alan R. Liss, New York, 327—340.

78) ISACSON, O., BJÖRKLUND, A., DUNNETT, S.B.: Conditions for neuronal survival and growth as assessed by the intracerebral transplantation technique in lesion models of the adult. CNS. GNC, 529—544.

79) ZIMMER, J. FINSEN, B., SØRENSEN, T., SUNDE, N.: Hippocampal transplants: synaptic organisation, their use in repair of neuronal circuits and mouse to rat xenografting. GNC, 545—564.

80) LINDSAY, R.M., RAISMAN, G., SEELEY, P.J.: Neural tissue transplants: studies using tissue culture manipulations, cell marking techniques and a plasma clot method to follow development of grafted neurons and glia. GNC, 585—603.

81) CRUTCHER, K.: A model of neuronal sprouting for examining the role of glia in axonal growth. GNC, 565—573.

82) SCHNELL, L., SCHWAB, M.E. (1990). Axonal regeneration in the rat spinal cord produced by an antibody against myelin associated neurite growth inhibitors. Nature 343, 269—272.

83) STOYKOVA, A., ALTHAUS, H.H., GEBICKE-HÄRTER, P.J., RITTNER, I., SCHWARTZ, P., NEUHOFF, V.: Culturing of neurons and oligodendrocytes bulk isolated from the same brain of young adult rats. GNC, 641—662.

AMINO ACID AVAILABILITY AND BRAIN FUNCTION IN HEALTH AND DISEASE

G. Huether

Max-Planck-Institut für experimentelle Medizin
Hermann-Rein-Straße 3, 3400 Göttingen, FRG

1. A STIMULATING CONCEPT: ATTRACTIVE SIMPLICITY

Until the early seventies, it was generally believed that normal brain function would be unaffected by the metabolic changes associated with the ingestion of different foods. This assumption was effectively undermined by the finding that the protein-to-carbohydrate ratio of a meal did affect the brain concentration of one particular neurotransmitter, serotonin. The following theory was proposed to explain this phenomenon: the ingestion of carbohydrates stimulates the release of insulin. This insulin release causes an increase in the uptake of amino acids into tissues. Because this increase is less for tryptophan, there is an increase in plasma of tryptophan relative to the other large neutral amino acids (LNAA: tryptophan, phenylalanine, tyrosine, valine, leucine and isoleucine). The LNAA in the blood compete for a common transport mechanism in order to cross the blood-brain barrier. The relative increase in plasma tryptophan results in an increased occupation of the transport mechanism with tryptophan and, therefore, an increase in the brain tryptophan levels. Because tryptophan is a rate limiting substrate for serotonin synthesis, the increase in brain tryptophan leads to an increase in brain serotonin.

Serotonin participates in the regulation of various important brain functions, including sleep, pain sensitivity, mood, aggression, and neuroendocrine secretion. Based on the initial finding that serotonin synthesis is limited by the availability of tryptophan, a flood of reports described the modulation of almost any of these brain functions by an increased or decreased availability of tryptophan (see Fig. 1).

The activity of other neurotransmitter systems was also found to be affected by the availability of their respective amino acid precursors for transmitter synthesis. For example the synthesis of two other monoamine transmitters, dopamine and epinephrine, was shown to be enhanced by an increased availability of their precursor amino acid, tyrosine, at least when neurons that release these transmitters were additionally stimulated. Consequently, another series of reports described the modulation of various brain functions after administration of tyrosine or its precursor, phenylalanine (see Fig. 1).

Even though neurochemists were the first to realize that the formation of amino acid derived transmitters in the brain can be affected by an altered supply of their respective precursor amino acids, the potential impact of this finding for the modulation of various brain functions and behaviour was soon recognized by neurobiologists and physicians,

NATO ASI Series, Vol. H 51
Cell to Cell Signals in Plants and Animals
Edited by V. Neuhoff and J. Friend

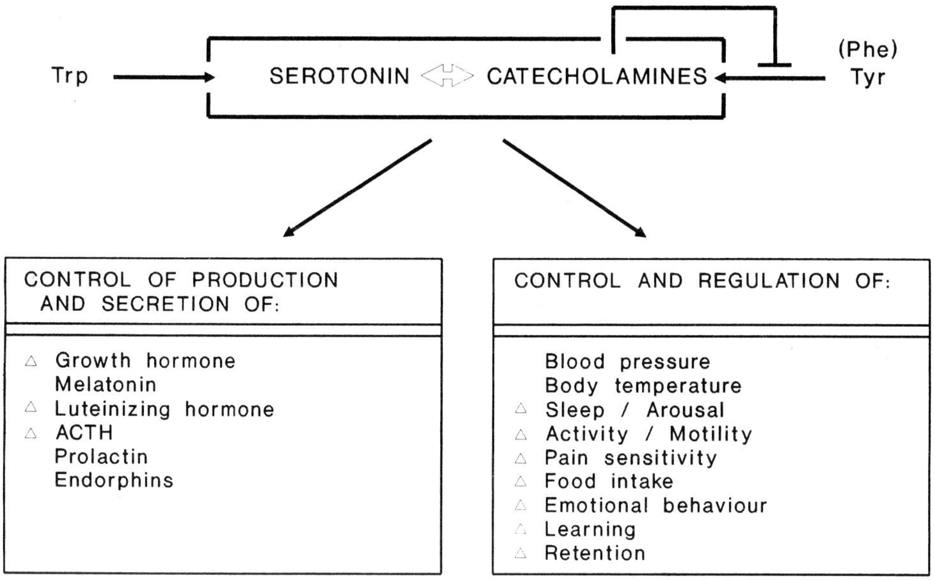

FIGURE 1
Central serotoninergic and catecholaminergic mechanisms participate in the regulation of many basic brain and body functions. Many of these functions have been shown to be affected by an increased or decreased availability of the precursor amino acids tryptophan and tyrosine, resp. phenylalanine (marked by triangles). For details and references see (1).

and workers from a growing number of different research disciplines became interested in various aspects of the relationship between amino acid availability and brain function.

For many neurobiologists the concept of precursor controlled synthesis of important neurotransmitters and modulators seemed to open new perspectives for a better understanding of normal brain function. The plasma concentrations of tyrosine and tryptophan, as well as of the other large neutral amino acids which compete with tyrosine or tryptophan for brain uptake are subject of wide variations. Their availability changes not only when people or animals receive various drugs or the amino acids themselves, but also in disease states affecting amino acid metabolism or under certain physiologic conditions, e.g. postprandially after the ingestion of a carbohydrate or protein rich meal, during hunger, exercise, stress or as a consequence of the release of various hormones. Much effort has been made to demonstrate that these variations often cause parallel changes in the levels of these amino acids within monoaminergic neurons, and therefore may influence both neurotransmitter synthesis and the activity of monoaminergic transmitter systems.

Psychiatrists became particularity interested in the concept of precursor-controlled synthesis of brain monoamines because a disturbed function of these systems was thought to be involved in many psychiatric deseases, e.g. depression, insomnia, anxiety or aggression.

A great number of investigations were initiated to study possible abnormalities in the availability of aromatic amino acids in psychiatric patients. A new therapeutic concept, amino acid precursor therapy, was introduced for the treatment of , e.g., depression and insomnia by oral tryptophan administration, and a lucrative market was discovered by pharmaceutical firms selling tryptophan as a natural tool for the support of serotoninergic mechanisms in all sorts of malaises. Similarily, tyrosine was offered as a natural tool for the support of catecholaminergic functions to prevent fatigue and to overcome exhaustion during stress.

Neuroendocrinologists were fascinated about the neuroendocrine effects of tryptophan and tyrosine, and neuroendocrine challenge tests were developed to study possible failures of cerebral monoaminergic transmission. New perspectives were opened also for nutritionists by the possible existence of a regulatory feedback loop for food intake control and macronutrient selection involving precursor dependent changes of the activity of serotoninergic systems which participate in the regulation of food intake. The activation of this system resulting from the increased tryptophan availability postpondial to the ingestion of a carbohydrate rich meal was thought to selectively reduce carbohydrate appetite during the next meal. For obvious reasons, a failure of this regulatory loop was assumed to play an important role in eating disorders and obesity. Finally, the introduction of a new artificial sweetener, the phenylalanine containing dipeptide aspartame, gave rise to controversies whether or not monoaminergic systems would be affected by the increased availability of tyrosine (in persons with normal liver phenylalanine hydroxylation), or by the limited supply of tyrosine (in persons with defective phenylalanine hydroxylation), respectively.

The discussions of the pros and cons of the concept of precursor controlled transmitter synthesis and its implications reached their summit in the mid-eighties and there was a substantial interest and urgent necessity for the scientists in these various fields to be given an opportunity to get together, to exchange their ideas, to critically assess the existing knowledge and to develop future research strategies. This opportunity was given in September 1987 at the Advanced Research Workshop "Amino Acid Availability and Brain Function in Health and Disease", which took place at the Max-Planck-Institute for Experimental Medicine in Göttingen, Germany. Leading scientists were invited to discuss the recent progress in this field, to address open questions, and to outline research strategies for the future.

2. THE TURNING POINT: UNEXPECTED COMPLEXITY

It was generally felt by the participants of this ARW that the research on the influence of altered amino acid availability on brain function at that time was entering a turning point with regard to three important aspects. Firstly, there existed certain long-standing controversies between individual groups, e.g., whether bound or free tryptophan would be more important for the regulation of serotonin synthesis, or whether carbohydrate or protein appetite is regulated via serotonergic mechanisms. However, it has since become apparent that these questions had to be answered on an "as well as" rather than an "either/or" basis. Hence, the Workshop did represent the end of certain scientific disagreements. Secondly, it was obvious that owing to the refinement of the experimental tools at our disposal (in vivo dialysis and voltammetry) we were entering an exciting phase of expansion and substantiation of our knowledge on the relationship between precursor availability and brain function. It seemed possible to measure now in the living animal what previously could only be assumed on a theoretical basis. And, as a third aspect, it became clear that the early enthusiasm about the perspectives of precursor treatments of various behavioural, mental, or neurologic dysfunctions was replaced by a more realistic judgment of the limits and problems that arise when rather mechanistic approaches are applied to a homeostatic system that is maintained by multiple feedback loops arranged in networks.

In other words, the proponents of the new concept agreed that many different mechanisms seem to be involved in the regulation of precursor availability, that individual transmitters seem to participate in the regulation of various brain functions, that certain assumptions were made on the basis of data obtained by inadequate methods and that things turned out to be more complicated than it was previously thought.

Obviously, the concept of precursor controlled transmitter synthesis, because of its appealing simplicity, invited many investigators to speculate much about its physiologic, pathogenetic and therapeutic implications. In view of these perspectives many investigators must have tended to ignore the critical points, the unsolved questions, and the unjustified assumptions of this concept. This is the only explanation of the fact that until now the most pressing questions related to the validity and the implications of the concept still await a definite and clear answer. Some examples may be helpful to illustrate this deficit:

- It has never been attempted to evaluate the relative importance of precursor availability for serotonin synthesis in relation to other processes involved in the regulation of the presynaptic releaseable serotonin pool.
- Until now, the impact of increased transmitter synthesis on actual transmitter release under normal physiologic conditions is not completely understood.
- Stress and metabolic strain affect the amino acid supply to the brain as well as the neuronal activity and the state of arousal in the brain. Such effects have never been adequately considered in animal experiments designed to study the influence of precursor treatments on transmitter formation and transmitter release.
- Even the most sophisticated methods, like microdialysis or in vivo voltammetry are inadequate to monitor the effects of altered transmitter synthesis on functional neurotransmission.
- There are indications that newly synthesized transmitters are preferentially released. It is unknown if neurons possess differentially releaseable transmitter pools and how they are regulated.
- Adaptation by counterregulation is one important mechanism how the brain reacts to long lasting changes of its sensory or nutritional input. The mechanisms by which the brain conterregulates an increased transmitter synthesis and release caused by long term precursor treatments have not been thouroughly investigated.
- Powerful pharmacologic tools are now available to characterize the effects of altered precursor availability on transmitter function and to demonstrate the involvement of even individual receptor subtypes. However, the pharmacological proof that the behavioural effects of, e.g. tryptophan administration, are indeed serotoninergic, is still the exception rather than the rule.
- Cell death and neurodegeneration is an increasing problem of neuropathology. It is not unlikely that changes in the release and the degradation, or in the compartmentation of neurotransmitters may have neurotoxic effects. It can not be excluded that such side effects are of relevance in precursor therapy.
- During brain development and early embryogenesis, neurotransmitters have been shown to play a particular role as morphogenetic signals and informational sub-

stances involved in the modulation of various properties and functions of embryonal cells or developing neurons. It is unknown if such non-transmitter functions of, e.g. monoamines, are of particular significance for the function of the adult brain and how such functions are modulated by an altered precursor supply.

These are only some of the fundamental problems with which the concept of precursor controlled neurotransmitter synthesis and brain function is confronted. In addition, there is a theoretical problem which a priori limits the value of this concept. Generally, in order for an increased precursor availability to affect not only transmitter synthesis but also the output of a neuron, its firing rate must not be tightly controlled by other mechanisms. Such situations are exceptions rather than the rule. In most cases transmitter release is regulated either by receptor-mediated feedback processes at the level of individual synapses or by multineuronal and multisynaptic feedback loops at the level of neuronal networks. A few exceptions might be anticipated in which such inhibitory feedback loops for the control of transmitter release will not be activated , so that precursor administration will be allowed to affect neurotransmission.

One example may be the permanently increased firing rate of surviving neurons in neurodegenerative disorders. These neurons (e.g. dopaminergic neurons of the substantia nigra in patients with Parkinson‘s disease) are more sensitive to the availability of their transmitter precursors than other neurons using the same transmitter. Other examples may be physiologic circumstances in which neurons undergo sustained elevations in firing frequencies (e.g. sympatho-adrenal cells in hemorrhagic shock) or neurons which lack multisynaptic feedback loops (e.g. peripheral neurons).

In conclusion, at closer examination of the complexity of the mechanisms involved in the regulation of neuronal firing rate and transmitter release, the early enthusiasm about the possible role and the perspectives of the control of neurotransmitter synthesis by precursor availability became gradually replaced by a more realistic judgement of the problems and the limitations of this concept.

3. A FADING CONCEPT: APPROACHING REALITY

The actual value of the concept of precursor dependent neurotransmitter synthesis must be measured by its contribution to a better understanding of the mechanisms and processes involved in the regulation of normal and disturbed brain function, and by the new approaches opened for the diagnosis and the therapy of brain dysfunctions. Of all possible implications of precursor availability on transmitter synthesis and brain function, the modulation of activity of serotoninergic pathways by the availability of tryptophan has been the most promising one.

The fluctuations of tryptophan availability associated with various physiologic states was assumed to affect the rate of serotonin synthesis in the brain. However, clear evidence for a specific modulation of the activity of serotoninergic mechanisms, and therefore, brain function by the altered availability of tryptophan resulting from, e.g. hunger, exercise, stress or hormonal changes is still lacking. A tryptophan induced activation of central serotoninergic mechanisms involved in food intake regulation after the ingestion of a carbohydrate rich meal was though to result in a preferential reduction of carbohydrate intake during the next meal. This simple concept of the regulation of macronutrient intake attracted many nutritionists involved in the management of eating disorders and obesity. A failure of this loop was assumed to play a role for carbohydrate craving of obese patients. However, carefully controlled animal studies revealed that the postulated loop does not exist or is only of minor importance compared to the role of sensory cues and learning (2). The concept of precursor-dependent transmitter synthesis stimulated an intense search for changes in the availability of tryptophan in psychiatric patients. Even though slight changes were observed, e.g. in certain subgroups of depressed patients, the general outcome of this strategy was rather discouraging. One interesting finding was that antidepressants, by inhibiting liver tryptophan-pyrrolase may cause a rise of circulating plasma tryptophan concentrations (3). But again, evidence is missing that this effect contributes to the therapeutic effect of antidepressants. The administration of tryptophan to experimental animals and normal humans at doses which cause an unphysiologic rise of circulating tryptophan concentrations has been shown to affect various brain functions, like mood, performance, sleep, food intake, pain sensitivity or the release of pituitary hormones (for references see (1)). However, it is still uncertain that these changes are actually caused by a precursor-induced activation of serotoninergic systems. Neuroactive tryptophan metabolites, like kynurenic acid or quinolinic acid are formed by the degradation of tryptophan via tryptophan- or indoleamine 2,3-dioxygenase in peripheral tissues and in the brain (4). In addition, the formation of other neuroactive indoleamines which, in contrast to serotonin can easily penetrate the blood brain barrier, e.g. 5-methoxytryptamine and melatonin, is stimulated in the gut and these substances are released into the circulation after oral or intravenous tryptophan administration (own unpublished observations). It is not unlikely, that the central effects of tryptophan are mediated by one or several of these metabolites.

Even though the peripheral metabolism and the mechanisms of action of orally administered tryptophan were not definitely solved, tryptophan was sold from the mid-sixties as a natural tool for the support of central serotoninergic mechanisms. It was used both as a drug for the treatment of depression and insomnia and as a food additive with no particular restrictions or care of possible side effects. At the end of 1989, unexpected side effects and even the death of some persons were reported after oral consumption or administration of tryptophan. Consequently tryptophan was taken off the market and neither pharmaceutical firms nor psychiatrists are interested any longer in this sort of precursor therapy. The unspecific activation of the immune system seen in some persons after tryptophan ingestion was explained by contaminations, including bacterial endotoxins, of the tryptophan preparations used. However, the protagonists of the tryptophan precursor therapy notoriously overlooked the close relationship between tryptophan metabolism and activation of the immune system. It has been known, for instance, that the extrahepatic metabolism of tryptophan proceeds via indoleamine-2,3-dioxygenase, an enzyme which is particularily active in tissues and cells involved in immunologic defense mechanisms. This enzyme is extremely sensitive to activation by interferon gamma. Infections, endotoxins or bacterial antigens cause a T-lymphocyte dependent, gamma-interferon mediated increase in the activity of this enzyme. Suchwise activated macrophages degrade tryptophan and release large amounts of very reactive tryptophan metabolites, e.g. 3-hydroxyanthranilic acid, which readily reacts with reactive oxygen species to form a lipophilic and membrane soluble, deleterious radical which seems to amplify the cytotoxiticy of activated macrophages (5). Under normal conditions, the overriding of this system is prevented by the limited availability of tryptophan. The daily intake of one or more grams of pure tryptophan by a person with an infection and an activation of the immune system, however, may support this defense system in a unforeseeable and fatal manner.

If this scenario is indeed a valid explanation of the observed nonspecific activation of the immune response in some persons after oral tryptophan intake, we can only hope that this time the complex interactions will be carefully clarified and evaluated before tryptophan is offered for the natural support of the immune system.

REFERENCES

1) HUETHER, G. (ed.). Amino Acid Availability and Brain Function in Health and Disease. NATO-ASI-Series, Series H: Cell Biology, Vol. 20, Springer Verlag, Berlin, 1988.

2) HOLDER, M.D., HUETHER, G. (1990). Role of prefeedings, plasma amino acid ratios and brain serotonin levels in carbohydrate and protein selection. Physiology and Behaviour, 47, 113—119.

3) BADAWY, A.A.B., EVANS, M. (1981). Inhibition of rat liver Trp-pyrrolase activity and elevation of brain Trp concentration by administration of antidepressants. Biochem. Pharmacol. 30, 1211—1216.

4) DURING, M.J., FREESE, A., HEYES, M.P., SWARTZ, K.J., MARKEY, S.P., ROTH, R.H. AND MARTIN, J.B. (1989). Neuroactive metabolites of L-tryptophan, serotonin and quinolinic acid, in striatal extracellular fluid. FEBS-Letters, 247, 438—444.

5) WERNER-FELMAYER, G, WERNER, E.R., FUCHS, D., HAUSEN, A., REIBNEGGER, G. AND WACHTER, H. (1989). Characteristics of interferon induced tryptophan metabolism in human cells in vitro. Biochimica et Biophysica Acta, 1012, 140—147.

INDEX

Bold face numerals indicate the pages where detailed comments can be found

A

D

E

F

I

K

L

M

N

O

U

V

Printing: Druckhaus Beltz, Hemsbach
Binding: Buchbinderei Schäffer, Grünstadt

NATO ASI Series H

Vol. 1: Biology and Molecular Biology of Plant-Pathogen Interactions. Edited by J. A. Bailey. 415 pages. 1986.

Vol. 2: Glial-Neuronal Communication in Development and Regeneration. Edited by H. H. Althaus and W. Seifert. 865 pages. 1987.

Vol. 3: Nicotinic Acetylcholine Receptor: Structure and Function. Edited by A. Maelicke. 489 pages. 1986.

Vol. 4: Recognition in Microbe-Plant Symbiotic and Pathogenic Interactions. Edited by B. Lugtenberg. 449 pages. 1986.

Vol. 5: Mesenchymal-Epithelial Interactions in Neural Development. Edited by J. R. Wolff, J. Sievers, and M. Berry. 428 pages. 1987.

Vol. 6: Molecular Mechanisms of Desensitization to Signal Molecules. Edited by T. M. Konijn, P. J. M. Van Haastert, H. Van der Starre, H. Van der Wel, and M. D. Houslay. 336 pages. 1987.

Vol. 7: Gangliosides and Modulation of Neuronal Functions. Edited by H. Rahmann. 647 pages. 1987.

Vol. 8: Molecular and Cellular Aspects of Erythropoietin and Erythropoiesis. Edited by I. N. Rich. 460 pages. 1987.

Vol. 9: Modification of Cell to Cell Signals During Normal and Pathological Aging. Edited by S. Govoni and F. Battaini. 297 pages. 1987.

Vol. 10: Plant Hormone Receptors. Edited by D. Klämbt. 319 pages. 1987.

Vol. 11: Host-Parasite Cellular and Molecular Interactions in Protozoal Infections. Edited by K.-P. Chang and D. Snary. 425 pages. 1987.

Vol. 12: The Cell Surface in Signal Transduction. Edited by E. Wagner, H. Greppin, and B. Millet. 243 pages. 1987.

Vol. 13: Toxicology of Pesticides: Experimental, Clinical and Regulatory Perspectives. Edited by L. G. Costa, C. L. Galli, and S. D. Murphy. 320 pages. 1987.

Vol. 14: Genetics of Translation. New Approaches. Edited by M. F. Tuite, M. Picard, and M. Bolotin-Fukuhara. 524 pages. 1988.

Vol. 15: Photosensitisation. Molecular, Cellular and Medical Aspects. Edited by G. Moreno, R. H. Pottier, and T. G. Truscott. 521 pages. 1988.

Vol. 16: Membrane Biogenesis. Edited by J. A. F. Op den Kamp. 477 pages. 1988.

Vol. 17: Cell to Cell Signals in Plant, Animal and Microbial Symbiosis. Edited by S. Scannerini, D. Smith, P. Bonfante-Fasolo, and V. Gianinazzi-Pearson. 414 pages. 1988.

Vol. 18: Plant Cell Biotechnology. Edited by M. S. S. Pais, F. Mavituna, and J. M. Novais. 500 pages. 1988.

Vol. 19: Modulation of Synaptic Transmission and Plasticity in Nervous Systems. Edited by G. Hertting and H.-C. Spatz. 457 pages. 1988.

Vol. 20: Amino Acid Availability and Brain Function in Health and Disease. Edited by G. Huether. 487 pages. 1988.

NATO ASI Series H

Vol. 21: **Cellular and Molecular Basis of Synaptic Transmission.** Edited by H. Zimmermann. 547 pages. 1988.

Vol. 22: **Neural Development and Regeneration. Cellular and Molecular Aspects.** Edited by A. Gorio, J.R. Perez-Polo, J. de Vellis, and B. Haber. 711 pages. 1988.

Vol. 23: **The Semiotics of Cellular Communication in the Immune System.** Edited by E.E. Sercarz, F. Celada, N.A. Mitchison, and T. Tada. 326 pages. 1988.

Vol. 24: **Bacteria, Complement and the Phagocytic Cell.** Edited by F.C. Cabello und C. Pruzzo. 372 pages. 1988.

Vol. 25: **Nicotinic Acetylcholine Receptors in the Nervous System.** Edited by F. Clementi, C. Gotti, and E. Sher. 424 pages. 1988.

Vol. 26: **Cell to Cell Signals in Mammalian Development.** Edited by S.W. de Laat, J.G. Bluemink, and C.L. Mummery. 322 pages. 1989.

Vol. 27: **Phytotoxins and Plant Pathogenesis.** Edited by A. Graniti, R.D. Durbin, and A. Ballio. 508 pages. 1989.

Vol. 28: **Vascular Wilt Diseases of Plants. Basic Studies and Control.** Edited by E.C. Tjamos and C.H. Beckman. 590 pages. 1989.

Vol. 29: **Receptors, Membrane Transport and Signal Transduction.** Edited by A.E. Evangelopoulos, J.P. Changeux, L. Packer, T.G. Sotiroudis, and K.W.A. Wirtz. 387 pages. 1989.

Vol. 30: **Effects of Mineral Dusts on Cells.** Edited by B.T. Mossman and R.O. Bégin. 470 pages. 1989.

Vol. 31: **Neurobiology of the Inner Retina.** Edited by R. Weiler and N.N. Osborne. 529 pages. 1989.

Vol. 32: **Molecular Biology of Neuroreceptors and Ion Channels.** Edited by A. Maelicke. 675 pages. 1989.

Vol. 33: **Regulatory Mechanisms of Neuron to Vessel Communication in Brain.** Edited by F. Battaini, S. Govoni, M.S. Magnoni, and M. Trabucchi. 416 pages. 1989.

Vol. 34: **Vectors as Tools for the Study of Normal and Abnormal Growth and Differentiation.** Edited by H. Lother, R. Dernick, and W. Ostertag. 477 pages. 1989.

Vol. 35: **Cell Separation in Plants: Physiology, Biochemistry and Molecular Biology.** Edited by D.J. Osborne and M.B. Jackson. 449 pages. 1989.

Vol. 36: **Signal Molecules in Plants and Plant-Microbe Interactions.** Edited by B.J.J. Lugtenberg. 425 pages. 1989.

Vol. 37: **Tin-Based Antitumour Drugs.** Edited by M. Gielen. 226 pages. 1990.

Vol. 38: **The Molecular Biology of Autoimmune Disease.** Edited by A.G. Demaine, J-P. Banga, and A.M. McGregor. 404 pages. 1990.

Vol. 39: **Chemosensory Information Processing.** Edited by D. Schild. 403 pages. 1990.

Vol. 40: **Dynamics and Biogenesis of Membranes.** Edited by J.A.F. Op den Kamp. 367 pages. 1990.

Vol. 41: **Recognition and Response in Plant-Virus Interactions.** Edited by R.S.S. Fraser. 467 pages. 1990.

NATO ASI Series H

Vol. 42: Biomechanics of Active Movement and Deformation of Cells.
Edited by N. Akkaş. 524 pages. 1990.

Vol. 43: Cellular and Molecular Biology of Myelination.
Edited by G. Jeserich, H. H. Althaus, and T. V. Waehneldt. 565 pages. 1990.

Vol. 44: Activation and Desensitization of Transducing Pathways.
Edited by T. M. Konijn, M. D. Houslay, and P. J. M. Van Haastert. 336 pages. 1990.

Vol. 45: Mechanism of Fertilization: Plants to Humans.
Edited by B. Dale. 710 pages. 1990.

Vol. 46: Parallels in Cell to Cell Junctions in Plants and Animals.
Edited by A. W. Robards, W. J. Lucas, J. D. Pitts, H. J. Jongsma, and D. C. Spray. 296 pages. 1990.

Vol. 47: Signal Perception and Transduction in Higher Plants.
Edited by R. Ranjeva and A. M. Boudet. 357 pages. 1990.

Vol. 48: Calcium Transport and Intracellular Calcium Homeostasis.
Edited by D. Pansu and F. Bronner. 456 pages. 1990.

Vol. 49: Post-Transcriptional Control of Gene Expression.
Edited by J. E. G. McCarthy and M. F. Tuite. 671 pages. 1990.

Vol. 50: Phytochrome Properties and Biological Action.
Edited by B. Thomas and C. B. Johnson, 1991.

Vol. 51: Cell to Cell Signals in Plants and Animals.
Edited by V. Neuhoff and J. Friend. 404 pages. 1991.